OXFORD TEXTS IN APPLIED AND ENGINEERING
MATHEMATICS

OXFORD TEXTS IN APPLIED AND ENGINEERING MATHEMATICS

* G. D. Smith: *Numerical Solution of Partial Differential Equations* 3rd Edition
* R. Hill: *A First Course in Coding Theory*
* I. Anderson: *A First Course in Combinatorial Mathematics* 2nd Edition
* D. J. Acheson: *Elementary Fluid Dynamics*
* S. Barnett: *Matrices: Methods and Applications*
* L. M. Hocking: *Optimal Control: An Intro to the Theory with Applications*
* D. C. Ince: *An Intro to Discrete Mathematics, Formal System Specification, and Z* 2nd Edition
* O. Pretzel: *Error-Correcting Codes and Finite Fields*
* P. Grindrod: *The Theory and Applications of Reaction-Diffusion Equations: Patterns and Waves* 2nd Edition

1. Alwyn Scott: *Nonlinear Science: emergence and dynamics of coherent structures*
2. D. W. Jordan and P. Smith: *Nonlinear ordinary differential equations: an introduction to dynamical systems* 3rd Edition
3. I. J. Sobey: *Introduction to interactive boundary layer theory*
4. A. B. Tayler: *Mathematical Models in Applied Mechanics* (reissue)
5. Ramdas Ram-Mohan: *Finite Element and Boundary Element Applications in Quantum Mechanics*
6. Lapeyre et al: *Introduction to Monte-Carlo Methods for Transport and Diffusion Equations*
7. Isaac Elishakoff & Yong Jin Ren: *Finite Element Methods for Structures with Large Stochastic Variations*
8. Alwyn Scott: *Nonlinear Science: emergence and dynamics of coherent structures* 2nd Edition

Titles marked with an asterix (*) appeared in the series Oxford Applied Mathematics and Computing Science Series, which has been folded into, and is continued by, the current series.

Nonlinear Science

Emergence and Dynamics of Coherent Structures

(Second Edition)

Alwyn Scott

Department of Mathematics, University of Arizona

This book has been printed digitally and produced in a standard specification in order to ensure its continuing availability

OXFORD
UNIVERSITY PRESS

Great Clarendon Street, Oxford OX2 6DP

Oxford University Press is a department of the University of Oxford.
It furthers the University's objective of excellence in research, scholarship, and education by publishing worldwide in

Oxford New York

Auckland Cape Town Dar es Salaam Hong Kong Karachi
Kuala Lumpur Madrid Melbourne Mexico City Nairobi
New Delhi Shanghai Taipei Toronto
With offices in
Argentina Austria Brazil Chile Czech Republic France Greece
Guatemala Hungary Italy Japan South Korea Poland Portugal
Singapore Switzerland Thailand Turkey Ukraine Vietnam

Oxford is a registered trade mark of Oxford University Press
in the UK and in certain other countries

Published in the United States
by Oxford University Press Inc., New York

© Alwyn Scott 2003

The moral rights of the author have been asserted

Database right Oxford University Press (maker)

Reprinted 2006

All rights reserved. No part of this publication may be reproduced, stored in a retrieval system, or transmitted, in any form or by any means, without the prior permission in writing of Oxford University Press, or as expressly permitted by law, or under terms agreed with the appropriate reprographics rights organization. Enquiries concerning reproduction outside the scope of the above should be sent to the Rights Department, Oxford University Press, at the address above

You must not circulate this book in any other binding or cover
And you must impose this same condition on any acquirer

ISBN 0-19-852852-3

This book is dedicated to

JOHN SCOTT RUSSELL (1808–1882)

PREFACE TO THE SECOND EDITION

Since *Nonlinear Science: Emergence and Dynamics of Coherent Structures* went to press in the autumn of 1998, several developments suggest that a second edition would be useful. First have been the experiences of teaching from the book, both by me and by friends and colleagues who have shared their questions and comments, noting typographical errors and suggesting ways in which the material might be better explained or more conveniently arranged. Second, I am now editing the forthcoming *Encyclopedia of Nonlinear Science*—an activity that corrects misconceptions and provides balanced perspectives on the structure of nonlinear science. Third, I have recently published a survey of neuroscience (*Neuroscience: A Mathematical Primer*), which motivates revisions of those portions of the present book that are devoted to nerve impulse dynamics. Finally, there have been significant advances in nonlinear science that are appropriate to discuss in a second edition, including new results on the nature of impulse propagation on myelinated nerves, progress in understanding the dynamics of nonlinear lattices (offering appreciations of oscillating self-localized modes or "breathers" and their interactions), and the advent of pump-probe measurements in the infrared range of the electromagnetic spectrum (leading to a spectroscopy of molecular crystals that is sensitive only to nonlinear effects, thereby providing direct indications of self-localization). All of these insights and more have been incorporated into the rewriting of this new edition.

As in the former edition, the first and last chapters can be read together as a single essay on how nonlinear science got started and where it seems to be going. It is suggested that introductory courses begin with a reading and discussions of these two chapters—the alpha and omega of the subject.

Related to the third chapter (The classical soliton equations), an appendix on elliptic functions has been added to aid those students who are unfamiliar with this important subject. The fourth chapter (Reaction-diffusion systems) is almost entirely rewritten to better introduce nerve impulse propagation and present a variety of useful nerve models, while detailed discussions of nerve impulse stability are moved to an appendix. In the fifth chapter (Nonlinear lattices), a detailed analysis of the Toda lattice is moved to an appendix, the sections on energy conserving nonlinear lattices are augmented to include recent developments in the area of "breathers" and "intrinsic localized modes," and the discussion of myelinated nerves is rewritten to include recent perspectives with implications for biological evolution. In Chapter 7 (Perturbation theory), some detailed derivations are moved to appendices, and the sections on nerve impulses are modified to complement the augmented discussion of Chapter 4.

In the eighth chapter (Quantum lattice solitons), a discussion of the quantum theory of differential pump-probe studies has been added and related to recent empirical and theoretical results with relevance for biological energy transfer. Finally, the problems have been revised, hints have been introduced for the solutions of some of the more difficult problems, all of the figures have been redrawn, and many new figures have been added.

First among those whose suggestions and insights have aided in making these revisions are Mads Peter Sørensen and Peter Leth Christiansen. Their contributions to the first edition are included and the lively discussions arising from our joint teaching activities at the Technical University of Denmark have been of great value. Particularly helpful feedback has been received from Leon Shohet, who uses the book in a course in nonlinear waves at the University of Wisconsin. Special thanks go also to Peter Hamm for introducing me to the merits of pump-probe spectroscopy and for several comments on the book. Others whose criticism and encouragement have enhanced the revisions include Ole Bang, Stephane Binczak, Leonor Cruzeiro-Hansson, Chris Eilbeck, Victor Enolskii, Chris Fall, Sergej Flach, Peter Vingaard Larsen, Robert MacKay, Jens Juul Rasmussen, Mario Salerno, Al Sievers, Alexey Ustinov, Jonathan Wattis, Yaroslav Zolotariuk, and many, many students. Sincere thanks are extended to all for their generous assistance.

With these improvements it is hoped that the second edition will continue to be a useful text for introductory courses on coherent nonlinear phenomena in applied science.

Lyngby ALWYN SCOTT
2003

PREFACE TO THE FIRST EDITION

This book introduces a revolution that has quietly transformed the realm of science over the past quarter century. Up to the mid-1970s, models of physical phenomena were usually assumed to be linear, or nearly so, allowing advantage to be taken of the convenient but often unrealistic property that complicated causes can be resolved into more simple components, the effects of which are treated separately. Over the past two decades, however, it has become increasingly evident that the assumption of linearity leads the theorist to miss qualitatively significant phenomena.

For a *nonlinear* system, it is now known, *the whole is greater than the sum of its parts*, leading to the *emergence* of new structures that are spatially or temporally coherent. Awareness of this is becoming recognized as a watershed in science, leading to a new understanding of its organization and structure.

Such emergent structures are very much *things*, having their own features, lifetimes, and peculiar ways of interacting. Since these interactions are also nonlinear, new dynamics give rise to yet other emergent structures appearing at higher levels of description. Thus the molecules of chemistry emerge from nonlinear interactions between the elements of atomic physics, providing a structural basis for the proteins and ribonucleic acids of biochemistry, and so on, up through the many levels of activity in a living organism. The intricacy of this hierarchical perspective reflects that of reality, and it should, I believe, be appreciated by all who would contribute to the biologically and socially oriented sciences of the coming century.

Although the present book emerges (if you please) from a lifetime spent with problems of nonlinear science, it is not intended to be a research monograph. Originally prepared as notes for courses in nonlinear science that are presented to advanced undergraduates at the Technical University of Denmark, the book is designed as an introduction to the study of nonlinear partial and difference—differential equations with special emphasis on emergent phenomena. While advanced topics are treated, an effort has been made to ease the ways through these sections by including appropriate preliminary material.

Over the past decade, several books have been used for these courses, but for a number of reasons none has been entirely satisfactory:

- Texts that do a proper job on the theory of nonlinear science are often weak on physical perspectives and realistic applications.
- Available books are usually research monographs in which the authors strive to communicate with their professional peers, neglecting the explanatory material that newcomers to the field require.

- Books on solitons tend to overlook the phenomena of nonlinear diffusion. Yet both nonlinear diffusion and energy conserving systems are important aspects of modern research in nonlinear partial differential equations, and some applications—such as the complex Ginsburg–Landau equation of nonlinear optics and also a (superconducting) Josephson transmission line—change from being soliton-like to nonlinear diffusion systems through the adjustment of experimental parameters, thereby associating the two behaviors. From a biological perspective, the emergence of impulses on a nerve fiber is of fundamental importance and a clear example of nonlinear diffusion.
- Finally, to master the many facets of nonlinear science, it is necessary to work problems. Without the experience of seeing the details in one's own handwriting, theoretical understanding is inadequate. Many of the exercises presented in this book encourage the reader to further develop facets of the theory that are introduced in the text. Others may suggest directions for research.

As a profession approaches maturity, its historical roots often become obscure. Unaware of the sources of basic ideas, scientists tend to reinvent the wheel, adding confusion to an already distended literature. Thus the book begins with a historical introduction to the emergent paradigm, attempting to put the early contributions into perspective by making annotated references to seminal papers that are sometimes forgotten.

The second chapter is also introductory. Since it is important to undertake the study of nonlinear systems with an adequate mastery of the established body of knowledge for linear systems, Chapter 2 is intended as a brief review of some key ideas. The student who has difficulty with this chapter should find time to re-examine the standard references and solve some of the problems. The chapter closes with two aspects of linear theory that are less familiar to intermediate students: stability theory and linear scattering theory. It is convenient to present the basic ideas of inverse scattering early in a series of lectures because the associated Gel'fand–Levitan equation plays a key role later on in the inverse scattering theory.

The book proper begins in Chapter 3 with an introduction to the concept of a *soliton*, the remarkable emergent entity that preserves its shape and speed under collisions with others of its kind. Basic soliton equations are related to the underlying physics, and their salient properties are described in some detail. In each case, an appropriate method is used to obtain localized and periodic traveling waves and N-soliton formulas, thereby introducing the reader to a variety of computational strategies and instilling a common sense attitude toward the investigation of new situations.

Chapter 4 discusses that neglected step-sister of soliton theory: nonlinear diffusion, a vast and vital subject with close ties to several aspects of biophysics—especially electrophysiology. The aims in this chapter are two-fold. First, the reader is encouraged to appreciate the scientific importance of nonlinear diffusion in excitable media and to see how it is related to soliton studies. Then directions of present and future research activities are indicated in studies of dynamics of single nerve cells and in the heart, with possible extensions to models of biological morphogenesis and evolution.

Although difference–differential equations (DDEs) are sometimes viewed as approximations to partial differential equations that are needed for numerical calculations, they have a strong claim to consideration on their own merits. Currently one of the most exciting research areas of condensed matter physics and physical chemistry is the study of local modes in anharmonic molecules and molecular crystals, and—in the classical (nonquantum) approximation—these are described by DDEs. The list of such problems is long and growing longer, including also: mechanical vibrations (nonlinear phonons), lattice polarons, dislocation dynamics, myelinated nerve fibers in vertebrate animals, biomolecular dynamics, and some aspects of the dynamics of the human brain. In Chapter 5, a variety of analytical techniques are introduced for the systems of this class, again in order to develop a flexible and optimistic attitude toward the study of related problems.

From a theoretical perspective, the inverse scattering transform provides the basis of soliton theory, underlying the unexpected properties of soliton bearing systems, and this theory is introduced in Chapter 6, with the aim of revealing its structure. The chapter begins with a discussion of inverse scattering for the linear Schrödinger equation—an application that is familiar to many students—paying particular attention to points that tend to confuse the novice. In the course of this discussion, the Gel'fand–Levitan equation is rederived from a different perspective than that of Chapter 2, reinforcing the basic ideas. An inverse scattering method (ISM) is then described in detail for the Korteweg–de Vries equation, and demonstrated through several physically motivated examples. Building upon this foundation, two-component (matrix) scattering operators are introduced as generalizations of the linear Schrödinger operator, and corresponding ISMs are applied to the most important nonlinear wave systems: the sine–Gordon equation and the nonlinear Schrödinger equation. Finally, it is shown how the ISM formulation can be used to generate countably infinite sets of independent conservation laws.

In the pre-soliton era, the study of nonlinear problems was largely limited to perturbation expansions about small amplitude linear approximations to the true solution. With the discovery of nonlinear traveling waves and N-soliton formulas, one can now start with an exact solution that is much closer to the final result, gaining thereby a significant increase in descriptive power. From a broader perspective, perturbation theory leads to nonlinear laws governing the interactions among emergent structures. This area of research is discussed in Chapter 7, with emphasis on a variety of analytic tools so the student becomes empowered to deal with the many such problems that are coming over the horizon.

Chapter 8 is devoted to a survey of lattice solitons and solitary waves from the perspective of quantum theory. Since—as was noted above—nonlinear science has many applications to molecules and molecular crystals, quantum behavior often needs to be considered. The primary aim in this chapter is to describe methods for formulating the quantum theory of nonlinear lattices that allow one to calculate energy eigenvalues and the structures of soliton wave functions. To ease the way, several introductory examples are treated, starting with the slightly anharmonic oscillator and models of molecular vibrations, before presenting a complete description of a quantum lattice soliton. Applications to problems of experimental interest are emphasized, and the chapter concludes with an evaluation of different procedures for solving quantized problems.

Experience has shown that the material presented here is more than enough for one semester; thus the instructor may wish to select some subset to present. For an introductory course, Chapters 1 through 5 and parts of 6 are found to be appropriate, while a graduate semester might review these ideas before concentrating on Chapters 6 through 8. The entire book can be comfortably covered in two semesters.

Although mentioned in several places, the phenomenon of deterministic chaos is not underscored in this book for three reasons. First of all, the present focus is on the emergence of coherent structures in nonlinear systems—a major component of nonlinear theory that should be emphasized in one university course. Second, the subject of chaos has been widely discussed in a number of books, several of which are directed to the literate public, leaving the impression with some that nonlinear science comprises only the theory of chaos. Finally, *an understanding of emergent phenomena is of fundamental importance in the study of living organisms*, whereas chaotic effects are of lesser interest in the biological realm.

The final chapter differs in character from the earlier parts of the book, being a speculative essay on the future directions of research in applied science. In the belief that we should think about what we do, these pages present some personal ideas about the nature of things to come. Noting that twentieth century science has been primarily directed toward the understanding of inanimate objects from a reductive perspective, it is predicted that the study of life will be a central issue in the coming century. Indeed, this shift in the center of gravity of scientific inquiry is already underway, transforming the nature of research in directions that are yet to be appreciated.

As the curtain rises on the science of the twenty-first century, the nonlinear phenomenon of emergence is destined to play a leading role. I hope this book will help those who wish to participate in the unfolding drama.

Lyngby and Tucson   ALWYN SCOTT
1998

ACKNOWLEDGEMENTS IN THE FIRST EDITION

Although many people have contributed to this book, thanks go first of all to Mads Peter Sørensen and Peter Leth Christiansen for their direct contributions and also for careful readings of the text and several preliminary versions that were tried out at the Technical University of Denmark. Mario Salerno, Chris Eilbeck, Lynne MacNeil, Carl Clausen, and many students have also read portions of the manuscript, making suggestions and corrections for which I am grateful. Figures were kindly provided by Carl Clausen, Kenneth Cole, Chris Eilbeck, François Fillaux, Dyan Louria, Art Winfree, and Anatol Zhabotinsky.

The understanding of nonlinear science has been a lifetime project, to which many students have contributed. I am grateful to have had the opportunity of working with Herb Aumann, Ole Bang, Abdel Benabdallah, Lisa Bernstein, Flora Chu, Leonor Cruzeiro, Henrik Feddersen, Tom Gabriel, Henriette Gilhøj, Mark Hayes, Sigmund Hoel, Jesper Halding Jensen, Wayne Johnson, Bill Keller, Aaron King, Steve Luzader, Carl Magee, Peter Miller, Virginia Muto, Bob Parmentier, Belur Prasanna, Bob Proebsting, Anita Rado, Kim Rasmussen, Mudita Reddy, Stan Reible, Paul Rissman, Mahendra Shah, and Han-Tzong Yuan.

Colleagues with whom I have worked directly on research projects related to the material presented here include: Antonio Barone, Irving Bigio, Jean-Guy Caputo, Giorgio Careri, Peter Leth Christiansen, Gianni Costabile, Chris Eilbeck, Victor Enol'skii, Filippo Esposito, Enrico Gratton, Mac Hyman, Scott Layne, Peter Lomdahl, Karl Lonngren, Dave McLaughlin, Jim Nordman, Bob Parmentier, Antonio Petraglia, Luigi-Maria Ricciardi, Mario Salerno, Bonaventura Savo, Henrik Smith, Uja Vota-Pinardi, and Ewan Wright, but many others have contributed less directly. Since its emergence in the early 1970s, the nonlinear science community has become rather like an extended family, providing mutual support and a wealth of ideas that are essential for a sound understanding of the field. Some of these people have been colleagues at Wisconsin, Naples, Los Alamos, Arizona, and the Technical University of Denmark, and for their collective influence I am particularly thankful.

Finally, I wish to express appreciation for the generous financial support that has been provided over the past 35 years by the National Science Foundation (USA), the National Institutes of Health (USA), The Consiglio Nazionale delle Ricerche (Italy), the European Molecular Biology Organization, the Department of Energy (USA), the Technical Research Council (Denmark), the Natural Science Research Council (Denmark), the Thomas B. Thriges Foundation, and the Fetzer Foundation.

CONTENTS

List of Figures		xxi
1	**THE BIRTH OF A PARADIGM**	1
1.1	From the Great Wave to the Great War	1
	1.1.1 Hydrodynamics	1
	1.1.2 Nonlinear diffusion	3
	1.1.3 Bäcklund transformation theory	6
	1.1.4 A theory of matter	7
1.2	Between the wars	8
1.3	Nonlinear research from 1945 to 1985	11
	1.3.1 Nerve studies	11
	1.3.2 Autocatalytic chemical reactions	12
	1.3.3 Solitons	14
	1.3.4 Local modes in molecules and molecular crystals	19
	1.3.5 Elementary particle research	20
1.4	Recent developments	21
	References	23
2	**LINEAR WAVE THEORY**	28
2.1	Dispersionless linear equations	28
2.2	Dispersive linear equations	30
2.3	The linear diffusion equation	31
2.4	Driven systems	33
	2.4.1 Green's method	33
	2.4.2 Fredholm's theorem	35
2.5	Stability	37
	2.5.1 General definitions	37
	2.5.2 Linear stability	38
	2.5.3 Signaling problems	39
2.6	Scattering theory	40
	2.6.1 Solutions of Schrödinger's equation	40
	2.6.2 Gel'fand–Levitan theory	43
	2.6.3 A reflectionless potential	48
2.7	Problems	48
	References	53

3 THE CLASSICAL SOLITON EQUATIONS — 55
- 3.1 The Korteweg–de Vries (KdV) equation — 57
  - 3.1.1 Long water waves — 57
  - 3.1.2 Solitary wave solutions — 58
  - 3.1.3 Periodic solutions — 59
  - 3.1.4 A Bäcklund transformation for KdV — 61
  - 3.1.5 N-soliton formulas — 67
- 3.2 The sine–Gordon (SG) equation — 71
  - 3.2.1 Long Josephson junctions — 71
  - 3.2.2 Solitary waves — 72
  - 3.2.3 Periodic waves — 74
  - 3.2.4 Nonlinear standing waves — 77
  - 3.2.5 Two-soliton solutions — 81
  - 3.2.6 More spatial dimensions — 85
- 3.3 The nonlinear Schrödinger (NLS) equation — 88
  - 3.3.1 Nonlinear wave packets — 88
  - 3.3.2 Modulated traveling-wave solutions of NLS(+) — 90
  - 3.3.3 Dark soliton solutions of NLS(−) — 92
  - 3.3.4 A BT for NLS(+) — 93
  - 3.3.5 Transverse phenomena — 95
- 3.4 Summary — 98
- 3.5 Problems — 98
- References — 106

4 REACTION-DIFFUSION SYSTEMS — 110
- 4.1 Simple reaction-diffusion equations — 111
  - 4.1.1 The Zeldovich–Frank-Kamenetsky (Z–F) equation — 111
  - 4.1.2 The Burgers equation — 116
- 4.2 The Hodgkin–Huxley (H–H) system — 117
  - 4.2.1 Space-clamped squid membrane dynamics — 118
  - 4.2.2 The H–H impulse — 124
- 4.3 Simplified nerve models — 127
  - 4.3.1 The Markin–Chizmadzhev (M–C) model — 127
  - 4.3.2 The FitzHugh–Nagumo (F–N) model — 129
  - 4.3.3 Morris–Lecar (M–L) models — 134
- 4.4 Stability analyses — 138
  - 4.4.1 The Z–F equation — 138
  - 4.4.2 The M–C model — 139
  - 4.4.3 The F–N model — 140
  - 4.4.4 The H–H and M–L systems — 143
- 4.5 Decremental conduction — 143
- 4.6 Nonuniform fibers — 147
  - 4.6.1 Tapered fibers — 147
  - 4.6.2 Leading-edge charge and impulse ignition — 149

		4.6.3	Dendritic logic	150
	4.7	More space dimensions		154
		4.7.1	Two-dimensional nonlinear diffusion	154
		4.7.2	Nonlinear diffusion in three dimensions	156
		4.7.3	Turing patterns	159
		4.7.4	Hypercycles	160
	4.8	Summary		161
	4.9	Problems		162
	References			171
5	**NONLINEAR LATTICES**			176
	5.1	Spring-mass lattices		177
		5.1.1	The Toda-lattice soliton	178
		5.1.2	Lattice solitary waves	179
		5.1.3	Existence of lattice solitary waves	180
		5.1.4	Intrinsic localized modes and intrinsic gap modes	182
	5.2	Lattices with nonlinear on-site potentials		185
		5.2.1	The discrete sine–Gordon equation	187
		5.2.2	Nonlinear Schrödinger lattices	190
		5.2.3	The discrete self-trapping equation	197
	5.3	Biological solitons		202
		5.3.1	Alpha-helix solitons in protein	202
		5.3.2	Self-trapping in globular proteins	205
		5.3.3	Solitons in DNA	207
	5.4	Nonconservative lattices		210
		5.4.1	Quasiharmonic lattices	210
		5.4.2	Myelinated nerves	215
		5.4.3	Emergence of form by replication	219
	5.5	Assemblies of neurons		221
	5.6	Summary		223
	5.7	Problems		224
	References			230
6	**INVERSE SCATTERING METHODS**			238
	6.1	Linear scattering revisited		240
		6.1.1	Scattering solutions, bound states, and upper half plane poles	240
		6.1.2	Why the upper half plane poles must be simple	242
		6.1.3	The Gel'fand–Levitan equation again	245
		6.1.4	Any questions?	249
	6.2	Inverse scattering method for KdV		250
		6.2.1	General description	250
		6.2.2	Some examples	252

		6.2.3	Reduction to Fourier analysis in the small amplitude limit	257
	6.3	Two-component scattering theory	258
		6.3.1	Linear theory	258
		6.3.2	ISMs for two-component scattering	264
	6.4	The sine–Gordon equation	266
	6.5	The nonlinear Schrödinger equation	271
	6.6	Conservation laws	273
		6.6.1	Conservation laws for the KdV equation	274
		6.6.2	Conserved densities for matrix scattering	276
	6.7	Summary	277
	6.8	Problems	278
	References	285

7 PERTURBATION THEORY — 287

	7.1	Perturbed matrices	288
	7.2	A damped harmonic oscillator	290
		7.2.1	Energy analysis	290
		7.2.2	Multiple time scales	291
	7.3	Energy analysis of soliton dynamics	293
		7.3.1	Korteweg–de Vries solitons	294
		7.3.2	Sine–Gordon solitons	296
		7.3.3	Nonlinear Schrödinger solitons	299
	7.4	More general soliton analyses	301
		7.4.1	Multiple scale analysis of an SG kink	301
		7.4.2	Variational analysis of an NLS soliton	306
	7.5	Multisoliton perturbation theory	309
		7.5.1	General theory	310
		7.5.2	Kink–antikink collisions	314
		7.5.3	Radiation from a fluxon	317
	7.6	Neural perturbations	319
		7.6.1	The FitzHugh–Nagumo system	320
		7.6.2	Electrodynamic (ephaptic) coupling of nerves	322
	7.7	Summary	326
	7.8	Problems	327
	References	335

8 QUANTUM LATTICE SOLITONS — 337

	8.1	Quantum oscillators	337
		8.1.1	A classical nonlinear oscillator	337
		8.1.2	The birth of quantum theory	339
		8.1.3	A quantum linear oscillator	342
		8.1.4	The rotating wave approximation	345
		8.1.5	The Born–Oppenheimer approximation	347

		8.1.6	Dirac's notation	350
		8.1.7	Pump-probe measurements	351
	8.2	Self-trapping in the dihalomethanes		353
		8.2.1	Classical analysis	354
		8.2.2	Quantum analysis	356
		8.2.3	Comparison with experiments	360
	8.3	Boson lattices		361
		8.3.1	The discrete self-trapping equation	361
		8.3.2	A lattice nonlinear Schrödinger equation	365
		8.3.3	Soliton wave packets	370
		8.3.4	The Hartree approximation	372
	8.4	More general quanta		377
		8.4.1	The Ablowitz–Ladik equation	377
		8.4.2	Salerno's equation	380
		8.4.3	A fermionic polaron model	381
		8.4.4	The Hubbard model	384
	8.5	Energy transport in protein		386
		8.5.1	Dynamic equations	386
		8.5.2	Experimental observations	390
		8.5.3	Recent comments	398
	8.6	A quantum lattice sine–Gordon equation		401
	8.7	Theoretical perspectives		403
		8.7.1	Number state method	403
		8.7.2	Quantum inverse scattering method	404
		8.7.3	QISM analysis of the DST dimer	406
		8.7.4	Comparison of the NSM and the QISM	407
	8.8	Summary		409
	8.9	Problems		409
	References			420
9	**LOOKING AHEAD**			424
	References			431
APPENDIX A CONSERVATION LAWS AND CONSERVATIVE SYSTEMS				433
	References			437
APPENDIX B MULTISOLITON FORMULAS				438
B.1	The KdV equation			438
B.2	The SG equation			438
B.3	The NLS equation			440
B.4	The Toda lattice			440
References				441

APPENDIX	**C**	**ELLIPTIC FUNCTIONS**	443
		References	447
APPENDIX	**D**	**STABILITY OF NERVE IMPULSES**	448
		References	454
APPENDIX	**E**	**PERIODIC TODA-LATTICE SOLITONS**	456
		References	457
APPENDIX	**F**	**ANALYTIC APPROXIMATIONS FOR LONG LATTICE SOLITARY WAVES**	458
		Reference	459
APPENDIX	**G**	**MULTIPLE-SCALE ANALYSIS OF A DAMPED-HARMONIC OSCILLATOR**	460
		References	462
APPENDIX	**H**	**GREEN FUNCTIONS FOR SOLITON RADIATION**	463
		References	467
INDEX			469

LIST OF FIGURES

1.1	A hydrodynamic soliton created in a wave tank by John Scott Russell in the 1830s.	2
1.2	Measurement of the change in membrane conductance (band) and membrane voltage (line) with time during the passage of a nerve impulse on a squid axon.	9
1.3	The planar structure of a benzene molecule, showing a local mode of the CH stretching oscillation.	11
1.4	Self-oscillatory ring waves in a two-dimensional chemical reaction diffusion system.	13
2.1	Sketch of the boundary conditions on $K(x, \xi)$ in the (x, ξ)-plane.	45
2.2	Contours in the complex k-plane for computing $B(x - \tau)$ from the inverse Fourier transform of Equation (2.38).	46
3.1	The third-degree polynomial, $P(\tilde{u})$, in Equation (3.11).	59
3.2	A periodic solution of Equation (3.20).	60
3.3	A plot of a 2-soliton collision: $-u_Z(x, t)$ from Equation (3.36).	64
3.4	(a) Superconducting strip-line, which supports TEM propagation in the x-direction. (b) An equivalent circuit model of the TEM mode.	71
3.5	A kink on a simple mechanical model of the SG equation that can be made from dressmaker pins and an elastic band.	74
3.6	Surface plot of a kink–kink solution of the SG equation from Equation (3.82) with $v = 1/2$.	82
3.7	Surface plot of a kink–antikink solution plotted from Equation (3.83) with $v = 1/2$.	82
3.8	Surface plot of a stationary breather from Equation (3.85) with $\omega = \pi/5$.	85
3.9	A plot of $|u_{\mathrm{nls2}}((x, t)|$ from Equation (3.113).	94
4.1	The dependence of the velocity of flame propagation (v) upon the diameter (d) of a candle.	111
4.2	Finding a traveling-wave solution of Equation (4.2).	113
4.3	Zero-velocity, pulse-like solutions of Equation (4.2).	116
4.4	(a) Sketch of a squid nerve axon. (b) A differential equivalent circuit of the axon.	118
4.5	From Equations (4.19), the stationary values of the switching variables: $n_0(V)$, $m_0(V)$, and $h_0(V)$ (upper panel), and the corresponding switching times: τ_n, τ_m, and τ_h (lower panel).	122

4.6	Schematic representation of a homoclinic trajectory in the (V, W, m, h, n) traveling-wave phase space defined in Equations (4.21).	125
4.7	A full-sized action potential (at $v = 18.8$ m/s) and an unstable threshold impulse (at 5.66 m/s) for the H–H axon at 18.5°C.	126
4.8	(a) Ionic currents in the M–C model as a function of the traveling-wave variable (ξ). (b) Structure of the associated nerve impulse.	128
4.9	The function $C(v)$ defined Equation (4.24) is plotted for parameter values corresponding to the standard H–H squid axon [70].	129
4.10	In the F–N system with $0 < \varepsilon \ll 1$, V and R are shown as functions of the traveling-wave variable ξ.	131
4.11	The homoclinic solution trajectory of Equations (4.27) corresponding to the traveling-wave impulse in Figure 4.10.	132
4.12	Propagation speeds for impulse solutions of a FitzHugh–Nagumo (F–N) system plotted against the temperature parameter ε.	133
4.13	Propagation speeds for periodic traveling-wave solutions of a F–N system plotted against the wavelength λ.	134
4.14	Plots of the initial ionic current (J_1) and the steady state current (J_{ss}) under the M–L model.	136
4.15	Amplitude of a traveling-wave impulse on an H–H axon plotted against a narcotization factor—η—reducing the maximum sodium and potassium conductances.	144
4.16	Propagation of a decremental impulse on an H–H axon narcotized by the factor $\eta = 0.25$.	145
4.17	A Purkinje cell of the human cerebellum. The dendrites of this cell spread over an area of about 0.25 mm×0.25 mm.	151
4.18	(a) Abrupt widening of a nerve fiber. (b) Branching region.	151
4.19	Spiral waves in an excitable chemical reagent.	156
4.20	A computer generated plot of a scroll ring.	158
4.21	Lichens growing on a wall in Scotland may be a biological example of spiral waves.	170
5.1	The upper figure shows a spring-mass lattice at rest with all values of y_n and $r_n = y_{n+1} - y_n$ equal to zero. The lower figure shows a compressive wave, corresponding to a Toda-lattice soliton (TLS) given in Equation (5.5).	177
5.2	(a) A plot of the acoustic mode dispersion relation for a uniform lattice given in Equation (5.14) with 10 masses and periodic boundary conditions. (b) A corresponding plot for two atoms in each of 10 unit cells, leading to a band of optical modes.	183
5.3	The relations between γ and ω for stationary solutions of DST with three lattice sites, M as in Equation (5.51), and $\varepsilon = 1$. Branch designations are explained in the text.	200

5.4	A short section of alpha helix. The dashed lines indicate relatively weak hydrogen bonds.	203
5.5	A cartoon of a short section of DNA double helix.	208
5.6	A unit cell of the two-dimensional tunnel diode array.	211
5.7	A 4×4 array ($N^2 = 16$) of the unit cells in Figure 5.6.	212
5.8	A Necker cube.	214
5.9	Equivalent circuit for a myelinated axon.	216
5.10	Leading-edge impulse velocity on a myelinated axon as a function of the discreteness parameter $D = 2\,\mathrm{mm/s}$.	219
6.1	Amplitudes of scattering solutions.	241
6.2	In the tank experiments of Figure 1.1, John Scott Russell observed that two or more solitary waves could be generated by increasing the quantity of water released.	257
7.1	A (superconducting) Josephson junction of length l with a fluxon (kink) traveling in the x-direction at velocity v.	298
7.2	The relationship between γ_c and α for kink-antikink annihilation, calculated from Equation (7.95).	317
7.3	First-order corrections to impulse speeds under the influence of mutual coupling. (Data courtesy of Steve Luzader [17].)	325
8.1	A simple nonlinear oscillator.	337
8.2	The first five eigenfunctions of a simple harmonic oscillator, plotted from Equation (8.24).	345
8.3	Sketch of a pump-probe experiment.	351
8.4	Energy eigenvalues of the QDNLS equation in the $f \to \infty$ limit, with $\varepsilon = 1$ and several values of γ.	368
8.5	Two quantum energy eigenvalues against propagation number for (a) the quantum Ablowitz–Ladik equation and (b) the fermionic polaron model with $\gamma < 2$.	379
8.6	Infrared absorption spectra of crystalline ACN in the Amide-I (CO stretching) region at three different temperatures.	391
8.7	Infrared absorption spectrum of crystalline ACN in the NH stretching region at two different temperatures.	400
8.8	The atomic structure of 4-methyl-pyridine.	401
C.1	Plots of the complete elliptic integral of the first kind $K(k)$ and the complete elliptic integral of the second kind $E(k)$ vs. k^2.	445
C.2	Plots of $\mathrm{sn}(\zeta, k)$ and $\mathrm{cn}(\zeta, k)$ as functions of $\zeta/K(k)$.	445
C.3	Plots of $\mathrm{dn}(\zeta, k)$ as functions of $\zeta/K(k)$. (Note that $\mathrm{dn}(\zeta, 0) = 1$.)	446

1
THE BIRTH OF A PARADIGM

"If I have seen further than others," said Isaac Newton, "it is because I have stood on the shoulders of giants." As we would not seem to see further than others because we stand on the *faces* of giants, this book begins with a historical introduction, noting salient contributions to nonlinear science that have been made over the past century and a half. More broadly, the present chapter is a study of the ways—constructive and otherwise—that the scientific community has met with new ideas.

1.1 From the Great Wave to the Great War

1.1.1 *Hydrodynamics*

Hydrodynamic solitary waves have been coursing up the fjords and firths of Europe since the dawn of time, but they were not noticed until 1834. In that year, a young Scottish engineer named John Scott Russell was engaged in an urgent endeavor. Britain's horse drawn canal boats were threatened by competition from the railroads, and he was conducting a series of experiments to measure the relationship between a boat's speed and its propelling force, with the aim of finding design parameters for conversion from horse-power to steam.

One late summer day, as chance would have it, a rope parted in his measurement apparatus and

the boat suddenly stopped—not so the mass of water in the channel which it had put in motion; it accumulated round the prow of the vessel in a state of violent agitation, then suddenly leaving it behind, rolled forward with great velocity, assuming the form of a large solitary elevation, a rounded, smooth and well defined heap of water, which continued its course along the channel without change of form or diminution of speed.

(See http://www.ma.hw.ac.uk/solitons/press.html for a recent photograph of this phenomenon.) Russell did not ignore his unexpected observation. Instead he "followed it on horseback, and overtook it still rolling on at a rate of some eight or nine miles an hour, preserving its original figure some thirty feet long and a foot to a foot and a half in height" until the wave became lost in the windings of the channel. And—as is described in his classic *Report on Waves* to the British Association for the Advancement of Science [83]—he continued to study the solitary wave in tanks (see Figure 1.1), canals, and the Firth of Forth over the following decade.

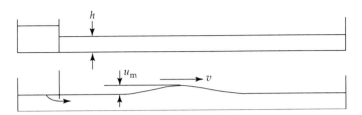

FIG. 1.1. *A hydrodynamic soliton created in a wave tank by John Scott Russell in the 1830s. In the upper diagram, a column of water is accumulated at the left-hand end of the tank. Release of this water by lifting a sliding panel (in the lower diagram) generates a solitary wave that travels to the right with velocity v. (Redrawn from reference [83].)*

During these studies, Russell showed the hydrodynamic solitary wave to be an independent dynamic entity that moves with constant shape and speed. A wave of height u_m and traveling in a channel of depth h was found to have a velocity given by the empirical expression

$$v \doteq \sqrt{g\,(h + u_\mathrm{m})}, \tag{1.1}$$

where g is the acceleration of gravity. Furthermore, he demonstrated that a sufficiently large initial mass of water would produce two or more independent solitary waves and that solitary waves cross each other "without change of any kind."

Russell's intuitive understanding of the emergence of solitary waves provided a scientific basis for his wave line hull, altering contemporary standards of ship design and leading him to become one of the great naval architects of the nineteenth century [33]. Although soon confirmed and extended in observations on the Canal de Bourgogne, near Dijon [8], subsequent scientific discussions of the hydrodynamic solitary wave focused on the mathematical correctness rather than the physical significance of Russell's results.

The collective judgment of nineteenth century science can be read from Horace Lamb's opus on hydrodynamics, which allots a mere 3 of 730 pages to the solitary wave [63]. Evidence that Russell maintained a broader and deeper appreciation of his discovery is provided by a posthumous work where—among several provocative ideas—he correctly estimated the height of the earth's atmosphere from Equation (1.1) and the fact that "the sound of a cannon (a large amplitude nonlinear wave) travels faster than the command (a low amplitude linear wave) to fire it" [84].

Building upon an earlier study by Joseph Boussinesq [16], Diederik Korteweg and Gustav de Vries published in 1895 a theory of shallow water waves that reduces Russell's problem to its essential features. Their main result was the nonlinear partial differential equation (PDE)

$$\frac{\partial u}{\partial t} + c\frac{\partial u}{\partial x} + \varepsilon\frac{\partial^3 u}{\partial x^3} + \gamma u\frac{\partial u}{\partial x} = 0, \qquad (1.2)$$

which came to play a key role in future theoretical developments [59]. In this equation, $c = \sqrt{gh}$ is the speed of small amplitude waves, the dispersive parameter $\varepsilon \equiv c(h^2/6 - T/2\rho g)$, the nonlinear parameter $\gamma \equiv 3c/2h$, and T and ρ are respectively the surface tension and the density of water. (Pego has noted that for zero surface tension Equation (1.2) is equivalent to a pair of equations studied by Boussinesq in reference [16, 76].)

Korteweg and de Vries showed that Equation (1.2) has exact traveling-wave solutions of the form $u(x,t) = \tilde{u}(x - vt)$, where $\tilde{u}(\cdot)$ is the "rounded, smooth and well defined heap" that was first observed by Russell, and v is the wave speed. If the dispersive term (ε) and the nonlinear term (γ) in Equation (1.2) are both zero, then the Korteweg–de Vries (KdV) equation becomes linear and has a traveling-wave solution for any shape of the pulse at the fixed speed c. In general Equation (1.2) is nonlinear, and a traveling-wave solution obtains for a particular shape that is found experimentally to be

$$u(x,t) \propto \operatorname{sech}^2[\kappa(x - vt)]. \qquad (1.3)$$

With this particular shape the effects of dispersion are in balance with those of nonlinearity over a continuous range of wave speeds. This balance can be pictured as

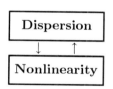

because the effect of dispersion is to spread out the energy of the pulse and the effect of nonlinearity is to draw it together. The solitary wave thus emerges as an independent dynamic entity, maintaining a dynamic balance between the *yin* of dispersion and the *yang* of nonlinearity.

Let us pause to reflect on this diagram, for we shall meet it again in this book. Note that the diagram represents a closed causal loop, biting its own tail like the serpentine *ouroboros* of mythology. Nonlinearity causes a wave to steepen, increasing the effect of dispersion. The ensuing balance between nonlinearity and dispersion leads to the emergence of a new thing: the solitary wave.

1.1.2 *Nonlinear diffusion*

Hydrodynamic solitary waves conserve energy, but there is another type of solitary wave that establishes a dynamic balance between nonlinear and dissipative effects. A simple example is provided by the ordinary candle, with which we are all familiar. In a process called "nonlinear diffusion," the heat from the flame diffuses into the wax, vaporizing it at the rate required to provide fuel for the flame. If the chemical energy stored in the wax is E (J/cm) and the power dissipated by the flame is P (J/s), the flame will propagate at the fixed velocity v (cm/s) for which

$$P = vE. \tag{1.4}$$

In other words, the flame digests energy at the same rate that it is eaten. The *yin* of thermal diffusion is balanced by the *yang* of nonlinear energy release according to the diagram

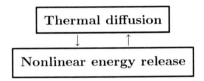

representing another closed causal loop leading to the emergence of another new thing: the flame.

Among the most important systems to display such nonlinear diffusion are the nerve fibers of animal organisms, but in the mid-nineteenth century electrophysiology was a subject steeped in mystery. Because it had been shown that Luigi Galvani's "animal electricity" was related to Alessandro Volta's "chemical electricity" and Ben Franklin's "atmospheric electricity," some wondered if the spark of life could be drawn from the heavens, as did Victor Frankenstein in Mary Shelley's tale of science gone amok. How might the conflicting claims of science and superstition be sorted out?

Important progress was made in 1850 by the German biophysicist Hermann Helmholtz, who first measured the velocity of nerve impulse propagation [48]. He did so over the objections of his father, a philosopher, who advised him not to bother with the experiment. Because a volitional act is identical to the will that directs it, the elder Helmholtz reasoned, it was absurd to try to observe a delay between the application of an electrical current and its muscular effect. Unfazed, young Helmholtz proceeded with the experiment, supplying a brief pulse of chemical electricity to the sciatic nerve of a frog and waiting for the resulting contraction of the leg muscle. The observed delay was unexpectedly long, leading to a velocity of about 27 m/s, which is close to the currently measured value.

Helmholtz missed the explanation of nerve impulse conduction (nonlinear diffusion) by supposing that its relatively low velocity implied that signals were carried by the motion of material particles, a notion that was to confuse neurophysiologists into the twentieth century. In retrospect, it is difficult to understand the failure of nineteenth century mathematical science to appreciate the descriptive power of nonlinear diffusion. It was certainly not a question of analytical ability, as Helmholtz's early study of hydrodynamic vortex motion was a prophetic contribution to later developments in solitary wave theory [49]. How could he have made such an elementary mistake in the case of nerve conduction?

The reason for this lapse may have been that linear diffusion is so unwavelike. Any pulse shaped solution of the linear diffusion equation

$$\frac{\partial^2 u}{\partial x^2} - \frac{\partial u}{\partial t} = 0$$

will fall with time near the peak (where it has negative curvature) and rise on the skirts (where it curves upward); thus it must spread out. To see how this qualitative effect is altered by the presence of nonlinearity, consider augmenting the linear diffusion equation to

$$\frac{\partial^2 u}{\partial x^2} - \frac{\partial u}{\partial t} = f(u). \tag{1.5}$$

For $f(u) = u(u-a)(u-1)$, an exact traveling-wave solution is [107]

$$\tilde{u}(x - vt) = \frac{1}{1 + \exp[(x-vt)/\sqrt{2}]},$$

propagating at the fixed velocity

$$v = (1 - 2a)/\sqrt{2}. \tag{1.6}$$

The first model of nerve impulse propagation was physical. In 1900, William Ostwald described experiments with metal wires in acid solutions, which are metastable because the metal is protected by an oxide layer [73]. When the oxide is disturbed (by scratching, say), however, a reacting region propagates along the wire with a characteristic shape and a speed of about a meter per second. But what determines this speed?

Computing the velocity of a nonlinear diffusion wave was described in 1906 by Robert Luther at the Main Meeting of the German Society for Applied Physical Chemistry in Dresden [67]. His subsequent publication is interesting today not only for the light it casts upon confusion surrounding the nerve problem at the end of the nineteenth century, but because it is the earliest known discussion of traveling waves in autocatalytic chemical reactions (in which a product of the reaction acts as a catalyst, stimulating its progress).

On this exciting morning, Luther began his lecture by recalling several current explanations for the phenomenon of nerve impulse conduction and then demonstrating a wave of chemical reaction, across which the color of a solution in a test tube changed. He suggested such an "autocatalytic process" as a means for nerve impulse conduction, noting that the propagation speed would be given by an expression of the form

$$v \propto \sqrt{D/\tau}, \tag{1.7}$$

where D is an ionic diffusion constant and τ is a reaction time. His audience included Walther Nernst (discoverer of the third law of thermodynamics) who expressed interest in the derivation of this equation[1] but doubted whether—in the context of homogeneous chemical reactions—it could explain the propagation speeds measured on nerve.

Although Equations (1.6) and (1.7) seem dissimilar in form, they are different aspects of the same general result. As we shall see in Chapter 4, the first provides the proportionality constant for the second, and the second indicates how the first depends upon physical parameters [90].

1.1.3 Bäcklund transformation theory

Although today's universities have much to offer, a modern graduate student might benefit from a year or two of study in Paris at the beginning of the twentieth century. The literature of this golden age of applied mathematics is difficult to come by in the United States nowadays, but aspects of it have been reviewed by George Lamb [62]. Of particular interest is the extension by Swedish mathematician Albert Bäcklund of Norwegian Sophus Lie's contact transformations, leading to a means for solving certain nonlinear PDEs [5]. His Bäcklund transformation (BT) can be motivated by requiring an expression of the form

$$du = P\,dx + Q\,dt,$$

to be an exact differential. For such a differential form, one must require that $u_x = P, u_t = Q$, and $P_t = Q_x$, where the subscripts are a convenient notation for partial derivatives. If P and Q can be found as functions of u and a known solution, u_0, then a new solution—let's call it u_1—can be generated by integrating the first-order pair

$$\frac{\partial u_1}{\partial x} = P(u_1, u_0, u_{0x}, u_{0t}),$$

$$\frac{\partial u_1}{\partial t} = Q(u_1, u_0, u_{0x}, u_{0t}).$$

Thus a known solution (u_0) is used to generate a new solution (u_1), after which the new solution is also known and can be used to generate another new solution.

It is straightforward to find a BT for any linear PDE, introducing a new eigenfunction into the total solution with each "turn of the Bäcklund crank." Only certain nonlinear PDEs are found to have BTs, but late nineteenth century

---

[1]The recorded discussion goes as follows. Nernst: "Who has derived this formula?" Luther: "Myself." N: "But this is not published yet?" L: "No, but it is a simple consequence of the differential equation." N: "Frankly, I cannot understand how one could obtain such high velocities, but one will look forward with even greater interest to the complete publication." Strangely, Luther never published his derivation [90].

applied mathematicians knew that these include

$$\frac{\partial^2 u}{\partial \xi \partial \tau} = \sin u, \qquad (1.8)$$

arising in research on the geometry of curved surfaces. The BT for Equation (1.8) is

$$\begin{aligned}\frac{\partial u_1}{\partial \xi} &= 2a \sin\left(\frac{u_1 + u_0}{2}\right) + \frac{\partial u_0}{\partial \xi}, \\ \frac{\partial u_1}{\partial \tau} &= \frac{2}{a} \sin\left(\frac{u_1 - u_0}{2}\right) - \frac{\partial u_0}{\partial \tau},\end{aligned} \qquad (1.9)$$

where a is an arbitrary constant. (To check this, one shows that if $u_{0,\xi\tau} = \sin u_0$ then $u_{1,\xi\tau} = \sin u_1$.)

A known solution of Equation (1.8) is clearly the vacuum: $u_0 = 0$. Starting from the vacuum, integration of Equations (1.9) gives

$$u_1(\xi, \tau) = 4 \arctan[\exp(a\xi + \tau/a)], \qquad (1.10)$$

a function that was familiar enough in the nineteenth century to have a name: the "gudermannian."[2] Generation of this function from the vacuum can be represented by the diagram

$$0 \xrightarrow{a} u_1,$$

and the subsequent development of a hierarchy of solutions by

$$0 \xrightarrow{a_1} u_1 \xrightarrow{a_2} u_2 \cdots \xrightarrow{a_N} u_N.$$

Each level in this diagram introduces an additional component (or nonlinear eigenfunction) with the gudermannian form. Direct integration to find higher level solutions becomes increasingly difficult, but algebraic techniques are available that do not require integration above the first level [62]. All this was known in the late-nineteenth century.

1.1.4 A theory of matter

Another thread of our story began in 1912 when Gustav Mie started to publish his "theory of matter" [69]. In this prescient though seemingly forgotten series of papers, Mie suggested a nonlinear augmentation of Maxwell's electromagnetic equations out of which elementary particles (e.g., the electron) would arise in a natural way. To this end, he defined a world function (Φ) as an energy functional depending upon electric field intensity ($\mathbf{E}$) and magnetic flux density ($\mathbf{B}$) and the four components of the electromagnetic potential ($\mathbf{A}, \phi$). Requiring Φ to

---

[2] For the German mathematician Christoph Gudermann (1798–1852), who was interested in the theory of elliptic functions.

be a function of the parameters $\eta \equiv \sqrt{\mathbf{E}^2 - \mathbf{B}^2}$ and $\chi \equiv \sqrt{\phi^2 - \mathbf{A}^2}$ insured relativistic invariance. The specific choice $\Phi = -\eta^2/2 + a\chi^6/6$ led to a static, spherically symmetric electric potential (ϕ) satisfying $r\phi'' + 2\phi' + ar\phi^5 = 0$ of the form

$$\phi(r) \approx \frac{(3r_0^2/a)^{1/4}}{\sqrt{r^2 + r_0^2}}.$$

Setting $4\pi(3r_0^2/a)^{1/4} = e$ (the electronic charge) yielded a spherically symmetric model for the electron with a radius of r_0 and electric potential

$$\phi(r) \to e/4\pi r$$

as $r \to \infty$. The Lorentz invariance that is built into the theory permits this solution to travel with any speed up to the limiting velocity of light with an appropriate Lorentz contraction.

Those who find Mie's ideas eccentric should note Albert Einstein comment that: "In the foundation of any consistent field theory, the particle concept must not appear in addition to the field concept. The whole theory must be based solely on partial differential equations and their singularity-free solutions" [31].

1.2 Between the wars

Strangely, research into the theory of BTs ceased to be of interest after the First World War. Although largely due to growing excitement with developments in (linear) quantum theory, part of the reason may have been the battlefield deaths of several French scientists who had been active in BT research [62]. Be that as it may, nonlinear research seemed—like the mythical American rustic Rip Van Winkle—to sleep for almost two decades until 1933 when a young Soviet physicist, Lev Landau, suggested the idea of a "polaron" as a means for an electron to move through a crystal lattice [64]. His basic idea was that a localized electronic wave packet would distort the lattice in its vicinity, forging a path for its progress, much as a child's marble works its way through a plate of spaghetti.

In the following year, inspired by Carl Anderson's discovery of positron-electron creation from cosmic radiation [80], the German physicist Max Born revisited Mie's nonlinear electromagnetics. Born was particularly interested in establishing a theory with solutions that are independent of the absolute electromagnetic potential so that it could be compatible with the requirements of quantum theory. Together with Leopold Infeld he eliminated the χ-dependence in Mie's functional formulation and chose instead a Lagrangian density of the form [15]

$$\mathcal{L} = E_0^2 \sqrt{1 + (\mathbf{H}^2 - \mathbf{E}^2)/E_0^2} - E_0^2,$$

where $\mathbf{H}$ is the magnetic field intensity and E_0 sets the field intensities at which nonlinearities arise. Thus, at low field amplitudes, this reduces to the classical Lagrangian density $\mathcal{L} = (\mathbf{H}^2 - \mathbf{E}^2)/2$. They found a spherically symmetric model electron with **E** everywhere finite although the electric displacement exhibits a singularity at the origin. Erwin Schrödinger was interested in Born's theory as early as 1935 [82] and continuing through the 1940s when—as founding director of the Dublin Institute for Advanced Studies—he attempted to move research in physics from the study of linear problems toward nonlinear science [72].

Entirely unrelated in the mid-1930s were the first mathematical studies of the nonlinear diffusion Equation (1.5) in the Soviet Union. Motivated by the English geneticist Ronald Fisher's analysis of genetic diffusion [39], Andrej Kolmogoroff and his colleagues considered the nonlinear function $f(u) = u(u-1)$ [58]. David Frank-Kamenetsky and Yakov Zeldovich, on the other hand, sought to understand flame propagation through a study of nonlinear diffusion with $f(u) = u(u - a)(u - 1)$ [107]. One wonders how much more rapidly neurophysiology would have advanced if these scientists had taken interest in the problem of nerve conduction. Mathematical input was clearly needed, for in 1938 the giant axon of the common squid had recently been recognized as a nerve fiber, and the American biophysicist Kenneth Cole published the first oscillograph of a nerve impulse, reproduced in Figure 1.2 [23]. Just as for the temperature of a candle flame, this classic image shows the voltage of the nerve membrane rising to a maximum value before returning to its resting level.

Also from the Soviet Union there emerged in 1938 yet another unrelated line of research that was motivated by a fundamental problem in solid state

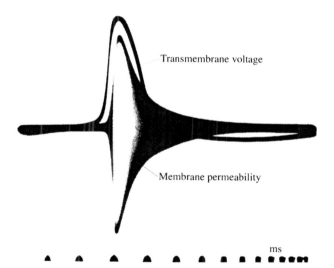

FIG. 1.2. *Measurement of the change in membrane conductance (band) and membrane voltage (line) with time during the passage of a nerve impulse on a squid axon. (The marks along the lower edge indicate time intervals of 1 ms.) (Courtesy of K.S. Cole.)*

physics: the relationship between dislocation dynamics and plastic deformation of a crystal. In this study, Yakov Frenkel and Tatiana Kontorova showed that a basic equation to describe dislocation motion is [42]

$$\frac{\partial^2 u}{\partial x^2} - \frac{\partial^2 u}{\partial t^2} = \sin u, \tag{1.11}$$

which can be viewed as a nonlinear augmentation of the linear wave equation ($u_{xx} - u_{tt} = 0$) in the same sense that Equation (1.5) is a nonlinear augmentation of the linear diffusion equation ($u_{xx} - u_t = 0$). Furthermore, Equation (1.11) is identical to Equation (1.8) after the independent variable transformation

$$\xi = (x+t)/2 \quad \text{and} \quad \tau = (x-t)/2. \tag{1.12}$$

The gudermannian solution of Equation (1.10) corresponds, under this transformation, to the propagation of a dislocation that is described by

$$u(x,t) = 4\arctan\left[\exp\left(\frac{x-vt}{\sqrt{1-v^2}}\right)\right]. \tag{1.13}$$

with velocity $v = (1-a^2)/(1+a^2)$. Frenkel and Kontorova would have been excited to read a work on BTs by Rudolf Steuerwald, which appeared in 1936 and included many of the analytic solutions of Equation (1.11) that were to become of great interest some 40 years later [92].

Another facet of early research in nonlinear science was the observation that vibrational energy in small molecules can be concentrated near particular interatomic bonds. An example of such a "local mode" in benzene (C_6H_6), with its hexagonal structure of six CH oscillators, is shown in Figure 1.3. By the mid-1920s, experimental evidence from measurements of infrared absorption and heats of dissociation suggested that the energy of the six CH stretching oscillators was localized on a single bond and not spread out over the entire molecule [13, 32]. In the jargon of modern nonlinear theory, such a local mode is called a "breather" that is pinned to the lattice. Nomenclature aside, the nature of the phenomenon was well understood by the German molecular spectroscopist Reinhard Mecke, who published a series of papers on the subject during the 1930s (see [79] for a bibliography of this work). Although a well established description of energy localization by nonlinearity, Mecke's research escaped the attention of those with budding interest in nonlinear science during the 1970s [68].

Between the wars, therefore, nonlinear research was fragmented in ways that are difficult to understand today. Old knowledge was largely unavailable for application to new problems. Fundamental relationships between nonlinear studies in electrodynamics, solid state physics, hydrodynamics, neurodynamics, genetic diffusion, flame propagation, physical chemistry, and applied mathematics were almost entirely overlooked. John Scott Russell's prophetic vision had been almost completely forgotten [84].

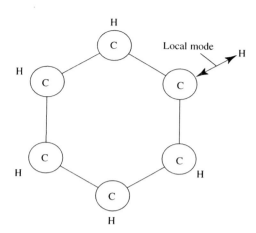

FIG. 1.3. *The planar structure of a benzene molecule, showing a local mode of the CH stretching oscillation.*

1.3 Nonlinear research from 1945 to 1985

If science had suffered from the spilling of its young blood during the First World War, this was not to be repeated during the century's second exercise in global slaughter. Poets and peasants could still be driven into the guns but not engineers and scientists, who were drafted for different tasks. North American research, in particular, was greatly strengthened by the war. Not only did several dozen of the very best European scientists emigrate to the United States, but the forced development of technological skills—in such diverse areas as electronics, microwaves, communications and control, physical chemistry, and the manipulation of elementary particles—left the scientific community in a state of high morale. In late 1945 anything seemed possible.

1.3.1 *Nerve studies*

Postwar research in electrophysiology was, so to speak, galvanized by the substantial increase in electronic measurement technology that had been developed in support of military communications systems and radar, and peacetime dividends were not long in coming. Based on the concept of nonlinear diffusion, a sound theoretical understanding of impulse propagation on the giant axon of the squid soon emerged. Briefly, the axon voltage V (shown as the line in Figure 1.2) obeys the PDE

$$\frac{\partial^2 V}{\partial x^2} - rc\frac{\partial V}{\partial t} = rj_\mathrm{i}, \tag{1.14}$$

where r is the electrical resistance per unit length of the axonal tube and c is the capacitance per unit length of the membrane; thus $1/rc$ is an electrical diffusion constant. The left-hand side of Equation (1.14) is a linear diffusion operator that is augmented by the nonlinear effects of the transmembrane ionic current, j_i, as was noted above for Equation (1.5). The main problem facing postwar researchers was to properly represent the dynamics of j_i, and a definitive answer to this question was provided in 1952 by the British electrophysiologists Alan Hodgkin and Andrew Huxley [52]. Their formulation introduced three additional variables to represent the opening and closing of ionic channels across the membrane, giving excellent predictions of the speed and shape of the impulse shown in Figure 1.2.

A simpler dynamical description for j_i was introduced in 1962 by Jinichi Nagumo and his colleagues at the University of Tokyo, based upon discussions with Richard FitzHugh [74]. In this formulation—usually called the FitzHugh–Nagumo system—ionic membrane current takes the form

$$j_\mathrm{i} = f(V) + R,$$
$$\frac{\partial R}{\partial t} = \varepsilon V, \qquad (1.15)$$

where R is a recovery variable and ε^{-1} is a time constant for the recovery process. In the limiting case $\varepsilon = 0$, R remains constant and Equation (1.14) reduces to Equation (1.5). In fact, R does remain relatively constant over the leading edge of a typical nerve impulse, so analysis of Equation (1.5) provides useful first approximations to both the shape of the leading edge and the speed of the impulse. Researchers in Nagumo's laboratory were also studying metal–acid nerve models in the early 1960s, as all who were fortunate enough to visit this lively group became aware, but the work was not published in English until 1976 [93].

Interestingly, a nerve impulse emerges as a singularity-free solution of the nonlinear PDE (1.14) just as Einstein has prescribed for the elementary particles of matter [31]. Following this observation, one might refer to the nerve impulse as an "elementary particle of thought." From this perspective, it becomes exciting to study the dynamic behavior of nerve impulses as independent dynamic entities. The many opportunities for impulse interactions include: (i) head-on collisions on a single fiber, (ii) front and rear interactions on a single fiber, (iii) impulse logic (Boolean AND or OR) at a dendritic branching region, (iv) parallel fiber (ephaptic) interactions, and more. By the mid-1970s mathematical neuroscience had become a vigorous area of research with implications for the global properties of the human brain [57, 86].

1.3.2 *Autocatalytic chemical reactions*

The study of oscillations and traveling waves in solutions of reacting chemicals has a checkered history. Publicly demonstrated in 1906 by Luther [67], as we have noted above, related oscillatory phenomena were also considered theoretically by Alfred Lotka in 1920 [66] and promptly reconfirmed experimentally [17]. Yet this area of research was dropped for some four decades under the mistaken notion that the observed data were at variance with the second law of thermodynamics.

The modern history of autocatalytic chemical reactions began in 1951 with the discovery by Boris Belousov in the Soviet Union of oscillations in the cerium ion catalyzed oxidation of citric acid by bromate ions [38]. Unfortunately, Belousov was unable to publish this discovery until 1958, and then only in an obscure proceedings that few have seen [10]. (Belousov died a disappointed man, although he was posthumously honored with the Lenin Prize in 1980.) Fortunately, this work was continued throughout the 1960s by Anatol Zhabotinsky, and the class of oscillatory, metal ion catalyzed oxidation of organic compounds by ionic bromate is now widely known as the Belousov–Zhabotinsky (BZ) reaction [108]. Essentially, this class of reactions involves the nonlinear diffusion of both excitatory and inhibitory molecules through the liquid phase in a manner that is similar to the FitzHugh–Nagumo system but with two differences: the inhibitory components can also diffuse, and the diffusion operators can act in two or three spatial dimensions.

In 1970, it was demonstrated by Albert Zaikin and Zhabotinsky that a two-dimensional BZ system can support periodically self-exciting ring waves, as shown in the temporal sequence of photographs reproduced in Figure 1.4 [105]. From these data, it is seen that colliding waves of concentration annihilate each other (leaving characteristic cusps in the wave fronts), and also that wave trains of higher frequency tend to displace those of lower frequency. As Zaikin and Zhabotinsky noted in their seminal publication, such oscillating chemical reactions "are interesting not only in themselves but as models of a number of important biological processes."

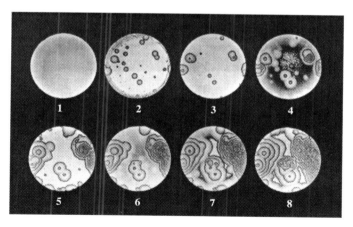

FIG. 1.4. *Self-oscillatory ring waves in a two-dimensional chemical reaction diffusion system. (The circular diameter is 100 mm and successive photographs are at intervals of 1 min.) (Courtesy of A.M. Zhabotinsky.)*

Shortly thereafter, US biologist Art Winfree showed that the BZ system can also produce spiral waves that continue to wind outward from a generating center without external stimulation [101]. (In his book *When Time Breaks Down*, Winfree notes that spiral waves were suggested by observations made a decade earlier in Nagumo's laboratory of two-dimensional waves on iron wire grids immersed in nitric acid [102].) After these two discoveries, the study of reaction and diffusion in a variety of chemical systems has become a subject of intense interest [38, 60].

From a more general perspective, nerve conduction and autocatalytic chemistry are examples from a broader field that Hermann Haken calls "synergetics," involving the application of modern tools of nonlinear analysis to a wide variety of problems in physics, chemistry, biology, sociology, and neurology [7, 45, 46].

1.3.3 Solitons

A striking development of the postwar period was the digital computer, which was hungry for appropriate applications. Enrico Fermi and Stan Ulam suggested one of the first scientific problems to be assigned to the 18,000 interconnected vacuum tubes of the Los Alamos MANIAC machine: the dynamics of energy equipartition in a slightly nonlinear mechanical system. The system they chose was a chain of 64 equal mass particles connected by slightly nonlinear springs, so from the linear perspective there were 64 normal modes of oscillation in the system. It was expected that if all the initial energy were put into a single mode, the slight nonlinearity would cause a gradual progress toward equal distribution of the energy among the normal modes in what could be considered a thermalized state. The computer results, obtained in collaboration with their young colleague, John Pasta, were surprising. If all the energy was originally in the mode of lowest frequency, for example, it returned almost entirely to that mode after a period of interaction with a few other low frequency modes [37]. In general, no thermalization was observed.[3]

Pasta has commented that Fermi believed this to be one of the most important problems he (Fermi) had studied. It certainly was one of the more puzzling. Pursuit of this mystery led Norman Zabusky and Martin Kruskal to approximate the nonlinear spring-mass system by the KdV equation, rederived for nonlinear wave propagation in a collisionless plasma. In 1965 they reported numerical observations that KdV solitary waves pass through each other with no change in shape or speed, and they coined the term "soliton" to suggest this particle-like property [104].

---

[3]Nick Metropolis told me that this discovery was accidental. Because on a short time scale the initial energy seemed to proceed toward a thermalized state (confirming the motivating intuition), the early computations were terminated before the energy returned to its original configuration. One afternoon they got to talking and unintentionally left MANIAC running, leading to their unexpected observation.

Zabusky and Kruskal were not the first to observe nondestructive interactions of energy conserving solitary waves. Apart from Russell's early tank experiments, J.K. Perring and T.R.H. Skyrme had studied solutions of Equation (1.11) with two solutions of the form in Equation (1.13) undergoing a collision. In 1962 they published numerical results showing perfect recovery of shapes and speeds after a collision and went on to discover an exact analytical description given by

$$u(x,t) = 4\arctan\left[\frac{v\sinh(x/\sqrt{1-v^2})}{\cosh(vt/\sqrt{1-v^2})}\right] \qquad (1.16)$$

for two solitary waves meeting at the origin with speeds of $\pm v$ [78]. (For a surface plot of this collision, look ahead to Figure 3.6.)

This result would not have surprised Bäcklund; it is merely the second (u_2) member of the hierarchy of solutions generated from the vacuum state by the BT defined in Equations (1.9). Nor would it have seemed unexpected to A. Seeger and his colleagues, who had noted in 1953 the connections between the nineteenth century work [92] and that of Frenkel and Kontorova [89]. But Perring and Skyrme were interested in Equation (1.11) as a nonlinear model for elementary particles of matter; thus the complete absence of scattering may have been disappointing. Throughout the 1960s, Equation (1.11) arose in a wide variety of problems including the propagation of domain walls in ferromagnetic and ferroelectric materials [9, 28, 35], self-induced transparency in nonlinear optics [61], and the propagation of magnetic flux quanta in long Josephson transmission lines [6, 85]. Eventually it became known as the "sine–Gordon" (SG) equation (a nonlinear version of the Klein–Gordon equation: $u_{xx} - u_{tt} = u$).

Perhaps the most important contribution by Zabusky and Kruskal in their 1965 paper was to recognize the relation between nondestructive soliton collisions and the riddle of Fermi–Pasta–Ulam (FPU) recurrence. The soliton solutions of the KdV equation were viewed as independent and localized dynamic entities, out of which more complex behavior could emerge, and from this perspective, they explained the FPU observations in the following manner. The initial condition $u(x,0)$ generates a family of solitons with different speeds, moving apart in the (x,t)-plane. Because the system studied was of finite length with perfect reflections at both ends, the solitons could not move infinitely far apart; instead they eventually reassembled in the (x,t)-plane and approximately recreated the initial condition $u(x,0)$ at a recurrence time t_0. If $u(x,0)$ puts all of the energy into a single mode of the system (or oscillatory configuration, like the shape of a vibrating piano string), then most of the energy reappears in this mode at t_0. By 1967, this insight had led Kruskal and his colleagues to devise a novel scheme for constructing a general solution to the KdV equation through a series of linear computations [43].

This procedure was soon expressed in the following general form by Peter Lax, for a general nonlinear wave equation

$$\frac{\partial u}{\partial t} + N(u) = 0, \tag{1.17}$$

where $N(u)$ denotes a nonlinear operator [65]. (For the KdV equation, $N(u) = cu_x + \varepsilon u_{xxx} + \gamma u u_x$.) Suppose one can find two linear operators, L and M, that depend upon $u(x,t)$—a solution of the nonlinear equation—and also satisfy the operator equation

$$\frac{\partial L}{\partial t} = ML - LM. \tag{1.18}$$

(Note that because $u(x,t)$ is embedded within Equation (1.18), Equation (1.17) is implied by Equation (1.18).)

Now L is viewed as an x-dependent linear scattering operator with potential $u(x,t)$, which is discussed in detail in Chapter 6. Its scattering data (bound state eigenvalues and reflection coefficient) are found from a study of

$$L\psi = \lambda\psi, \tag{1.19}$$

where $\psi(x,t)$ is the dependent variable in this "associated linear problem." If the time dependence of the scattered wave is taken to be

$$\frac{\partial \psi}{\partial t} = M\psi, \tag{1.20}$$

then the discrete (bound state) eigenvalues of Equation (1.19) are independent of time ($d\lambda/dt = 0$). Thus the solution of the nonlinear initial value problem

$$u(x,0) \longrightarrow u(x,t)$$

can be effected through the following three linear computations:

1. *Direct problem.* Calculate the scattering data of L at $t = 0$ from $u(x,0)$.
2. *Time evolution of the scattering data.* Use the asymptotic form of Equation (1.20) at $x = \pm\infty$ to find the time variation of the reflection coefficient. As noted, the bound state eigenvalues are independent of time.
3. *Inverse problem.* From the scattering data of L as a function of time, calculate the scattering potential $u(x,t)$.

Each bound state eigenfunction of Equation (1.19) is related to a particular soliton in the general solution, and the associated eigenvalue (λ_j) corresponds directly to the velocity (v_j) of that soliton. In the small amplitude (linear) limit, there are no solitons present in the solution of Equation (1.17) and no bound states in Equation (1.19). In this case, the analysis reduces to the Fourier transform method for linear PDEs, to be reviewed in the next chapter. Thus the technique of Kruskal and his colleagues has become known as the "inverse scattering method" (ISM), wherein solitons play the role of generalized Fourier components.

If $u(x,0)$ is made increasingly large, two or more solitons appear in the solution, just as had been reported by Russell for a sufficiently large initial mass of water 123 years before [83].

Another important development of the 1960s was the discovery by Morikazu Toda of exact 2-soliton interactions on a nonlinear spring-mass system with the spring potential having the form [97, 98]

$$\text{potential} = \frac{a}{b}e^{-br} + ar, \tag{1.21}$$

where r is the extension of a nonlinear spring from its equilibrium value. In the limit $a \to \infty$ and $b \to 0$ with ab finite, this reduces to the quadratic potential of a linear spring. In the limit $a \to 0$ and $b \to \infty$ with ab finite, it describes the interaction between hard spheres. Thus it was known by 1967 that solitons were not limited to PDEs; local solutions of difference differential equations could also exhibit the properties of unchanging shapes and speeds after collisions.

Up to this point we have been concentrating attention on the dynamics of individual solitary waves. In 1965, however, Gerry Whitham began a series of papers that describes the dynamics of periodic traveling waves [100]. Such solutions take the general form

$$u(x,t) = \tilde{u}(kx + \omega t) \equiv \tilde{u}(\theta),$$

and include waves that are periodic but not sinusoidal (called "cnoidal waves") that were obtained as solutions of Equation (1.2) by Korteweg and de Vries. Whitham assumed the solution to be locally periodic, though not sinusoidal, and assumed that the dynamics of the underlying PDE could be derived from a Lagrangian density function, $\mathcal{L}$. From this he obtained an averaged Lagrangian over a period of the wave

$$\langle \mathcal{L} \rangle \equiv \frac{1}{2\pi} \int_0^{2\pi} \mathcal{L}\, d\theta,$$

as a function of the local values of the wave number ($k = \theta_x$), frequency ($\omega = \theta_t$), and amplitude (A). Thus $\langle \mathcal{L} \rangle = \langle \mathcal{L} \rangle(A, \theta_x, \theta_t)$, for which he proposed an "averaged Lagrangian principle"

$$\delta \iint \langle \mathcal{L} \rangle(A, \theta_x, \theta_t) dx\, dt = 0,$$

based on the functions $A(x,t)$ and $\theta(x,t)$. Under this formulation, slow variations in A, k, and ω are governed by the PDE system

$$\frac{\partial \langle \mathcal{L} \rangle}{\partial A} = 0$$
$$\frac{\partial}{\partial x} \frac{\partial \langle \mathcal{L} \rangle}{\partial k} + \frac{\partial}{\partial t} \frac{\partial \langle \mathcal{L} \rangle}{\partial \omega} = 0 \qquad (1.22)$$
$$\frac{\partial k}{\partial t} - \frac{\partial \omega}{\partial x} = 0,$$

where the first of these equations is obtained by variation of $\langle \mathcal{L} \rangle$ with respect to A, the second by variation with respect to θ, and the last is a conservation law for wave crests (or solitons). The first of Equations (1.22) can be viewed as a nonlinear dispersion relation

$$D(A, k, \omega) = 0,$$

depending upon the local value of the wave amplitude.

The events noted above are but salient features of a growing panorama of nonlinear wave activities that became gradually less parochial during the 1960s. Solid state physicists began to see relationships between their solitary waves (magnetic domain walls, self-shaping pulses of light, quanta of magnetic flux, polarons, etc.) and those from classical hydrodynamics, and applied mathematicians began to suspect that the ISM might be used for a broader class of nonlinear wave equations. It was amid this excitement that Alan Newell and his colleagues organized a research workshop for three and a half weeks during the summer of 1972 in upstate New York, where the participants "ranged over a wide spectrum of ages (from graduate students to senior scientists), background interests (biology, electrical engineering, geology, geophysics, mathematics, physics) and countries of origin (United States, Canada, Great Britain, Australia)" [75]. It was a seminal meeting for soliton research in the English speaking world. Many cross-disciplinary bonds of collaboration and friendship were first formed, and a sense of the existence of nonlinear science as a broad and vigorous professional activity was established.

Interestingly, one of the most significant contributions to Newell's conference arrived by mail from the Soviet Union [106]. Vladimir Zakharov and Alexey Shabat had shown that Lax operators (L and M) exist for the nonlinear Schrödinger (NLS) equation

$$i\frac{\partial u}{\partial t} + \varepsilon \frac{\partial^2 u}{\partial x^2} + \gamma |u|^2 u = 0, \qquad (1.23)$$

and proceeded in elegant ways to develop the structure of the ISM and derive an infinite set of conservation laws corresponding to those previously found for KdV by Robert Miura [70, 71]. This equation had arisen in practice to describe nonlinear envelope waves in hydrodynamics [11], nonlinear optics [55], nonlinear acoustics [96], and plasma waves [54, 95]. Thus, many left Newell's conference realizing that four of the most fundamental nonlinear equations (KdV, SG, NLS, and Toda's nonlinear spring-mass system) displayed solitary wave behavior with the special properties that had led Zabusky and Kruskal to coin the term soliton.

Within the next two years, basic ingredients (Lax operators) for the ISM had been constructed for the SG equation [1, 94] and the Toda lattice [40, 41, 50]. Following the discovery of a BT for the KdV equation [99], the equivalence of the BT and the ISM for a wide class of nonlinear wave systems was established [2, 3]. Those were memorable days.

1.3.4 Local modes in molecules and molecular crystals

Between 1950 and 1970, the local mode interpretation of infrared molecular spectra fell into disuse as physical chemists turned to spectral assignments based on linear theories. The argument for such a shift away from the earlier nonlinear perspective went like this. Molecular vibrations are governed by quantum theory, which is linear. Because the quantum energy operator shares the symmetry of the molecule (six-fold rotations for benzene), the quantum mechanical eigenfunction must have the same symmetry. Thus the energy of an eigenfunction cannot be localized on an individual bond, as is indicated in Figure 1.3.

Although correct as far as it goes, this argument is incomplete. The physical chemist observes wave packets of eigenfunctions that are composed of several eigenfunctions, spread over an energy range ΔE. In an experimental observation of a local mode, as Lisa Bernstein has shown, the energy range of the corresponding wave packet is [12]

$$\Delta E \propto 1/(n-1)!,$$

where n is the principal quantum level of the oscillation. Because an initially localized wave packet will remain so for a time of order

$$\tau \sim \frac{\pi \hbar}{\Delta E} \propto (n-1)!$$

(where $\hbar \doteq 1.05 \times 10^{-34}$ Js is Planck's constant), it is not difficult to find experimental situations where τ is larger than the time of measurement. At $n \sim 20$ in a typical CH stretching oscillation, for example, τ exceeds the age of the universe. For such times, the wave packet is effectively localized.

As local modes were rediscovered by physical chemists in the 1970s, such insights came to light, but the original contributions of Mecke and his coworkers were almost completely forgotten [68, 79]. And although local modes in molecules are important examples of discrete solitons, the field of research was not appreciated by the soliton community until the mid-1980s [88], when spectroscopic evidence for local modes in molecular crystals began to appear [21].

The theory of local modes of vibration in molecular crystals is related to that for Landau's polaron, which was widely studied in the Soviet Union in the late 1940s and early 1950s [77]. In 1973, it was suggested that a polaronic mechanism might facilitate electron transport in biomolecules (proteins) [56], and a related means was proposed by the Ukrainian physicist Alexander Davydov for the storage and transport of infrared vibrational energy [25].

Davydov's theory was formulated for the alpha-helix regions of proteins, where chains of amino acids (each containing a strong carbon–oxygen (CO) stretching resonance) are wound into a helix. If CO vibrational energy is localized, it distorts the coil, which in turn acts as a potential well, preventing the dispersion of vibrational energy [26].

Interestingly, it was also in 1973 that the Italian experimental physicist Giorgio Careri first presented spectroscopic measurements on molecular crystals of acetanilide, a hydrogen bonded amide structure that models natural protein [20]. (During the nineteenth century, acetanilide was sold as a cure for headaches, but because users occasionally died, it lost out to aspirin.) Careri's studies showed that the spectral line of the CO stretching resonance was split, with an "anomalous" lower energy band sharing the total oscillator strength. After a decade of unsuccessful attempts to assign this anomalous resonance in a conventional manner, he turned to a polaronic assignment that was inspired by the work of Davydov [21].

Just as with the rediscovery of local modes in molecules, many physical chemists viewed Careri's suggestion of local modes in acetanilide with mistrust, and a substantial amount of experimental work was undertaken to challenge it. Viewed in retrospect, these efforts strongly supported the assignment of the anomalous CO stretching resonance in acetanilide to a polaronic mechanism [87], and recent optical pulse-probe measurements—that respond only to nonlinear features—confirm this assignment [29, 30].

Thus certain spectral lines in hydrogen bonded molecular crystals provide evidence of self-localized states, as described by the following diagram.

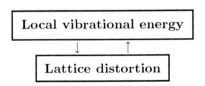

Opportunities for applications of such local modes as oscillators, mixers, and detectors in the terahertz (far infrared) range of the electromagnetic spectrum are yet to be explored.

1.3.5 *Elementary particle research*

Einstein's conviction that a consistent theory for particle physics must be based on localized solutions of nonlinear PDEs [31] was shared by some of his most distinguished colleagues. In addition to the above mentioned work by Born, Infeld, and Schrödinger in this area, both Werner Heisenberg [47] and Louis de Broglie [18, 19] have suggested nonlinear field theories that—in their simplest representations—can be viewed as the augmentation of linear field equations by a nonlinear term of the form $|u|^2 u$ as in Equation (1.23).

The ideas of de Broglie and David Bohm [14] are more closely related to those of the ISM. In their "theory of the double solution"—which has possibilities for further research [53]—the real particle is a localized solution of a nonlinear equation with the form

$$u = U e^{i\theta'}.$$

Associated with this localized nonlinear solution is the solution of a corresponding linear equation

$$\psi = \Psi e^{i\theta}$$

with

$$\theta = \theta' \qquad (1.24)$$

except in a small region surrounding the real particle. The function ψ is taken to be a solution of Schrödinger's quantum mechanical wave equation, and the phase condition of Equation (1.24) allows the particle to be guided by ψ. Similarly, the nonlinear solution of Equation (1.17) is guided through space-time by the linear asymptotic solution of Equations (1.19) and (1.20).

During the 1960s, several investigators considered the SG equation of Equation (1.11) as a two-dimensional field theory for elementary particles [27, 34, 51, 78, 81, 91], and this work gained momentum in the early 1970s when it became known that special properties of the SG equation allow the corresponding quantum problem to be solved, showing that certain qualitative properties of the classical solution survive quantization [24, 36, 44]. In particular, the classical field energy was found to be a useful first approximation for the soliton mass, with quantum effects coming in as second-order corrections. Following this discovery, there was a dramatic rise in research related to the nonlinear theory of elementary particles [4, 103].

1.4 Recent developments

Since the middle of the 1980s, a new paradigm—the emergent coherent structure—has become established in most areas of applied science. No longer is one surprised to find spatially localized regions of energy or activity that are stable in time, possessing the essential properties of objects. No more is nonlinear diffusion confused with its linear cousin. Nor is it unusual for a physical chemist to discuss science with an oceanographer or for a condensed matter physicist to sit down for a beer with a biologist. The proliferation of interdisciplinary research centers and conferences on nonlinear science shows that workers in diverse areas are learning to learn from each other [22].

Perhaps the most important implication of the new paradigm is that coherent structures must be treated as independent dynamic entities. Each emergent structure has its own regularities of propagation and interaction, which must be respected.

Current nonlinear studies that focus on understanding the details of such object-like behavior include the following:

1. *Optical fibers* where solitons of the nonlinear Schrödinger equation—Equation (1.23)—provide a means to carry bits of information over many thousands of kilometers.
2. *Nerve fibers* on which solitary wave solutions of the Hodgkin–Huxley equations of nonlinear diffusion—Equation (1.14)—carry motor signals to our muscles and provide a basis for the myriad internal computations of our brains.
3. *Autocatalytic chemical reactions* in one-, two-, and three-spatial dimensions that display a large and substantially unexplored variety of stable dynamic forms.
4. *Superconducting (Josephson) transmission lines* where SG solitons of Equation (1.11) (which are also quantum units of magnetic flux) provide the basis for oscillators and detectors in the submillimeter range of the electromagnetic spectrum.
5. *The surface of an ocean*, which carries energy in the form of KdV solitons—as in Equation (1.3)—from the site of a submerged earthquake to ravage the shores of a distant land.
6. *Solid state physics* where crystal dislocations and ferromagnetic or ferroelectric domain walls (described by the SG equation) play key roles in determining the global properties of materials.
7. *Local modes* in molecules and molecular crystals, which are becoming understood through recent developments in the quantum theory of nonlinear lattices. Such modes are important because they suggest means for laser catalysis of chemical reactions. Local modes in molecular crystals may lead to possibilities for generation and detection devices in the far infrared (terahertz) range of the electromagnetic spectrum.
8. *Biochemical solitons*, which are closely related to the local modes of molecular crystals, may play functional roles in the dynamics of enzymes and DNA.

This list can be extended from the basic fields of matter (underlying the properties of the elementary particles) through biomolecules; dynamics of the brain; biological organisms and the growth of cities, where the development of form (morphogenesis) is influenced by nonlinear diffusion; population dynamics and genetic diffusion; vortex structures of tornadoes, weather fronts and Jupiter's Great Red Spot; to the structures of black holes, galaxies, and the universe itself.

Such was the vision of nonlinear science that began to emerge as John Scott Russell first observed a solitary wave on the "happiest day of [his] life" in August of 1834 and was expressed in his posthumous work *The Wave of Translation in the Oceans of Water, Air and Ether* [84]. The aim of this book is to introduce Russell's vision to the reader.

REFERENCES

1. M J Ablowitz, D J Kaup, A C Newell, and H Segur. The initial value solution for the sine–Gordon equation. *Phys. Rev. Lett.* 30 (1973) 1262–1264.
2. M J Ablowitz, D J Kaup, A C Newell, and H Segur. Nonlinear evolution equations of physical significance. *Phys. Rev. Lett.* 31 (1973) 125–127.
3. M J Ablowitz, D J Kaup, A C Newell, and H Segur. The inverse scattering transform Fourier analysis for nonlinear problems. *Stud. Appl. Math.* 53 (1974) 294–315.
4. A Actor. Classical solutions of $SU(2)$ Yang–Mills theories. *Rev. Mod. Phys.* 51 (1979) 461–525.
5. A V Bäcklund. Zur Theorie der Flächentransformationen. *Math. Ann.* 19 (1882) 387–422.
6. A Barone, F Esposito, C J Magee, and A C Scott. Theory and applications of the sine–Gordon equation. *Riv. Nuovo Cimento* 1 (1971) 227–267.
7. E Basar, H Flohr, H Haken, and A J Mandell (eds.). *Synergetics of the Brain*. Springer-Verlag, Berlin, 1983.
8. M H Bazin. Rapport aux remous et a la propagation des ondes. Report to the Academy of Sciences, Paris, 1865.
9. C P Bean and R W deBlois. Ferromagnetic domain wall as a pseudorelativistic entity. *Bull. Am. Phys. Soc.* 4 (1959) 53.
10. B P Belousov. A periodic reaction and its mechanism. *Sbornik Referatov po Radiatsionni Meditsine*. Medgiz, Moscow (1959) 145. See reference [38] for the first English translation.
11. D J Benney and A C Newell. The propagation of nonlinear wave envelopes. *J. Math. Phys.* 46 (1967) 133–139.
12. L Bernstein, J C Eilbeck, and A C Scott. The quantum theory of local modes in a coupled system of nonlinear oscillators. *Nonlinearity* 3 (1990) 293–323.
13. R T Birge and H Sponer. The heat of dissociation of non-polar molecules. *Phys. Rev.* 28 (1926) 259–283.
14. D Bohm. *Causality and Chance in Modern Physics*. Routledge & Kegan Paul, London, 1957.
15. M Born and L Infeld. Foundations of a new field theory. *Proc. R. Soc.* (London) A 144 (1934) 425–451.
16. J Boussinesq. Théorie des ondes et des remous qui se propagent le long d'un canal rectangulaire horizontal, en communiquant au liquide contenu dans ce canal des vitesses sensiblement pareilles de la surface au fond. *J. Math. Pures Appl.* (ser. 2) 17 (1872) 55–108.
17. W C Bray. A periodic reaction in homogeneous solution and its relation to catalysis. *J. Am. Chem. Soc.* 43 (1921) 1262–1267.
18. L de Broglie. *Nonlinear Wave Mechanics*. Elsevier, Amsterdam, 1960.
19. L de Broglie. *Introduction to the Vigier Theory of Elementary Particles*. Elsevier, Amsterdam, 1963.

20. G Careri. Search for cooperative phenomena in hydrogen-bonded amide structures. In *Cooperative Phenomena*, H Haken and M Wagner, eds., Springer-Verlag, Berlin, 1973.
21. G Careri, U Buontempo, F Galluzzi, A C Scott, E Gratton, and E Shyamsunder. Spectroscopic evidence for Davydov-like solitons in acetanilide. *Phys. Rev. B* 30 (1984) 4689–4702.
22. P L Christiansen, M P Sørensen, and A C Scott. *Nonlinear Science at the Dawn of the 21st Century*. Springer-Verlag, Berlin, 2000.
23. K S Cole and H J Curtis. Electrical impedance of nerve during activity. *Nature* 142 (1938) 209.
24. R F Dashen, B Hasslacher, and A Neveu. Particle spectrum in model field theories from semiclassical functional integral techniques. *Phys. Rev. D* 11 (1975) 3424–3450.
25. A S Davydov. The theory of contraction of proteins under their excitation. *J. Theor. Biol.* 38 (1973) 559–569.
26. A S Davydov. *Solitons in Molecular Systems*. 2nd edition. Reidel, Dordrecht, 1991.
27. G H Derrick. Comments on nonlinear equations as models for elementary particles. *J. Math. Phys.* 5 (1964) 1252–1254.
28. W Döring. Über die Trägheit der Wände zwischen weisschen Bezirken. *Z. Naturforsch.* 31 (1948) 373–379.
29. J Edler and P Hamm. Self-trapping of the amide-I band in a peptide model crystal. *J. Chem. Phys.* 117 (2002) 2415–2424.
30. J Edler, P Hamm, and A C Scott. Femtosecond study of self-trapped vibrational excitons in crystalline acetanilide. *Phys. Rev. Lett.* 88 (2002) 067403-1–067403-4.
31. A Einstein. *Ideas and Opinions*. Crown, New York, 1954.
32. J W Ellis. Heats of linkage of C-H and N-H bonds from vibration spectra. *Phys. Rev.* 33 (1929) 27–36.
33. G S Emmerson. *John Scott Russell—A Great Victorian Engineer and Naval Architect*. John Murray, London, 1977.
34. U Enz. Discrete mass, elementary length, and a topological invariant as a consequence of a relativistic invariant variational principle. *Phys. Rev.* 131 (1963) 1392–1394.
35. U Enz. Die Dynamik der blochschen Wand. *Helv. Phys. Acta* 37 (1964) 245–251.
36. L D Faddeev. Hadrons from leptons? *JETP Lett.* 21 (1975) 64–65.
37. E Fermi, J R Pasta, and S M Ulam. Studies of nonlinear problems. Los Alamos Scientific Laboratory Report No. LA–1940 (1955). Reprinted in reference [75].
38. R J Field and M Burger. *Oscillations and Traveling Waves in Chemical Systems*. John Wiley, New York, 1985.
39. R A Fisher. The wave of advance of advantageous genes. *Ann. Eugen.*, now *Ann. Hum. Gen.* 7 (1937) 355–369.
40. H Flaschka. The Toda lattice. I. Existence of integrals. *Phys. Rev. B* 9 (1974) 1924–1925.
41. H Flaschka. On the Toda lattice. II. Inverse scattering solution. *Prog. Theor. Phys.* 51 (1974) 703–716.

REFERENCES

42. J Frenkel and T Kontorova. On the theory of plastic deformation and twinning. *J. Phys. (USSR)* 1 (1939) 137–149.
43. C S Gardner, J M Greene, M D Kruskal, and R M Miura. Method for solving the Korteweg–de Vries equation. *Phys. Rev. Lett.* 19 (1967) 1095–1097.
44. J Goldstone and R Jakiw. Quantization of nonlinear waves. *Phys. Rev. D* 11 (1975) 1486–1498.
45. H Haken. *Synergetics*. 3rd edition. Springer-Verlag, Berlin, 1983.
46. H Haken. *Advanced Synergetics*. Springer-Verlag, Berlin, 1983.
47. W Heisenberg. *Introduction to the Unified Field Theory of Elementary Particles*. John Wiley, New York, 1966.
48. H Helmholtz. Messungen über den zeitlichen Verlauf der Zuckung animalischer Muskeln und die Fortpflanzungsgeschwindigkeit der Reizung in den Nerven. *Arch. Anat. Physiol.* (1850) 276–364.
49. H Helmholtz. On the integrals of hydrodynamic equations that express vortex motions. *Phil. Mag. Suppl.* 33 (1867) 485–510.
50. M Hénon. Integrals of the Toda lattice. *Phys. Rev. B* 9 (1974) 1921–1923.
51. R H Hobart. On the instability of a class of unitary field models. *Proc. Phys. Soc.* 82 (1963) 201–203.
52. A L Hodgkin and A F Huxley. A quantitative description of membrane current and its application to conduction and excitation in nerve. *J. Physiol.* 117 (1952) 500–544.
53. P R Holland. *The Quantum Theory of Motion*. Cambridge University Press, Cambridge, 1993.
54. Y H Ichikawa, T Imamura, and T Tanuiti. Nonlinear wave modulation in collisionless plasma. *J. Phys. Soc. Jpn.* 33 (1972) 189–197.
55. V I Karpman and E M Kruskal. Modulated waves in a nonlinear dispersive media. *Sov. Phys. JETP* 28 (1969) 277–281.
56. G Kemeny and I M Goklany. Polarons and conformons. *J. Theor. Biol.* 40 (1973) 107–123.
57. B I Khodorov. *The Problem of Excitability*. Plenum, New York, 1974.
58. A Kolmogoroff, I Petrovsky, and N Piscounoff. Étude de l'équation de la diffusion avec croissance de la quantité de matière et son application a un problème biologique. *Bull. Univ. Moscow, Sér. Int.* A1 (1937) 137–149.
59. D J Korteweg and H de Vries. On the change of form of long waves advancing in a rectangular canal, and on a new type of long stationary waves. *Philos. Mag.* 39 (1895) 422–443.
60. V I Krinsky. *Self-Organization: Autowaves and Structures Far from Equilibrium*. Springer-Verlag, Berlin, 1984.
61. G L Lamb, Jr. Analytical description of ultrashort pulse propagation in a resonant medium. *Rev. Mod. Phys.* 43 (1971) 99–124.
62. G L Lamb, Jr. Bäcklund transforms at the turn of the century. In *Bäcklund Transforms*, R M Miura, ed., Springer Math. Series No. 515, Springer-Verlag, New York, 1976.
63. H Lamb. *Hydrodynamics*. 6th edition, Dover, New York, 1932.
64. L D Landau. Über die Bewegung der Elektronen im Kristallgitter. *Phys. Z. Sowjetunion* 3 (1933) 664–665.

65. P D Lax. Integrals of nonlinear equations of evolution and solitary waves. *Commun. Pure Appl. Math.* 21 (1968) 467–490.
66. A J Lotka. Undamped oscillations derived from the law of mass action. *J. Am. Chem. Soc.* 42 (1920) 1595–1599.
67. R Luther. Räumliche Fortpflanzung chemischer Reaktionen. *Z. Elektrochem.* (32) 12 (1906) 596–600. English translation in *J. Chem. Ed.* 64 (1987) 740–742.
68. W Lüttke and G A A Nonnenmacher. Reinhard Mecke (1895–1969): Scientific work and personality. *J. Mol. Struct.* 347 (1995) 1–18.
69. G Mie. Grundlagen einer Theorie der Materie. *Ann. Phys.* 37 (1912) 511–534; 39 (1912) 1–40; 40 (1913) 1–66.
70. R M Miura. Korteweg–de Vries equation and generalizations. *J. Math. Phys.* 9 (1968) 1202–1204.
71. R M Miura, C S Gardner, and M D Kruskal. Korteweg–de Vries equation and generalizations. II. Existence of conservation laws and constants of motion. *J. Math. Phys.* 9 (1968) 1204–1209.
72. W Moore. *Schrödinger: Life and Thought.* Cambridge University Press, Cambridge, 1989.
73. W Ostwald. Periodische Erscheinungen bei der Auflösung des Chrom in Säuren. *Zeit. Phys. Chem.* 35 (1900) 204–256.
74. J Nagumo, S Arimoto, and S Yoshizawa. An active pulse transmission line simulating nerve axon. *Proc. IRE* 50 (1962) 2061–2070.
75. A C Newell (ed.). *Nonlinear Wave Motion.* AMS Lectures in Applied Math. Vol. 15, 1974.
76. R Pego. Origin of the KdV equation. *Not. AMS* 45 (1998) 358.
77. I Pekar. *Untersuchungen über die Elektronentheorie der Kristalle.* Akademie-Verlag, Berlin, 1954.
78. J K Perring and T R H Skyrme. A model unified field equation. *Nucl. Phys.* 31 (1962) 550–555.
79. M Quack. Spectra and dynamics of coupled vibrations in polyatomic molecules. *Annu. Rev. Phys. Chem.* 41 (1990) 839–874.
80. F K Richtmyer and E H Kennard. *Introduction to Modern Physics.* McGraw-Hill, New York, pp. 598–603.
81. G Rosen. Particlelike solutions to nonlinear scalar wave theories. *J. Math. Phys.* 6 (1965) 1269–1272.
82. E Schrödinger. Contributions to Born's new theory of the electromagnetic field. *Proc. R. Soc. (London) A* 150 (1935) 465–477.
83. J Scott Russell. *Report on Waves.* 14th meeting of the British Association for the Advancement of Science (BAAS), 1844.
84. J Scott Russell. *The Wave of Translation in the Oceans of Water, Air and Ether.* London, 1885.
85. A C Scott. A nonlinear Klein–Gordon equation. *Am. J. Phys.* 37 (1969) 52–61.
86. A C Scott. *Neuroscience: A Mathematical Primer.* Springer-Verlag, New York, 2002.
87. A C Scott. Davydov's soliton. *Phys. Rep.* 217 (1992) 1–67.
88. A C Scott, P S Lomdahl, and J C Eilbeck. Between the local mode and normal mode limits. *Chem. Phys. Lett.* 113 (1985) 29–36.

REFERENCES

89. A Seeger, H Donth, and A Kochendörfer. Theorie der Versetzungen in eindimensionalen Atomreihen. *Z. Phys.* 134 (1953) 173–193.
90. K Showalter and J J Tyson. Luther's discovery and analysis of chemical waves. *J. Chem. Ed.* 64 (1987) 742–744.
91. T H R Skyrme. A nonlinear theory of strong interactions. *Proc. R. Soc.* (London) A 247 (1958) 260–278.
92. R Steuerwald. Über Enneper'sche Flächen und Bäcklund'sche Transformation. *Abh. Bayerischen Akad. Wiss.* München (1936) 1–105.
93. R Suzuki. Electrochemical neuron model. *Adv. Biophys.* 9 (1976) 115–156.
94. L A Takhtajan and L D Faddeev. Essentially nonlinear one-dimensional model of classical field theory. *Theor. Math. Phys.* 21 (1974) 1046–1057.
95. T Tanuiti and H Washimi. Self trapping and instability of hydromagnetic waves along the magnetic field in a cold plasma. *Phys. Rev. Lett.* 21 (1968) 209–212.
96. F Tappert and C M Varma. Asymptotic theory of self trapping of heat pulses in solids. *Phys. Rev. Lett.* 25 (1970) 1108–1111.
97. M Toda. Vibration of a chain with nonlinear interactions. *J. Phys. Soc. Jpn.* 22 (1967) 431–436.
98. M Toda. Wave propagation in anharmonic lattices. *J. Phys. Soc. Jpn.* 23 (1967) 501–506.
99. H D Wahlquist and F B Estabrook. Bäcklund transformation for solutions of the Korteweg–de Vries equation. *Phys. Rev. Lett.* 31 (1973) 1386–1390.
100. G B Whitham. A general approach to linear and nonlinear waves using a Lagrangian. *J. Fluid Mech.* 22 (1965) 273–283.
101. A T Winfree. Spiral waves of chemical activity. *Science* 175 (1972) 634–636.
102. A T Winfree. *When Time Breaks Down: The Three-Dimensional Dynamics of Electrochemical Waves and Cardiac Tissue.* Princeton University Press, Princeton, 1987.
103. Y Yang. The Lee–Weinberg magnetic monopole of unit charge: existence and uniqueness. *Physica D* 117 (1998) 215–240.
104. N J Zabusky and M D Kruskal. Interactions of solitons in a collisionless plasma and the recurrence of initial states. *Phys. Rev. Lett.* 15 (1965) 240–243.
105. A N Zaikin and A M Zhabotinsky. Concentration wave propagation in two-dimensional liquid-phase self-oscillating systems. *Nature* 225 (1970) 535–537.
106. V E Zakharov and A B Shabat. Exact theory of two-dimensional self-focusing and one-dimensional self-modulation of waves in nonlinear media. *Sov. Phys. JETP* 34 (1972) 62–69.
107. YaB Zeldovich and DA Frank-Kamenetsky. Toward a theory of uniformly propagating flames. In Russian. *Dokl. Akad. Nauk SSSR* 19 (1938) 693–697.
108. A M Zhabotinsky. Periodic movement in the oxidation of malonic acid in solution (Investigation of the kinetics of Belousov's reaction.) In Russian. *Biofizika* 9 (1964) 306–311.

2
LINEAR WAVE THEORY

In this book are presented a variety of methods for solving nonlinear partial differential equations (PDEs); thus it is appropriate to begin with a brief review of some techniques that can be used for solving linear PDEs. (One should learn to walk before trying to run.)

What is so special about linear equations? If $u_1(x,t)$ and $u_2(x,t)$ are both solutions of a linear equation, then

$$u(x,t) = Au_1(x,t) + Bu_2(x,t)$$

is also a solution, where A and B are arbitrary constants. "The whole is equal to the sum of its parts" is the way that some express this property. Thus the task of finding the general solution for a linear PDE can be broken up into a collection of simpler problems. In the more general realm of nonlinear systems, such a resolution is not possible because

$$u(x,t) \neq Au_1(x,t) + Bu_2(x,t)$$

and the whole is often greater than the sum of its parts. Thus nonlinearity becomes the source of emergent phenomena. As a rule, linear models tend to be more appropriate for the study of inanimate systems, where the threads of causality are more easily sorted out, whereas biological systems are usually nonlinear.

The plan for this chapter is first to apply some basic analytical techniques to linear limits of the nonlinear PDEs displayed in Chapter 1, augmented with a selection of practice problems. This brief review is followed by introductions to linear stability theory and linear scattering theory.

2.1 Dispersionless linear equations

Perhaps the most simple of all linear PDEs is obtained by setting ε and γ to zero in the Korteweg–de Vries (KdV) Equation (1.2) to obtain

$$\frac{\partial u}{\partial t} + c\frac{\partial u}{\partial x} = 0, \tag{2.1}$$

where $-\infty < x < \infty$. This result is reached by assuming the x-dependence of u is weak enough to neglect the dispersive (u_{xxx}) term and the amplitude is small enough to neglect the nonlinear (uu_x) term. The most general solution of Equation (2.1) is the traveling-wave

$$u(x,t) = \tilde{u}(x - ct),$$

and if the initial condition is $u(x,0) = f(x)$, then the solution is

$$u(x,t) = f(x - ct).$$

Thus solutions of Equation (2.1) propagate at speed c with any shape.

Consider next the second-order linear equation

$$\frac{\partial^2 u}{\partial x^2} - \frac{1}{c^2}\frac{\partial^2 u}{\partial t^2} = 0, \tag{2.2}$$

where $-\infty < x < \infty$, which is often called the wave equation. Assuming the solution to be a traveling-wave with velocity v,

$$u(x,t) = \tilde{u}(x - vt)$$

requires that

$$\tilde{u}''\left(1 - \frac{v^2}{c^2}\right) = 0.$$

For $\tilde{u}'' \neq 0$, it is clear that the traveling-wave velocity must take one of the two values

$$v = \pm c.$$

Thus there are two traveling-wave solutions and the general solution has the form

$$u(x,t) = f(x - ct) + g(x + ct),$$

where the functions $f(\cdot)$ and $g(\cdot)$ are arbitrary beyond the requirement that they possess second derivatives.

If the initial conditions for Equation (2.2) are

$$u(x,0) = F(x)$$
$$\frac{\partial u}{\partial t}(x,0) = G(x),$$

then

$$f(x) + g(x) = F(x)$$
$$-cf'(x) + cg'(x) = G(x),$$

and the solution is

$$u(x,t) = \frac{1}{2c}\{[cF(x - ct) - H(x - ct)] + [cF(x + ct) + H(x + ct)]\},$$

where

$$H(z) \equiv \int G(z)\,dz.$$

2.2 Dispersive linear equations

Let us now consider an approximation to the KdV equation with only the nonlinear term neglected by setting γ to zero in Equation (1.2) (or assuming the wave amplitude to be small). The resulting linear equation is

$$\frac{\partial u}{\partial t} + c\frac{\partial u}{\partial x} + \varepsilon\frac{\partial^3 u}{\partial x^3} = 0. \tag{2.3}$$

Guided by previous results, we again try a traveling-wave solution of the form $u(x,t) = \tilde{u}(x - vt)$, requiring $\tilde{u}(\cdot)$ to satisfy the ordinary differential equation (ODE)

$$(c - v)\tilde{u}' + \varepsilon\tilde{u}''' = 0, \tag{2.4}$$

for some value of the assumed traveling-wave velocity v. If $v = c$, then $\tilde{u}''' = 0$, implying

$$\tilde{u}(x - ct) = A + B(x - ct) + C(x - ct)^2,$$

where A, B, and C are arbitrary constants. For the solution to be bounded as $x \to +\infty$, both B and C must be set to zero, leaving the result that

$$u(x,t) = A.$$

This is certainly a solution—it represents the undisturbed surface of the water—but it is not very interesting.

To proceed further, assume that the traveling-wave speed is not equal to c. Then Equation (2.4) permits solutions of the form

$$u(x,t) = e^{ikx - i\omega t},$$

where $v = \omega/k$, and from Equation (2.3)

$$\omega(k) = ck - \varepsilon k^3. \tag{2.5}$$

Taking elementary components of the solution to be exponential functions of this sort is equivalent to using the Fourier transform[1] technique [6]. Thus if the initial condition is $u(x,0) = f(x)$, then

$$u(x,0) = \frac{1}{2\pi}\int_{-\infty}^{\infty} F(k)e^{ikx}dk, \tag{2.6}$$

where

$$F(k) = \int_{-\infty}^{\infty} f(x)e^{-ikx}dx. \tag{2.7}$$

[1] After French mathematician and physicist Jean Baptiste Joseph Fourier (1768–1830).

For $t > 0$, each Fourier component of the solution will evolve as an exponential traveling-wave. Thus

$$u(x,t) = \frac{1}{2\pi} \int_{-\infty}^{\infty} F(k) e^{ikx - i\omega(k)t} dk, \qquad (2.8)$$

where $\omega(k)$ is defined by the dispersion relation of Equation (2.5). It relates the radian frequency

$$\omega = 2\pi/T$$

to the propagation number

$$k = 2\pi/\lambda,$$

where T is the temporal period and λ is the spatial period.

2.3 The linear diffusion equation

As was mentioned in the previous chapter, solutions of the simple linear diffusion equation

$$\frac{\partial u}{\partial t} - \frac{\partial^2 u}{\partial x^2} = 0 \qquad (2.9)$$

are unwavelike in character. To see this, note that any pulse shaped solution has negative curvature ($\partial^2 u/\partial x^2 < 0$) near its maximum, so u must be decreasing in this region. Similarly, the second derivative with x is positive where $u \to 0$, so it must increase on the skirts. All a pulse shaped initial condition can do, therefore, is spread itself out, like a puff of smoke on a still morning or a dab of honey on a piece of bread.

A constraint on these diffusive dynamics arises because Equation (2.9) is a conservation law as defined in Equation (A.1) of Appendix A. Thus $D = u$ is the density and $F = -\partial u/\partial x$ is the flow of the conserved quantity

$$Q = \int_{-\infty}^{\infty} u \, dx = \text{constant}. \qquad (2.10)$$

To determine how $u(x,t)$ evolves from the initial condition

$$u(x,0) = f(x), \qquad (2.11)$$

it is convenient to carry through a computation using both the above defined Fourier transform and the closely related Laplace transform.[2] (As the use of both Fourier and Laplace transforms appears in standard undergraduate curricula of applied science, it is assumed that they are familiar to the reader. If this is not

---

[2] Named for Pierre Simon de Laplace (1749–1827), a French mathematician and astronomer. This method of solving differential equations was discovered by Oliver Heaviside (1850–1925), a self-taught English physicist and electromagnetic theorist [5].

the case, some of the practice problems should be studied along with references [2, 5, 6, 8, 11, 12].)

Defining the Laplace transform of $u(x,t)$ as

$$U(x,s) \equiv \int_0^\infty u(x,t)e^{-st}dt,$$

the Laplace transform of Equation (2.9) is

$$\left(s - \frac{\partial^2}{\partial x^2}\right)U(x,s) = f(x).$$

Denoting the Fourier transform of $U(x,s)$ as $\mathcal{U}(k,s)$, we have

$$\mathcal{U}(k,s) = \frac{F(k)}{s+k^2}.$$

Taking the inverse Laplace transform of $\mathcal{U}(k,s)$ gives

$$\tilde{U}(k,t) = F(k)e^{-k^2 t},$$

and finally an inverse Fourier transform of $\tilde{U}(k,t)$ yields

$$u(x,t) = \int_{-\infty}^\infty f(\xi) \frac{\exp[-(x-\xi)^2/4t]}{2\sqrt{\pi t}} d\xi. \tag{2.12}$$

Equation (2.12) expresses $u(x,t)$ as a convolution of $f(x)$ with the function

$$g(x,t) = \frac{\exp(-x^2/4t)}{2\sqrt{\pi t}}. \tag{2.13}$$

This function is a natural solution of the linear diffusion equation which satisfies the constraint of Equation (2.10).

As $t \to 0$ from positive values,

$$\frac{\exp[-(x-\xi)^2/4t]}{2\sqrt{\pi t}} \to \delta(x-\xi), \tag{2.14}$$

where $\delta(x)$ is Dirac's delta function defined by the properties

$$\delta(x) = 0 \quad \text{for } x \neq 0$$

and

$$\int_{-\infty}^\infty f(\xi)\delta(x-\xi)d\xi = f(x).$$

Thus the expression for $u(t)$ in Equation (2.12) satisfies the initial condition of Equation (2.11).

2.4 Driven systems

In the sections above, we have considered the dynamics of some linear PDEs that are evolving with time from prescribed initial conditions. Source terms on the right-hand sides of Equations (2.1), (2.2), (2.3), and (2.9) have all been zero. Such source terms represent causes in cause–effect relationships, and they also appear in connection with the study of perturbation theories for nonlinear waves in Chapter 7.

2.4.1 Green's method

As a motivating example, consider Equation (2.9) augmented with a source term and a term describing absorption, so

$$\frac{\partial u}{\partial t} - \frac{d^2 u}{dx^2} + a^2 u = f(x)$$

with $-\infty < x < \infty$. Physically, this equation can model an infinitely long heat-conducting rod that is partially wrapped in insulation, allowing some loss of heat. Thus u is the heat per unit length (or temperature), $a^2 u$ describes the loss of heat to the surroundings, and $f(x)$ is a time independent source of heat input along the rod.

In the steady state, all variables are assumed to be independent of time so the equation reduces to

$$-\frac{d^2 u}{dx^2} + a^2 u = f(x). \tag{2.15}$$

To solve Equation (2.15), notice that the source term can be represented in terms of Dirac's delta functions as

$$f(x) = \int_{-\infty}^{\infty} f(\xi)\delta(x - \xi)\, d\xi,$$

so it is of interest to study the related problem

$$-\frac{d^2 g}{dx^2} + a^2 g = \delta(x),$$

where $g(x) \to 0$ as $x \to \pm\infty$. The solution of this auxiliary problem

$$g(x) = \frac{e^{-a|x|}}{2a}$$

is called the "Green function" for Equation (2.15) [8, 12].

Because the source can be represented in terms of delta functions, the solution can be represented in terms of delta function responses—or the Green function—as

$$u(x) = \int_{-\infty}^{\infty} f(\xi)\frac{e^{-a|x-\xi|}}{2a}\, d\xi.$$

This approach to the solution of driven systems has been widely employed throughout applied science since its introduction in an 1828 essay by George Green as a means for solving problems in the newly formulated fields of electrostatics and magnetostatics[3] [4].

Green's method is not limited to problems with space as the independent variable. Electrical engineers, for example, often use the approach to solve time dependent problems, albeit with different jargon. Thus Dirac's delta function is called an "impulse function," and the Green function is termed the "impulse response."

As a simple example of such an application, consider the system

$$\frac{du}{dt} + au = f(t), \tag{2.16}$$

where $u(0) = 0$ and $f(t) = 0$ for $t \leq 0$. The corresponding Green function (or impulse response) is constructed to satisfy

$$\frac{dg}{dt} + ag = \delta(t), \tag{2.17}$$

with the additional requirement that

$$g(t) = 0 \quad \text{for } t < 0.$$

The reason for this extra constraint on a time dependent Green function is that time is a causal variable, implying that a response cannot precede its cause. Thus Equation (2.17) has the solution

$$g(t) = \begin{cases} 0 & \text{for } t < 0 \\ e^{-at} & \text{for } t \geq 0, \end{cases}$$

and the solution to Equation (2.16) can be written

$$u(t) = \int_{-\infty}^{\infty} f(\tau) \, g(t-\tau) d\tau$$

$$= \int_{0}^{t} f(\tau) \, e^{-a(t-\tau)} d\tau$$

$$= e^{-at} \int_{0}^{t} f(\tau) \, e^{a\tau} d\tau.$$

Green's method can also be used to solve problems in which both space and time appear as independent variables. As an example, consider the system

$$\frac{\partial u}{\partial t} - \frac{\partial^2 u}{\partial x^2} = f(x,t), \tag{2.18}$$

---

[3] George Green (1793–1841) was a self-taught miller's son from Nottingham, who published his work privately. Dirk Struik has described his book as "the beginning of mathematical physics in England."

defined on the half plane $-\infty < x < +\infty$ and $t \geq 0$, with the initial condition $u(x,0) = 0$. Resolving $f(x,t)$ into delta function components as

$$f(x,t) = \int_0^\infty d\tau \int_{-\infty}^\infty d\xi\, \delta(x-\xi)\delta(t-\tau)f(\xi,\tau),$$

where

$$\delta(x)\delta(t) = 0 \quad \text{for } x \neq 0 \quad \text{or } t \neq 0,$$

we seek a Green function that solves

$$\frac{\partial g}{\partial t} - \frac{\partial^2 g}{\partial x^2} = \delta(x)\delta(t),$$

with the causality requirement

$$g(x,t) = 0 \quad \text{for } t < 0.$$

From the results of Section 2.3, we note that such a Green function is

$$g(x,t) = \begin{cases} \dfrac{\exp(-x^2/4t)}{2\sqrt{\pi t}} & \text{for } t \geq 0 \\ 0 & \text{for } t < 0. \end{cases}$$

Thus the desired solution to Equation (2.18) is

$$u(x,t) = \int_0^t d\tau \int_{-\infty}^\infty d\xi\, f(\xi,\tau) \frac{\exp[-(x-\xi)^2/4(t-\tau)]}{2\sqrt{\pi(t-\tau)}}.$$

2.4.2 Fredholm's theorem

Appealing as Green's method is, one must be careful in using it because the stated problem may not have a solution. Consider the general system

$$Lu(x) = f(x), \tag{2.19}$$

where $-\infty < x < \infty$, L is a linear operator, and $f(x)$ is a source. In terms of an inner product that can be defined for real functions as

$$(v(x), w(x)) \equiv \int_{-\infty}^\infty v(x)w(x)\,dx,$$

the adjoint $(L^\dagger)$ of the operator L satisfies the condition

$$(Lv, w) = (v, L^\dagger w).$$

All functions φ for which

$$L^\dagger \varphi = 0 \tag{2.20}$$

are said to "span the null space" of $L^\dagger$, indicated as $\mathcal{N}(L^\dagger)$.

Now make two assumptions:

- Equation (2.19) has a solution.
- The inner product $(f, \varphi) \neq 0$ for some φ in the null space of $L^\dagger$.

Then the calculation

$$0 \neq (f, \varphi) = (Lu, \varphi) = (u, L^\dagger \varphi) = (u, 0) = 0$$

leads to the contradiction $0 \neq 0$. Thus a necessary condition for Equation (2.19) to have a solution is that the inner product of the source with any function that satisfies Equation (2.20) must be zero. In other words, it is necessary that

$$f(x) \perp \mathcal{N}(L^\dagger) \qquad (2.21)$$

for Equation (2.19) to have a solution, where "$\perp$" means "is orthogonal to."

Fredholm's theorem[4] states that condition (2.21) is both necessary and sufficient for Equation (2.19) to have a solution [12]. Equations that satisfy Fredholm's theorem are variously said to be "solvable," "compatible," or "consistent."

As a simple example of this requirement, let us reconsider Equation (2.15) assuming that the insulation is perfect ($a = 0$), and the rod is ring shaped (and of unit length) so all functions are periodic. Because

$$\left(\frac{d^2}{dx^2}\right)^\dagger = \left(\frac{d^2}{dx^2}\right)$$

and the boundary conditions are periodic, the adjoint null space is spanned by

$$\varphi = 1.$$

Thus the solvability condition is

$$\int_0^1 f(x)\,dx = 0.$$

In physical terms, this means that the net heat input to a perfectly insulated, ring shaped rod must be zero in the steady state.

Another example is the driven harmonic oscillator

$$\frac{d^2 x}{dt^2} + \omega^2 x = f(t).$$

Because the operator

$$\frac{d^2}{dt^2} + \omega^2$$

---

[4] For Erik Fredholm (1866–1927), a Swedish mathematician who established the theory of integral equations.

is self-adjoint, Fredholm's theorem requires

$$(f(t), \cos \omega t) = 0 \quad \text{and} \quad (f(t), \sin \omega t) = 0$$

for a solution to exist. These conditions forbid components of $f(t)$ at frequency ω, thereby avoiding the unbounded growth associated with resonance.

2.5 Stability

In the analysis of dynamic problems, it is often important to consider questions of stability, as unstable solutions are unlikely to be observed in nature. Here are briefly stated some of the basic concepts of stability theory for future reference.

2.5.1 *General definitions*

As an example of the ways that concepts of stability can enter into wave problems, consider the sine–Gordon (SG) equation

$$\frac{\partial^2 u}{\partial x^2} - \frac{\partial^2 u}{\partial t^2} = \sin u, \qquad (2.22)$$

defined on the half plane $-\infty < x < \infty$ and $t \geq 0$. This system evidently has stationary solutions

$$u_0 = 0 \quad \text{and} \quad u_\pi = \pi.$$

Consider now small changes Δ from these stationary solutions so

$$u(x,t) = 0 + \Delta(x,t)$$

in the first case and

$$u(x,t) = \pi + \Delta(x,t)$$

in the second. The stability of a stationary solution is related to the behavior of $\Delta(x,t)$, which can evolve with time in three ways [7, 13]:

1. *Asymptotic stability.* If

$$\Delta(x,t) \to 0 \quad \text{as} \quad t \to \infty$$

   for all initial conditions $\Delta(x,0)$ in a region around a stationary solution, that solution is said to be asymptotically stable. In other words, all solutions sufficiently close to the stationary solution converge to it.
2. *Instability.* A stationary solution is unstable if any one of the neighboring solutions grows away from it. Formally, this can be expressed by assuming that $\Delta(x,0) < \epsilon$ and showing that $\Delta(x,t)$ becomes larger than some fixed $\delta > 0$ for all $\epsilon > 0$ as $t \to \infty$.
3. *Neutral stability.* A stationary solution that is neither unstable nor asymptotically stable is said to be neutrally stable.

2.5.2 Linear stability

For nonlinear systems, such definitions can be treacherous. Think of a new (unsharpened) pencil standing on its flat end. Is this situation stable or not? If the room is quiet and the table is even, a practical person might conclude that it is asymptotically stable, although the "region around" the stationary solution is uncomfortably small. Given some classroom turmoil, however, introducing noise or gusts of moving air, one might conclude otherwise.

Thus motivated, it is interesting to apply the stability definitions in sufficiently small regions of the variable $\Delta(x,t)$ that it can be assumed to obey a linear equation. As an example, consider the SG solution $u_\pi = \pi$. From Equation (2.22),

$$\frac{\partial^2 \Delta}{\partial x^2} - \frac{\partial^2 \Delta}{\partial t^2} = \sin(\pi + \Delta)$$
$$\doteq -\Delta, \tag{2.23}$$

for $|\Delta| \ll 1$. If $\Delta(x,0)$ is chosen to be independent of x,

$$\Delta(x,t) = Ae^t + Be^{-t}, \tag{2.24}$$

evidently satisfying the condition for instability in the linear approximation. Thus the stationary SG solution

$$u_\pi = \pi$$

is linearly unstable.

For the solution $u_0 = 0$, on the other hand,

$$\frac{\partial^2 \Delta}{\partial x^2} - \frac{\partial^2 \Delta}{\partial t^2} \doteq \Delta$$

for Δ sufficiently small. Assuming that

$$\Delta(x,0) = f(x) \quad \text{and} \quad \frac{\partial \Delta}{\partial t}(x,0) = 0,$$

general solutions of this linear approximation can be constructed from Fourier components of the form

$$e^{ikx - i\omega(k)t},$$

where

$$\omega^2 = 1 + k^2 \tag{2.25}$$

is the linear dispersion equation.

Because $\Delta(x,0)$ can be constructed as the Fourier integral

$$\Delta(x,0) = \frac{1}{2\pi} \int_{-\infty}^{\infty} F(k) e^{ikx} dk,$$

where $F(k)$ is the Fourier transform of $f(x)$, then

$$\Delta(x,t) = \frac{1}{2\pi} \int_{-\infty}^{\infty} F(k)e^{ikx} \cos\left(\sqrt{k^2 + 1}\,t\right) dk, \qquad (2.26)$$

implying neutral stability of $u_0 = 0$ in the linear approximation.

2.5.3 Signaling problems

Solving the linear dispersion relation of Equation (2.25) for k as a function of ω gives

$$k = \pm\sqrt{\omega^2 - 1}.$$

For $\omega < 1$, k becomes an imaginary number and

$$e^{ikx} = e^{\pm x\sqrt{1-\omega^2}}. \qquad (2.27)$$

This formulation is of interest for the domain

$$0 \leq x < +\infty$$
$$0 \leq t < +\infty.$$

In such a signaling problem, the initial and boundary conditions

$$\Delta(x,0) = 0$$
$$\Delta(0,t) = f(t)$$

can be imposed, placing the frequency ω under the control of an experimenter. Three possibilities then emerge.

1. *Evanescent waves.* From Equation (2.27), it follows that for $\omega < 1$ one of the natural solutions decays in the $+x$-direction, while the other grows. If the wave system has no sources of energy, solutions must decay with distance from the source, and such waves are called evanescent. Only the solution that decays with increasing x is excited by the boundary conditions indicated above. The solution that grows with x is needed for boundary conditions of the sort

$$\Delta(0,t) = f(t)$$
$$\Delta(L,t) = g(t),$$

    where $0 \leq x \leq L$.
2. *Convective instability.* For wave systems with sources of energy, it is possible to have localized (pulse-like) solutions that grow in the direction of propagation, and such solutions are said to be convectively unstable [10].

Convectively unstable systems can be used both as traveling-wave amplifiers and as oscillators (with appropriate boundary conditions at $x = 0$ and $x = L$).

3. *Absolute instability.* For Equation (2.23), the dispersion relation is

$$\omega^2 = k^2 - 1.$$

In the Fourier transform corresponding to Equation (2.26), this implies exponential growth with time at all values of x, as we have seen in Equation (2.24). Such behavior is an example of absolute instability. Absolutely unstable systems cannot be used as amplifiers, but with appropriate boundary conditions at $x = 0$ and $x = L$, they may be employed as oscillators.

2.6 Scattering theory

Most of us have childhood memories of tossing stones into a pond and watching the waves they generate. Sometimes these waves impinge upon a floating object and are reflected, leading one to ask what the scattered waves might tell about the structure of the object that reflected them. From a general perspective, this is called the inverse problem, which has many important applications, including the use of sound waves for mineral prospecting, target identifications in radar and sonar systems, structure determinations in x-ray diffraction studies, and the interpretation of electrocardiograms, among others. Usually only partial information about the scattering object can be obtained, but in this section we consider a situation in which knowledge of the scattered waves is sufficient to reconstruct the scattering object.

2.6.1 Solutions of Schrödinger's equation

In Chapter 8, an important role is played by the homogeneous linear ODE

$$-\frac{d^2\psi}{dx^2} + u(x)\psi = \lambda\psi, \qquad (2.28)$$

called the time independent Schrödinger equation.[5] In the quantum mechanical interpretation of this equation, λ is the energy of a particle moving in the x-direction, $u(x)$ is a potential energy through which the particle moves, and $\psi(x, \lambda)$ is a wave function that guides its motion. Leaving aside the question of its physical interpretation, Equation (2.28) has two distinct types of finite solutions.

1. *Scattering solutions.* Assuming that $u(x) \to 0$ as $x \to \pm\infty$ and defining

$$k^2 \equiv \lambda > 0,$$

---

[5] After the Austrian physicist Erwin Schrödinger (1887–1961), who discovered it.

$\psi(x)$ can have the following asymptotic behavior far from the origin:

$$\psi(x,k) \sim \begin{cases} e^{-ikx} + b(k)e^{+ikx} & \text{as } x \to +\infty, \\ a(k)e^{-ikx} & \text{as } x \to -\infty. \end{cases} \qquad (2.29)$$

This is called a scattering solution, for which $\exp(-ikx)$ can be viewed as an incident wave, approaching the scattering potential $u(x)$ from $x = +\infty$. Then $a(k)$ is the transmission coefficient and $b(k)$ is the reflection coefficient. The scattering solution requires that $\lambda > 0$ so k takes all real values between $-\infty$ and $+\infty$.

2. *Bound states.* If $u(x)$ is sufficiently negative near the origin of the x-axis and $u(x) \to 0$ as $x \to \pm\infty$, analysis of Equation (2.28) in the $(\psi, d\psi/dx)$ phase plane shows that for certain negative eigenvalues ($\lambda = \lambda_n$), there are corresponding bound states—ψ_n—falling exponentially to zero as $x \to \pm\infty$. Because the eigenvalues are necessarily negative for this asymptotic behavior, it is convenient to write

$$-\lambda_n \equiv \kappa_n^2,$$

so small amplitude, evanescent solutions of Schrödinger's equation are of the form

$$\psi_n(x) \propto \begin{cases} e^{-\kappa_n x} & \text{as } x \to +\infty, \\ e^{+\kappa_n x} & \text{as } x \to -\infty. \end{cases} \qquad (2.30)$$

Equations (2.29) are obtained from analyzing Schrödinger's equation with $k^2 = \lambda > 0$ (so k is a real number running from $-\infty$ to $+\infty$), while the derivation of Equations (2.30) assumes $\lambda < 0$ (permitting bound state solutions only at certain discrete values of λ); yet the two results are related in an interesting way. If the scattering solution is extended from real values of k into the upper half of the k-plane, $b(k)$ and $a(k)$ are found to have simple poles at $k = +i\kappa_n$, where ($n = 1, 2, \ldots, N$), and Equations (2.29) evidently take the form of Equations (2.30). Of particular interest are the residues of these upper half plane poles of $b(k)$: $\{r_n\}$.

In making the analytic continuation of the scattering problem (defined on the real k-axis) to the bound state problem (defined in the upper half of the complex k-plane), consider that the bound states have finite square integrals (whereas the scattering states do not). Thus it is customary to normalize the bound states as

$$\int_{-\infty}^{\infty} \psi_n^2 dx = 1, \qquad (2.31)$$

whereupon the asymptotic behavior of ψ_n as $x \to +\infty$ is fixed as

$$\psi(x) \sim c_n e^{-\kappa_n x}. \qquad (2.32)$$

The items $b(k)$ ($-\infty < k < +\infty$), κ_n, and r_n ($n = 1, 2, \ldots, N$) are collectively referred to as the scattering data for a potential $u(x)$. From a knowledge of $u(x)$, it is a straightforward, if sometimes tedious, matter to compute the corresponding scattering data. This computation is called the direct problem.

At this point, it is appropriate to consider an example, and perhaps the most simple is

$$u(x) = \alpha \delta(x),$$

where α is a real constant that may be either positive or negative. With this potential in Equation (2.28), integration across the origin implies

$$\psi_x(+\epsilon) - \psi_x(-\epsilon) = \int_{-\epsilon}^{+\epsilon} [\alpha \delta(x) - \lambda] \psi(x)\, dx,$$

where ϵ can be made as small as we please. Thus

$$\psi_x(0^+) - \psi_x(0^-) = \alpha \psi(0),$$

implying that scattering solutions have the form given in Equations (2.29) at all values of x except at the origin where the slope of ψ jumps by α times its magnitude. Because the term $d^2\psi/dx^2$ appears in Equation (2.28), $\psi(x)$ must be continuous at the origin, so

$$\psi(0) = 1 + b(k) = a(k),$$

and the discontinuity condition on the derivative requires that

$$-ik + ikb - (-ika) = \alpha(1 + b).$$

Thus

$$b(k) = \frac{\alpha}{2ik - \alpha} = a(k) - 1 \tag{2.33}$$

is the reflection coefficient for real values of k and also for real values of α, which defines the scattering states.

What about the bound states? To answer this question, recall that the scattering states are

$$\psi(x, k) = \begin{cases} e^{-ikx} + b(k)e^{+ikx} & \text{for } x > 0, \\ a(k)e^{-ikx} & \text{for } x < 0, \end{cases} \tag{2.34}$$

where k takes any real value between $+\infty$ and $-\infty$. But if k is allowed to be a complex number, Equation (2.33) implies that both the reflection coefficient and the transmission coefficient have poles at

$$k = -i\alpha/2.$$

To appreciate the physical meaning of such a pole in the context of Equation (2.34), consider that the incident wave to be a cause, whereas the reflected wave and the transmitted waves are effects. Thus the pole indicates a complex value of k where the cause becomes negligibly small with respect to its effects. In other words, the solution becomes

$$\psi(x) = \begin{cases} c_1 e^{+\alpha x/2} & \text{for } x > 0, \\ c_1 e^{-\alpha x/2} & \text{for } x < 0. \end{cases} \quad (2.35)$$

Now this is not very interesting for $\alpha > 0$ because it diverges as $x \to \pm\infty$, but for $\alpha < 0$, it corresponds to a bound state with

$$\kappa_1 = -\alpha/2.$$

Because there is only one such bound state, $N = 1$, and the normalization of Equation (2.31) is satisfied for

$$c_1 = \sqrt{-\alpha/2}.$$

Finally writing $b(k)$ in the form $r_1/(k - i\kappa_1)$ implies that

$$r_1 = -i\alpha/2 = ic_1^2. \quad (2.36)$$

To summarize: in this example, we have assumed a scattering potential of the form $u(x) = \alpha\delta(x)$ and considered two possibilities: (i) $\alpha > 0$. In this case, there are no bound states, and Equation (2.33) comprises all of the scattering data. (ii) $\alpha < 0$. Now in addition to Equation (2.33) the scattering data include a single bound state ($N = 1$) with $\kappa_1 = -\alpha/2$ and $r_1 = -i\alpha/2$.

Thus the direct problem is solved by computing the scattering data from the potential. Next we ask: Is it possible to calculate the potential in Schrödinger's equation from its scattering data?

2.6.2 Gel'fand–Levitan theory

A means for calculating $u(x)$ in Equation (2.28) from its scattering data (solving the inverse problem) was developed by Gel'fand and Levitan in 1951 [3]. Here we sketch a plausibility argument for this procedure, following a particularly lucid discussion by Whitham [1, 14]. A more detailed derivation of these results is presented in Chapter 6.

The first move is to view the scattering solution $\psi(x, k)$ as a Fourier transform of $\varphi(x, \tau)$ with respect to a causal variable τ. Thus[6]

$$\psi(x, k) = \int_{-\infty}^{\infty} \varphi(x, \tau) e^{+ik\tau} d\tau \quad (2.37)$$

and

$$\varphi(x, \tau) = \frac{1}{2\pi} \int_{-\infty}^{\infty} \psi(x, k) e^{-ik\tau} dk. \quad (2.38)$$

---

[6]Note that this definition of the transform differs slightly from Equations (2.6) and (2.7).

To remind ourselves of this causal property, τ is called a "pseudotime."

As is usual in Fourier transform theory with respect to a causal variable, the path of integration in Equation (2.38) is assumed to be deformed as shown in Figure 2.2 so all singularities of $\psi(x, k)$ in the upper half of the complex k-plane are included in the integration over the lower half (which is equivalent to using a Laplace transform). Under this transformation, Schrödinger's equation becomes

$$\frac{\partial^2 \varphi}{\partial x^2} - \frac{\partial^2 \varphi}{\partial \tau^2} - u(x)\varphi = 0, \qquad (2.39)$$

which is a linear, hyperbolic, homogeneous wave equation with the spectral parameter (k) replaced by a causal pseudotime.

As $x \to +\infty$, a scattering solution of Equation (2.39) takes the form

$$\varphi(x, \tau) \sim \varphi_\infty(x, \tau) \equiv \delta(x + \tau) + B(x - \tau), \qquad (2.40)$$

where $\delta(x + \tau)$ is an incident impulse, and $B(x - \tau)$ represents the reflection from $u(x)$.

Now we try to construct the solution to Equation (2.39) as

$$\varphi(x, \tau) = \varphi_\infty(x, \tau) + \int_x^\infty K(x, \xi)\, \varphi_\infty(\xi, \tau)\, d\xi, \qquad (2.41)$$

where $K(x, \xi)$ is a kernel that is not only undetermined, but may not even exist! Being optimistic, however, we substitute Equation (2.41) into (2.39), noting that

$$\frac{\partial^2 \varphi_\infty(\xi, \tau)}{\partial \tau^2} = \frac{\partial^2 \varphi_\infty(\xi, \tau)}{\partial \xi^2}$$

and integrating by parts, to find

$$\int_x^\infty \left[\frac{\partial^2 K(x, \xi)}{\partial x^2} - \frac{\partial^2 K(x, \xi)}{\partial \xi^2} - u(x) K(x, \xi) \right] \varphi_\infty(\xi, \tau)\, d\xi$$

$$- \left[2 \frac{d}{dx} K(x, x) + u(x) \right] \varphi_\infty(x, \tau)$$

$$- K(x, \xi) \frac{\partial \varphi_\infty(\xi, \tau)}{\partial \xi} \bigg|_{\xi = +\infty} + \frac{\partial K(x, \xi)}{\partial \xi} \varphi_\infty(\xi, \tau) \bigg|_{\xi = +\infty}$$

$$= 0.$$

Evidently this condition is satisfied by requiring

$$\frac{\partial^2 K(x, \xi)}{\partial x^2} - \frac{\partial^2 K(x, \xi)}{\partial \xi^2} - u(x) K(x, \xi) = 0,$$

$$u(x) = -2 \frac{d}{dx} K(x, x),$$

$$K(x, \xi) \to 0 \quad \text{as } \xi \to +\infty,$$

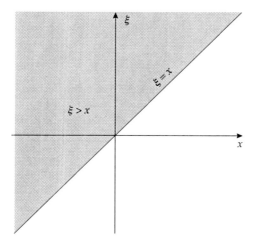

FIG. 2.1. *Sketch of the boundary conditions on $K(x,\xi)$ in the (x,ξ)-plane.*

and
$$\frac{\partial K(x,\xi)}{\partial \xi} \to 0 \quad \text{as } \xi \to +\infty.$$

That these requirements can be satisfied is seen from Figure 2.1. In the integral of Equation (2.41), $K(x,\xi)$ needs to be defined only over the region where $\xi \geq x$, indicated by shading in the figure. Along the lower boundary of this region, $x = \xi$, and $K(x,\xi)$ is determined by the condition

$$u(x) = -2\frac{d}{dx}K(x,x). \tag{2.42}$$

In the upper reaches of the shaded region, $\xi \to \infty$, and both $K(x,\xi)$ and $\partial K(x,\xi)/\partial \xi$ go to zero. Thus the problem is well posed and some $K(x,\xi)$ exists for $\xi > x$. How to find it?

Recall that the problem is causal in the pseudotime variable τ, and the source for the scattering is $\delta(x+\tau)$. Because an effect cannot precede its cause, $\varphi(x,\tau)$ must be zero for $x + \tau < 0$.

Imposing this constraint on Equation (2.41) and substituting the expression for $\varphi_\infty(x,\tau)$ from Equation (2.40), it follows that

$$B(x-\tau) + K(x,-\tau) + \int_x^\infty K(x,\xi)\, B(\xi-\tau)\, d\xi = 0 \tag{2.43}$$

for $x + \tau < 0$. This is the Gel'fand–Levitan equation for finding the potential in Schrödinger's equation from observations of its scattering [3]. It is a linear integral equation, with $B(x-\tau)$ assumed to be given [12]. Solving for $K(x,-\tau)$ allows $u(x)$ to be computed from Equation (2.42).

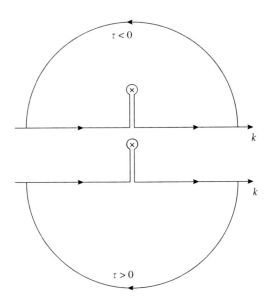

FIG. 2.2. *Contours in the complex k-plane for computing $B(x - \tau)$ from the inverse Fourier transform of Equation (2.38).*

The remaining task is to determine $B(x-\tau)$ from the scattering data: (i) $b(k)$ ($-\infty < k < +\infty$), and (ii) $\{\kappa_n\}$ and $\{r_n\}$ ($n = 1, 2, \ldots, N$). This can be done using the (causal) inverse Fourier transform defined in Equation (2.38) with the contours indicated in Figure 2.2.

To insure causality with respect to τ, it is convenient to assume $u(x) = 0$ for $x \geq 0$. Because the reflected wave of the scattering solution is $b(k)\exp(ikx)$, Equation (2.38) implies

$$B(x - \tau) = \frac{1}{2\pi} \int_{-\infty}^{\infty} b(k) e^{ik(x-\tau)} dk$$
$$= 0 \quad \text{for } \tau < 0,$$

if two conditions are satisfied. First, all singularities must be excluded from the integration contour for $\tau < 0$, as is indicated in the upper half of Figure 2.2. Second, the integral along the upper semicircular arc must go to zero as its radius becomes large. As $x - \tau > 0$ for $x \geq 0$ and $\tau < 0$, this second condition is satisfied.

Evidently $B(x - \tau)$ has two distinct components:

$$B(x - \tau) = B_1(x - \tau) + B_2(x - \tau),$$

where B_1 is generated by the scattering solutions, and B_2 is generated by the bound states.

As the reflected wave of the scattering solution is $b(k)\exp(ikx)$, the first of these components is computed directly from Equation (2.38) as

$$B_1(x-\tau) = \frac{1}{2\pi} \int_{-\infty}^{\infty} b(k)\, e^{ik(x-\tau)} dk, \tag{2.44}$$

where the path of integration is along the real k-axis, making no excursion into the upper half of the k-plane. From Figure 2.2, it is evident that this integral is generated by singularities of $b(k)$ in the lower half of the k-plane.

$B_2(x-\tau)$, on the other hand, arises from poles of $b(k)$ in the upper half of the k-plane, where

$$k = +i\kappa_n.$$

Such singularities give rise to bound states, and they are included within the $\tau > 0$ contour in order to preserve the causality with respect to pseudotime upon which the derivation depends. Because the integration is around a closed path in the clockwise direction (for $\tau > 0$), Cauchy's theorem[7] implies that the total contribution of the bound states to $B(x-\tau)$ is

$$B_2(x-\tau) = -i \sum_{n=1}^{N} r_n\, e^{-\kappa_n(x-\tau)}. \tag{2.45}$$

In a more conventional notation, the Gel'fand–Levitan equation can be written as[8]

$$B(x+y) + K(x,y) + \int_x^{\infty} K(x,\xi)\, B(\xi+y) d\xi = 0, \tag{2.46}$$

where $y > x$ and

$$B(x+y) = \frac{1}{2\pi} \int_{-\infty}^{\infty} b(k)\, e^{ik(x+y)} dk - i\sum_{n=1}^{N} r_n\, e^{-\kappa_n(x+y)}.$$

Because the relation between r_n and c_n found in Equation (2.36) holds in general, the second of these equations can also be written

$$B(x+y) = \frac{1}{2\pi} \int_{-\infty}^{\infty} b(k)\, e^{ik(x+y)} dk + \sum_{n=1}^{N} c_n^2\, e^{-\kappa_n(x+y)},$$

where the c_n are bound state normalization constants defined as in Equations (2.31) and (2.32).

---

[7] After the French mathematician Augustin Louis Cauchy (1789–1857).
[8] The Gel'fand–Levitan equation might more properly be called the Gel'fand–Levitan–Marchenko equation, as the original analysis was for $0 \leq x < +\infty$ [3, 9].

2.6.3 A reflectionless potential

As a simple example of the ideas developed in the previous section, assume

$$B(x+y) = \gamma e^{-\kappa(x+y)}.$$

Upon substituting into Equation (2.46), one finds

$$\gamma e^{-\kappa(x+y)} + K(x,y) + \gamma e^{-\kappa y} \int_x^\infty K(x,\xi) e^{-\kappa \xi} d\xi = 0;$$

thus $K(x,y)$ is proportional to $\exp(-\kappa y)$. Writing it as

$$K(x,y) = K_0(x) e^{-\kappa y},$$

the Gel'fand–Levitan equation can be solved for

$$K(x,y) = -\frac{\gamma e^{-\kappa(x+y)}}{1 + (\gamma/2\kappa)e^{-2\kappa x}}.$$

From Equation (2.42), the corresponding Schrödinger potential is

$$u(x) = \frac{-4\kappa\gamma}{[e^{\kappa x} + (\gamma/2\kappa)e^{-\kappa x}]^2}.$$

Choosing $\gamma = 2\kappa$ reduces this expression to

$$u(x) = -2\kappa^2 \operatorname{sech}^2 \kappa x,$$

and from the fact that $b(k) = 0$ was assumed in its construction, it is evident that this is a reflectionless potential in Schrödinger's equation.

The alert reader will not fail to notice that the shape of this potential is similar to Russell's solitary wave as given in Equation (1.3).

2.7 Problems

1. Discuss the solutions of Equation (1.2) when both $\varepsilon = 0$ and $\gamma = 0$.

2. Use qualitative arguments to show that solutions of the linear diffusion equation

$$u_{xx} - u_t = 0$$

are never wavelike. [Hint: consider the sign of the second derivative.]

3. Consider the linear wave equation

$$cu_x + u_t = 0,$$

with the initial condition

$$u(x,0) = f(x).$$

Solve this problem by taking: (i) the Laplace transform of $u(x,t)$ to obtain $U(x,s)$, (ii) the Fourier transform of $U(x,s)$ to obtain $\mathcal{U}(k,s)$, (iii) the inverse Laplace transform of $\mathcal{U}(k,s)$ to obtain $\tilde{U}(k,t)$, and finally (iv) the inverse Fourier transform of $\tilde{U}(k,t)$ to obtain $u(x,t)$. Discuss your result.

4. Use a Laplace transform on t and then a Fourier transform on x to find the solution of
$$u_{xx} - u_{tt} = 0$$
for $-\infty < x < +\infty$, with the initial conditions
$$u(x,0) = f(x) \quad \text{and} \quad u_t(x,0) = 0.$$

5. (a) Use the combined Fourier–Laplace transform method of Problems 3 and 4 to solve the problem
$$iu_t + u_{xx} = 0,$$
with $-\infty < x < +\infty$ and the initial condition $u(x,0) = f(x)$.

(b) How is the initial velocity of a wave packet determined by $f(x)$?

6. Use the combined Fourier–Laplace transform method to solve the problem
$$u_{xx} - u_{tt} = u,$$
with $-\infty < x < +\infty$ and the initial conditions
$$u(x,0) = f(x) \quad \text{and} \quad u_t(x,0) = g(x).$$
[Hint: note that the dispersion relation $\omega = \omega(k)$ is double valued.]

7. In Equation (2.8), assume that $u(x,0)$ is a localized pulse with a well defined maximum value so its Fourier transform has a maximum value at $k = k_0$. Show that the speed of propagation of the pulse is $c - 3\varepsilon k_0^2$.

8. Find a simple dependent variable transformation that reduces a linear diffusion equation of the form
$$w_t - w_{xx} + w = 0$$
to
$$u_t - u_{xx} = 0.$$

9. Show that the natural solution of a linear diffusion equation given in Equation (2.13) satisfies the condition
$$\int_{-\infty}^{\infty} g\,dx = \text{constant}.$$
[Hint: see Appendix A.]

10. Verify the limit for the natural solution of a linear diffusion equation expressed in Equation (2.14).

11. Show that
$$g(x) = \frac{e^{-a|x|}}{2a}$$
is a solution of
$$-\frac{d^2 g}{dx^2} + a^2 g = \delta(x),$$
over $-\infty < x < \infty$.

12. Consider the problem
$$\frac{\partial u}{\partial t} - \frac{\partial^2 u}{\partial x^2} = f(x, t),$$
defined on the half plane $-\infty < x < +\infty$ and $t \geq 0$ with the initial condition
$$u(x, 0) = 0$$
and the requirement that $f(x, t) = 0$ for $t < 0$. Show by direct substitution that
$$u(x, t) = \int_0^t d\tau \int_{-\infty}^{\infty} d\xi \, f(\xi, \tau) \frac{\exp[-(x - \xi)^2/4(t - \tau)]}{2\sqrt{\pi(t - \tau)}}.$$

13. Solve the previous problem for the initial condition
$$u(x, 0) = F(x).$$

14. Show that the operator
$$\frac{d^2}{dx^2}$$
with periodic boundary conditions is self-adjoint.

15. Consider the matrix equation
$$M\mathbf{u} = \mathbf{f},$$
where M is an $n \times n$ square matrix with real coefficients, and $\mathbf{u}$ and $\mathbf{f}$ are column vectors with n real components. With respect to the vector inner product
$$(\mathbf{v}, \mathbf{w}) \equiv \sum_{j=1}^{n} v_j w_j,$$
the adjoint $(M^\dagger)$ is defined as
$$(M\mathbf{v}, \mathbf{w}) = (\mathbf{v}, M^\dagger \mathbf{w}).$$

(a) Show that the adjoint of M is equal to its transpose.
(b) Show that a necessary condition for the matrix equation to have a solution is
$$(\mathbf{f}, \mathbf{v}) = 0$$

for all vectors **v** that satisfy
$$M^\dagger \mathbf{v} = \mathbf{0}.$$

16. The matrix equation
$$\begin{bmatrix} 1 & 1 & 1 \\ 2 & -1 & 1 \\ 1 & -2 & 0 \end{bmatrix} \begin{bmatrix} u_1 \\ u_2 \\ u_3 \end{bmatrix} = \begin{bmatrix} f_1 \\ f_2 \\ f_3 \end{bmatrix}$$
is an example of the previous problem. With **v** defined as the solution of
$$\begin{bmatrix} 1 & 2 & 1 \\ 1 & -1 & -2 \\ 1 & 1 & 0 \end{bmatrix} \begin{bmatrix} v_1 \\ v_2 \\ v_3 \end{bmatrix} = \begin{bmatrix} 0 \\ 0 \\ 0 \end{bmatrix},$$
find the conditions on **f** for the original matrix equation to be solvable. Discuss your result.

17. Consider the problem
$$-\frac{d^2 u}{dx^2} = f(x),$$
where $0 \le x \le 1$, $u(0) = a$, and $u(1) = b$. If the solution is expressed as
$$u(x) = \int_0^1 f(\xi) g(x, \xi) d\xi,$$
find $g(x, \xi)$.

18. Is the solution $u_0 = 0$ of the SG equation stable or asymptotically stable from the perspective of a linear approximation.

19. Consider the system
$$\frac{dx}{dt} = \alpha x (1 - x/x_0) - \beta xy$$
$$\frac{dy}{dt} = -\gamma y + \delta xy,$$
proposed by Volterra to represent the interaction of a predator species (y), with its prey (x) having a maximum sustainable population of x_0.

   (a) From a sketch of trajectories in the (x, y)–phase plane, find all stationary solutions ($dx/dt = 0$ and $dy/dt = 0$) of this system with $x \ge 0$ and $y \ge 0$.

   (b) Determine the stability of these stationary solutions from the perspective of a linear analysis.

20. Show that the assumptions of Equation (1.20) and Equation (1.18) imply that the discrete eigenvalues of Equation (1.19) are independent of time.

21. The Wronskian $W(\psi; \phi)$ of the functions $\psi(x)$ and $\phi(x)$ is defined as
$$W(\psi; \phi) \equiv \psi \phi_x - \phi \psi_x.$$

If ψ and ϕ are two solutions of Schrödinger's equation (2.28) for the same value of λ, show that $W(\psi; \phi)$ is independent of x.

22. (a) If the scattering solution $\psi(x, k)$ in Equation (2.29) satisfies Schrödinger's equation, show that $\psi^*(x, k)$ does also.

    (b) From this observation, evaluate the Wronskian $W(\psi; \psi^*)$ to show that
    $$|a|^2 + |b|^2 = 1.$$

    (c) Give a physical interpretation of this result.

23. (a) Sketch solution trajectories of Equation (2.28) in the $(\psi, d\psi/dx)$ phase plane under the assumptions: $u(x)$ is negative near $x = 0$, and $u(x) \to 0$ as $x \to \pm\infty$.

    (b) Show that there are a finite number of bound solutions, going to zero as $x \to \pm\infty$.

    (c) Relate the eigenvalue (λ_n) of each solution to the number of times that the corresponding phase plane trajectory loops around the origin.

24. Consider Schrödinger's equation (2.28) with a potential function of the form
    $$u(x) = \begin{cases} \alpha & 0 < x < 1, \\ 0 & x < 0 \text{ and } x > 1, \end{cases}$$
    where α is a real number that can take any positive or negative value. (Note that for $\alpha > 0$, this is a potential barrier, while for $\alpha < 0$, it is a potential well.)

    (a) Solve the scattering problem for $\alpha > 0$. [Hint: match boundary conditions at discontinuities.]

    (b) Solve the scattering problem for $\alpha < 0$.

    (c) Extend the results of part (b)—by analytic continuation—into the upper half of the complex k-plane to find the conditions on α for the existence of bound states.

    (d) Check Equation (2.36) for these bound states.

25. Consider the reflected wave $b(k) \exp(ikx)$ to be the causal Fourier (Laplace) transform of $B(x - \tau)$ in Equation (2.43), and assume that $u(x) = 0$ for $x \geq 0$.

    (a) Sketch incident and scattered waves on the (x, τ) plane, showing that $B(x - \tau) = 0$ for $\tau < 0$.

    (b) In evaluating the inverse transform
    $$B(x - \tau) = \frac{1}{2\pi} \int_{-\infty}^{\infty} b(k) e^{ik(x-\tau)} dk$$
    for $x \geq 0$, show that $x - \tau < 0$ implies $\tau > 0$.

(c) Use the results of (a) and (b) to check that integrals on the large semi-circles of Figure 2.2, go to zero at large radius for $x = 0$ and the indicated values of τ.

(d) From (c), show that causality in τ is satisfied by deformations of the integration paths as indicated in Figure 2.2.

26. Verify the details of the complex integration leading to the expression for $B_2(x-\tau)$ in Equation (2.45).

27. Assume that the reflection coefficient for scattering states in Schrödinger's equation (2.28) is
$$b(k) = -1/(2ik + 1).$$

(a) Use this information to find $u(x)$.

(b) If the potential function
$$u(x) = -\delta(x)$$
in Schrödinger's equation, find $B(x+y)$ in the Gel'fand–Levitan equation.

28. Assuming that
$$B(x+y) = \gamma_1 e^{-\kappa_1(x+y)} + \gamma_2 e^{-\kappa_2(x+y)}$$
in Equation (2.46), find the corresponding reflectionless potential of Schrödinger's equation (2.28).

REFERENCES

1. G N Balanis. The plasma inverse problem. *J. Math. Phys.* 13 (1972) 1001–1005.
2. P P G Dyke. *An Introduction to Laplace Transforms and Fourier Series.* Springer-Verlag, New York, 1999.
3. I M Gel'fand and B M Levitan. On the determination of a differential equation from its spectral function. *Am. Math. Transl.* (2) 1 (1951) 253–304.
4. G Green. *An Essay on the Application of Mathematical Analysis to the Theories of Electricity and Magnetism.* Printed for the author by T Wheelhouse, Nottingham, 1828.
5. O Heaviside. *Electromagnetic Theory.* Dover Publications Inc., New York, 1950.
6. F B Hildebrand. *Advanced Calculus for Applications.* 2nd edition. Prentice-Hall, Englewood Cliffs, New Jersey, 1976.
7. W Hurewicz. *Lectures on Ordinary Differential Equations.* John Wiley & Sons, New York, 1958.
8. J P Keener. *Principles of Applied Mathematics, Transformation and Approximation.* Perseus Publishing, Cambridge, Massachusetts, 2000.
9. V A Marchenko. On the reconstruction of the potential energy from phases of the scattered waves. *Dokl. Akad. Nauk SSSR* 104 (1955) 695–698.
10. A C Scott. *Active and Nonlinear Wave Propagation in Electronics.* John Wiley & Sons, New York, 1970.

11. J L Schiff. *The Laplace Transform: Theory and Applications.* Springer-Verlag, New York, 1999.
12. I Stackgold. *Green's Functions and Boundary Value Problems.* John Wiley & Sons, New York, 1979.
13. G Strang. *Introduction to Applied Mathematics.* Wellesley-Cambridge Press, Wellesley, 1986.
14. G B Whitham. *Linear and Nonlinear Waves.* John Wiley & Sons, New York, 1974.

3
THE CLASSICAL SOLITON EQUATIONS

An exciting development in nonlinear science during the 1970s was the gradual realization that certain nonlinear partial differential equations display a variety of exact solutions. These include not only solitary waves—known since the nineteenth century studies of Russell, Bazin, and Boussinesq—but solutions involving an arbitrary number of solitary waves of varying speeds and amplitudes undergoing a collective collision. In such events, the solitary waves are now called "solitons," emphasizing their particle-like character, because they leave the interaction region of space-time with exactly the same speeds and amplitudes that they had upon entry [61]. (As was noted in the introduction, Russell observed hydrodynamic solitary waves passing through each other without change, but he provided no analytic explanation for the phenomenon.)

The first and still most significant soliton systems arose prior to the 1970s in the context of outstanding problems in applied science. Foremost among these are the following three:

1. *The Korteweg–de Vries (KdV) equation.* Proposed in 1895 as a model for Russell's hydrodynamic solitary wave [36], this equation can be normalized as

$$\frac{\partial u}{\partial t} - 6u\frac{\partial u}{\partial x} + \frac{\partial^3 u}{\partial x^3} = 0. \qquad (3.1)$$

Interestingly, *Physical Review Letters* was reluctant to publish the seminal paper by Zabusky and Kruskal (where the term "soliton" was coined) because some believed that the KdV equation has little to do with physics. The authors dealt with this concern by deriving Equation (3.1) for ion-acoustic waves in a plasma medium [70, 75], but its scientific significance is more broad. In fact, the KdV system arises whenever one studies unidirectional propagation of long waves in a dispersive energy conserving medium at the lowest order of nonlinearity, a generic property that led to several early applications, including: the anharmonic lattice [74], pressure waves in liquid-gas bubble mixtures [72], rotating flow of a liquid through a tube [42], and thermally excited phonon packets in low temperature nonlinear crystals [68]. Many such applications are described in the recent book by Naugolnykh and Ostrovsky [48].

The author thanks P. L. Christiansen for helping to plan this chapter and for contributions to Section 3.2.4.

2. *The sine–Gordon (SG) equation.* Augmenting the linear wave equation with one of the elementary functions leads to

$$\frac{\partial^2 u}{\partial x^2} - \frac{\partial^2 u}{\partial t^2} = \sin u, \qquad (3.2)$$

where normalizing units have been used to measure x, t, and u. As was noted in Chapter 1, this equation has many physical applications [2], including the propagation of crystal defects [27] and domain walls in ferromagnetic and ferroelectric materials [22], a one-dimensional model for elementary particles [24,55,64], the propagation of splay waves on a biological (lipid) membrane [25], self-induced transparency of short optical pulses [37], and the propagation of quantum units of magnetic flux (called fluxons) on long Josephson (superconducting) transmission lines [62]. All of these applications had been considered before the special properties of the SG equation were rediscovered in the 1970s.

3. *The nonlinear Schrödinger (NLS) equation.* Again in normalized form, this equation can be written as

$$\mathrm{i}\frac{\partial u}{\partial t} + \frac{\partial^2 u}{\partial x^2} + 2|u|^2 u = 0, \qquad (3.3)$$

which arose as a model for packets of hydrodynamic waves on deep water [7]. It takes its name from the fact that the small amplitude approximation is the equation that Erwin Schrödinger proposed in January of 1926 for the propagation of a quantum wave packet in free space. (When one wishes to distinguish different degrees of nonlinearity, NLS is sometimes called the "cubic Schrödinger" equation.) Since the mid-1970s, the NLS equation has been important as the fundamental description of nonlinear pulses on an optical fiber [29], but its potential significance is even greater. Like KdV, NLS is a generic equation, arising whenever one studies unidirectional propagation of wave packets in a dispersive energy conserving medium at the lowest order of nonlinearity. Again this property led to many early physical applications, including: two-dimensional self-focusing of a plane wave [33, 66], one-dimensional self-modulation of a monochromatic wave [29, 32, 67], propagation of a heat pulse in a solid [68], and Langmuir waves in plasmas [28, 31].

In addition to its solitons, each of these three equations supports nonlinear periodic waves and exact N-soliton solutions, describing the above mentioned collective collisions. It is the primary aim of this chapter to explore solutions in some detail and to present various analytical methods for finding them. In each case, we shall sketch physical applications before entering the thicket of mathematical analysis. (Excellent discussions of physical models for these three equations are included in the recent book by Remoissenet [54].)

3.1 The Korteweg–de Vries (KdV) equation

3.1.1 Long water waves

Korteweg and de Vries studied the dynamics of the long water waves of low amplitude that were observed by Russell, concluding that propagation along the canal can be approximately described by the equation

$$\frac{1}{c}\frac{\partial u}{\partial t} + \frac{\partial u}{\partial x} + \frac{3}{2h}u\frac{\partial u}{\partial x} + \frac{h^2}{6}\frac{\partial^3 u}{\partial x^3} \doteq 0, \tag{3.4}$$

where

- u (in centimeters) is the vertical displacement of the water surface from its resting level,
- x (in centimeters) is the distance along the canal,
- t (in seconds) is time, and
- h (in centimeters) is the resting depth of the water in the canal.

In the derivation of Equation (3.4), four assumptions are made: (i) inviscid water, implying no loss of energy by the wave, (ii) small amplitude, expressed as the condition that $u \ll h$, (iii) slow spatial variation, expressed as the condition that $hu_x \ll u$, and (iv) no surface tension.

From assumptions (ii) and (iii), it is evident that the first two terms of Equation (3.4) are larger than the remaining two; thus to an even lower approximation

$$\frac{\partial u}{\partial t} + c\frac{\partial u}{\partial x} \approx 0.$$

This is satisfied by a wave of arbitrary shape, propagating in the $+x$-direction at the fixed speed

$$c = \sqrt{gh},$$

where $g = 980 \, \text{cm/s}^2$ is the acceleration of gravity.

The third and fourth terms on the left-hand side of Equation (3.4) are taken to be of the same order of smallness, which can be expressed as

$$\varepsilon \sim \left|\frac{u}{h}\right| \sim \left|h^2\frac{\partial^2}{\partial x^2}\right|.$$

Thus the symbol "$\doteq$" indicates that terms of higher order in ε have been neglected in the derivation of Equation (3.4). As was noted in Chapter 1, the third term in Equation (3.4) makes the system nonlinear, whereas the fourth term introduces dispersion. It is the balance between these two effects that results in the emergence of a solitary wave.

Although it is possible to analyze Equation (3.4) as it stands, a more convenient form is obtained by eliminating the $\partial u/\partial x$ term via a (Galilean) transformation of the independent variables of the form

$$u(x,t) \to \tilde{u}(\zeta, \tau),$$

where
$$\zeta = x - ct$$
$$\tau = t,$$

implying that
$$\frac{\partial u}{\partial x} = \frac{\partial \tilde{u}}{\partial \zeta}$$
$$\frac{\partial u}{\partial t} = \frac{\partial \tilde{u}}{\partial \tau} - c\frac{\partial \tilde{u}}{\partial \zeta}.$$

Under this transformation, Equation (3.4) becomes
$$\frac{\partial \tilde{u}}{\partial \tau} + \frac{3c}{2h}\tilde{u}\frac{\partial \tilde{u}}{\partial \zeta} + \frac{h^2 c}{6}\frac{\partial^3 \tilde{u}}{\partial \zeta^3} \doteq 0. \tag{3.5}$$

By choosing appropriate units for $\tilde{u}$ and ζ, Equation (3.5) takes the form of Equation (3.1).

3.1.2 Solitary wave solutions

Consider Equation (3.1)
$$u_t - 6uu_x + u_{xxx} = 0$$
with $-\infty < x < +\infty$ and $-\infty < t < +\infty$. Introducing the traveling-wave assumption $u(x,t) = \tilde{u}(x - vt)$ (where $\tilde{u}$ is an undetermined function) leads to the ordinary differential equation
$$(v + 6\tilde{u})\tilde{u}' - \tilde{u}''' = 0. \tag{3.6}$$

Here prime denotes differentiation with respect to the traveling-wave variable $\xi \equiv x - vt$. Integrating once with respect to ξ gives
$$(v + 3\tilde{u})\tilde{u} - \tilde{u}'' = \frac{A}{2}$$
and again (after multiplication by $\tilde{u}'$)
$$(v + 2\tilde{u})\tilde{u}^2 - (\tilde{u}')^2 = A\tilde{u} + B, \tag{3.7}$$
where A and B are integration constants.

For boundary conditions $\tilde{u}$ and $\tilde{u}' \to 0$ as $\xi \to \pm\infty$, Equation (3.7) becomes
$$(\tilde{u}')^2 = \tilde{u}^2(2\tilde{u} + v)$$
or
$$\xi - x_0 = \pm \int^{\tilde{u}} \frac{dy}{y\sqrt{2y + v}}. \tag{3.8}$$

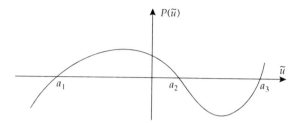

FIG. 3.1. *The third-degree polynomial, $P(\tilde{u})$, in Equation (3.11).*

This integral is readily evaluated to obtain the traveling-pulse solution

$$u(x,t) = -\frac{v}{2}\text{sech}^2\left[\frac{\sqrt{v}}{2}(x - vt - x_0)\right], \qquad (3.9)$$

describing the "Wave of Translation" discovered by Russell in 1834 and shown in Figure 1.1.

This expression represents a bounded solution of the KdV equation for all values of the velocity $0 \leq v < +\infty$. As the velocity increases, the pulse becomes more narrow and its amplitude increases. The integration constant x_0 locates the center of the pulse at time $t = 0$.

The solitary pulse of Equation (3.9) is the basic building block from which exact N-soliton solutions of the KdV equation can be constructed. Before taking up such solutions, however, let's consider the possibility of periodic traveling-wave solutions, extending over the range $-\infty < \xi < +\infty$.

3.1.3 *Periodic solutions*

For periodic solutions, the arbitrary constants, A and B, in the ordinary differential equation (3.7) for the traveling-wave $\tilde{u}$ are not zero. The equation may be rewritten as

$$\tilde{u}' = \pm\sqrt{P(\tilde{u})} \qquad (3.10)$$

with the third-degree polynomial

$$P(\tilde{u}) = 2\left(\tilde{u}^3 + v\tilde{u}^2/2 - A\tilde{u}/2 - B/2\right)$$
$$\equiv 2(\tilde{u} - a_1)(\tilde{u} - a_2)(\tilde{u} - a_3). \qquad (3.11)$$

If v, A, and B have such values that the three roots are real with $a_3 > a_2 > a_1$, it is shown in Appendix C that

$$\tilde{u}(\xi) = a_2 - (a_2 - a_1)\text{cn}^2\left[(\xi - \xi_1)\sqrt{(a_3 - a_1)/2}, k\right], \qquad (3.12)$$

where $\text{cn}(\zeta, k)$ is a Jacobi elliptic function of modulus k and

$$k^2 = (a_2 - a_1)/(a_3 - a_1).$$

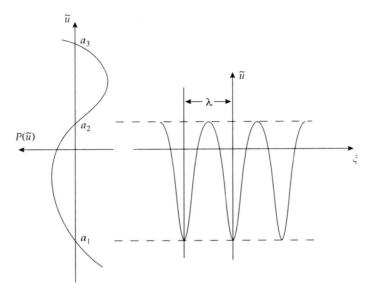

FIG. 3.2. *A periodic solution of Equation (3.12).*

In Figure 3.2, such a periodic wave is sketched (with $\xi_1 = 0$) and related to the underlying polynomial $P(\tilde{u})$. From this sketch, the following points should be noted.

1. Although the elliptic functions introduced in Appendix C and used to construct Equation (3.12) are precise and convenient, the qualitative nature of $\tilde{u}(\xi)$ can be obtained from inspection of Figure 3.2. Thus a real and bounded solution of Equation (3.10) can exist only for $a_1 \leq \tilde{u} \leq a_2$, where $P(\tilde{u})$ is positive. This solution has turning points at a_1 and a_2, where $P(\tilde{u}) = 0$.
2. For $k = 0$, $\text{cn}(\zeta, 0) = \cos(\zeta)$, and the problem is reduced to a linear (sinusoidal) limit.
3. For $0 < k < 1$, $\text{cn}(\zeta, k)$ retains the qualitative character of a cosine function (see Figure C.2) but with a period, $4K(k)$, that increases without bound as $k \to 1$. This is the solitary-wave limit treated in the previous section.
4. Because $\text{cn}(\zeta, k)$ appears as the square in Equation (3.12), $\tilde{u}(\xi)$ has the period
$$\lambda = 2K(k)/\sqrt{(a_3 - a_1)/2}, \qquad (3.13)$$
as indicated in Figure 3.2.
5. From the definition of the polynomial in Equation (3.11), it is seen that
$$v = -2(a_1 + a_2 + a_3).$$

Thus specifying the maximum value of a periodic wave (a_2), the minimum value (a_1), and the traveling-wave velocity (v) fixes all of the parameters in the problem.

As was noted in Chapter 1, a periodic nonlinear wave may be modulated to allow the wave's speed (v), wavelength (λ), and amplitude ($a_2 - a_1$) to be slowly varying functions of space and time. Whitham has shown how an averaged Lagrangian formulation of the problem can be used to obtain dynamic equations relating these slow variations [71].

3.1.4 A Bäcklund transformation for KdV

Having explored the traveling-wave solutions of KdV, we are now ready to consider the construction of analytic expressions for N-soliton collisions. One way to obtain such solutions is through the repeated application of a Bäcklund transformation (BT) with the general form

$$w_{1,x} = F(w_0, w_1, w_{0,x}, w_{0,t}; \lambda)$$
$$w_{1,t} = G(w_0, w_1, w_{0,x}, w_{0,t}; \lambda), \quad (3.14)$$

where $w_0(x,t)$ is a known solution of the system under study and $w_1(x,t)$ is a new solution to be generated by the transformation. In principle, this process can be repeated an indefinite number of times, with the parameter λ characterizing the added feature of the new solution.

To show that such functions (F and G) exist, it is necessary to prove that if w_0 is a solution of the equation of interest, then w_1, generated by integration of Equations (3.14), is also a solution of that equation. For linear PDEs, interestingly, there is always such a BT that has this property, which introduces a new eigenfunction into the solution with each application. It is for nonlinear PDEs that difficulties arise.

For the KdV equation, a BT was first obtained by Wahlquist and Estabrook in 1973, and their results are presented here, following the lucid discussion by Drazin and Johnson [21, 69].

In 1968, Miura showed that if $v(x,t)$ is a solution to the modified Korteweg–de Vries (MKdV) equation

$$v_t - 6v^2 v_x + v_{xxx} = 0, \quad (3.15)$$

then

$$u \equiv v^2 + v_x \quad (3.16)$$

is a solution to the KdV equation (3.1) [46]. To see this, note that substitution of Equation (3.16) into KdV yields an equation that can be rewritten as

$$\left(2v + \frac{\partial}{\partial x}\right)(v_t - 6v^2 v_x + v_{xxx}) = 0. \quad (3.17)$$

This implies two observations:

1. If v is a solution of
$$v_t - 6(v^2 + \lambda)v_x + v_{xxx} = 0, \tag{3.18}$$
then
$$u = \lambda + v^2 + v_x \tag{3.19}$$
is a solution of the KdV equation.
2. If v is a solution to Equation (3.18), then $-v$ is also a solution to this equation.

From Equation (3.19) are obtained two different solutions to the KdV equation:
$$u_0 = \lambda + v^2 + v_x$$
$$u_1 = \lambda + v^2 - v_x.$$

Adding and subtracting these equations yields
$$u_0 - u_1 = 2v_x$$
$$u_0 + u_1 = 2(\lambda + v^2). \tag{3.20}$$

It is now convenient to introduce a change of dependent variable as
$$u_0 = \frac{\partial w_0}{\partial x} \quad \text{and} \quad u_1 = \frac{\partial w_1}{\partial x}, \tag{3.21}$$
where w_0 and w_1 are defined up to an arbitrary function of time. Equations (3.20) can then be rewritten as
$$w_0 - w_1 = 2v$$
and
$$(w_1 + w_0)_x = 2\lambda + (w_1 - w_0)^2/2. \tag{3.22}$$

Using these results, Equation (3.18) becomes
$$(w_1 - w_0)_t - 3(w_{1,x}^2 - w_{0,x}^2) + (w_1 - w_0)_{xxx} = 0. \tag{3.23}$$

Equations (3.22) and (3.23) are recognized as a BT, with the structure indicated in Equations (3.14). [Although Equation (3.23) may not, at first glance, appear to have this structure, its dependence on $w_{1,x}$ and $w_{1,xxx}$ is readily eliminated by substitution from Equation (3.22).]

The first step in checking a BT is to see if it can correctly generate an elementary solution (soliton) from the vacuum. To this end, assume $w_0 = 0$ (the vacuum solution), whereupon Equation (3.22) becomes
$$w_{1,x} = 2\lambda + w_1^2/2. \tag{3.24}$$

Upon setting
$$\lambda = -\kappa^2 < 0,$$
integration leads to the intermediate result
$$w_1(x,t) = -2\kappa \tanh[\kappa x + f(t)],$$
with κ a real number and $f(t)$ an arbitrary function of time.

To fix the time dependence of w_1, return to Equation (3.23), which implies that
$$w_{1,t} = 3w_{1,x}^2 - w_{1,xxx}. \tag{3.25}$$
Differentiating Equation (3.24), gives
$$w_{1,xx} = w_1 w_{1,x}$$
$$w_{1,xxx} = w_{1,x}^2 + w_1 w_{1,xx},$$
so Equation (3.25) becomes
$$w_{1,t} = 2w_{1,x}^2 - w_1 w_{1,xx}$$
$$= 2w_{1,x}(w_{1,x} - w_1^2/2)$$
or from Equation (3.24)
$$w_{1,t} + 4\kappa^2 w_{1,x} = 0.$$

This latter result indicates that w_1 is a traveling wave, with the functional dependence
$$w_1(x,t) = w_1(x - 4\kappa^2 t)$$
$$= -2\kappa \tanh[\kappa(x - 4\kappa^2 t - x_0)]. \tag{3.26}$$
Because $w_{1,x} \equiv u_1$, this implies
$$u_1(x,t) = -2\kappa^2 \text{sech}^2[\kappa(x - 4\kappa^2 t - x_0)] \tag{3.27}$$
as an expression for the soliton that can be compared with the previous result in Equation (3.9). Evidently the two results are identical if the traveling-wave speed
$$v = 4\kappa^2 = -4\lambda.$$

Thus a BT can be used to generate a soliton from the vacuum, an operation that can be expressed by the diagram
$$0 \xrightarrow{\lambda} u_1.$$

Although comforting, this achievement is not overly impressive because we already had a formula for the KdV soliton, and it is possible to find a BT that will generate any solitary wave from the vacuum. To show its worth, a BT must generate a solution with more than one soliton.

Let us now construct a KdV solution with two soliton components. The first such solution, published by Zabusky in 1968, has the form [73]

$$u_Z(x,t) = -12 \frac{3 + 4\cosh(2x - 8t) + \cosh(4x - 64t)}{[3\cosh(x - 28t) + \cosh(3x - 36t)]^2}. \qquad (3.28)$$

This is an exact solution of Equation (3.1), describing the collision of two solitons near the origin of the (x,t)-plane. Thus as $t \to -\infty$,

$$u_Z(x,t) \sim -8\operatorname{sech}^2\left(2x - 32t + \tfrac{1}{2}\log 3\right) - 2\operatorname{sech}^2\left(x - 4t - \tfrac{1}{2}\log 3\right), \qquad (3.29)$$

the sum of two solitons on a collision course. As $t \to +\infty$, on the other hand, Equation (3.28) becomes

$$u_Z(x,t) \sim -8\operatorname{sech}^2\left(2x - 32t - \tfrac{1}{2}\log 3\right) - 2\operatorname{sech}^2\left(x - 4t + \tfrac{1}{2}\log 3\right), \qquad (3.30)$$

the sum of the same solitons moving away from the collision. As a result of the collision, the trajectory of the faster ($v = 16$) soliton is advanced by a distance of $\log 3$, while that of the slower ($v = 4$) soliton is retarded by a distance of $\log 3$. All of these facts, and more, are contained in Equation (3.28), which is plotted in Figure 3.3.

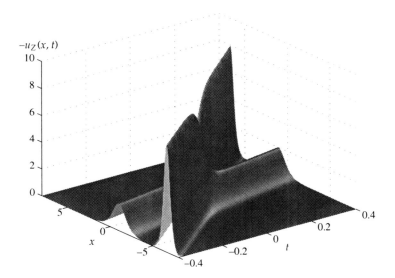

FIG. 3.3. *A plot of a 2-soliton collision: $-u_Z(x,t)$ from Equation (3.36).*

To get this result from our BT, we might consider letting w_0 be one of the incoming solitons in Figure 3.3, say

$$w_1 = -2\tanh(x - 4t),$$

which is generated from the vacuum by the transformation

$$w_{1,x} = -2 + w_1^2/2$$
$$w_{1,t} = 3w_{1,x}^2 - w_{1,xxx}.$$

Then setting $\lambda = -4$, corresponding to the speed ($v = 16$) of the other soliton, the 2-soliton collision should be generated by integrating

$$(w_{12} + w_1)_x = -8 + (w_{12} - w_1)^2/2$$
$$(w_{12} + w_1)_t = 3(w_{12,x}^2 - w_{1,x}^2) - (w_{12} - w_1)_{xxx}. \tag{3.31}$$

These equations are represented by the diagram

$$0 \xrightarrow{-1} w_1 \xrightarrow{-4} w_{12},$$

where $w_{12,x}$ is expected to equal $u_Z(x,t)$ as defined in Equation (3.28). A moment of reflection, however, should convince the reader that this integration will be difficult to carry through. Is it possible to find the 2-soliton formula with less difficulty?

There is a better way, and it is based on the permutative property of successive BTs, established for the SG equation in the nineteenth century [23]. In the context of inverse scattering transform theory, to be discussed in Chapter 6, each new soliton introduces a corresponding factor into the transmission coefficient of an associated scattering problem. Because the solution to the nonlinear problem can be constructed from the transmission coefficient—the factors of which can be arranged in any order—a succession of BTs performed in any order will produce the same result [26, 39]. This useful property can be expressed by the diagrams

$$w_0 \xrightarrow{\lambda_1} w_1 \xrightarrow{\lambda_2} w_{12} = w_{21},$$

and

$$w_0 \xrightarrow{\lambda_2} w_2 \xrightarrow{\lambda_1} w_{21} = w_{12}.$$

The first of the above diagrams represents the pair of equations

$$(w_1 + w_0)_x = 2\lambda_1 + (w_1 - w_0)^2/2$$
$$(w_{12} + w_1)_x = 2\lambda_2 + (w_{12} - w_1)^2/2,$$

and the second represents

$$(w_2 + w_0)_x = 2\lambda_2 + (w_2 - w_0)^2/2$$
$$(w_{21} + w_2)_x = 2\lambda_1 + (w_{21} - w_2)^2/2.$$

Because

$$w_{12} = w_{21},$$

the x-derivatives can be eliminated from these four equations, giving the algebraic expression

$$w_{12} = w_0 - \frac{4(\lambda_1 - \lambda_2)}{w_1 - w_2}. \tag{3.32}$$

After contemplating the direct integration of Equations (3.31) with respect to both x and t, Equation (3.32) is a remarkably simple result. Consider two of its implications.

1. Setting $\lambda_1 = \lambda_2$ in Equation (3.32) implies that $w_{12} = w_0$. Because soliton speeds are directly related to the λs, we see that we cannot create two different solitons with the same speed. This is consistent with the traveling-wave results from Sections 3.1.2 and 3.1.3, where it was found that there could be either a single soliton or an infinite number (a periodic traveling-wave), but not a finite number.
2. In order to avoid a singularity of w_{12} in the (x,t)-plane, it is necessary that w_1 be nowhere equal to w_2. Because there are always values of x and t where components of the form

$$-2\kappa \tanh[\kappa(x - 4\kappa^2 t - x_0)]$$

(with different values of κ) intersect, this seems to present a problem, but it can be overcome by the following trick. Although in the derivation of the single soliton, given in Equations (3.26) and (3.27), we quite properly ignored singular solutions such as

$$-2\kappa \coth[\kappa(x - 4\kappa^2 t - x_0)],$$

choosing (say) w_1 as the "tanh" and w_2 as the "coth" (with $\kappa_2 > \kappa_1$) insures the regularity of w_{12}.

Thus to construct Zabusky's 2-soliton solution—$u_Z(x,t)$—of Equation (3.28) from Equation (3.32), one can assume

$$w_0 = 0$$
$$\lambda_1 = -1$$
$$\lambda_2 = -4$$
$$w_1 = -2\tanh(x - 4t)$$
$$w_2 = -4\coth(2x - 32t),$$

whereupon
$$w_{12} = -\frac{12}{4\coth(2x-32t) - 2\tanh(x-4t)} \qquad (3.33)$$
and $u_Z(x,t) = \partial w_{12}/\partial x$.

3.1.5 N-soliton formulas

At this point in the discussion, the reader should be convinced that the KdV equation has exact N-soliton solutions, and it is not so easy to find them. In 1971 Hirota developed an ingenious scheme for constructing such formulas, which is summarized in this section [21, 30, 71].

Convenient for Hirota's approach is the introduction of bilinear differential operators, defined as

$$D_t^m D_x^n (a \cdot b) \equiv \left[\frac{\partial}{\partial t} - \frac{\partial}{\partial t'}\right]^m \left[\frac{\partial}{\partial x} - \frac{\partial}{\partial x'}\right]^n a(x,t)b(x',t')\bigg|_{\substack{x=x' \\ t=t'}}. \qquad (3.34)$$

Several of the salient properties of this operator are presented in Problem 19. (To become familiar with the notation, the reader should pause at this point and work them through.) This formulation may seem less strange if one notes that the standard expression for the multiple derivative of a product can be written

$$\frac{\partial^{m+n}}{\partial x^m \partial t^n}(a \cdot b) = \left[\frac{\partial}{\partial t} + \frac{\partial}{\partial t'}\right]^m \left[\frac{\partial}{\partial x} + \frac{\partial}{\partial x'}\right]^n a(x,t)b(x',t')\bigg|_{\substack{x=x' \\ t=t'}}.$$

As it stands, the KdV equation has terms of different powers of the dependent variables (u), but this property can be changed by substituting

$$u = -2\frac{\partial^2}{\partial x^2}\log f, \qquad (3.35)$$

a transformation that may be written in two steps. Letting $u = w_x$ in Equation (3.1) and integrating with respect to x we get

$$w_t - 3(w_x)^2 + w_{xxx} = 0 \qquad (3.36)$$

for suitable boundary conditions at infinity. Introducing $w = -2f_x/f$ into Equation (3.36) yields

$$ff_{xt} - f_x f_t + ff_{xxxx} - 4f_x f_{xxx} + 3f_{xx}^2 = 0. \qquad (3.37)$$

While this equation may not appear to be a simplification of KdV, it has several useful properties.

1. If $f(x,t)$ is a solution of Equation (3.37), then so also is $\alpha f(x,t)$, where α is a constant. Thus although nonlinear, Equation (3.37) shares a useful property of linear systems.

2. If $f(x,t)$ is a solution of Equation (3.37), then so is $\beta(t)f(x,t)$, where $\beta(t)$ is any differentiable function of time.
3. Equation (3.37) can be rearranged to

$$f(f_t + f_{xxx})_x - f_x(f_t + f_{xxx}) + 3(f_{xx}^2 - f_x f_{xxx}) = 0,$$

emphasizing the role of the linear operator

$$\frac{\partial}{\partial t} + \frac{\partial^3}{\partial x^3}.$$

4. In terms of Hirota's bilinear differential operators, Equation (3.37) can be written

$$D_x(D_t + D_x^3)(f \cdot f) = 0.$$

5. In this formulation, the single-soliton solution of KdV

$$u(x,t) = -\frac{v}{2}\operatorname{sech}^2\left[\frac{\sqrt{v}}{2}(x - vt - x_0)\right]$$

takes the simple form

$$f(x,t) = 1 + e^{-\sqrt{v}(x-vt-x_0)}. \tag{3.38}$$

With optimism that stems from the above properties and the knowledge that N-soliton solutions exist, we now proceed as follows. First define the bilinear differential operator

$$B \equiv D_x(D_t + D_x^3)$$

for typographic convenience, and note that

$$B(a_1 \cdot b_1 + a_2 \cdot b_2) = B(a_1 \cdot b_1) + B(a_2 \cdot b_2).$$

Next take f to be

$$f(x,t) = 1 + \sum_{n=1}^{\infty} \varepsilon^n f_n(x,t), \tag{3.39}$$

and require

$$B(f \cdot f) = 0 \tag{3.40}$$

at each power of ε.

Interestingly, the terms up to ε^N generate an N-soliton solution. To see how this goes, consider the relevant equations with increasing orders of ε.

Single soliton

The order ε equation

$$B(f_1 \cdot 1 + 1 \cdot f_1) = B(f_1 \cdot 1) + B(1 \cdot f_1) = 0$$

can be written as

$$\frac{\partial}{\partial x}\left(\frac{\partial f_1}{\partial t} + \frac{\partial^3 f_1}{\partial x^3}\right) = 0, \qquad (3.41)$$

with the solution

$$f_1 = e^{\theta_1}$$
$$\theta_1 = k_1 x - k_1^3 t + \alpha_1.$$

The order ε^2 equation is then

$$2\left(\frac{\partial^2}{\partial x \partial t} + \frac{\partial^4}{\partial x^4}\right) f_2 = -B(f_1 \cdot f_1)$$
$$= 0,$$

which is satisfied for $f_2 = 0$, implying $f_3 = f_4 = \cdots = 0$. Because the series for f truncates, one can set $\varepsilon = 1$, whereupon

$$f = 1 + e^{\theta_1}.$$

This corresponds to the single-soliton solution of Equation (3.38) with $k_1 = -\sqrt{v}$ and $\alpha_1 = \sqrt{v} x_0$.

Two-solitons

In Equation (3.41), assume that

$$f_1 = e^{\theta_1} + e^{\theta_2},$$

where $\theta_i = k_i x - k_i^3 t + \alpha_i (i = 1, 2)$. Now the order ε^2 equation becomes

$$2\left(\frac{\partial^2}{\partial x \partial t} + \frac{\partial^4}{\partial x^4}\right) f_2 = -B\left(e^{\theta_1} \cdot e^{\theta_2}\right) - B\left(e^{\theta_2} \cdot e^{\theta_1}\right)$$
$$= -2\left[(k_1 - k_2)(k_2^3 - k_1^3) + (k_1 - k_2)^4\right] e^{\theta_1 + \theta_2},$$

having the solution

$$f_2 = \left(\frac{k_1 - k_2}{k_1 + k_2}\right)^2 e^{\theta_1 + \theta_2}.$$

With this f_2, the order ε^3 equation becomes

$$2\left(\frac{\partial^2}{\partial x \partial t} + \frac{\partial^4}{\partial x^4}\right) f_3 = -B(f_1 \cdot f_2 + f_2 \cdot f_1), \qquad (3.42)$$

and a detailed computation shows that the right-hand side is zero. Thus it is possible to assume $f_3 = f_4 = \cdots = 0$ and set $\varepsilon = 1$, whereupon

$$f = 1 + e^{\theta_1} + e^{\theta_2} + \left(\frac{k_1 - k_2}{k_1 + k_2}\right)^2 e^{\theta_1 + \theta_2},$$

corresponding to a 2-soliton solution of KdV.

N-solitons

In the general case, assume in Equation (3.41) that

$$f_1 = e^{\theta_1} + e^{\theta_2} + \cdots + e^{\theta_N},$$

where $\theta_i = k_i x - k_i^3 t + \alpha_i$ with $i = 1, 2, \ldots, N$. Then Equation (3.40) implies

$$2\left(\frac{\partial^2}{\partial x \partial t} + \frac{\partial^4}{\partial x^4}\right) f_1 = 0$$

$$2\left(\frac{\partial^2}{\partial x \partial t} + \frac{\partial^4}{\partial x^4}\right) f_2 = -B(f_1 \cdot f_1)$$

$$2\left(\frac{\partial^2}{\partial x \partial t} + \frac{\partial^4}{\partial x^4}\right) f_3 = -B(f_2 \cdot f_1 + f_1 \cdot f_2)$$

$$2\left(\frac{\partial^2}{\partial x \partial t} + \frac{\partial^4}{\partial x^4}\right) f_4 = -B(f_3 \cdot f_1 + f_2 \cdot f_2 + f_1 \cdot f_3)$$

$$\vdots$$

$$2\left(\frac{\partial^2}{\partial x \partial t} + \frac{\partial^4}{\partial x^4}\right) f_N = -B(f_{N-1} \cdot f_1 + \cdots + f_1 \cdot f_{N-1})$$

$$2\left(\frac{\partial^2}{\partial x \partial t} + \frac{\partial^4}{\partial x^4}\right) f_{N+1} = 0$$

$$2\left(\frac{\partial^2}{\partial x \partial t} + \frac{\partial^4}{\partial x^4}\right) f_{N+2} = 0$$

$$\vdots$$

Thus the computational task is to solve these equations for $f_2, f_3, \ldots, f_N$, set $\varepsilon = 1$, and then construct the N-soliton solution of KdV as

$$u_N(x, t) = -2\frac{\partial^2}{\partial x^2} \log(1 + f_1 + f_2 + \cdots + f_N),$$

which is recorded as Equation (B.1) of Appendix B.

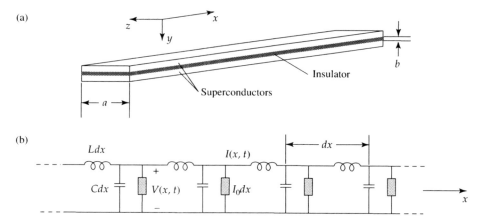

FIG. 3.4. *(a) Superconducting strip-line, which supports TEM propagation in the x-direction. (b) An equivalent circuit model of the TEM mode.*

3.2 The sine–Gordon (SG) equation

3.2.1 Long Josephson junctions

Of the several applications of the SG equation mentioned in the introduction to this chapter, the most widely used is as a model for the propagation of transverse electromagnetic (TEM) waves on a superconducting (or "strip-line") transmission system. Such a system is shown in Figure 3.4, indicating propagation in the x-direction, with the electric field (E) oriented in the y-direction and the magnetic field (H) in the z-direction.

From the equivalent circuit for the TEM mode shown in Figure 3.4, the voltage (V) across the insulating region and the longitudinal current (I) are related by conservation laws for magnetic flux and electric charge (or Kirchhoff's voltage and current laws[1]) as [59]

$$\frac{\partial V}{\partial x} = -L\frac{\partial}{\partial t}$$
$$\frac{\partial I}{\partial x} = -C\frac{\partial V}{\partial t} - I_0 \sin\phi. \qquad (3.43)$$

In these equations, L is the inductance per unit length of the strip-line, and C is the capacitance per unit length.

The term $I_0 \sin\phi$ in the second equation represents quantum mechanical (Josephson) tunneling of superconducting electron pairs through the insulating barrier indicated in Figure 3.4 [3]. Within this term, ϕ represents the change in phase of the superconducting wave function across the barrier, and it is related

---

[1] After German physicist Gustav Robert Kirchhoff (1824–1887).

to the voltage by
$$\frac{\partial \phi}{\partial t} = +\frac{2e}{\hbar}V, \tag{3.44}$$

e being the electronic charge and $\hbar \equiv h/2\pi$ being Planck's constant.

Substituting Equation (3.44) into the first of Equations (3.43) and integrating with respect to time, one finds
$$\frac{\partial \phi}{\partial x} = -\frac{2eL}{\hbar}I, \tag{3.45}$$

up to an arbitrary constant that is set to zero.

Thus the second of Equations (3.43) can be written as
$$\frac{\partial^2 \phi}{\partial x^2} - \frac{1}{c^2}\frac{\partial^2 \phi}{\partial t^2} = \frac{1}{\lambda_J^2}\sin\phi, \tag{3.46}$$

where
$$c = 1/\sqrt{LC} \tag{3.47}$$

is called the Swihart velocity [65] and
$$\lambda_J = \sqrt{\frac{\hbar}{2eLI_0}}. \tag{3.48}$$

Measuring distance in units of λ_J and time in units of λ_J/c normalizes Equation (3.46) to (3.2).

3.2.2 Solitary waves

Consider the normalized SG equation (3.2)
$$u_{xx} - u_{tt} = \sin u$$

with $-\infty < x < +\infty$ and $-\infty < t < +\infty$. Introducing the traveling-wave assumption $u(x,t) = \tilde{u}(x - vt)$, gives the ODE
$$(1 - v^2)\tilde{u}'' = \sin \tilde{u}. \tag{3.49}$$

Multiplying by $\tilde{u}'$ and integrating with respect to the traveling-wave variable $\xi = x - vt$ leads to
$$\tfrac{1}{2}(1 - v^2)(\tilde{u}')^2 = A - \cos \tilde{u}, \tag{3.50}$$

where A is a constant of integration.

Setting $A = 1$ and imposing the boundary conditions $\tilde{u} \to 0 \pmod{2\pi}$ and $\tilde{u}' \to 0$ as $\xi \to \pm\infty$, one obtains the solitary wave
$$u(x,t) = 4\arctan\left[\exp\left(\pm\frac{x - vt - x_0}{\sqrt{1 - v^2}}\right)\right]. \tag{3.51}$$

Although this gudermannian solution is a soliton, it differs in several ways from the KdV soliton, considered in the previous section.

1. *Kinks and antikinks.* Instead of being a true pulse, Equation (3.51) represents a monotonic level change of magnitude 2π as ξ increases from $-\infty$ to $+\infty$. It is customary to call the "+" solution a kink and the "−" solution an antikink.
2. *Lorentz invariance.*[2] Like a relativistic particle, an SG kink exhibits Lorentz contraction, becoming more narrow as $v \to \pm 1$. This property is a result of the fact that the SG equation is invariant under the independent variable transformation $(x, t) \to (\xi, \tau)$, where

$$\xi = (x - vt)/\sqrt{1 - v^2} \quad \text{and} \quad \tau = (t - vx)/\sqrt{1 - v^2}. \tag{3.52}$$

3. *Kink-antikink annihilation.* Fixed by the boundary conditions at $x = \pm\infty$, the total change of level in a solution

$$u(+\infty, t) - u(-\infty, t)$$

is a constant of the motion. For a collision between a kink and an antikink, this difference is zero, allowing the two wave components to annihilate each other and release their energy as radiation of the (linear) Klein–Gordon equation[3] $u_{xx} - u_{tt} \approx u$. This phenomenon is influenced by dissipative effects, to be studied in Chapter 7, and it is qualitatively similar to the annihilation of a particle by its antiparticle (an electron by a positron, for example, to produce gamma radiation) in the realm of high energy physics.

4. *Tachyons.* In addition to the waves described by Equation (3.51), where $|v| < 1$, there are other kink-like solutions of the form

$$u(x, t) = 4 \arctan\left[\exp\left(\pm \frac{x - vt - x_0}{\sqrt{v^2 - 1}}\right)\right] + \pi, \tag{3.53}$$

which require $|v| > 1$. These correspond to conjectured faster than light particles called "tachyons," but they are unstable.

All of the above phenomena (except tachyons) can be observed on mechanical models of the SG equation [58], a simple example of which is sketched in Figure 3.5. Such a model is readily constructed by sticking dressmaker pins into an elastic band.

---

[2]For the Dutch physicist Hendrik Antoon Lorentz (1853–1928), who observed that the electromagnetic equations are invariant to this transformation.
[3]After O. Klein and W. Gordon, who independently proposed it in the 1920s as a quantum wave equation. The term "sine–Gordon" was chosen as a pun on the name of this earlier equation.

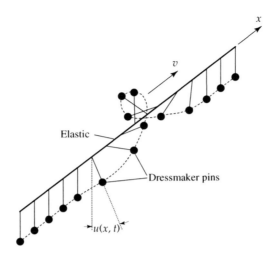

FIG. 3.5. *A kink on a simple mechanical model of the SG equation that can be made from dressmaker pins and an elastic band. (The pins are rotating clockwise in the direction of propagation.)*

It is of interest to express the kink (or antikink) wave of Equation (3.51) in the context of the Josephson transmission line introduced in the previous section. The first of Equations (3.43) is an exact conservation law for magnetic flux

$$\Phi = \int_{-\infty}^{\infty} LI\,dx,$$

where I is related to the x-derivative of ϕ (or u) by Equation (3.45). Because ϕ changes by 2π upon integrating across a kink, each carries an amount of magnetic flux

$$\Phi_0 = \frac{2\pi\hbar}{2e} = \frac{h}{2e} \doteq 2.064 \times 10^{-15}\,\text{Vs},$$

called the flux quantum.

Studies of the dynamics of such flux quanta, often termed fluxons and antifluxons instead of kinks and antikinks, are presently an active area of applied soliton research, relating a wide range of experimental, theoretical, and numerical studies [3, 51].

3.2.3 Periodic waves

To obtain the kink solution in Equation (3.51), A (the first integration constant) was assumed to be unity, requiring that $|v| < 1$. In this section, we study periodic solutions

$$u(x,t) = \tilde{u}(\theta) = \tilde{u}(\theta + 2\pi) \pmod{2\pi}, \tag{3.54}$$

where θ is a traveling-wave variable of the form

$$\theta = \omega t - \beta x,$$

and the local value of the wave speed is

$$v = \frac{\omega}{\beta}.$$

Under this notation, consider the following solutions:

1. $A > 1$ and $|v| < 1$: In this case, Equation (3.50) can be written

$$\sqrt{\beta^2 - \omega^2} \int^{\tilde{u}} \frac{dy}{\sqrt{2(A - \cos y)}} = \theta, \qquad (3.55)$$

indicating that $\tilde{u}$ is a monotone increasing function of θ. (In the context of electromagnetic waves on a Josephson transmission line, this solution represents a one-dimensional array of fluxons, propagating at a subcritical speed.)

2. $1 > A > -1$ and $|v| < 1$: This case is also governed by Equation (3.55), implying a solution that oscillates periodically about $\tilde{u} = \pi$.

3. $1 > A > -1$ and $|v| > 1$: Now Equation (3.50) can be written as

$$\sqrt{\omega^2 - \beta^2} \int^{\tilde{u}} \frac{dy}{\sqrt{2(\cos y - A)}} = \theta, \qquad (3.56)$$

implying periodic oscillations about $\tilde{u} = 0$.

4. $A < -1$ and $|v| > 1$: Equation (3.56) indicates that u is a monotone increasing function of θ, representing a one-dimensional array of tachyons, propagating at supercritical speed.

These solutions are, by construction, strict traveling-waves. As noted in the introduction, Whitham has shown how an averaged Lagrangian formulation can be used to obtain dynamic equations of slow modulations of these waves [71].

A Lagrangian density for Equation (3.2) is

$$\mathcal{L} = (u_t^2 - u_x^2)/2 + \cos u, \qquad (3.57)$$

which can be averaged over a period in $\theta = \omega t - \beta x$ to obtain

$$\langle \mathcal{L} \rangle = \frac{1}{2\pi} \int_0^{2\pi} \left[\frac{1}{2}(\omega^2 - \beta^2)\tilde{u}_\theta^2 + \cos \tilde{u} \right] d\theta. \qquad (3.58)$$

The integral can be simplified by noting that Equation (3.50) becomes

$$\tfrac{1}{2}(\beta^2 - \omega^2)\tilde{u}_\theta^2 + \cos \tilde{u} = A, \qquad (3.59)$$

and substituting this expression for $\cos \tilde{u}$ into Equation (3.58), implies

$$\langle \mathcal{L} \rangle = \frac{1}{2\pi} \int_0^{2\pi} (\omega^2 - \beta^2) \tilde{u}_\theta^2 \, d\theta + A$$

$$= \frac{1}{2\pi}(\omega^2 - \beta^2) \int_0^{2\pi} \tilde{u}_\theta \, d\tilde{u} + A.$$

Thus

$$\langle \mathcal{L} \rangle(\omega, \beta, A) = \frac{1}{2\pi} \sqrt{\omega^2 - \beta^2} \oint \sqrt{2(\cos y - A)} \, dy + A$$

is an expression for an averaged Lagrangian density that governs the three parameters (ω, β, A) of a slow modulation.

Three equations relating these parameters are obtained as follows. First, from variation of $\langle \mathcal{L} \rangle$ with A

$$\sqrt{\omega^2 - \beta^2} \oint \frac{dy}{\sqrt{2(\cos y - A)}} = 2\pi, \qquad (3.60)$$

a nonlinear dispersion equation. Second, variation of $\langle \mathcal{L} \rangle$ with θ implies

$$\frac{\partial}{\partial t} \frac{\partial \langle \mathcal{L} \rangle}{\partial \omega} - \frac{\partial}{\partial x} \frac{\partial \langle \mathcal{L} \rangle}{\partial \beta} = 0. \qquad (3.61)$$

(For an explanation of this variational calculation, refer to Appendix A.) Finally, for θ to be defined as a function of x and t, there is the consistency relation

$$\frac{\partial \beta}{\partial t} + \frac{\partial \omega}{\partial x} = 0. \qquad (3.62)$$

These three equations can be written more compactly by introducing the notation

$$I(A) \equiv \frac{1}{2\pi} \oint \sqrt{2(\cos y - A)} \, dy. \qquad (3.63)$$

Substituting into Equation (3.60), squaring, and differentiating with respect to x, we find

$$(I')^2 (\omega \omega_x - \beta \beta_x) + (\omega^2 - \beta^2) I' I'' A_x = 0. \qquad (3.64)$$

Also Equation (3.61) becomes

$$(\omega^2 - \beta^2) I' (\omega A_t + \beta A_x) + \omega I (\beta \beta_t + \omega \beta_x) - \beta I (\beta \omega_t + \omega \omega_x) = 0. \qquad (3.65)$$

In matrix form, Equations (3.64), (3.65), and (3.62) can be written

$$\begin{bmatrix} \omega^2 I & -\omega \beta I & \beta(\omega^2 - \beta^2) I' \\ -\beta (I')^2 & \omega (I')^2 & (\omega^2 - \beta^2) I' I'' \\ 0 & 1 & 0 \end{bmatrix} \frac{\partial}{\partial x} \begin{bmatrix} \beta \\ \omega \\ A \end{bmatrix}$$

$$+ \begin{bmatrix} \omega \beta I & -\beta^2 I & \omega(\omega^2 - \beta^2) I' \\ 0 & 0 & 0 \\ 1 & 0 & 0 \end{bmatrix} \frac{\partial}{\partial t} \begin{bmatrix} \beta \\ \omega \\ A \end{bmatrix} = 0, \qquad (3.66)$$

which is a nonlinear system for the slow variations of the parameters $\beta(x,t)$, $\omega(x,t)$, and $A(x,t)$.

The characteristic velocity (v_e) of this system is determined by

$$v_e \frac{\partial}{\partial x} \begin{bmatrix} \beta \\ \omega \\ A \end{bmatrix} = \frac{\partial}{\partial t} \begin{bmatrix} \beta \\ \omega \\ A \end{bmatrix}$$

and found to be

$$v_e = \frac{\beta \pm \sqrt{\alpha}\omega}{\omega \pm \sqrt{\alpha}\beta}, \tag{3.67}$$

where $\alpha \equiv -II''/(I')^2$.

Of particular interest is the type 3 solution given by Equation (3.56), which represents a periodic traveling-wave oscillating about $\tilde{u} = 0$. In the linear limit, $A \to 1$, $\alpha \to 0$, and Equation (3.60) reduces to

$$\omega^2 - \beta^2 = 1.$$

Thus

$$v_e = \beta/\omega = d\omega/d\beta$$

is the linear group velocity. Evidently Equations (3.66) represent a nonlinear envelope wave, or wave of modulation, that rides over the carrier wave, defined in Equation (3.54), with velocity $v_c = \omega/\beta$.

If $II'' > 0$, $\sqrt{\alpha}$ is imaginary, and Equation (3.67) no longer implies a real envelope velocity—a criterion for modulational instability that was first noted in the context of nonlinear optics by Ostrovsky [50]. Whitham has shown that this is so for the periodic waves of type 3 [71], implying an instability that has been studied in the context of deep water waves by Benjamin and Feir [6]. In this case, a wave of constant amplitude is unstable to small changes in the modulation, and this "Benjamin–Feir instability" can lead to the formation of envelope wave solitons that will be studied in Section 3.3.1. (For further perspectives on type 3 solutions, see Problems 31 and 32.)

3.2.4 Nonlinear standing waves

In our development of traveling-wave solutions to the SG equation, we have assumed $-\infty < x < +\infty$, but a real superconducting strip-line (see Figure 3.4) has a finite length in the x-direction. Thus it is of interest to consider the system

$$u_{xx} - u_{tt} = \sin u \tag{3.68}$$

with

$$0 \leq x \leq l$$
$$0 \leq t < +\infty$$
$$u_x(0,t) = u_x(l,t) = 0,$$

where the boundary conditions indicate that the x-directed current (being proportional to u_x) is zero at the two open-circuit ends of the strip-line structure. The system of Equations (3.68) describes a nonlinear resonator, for which it is possible to construct a variety of exact solutions that have been experimentally observed [51].

To solve a linear problem of this sort, one would use the method of separation of variables by seeking solutions of the form $u(x,t) = X(x)\,T(t)$ and finding ODEs for $X(x)$ and $T(t)$. While this method is not generally effective for nonlinear PDEs, the SG equation has some interesting solutions of the class [38]

$$u(x,t) = 4\arctan[X(x)\,T(t)], \tag{3.69}$$

evidently including the kinks and antikinks of Equation (3.51).

Substituting Equation (3.69) into (3.68) yields

$$(1 + X^2T^2)(X''T - XT'' - XT)$$
$$- 2XT[(X')^2T^2 - X^2(T')^2 - X^2T^2] = 0. \tag{3.70}$$

Now assume $X(x)$ and $T(t)$ to be elliptic functions which satisfy equations of the form

$$(X')^2 = a_1 X^4 + a_2 X^2 + a_3 \tag{3.71}$$

and

$$(T')^2 = b_1 T^4 + b_2 T^2 + b_3, \tag{3.72}$$

where $a_1, a_2, a_3, b_1, b_2,$ and b_3 are constants. From Equations (3.71) and (3.72), it follows that

$$X'' = 2a_1 X^3 + a_2 X \tag{3.73}$$

and

$$T'' = 2b_1 T^3 + b_2 T. \tag{3.74}$$

Inserting Equations (3.71)–(3.74) into (3.70), terms of order X^5T^3 and X^3T^5 drop out, leaving

$$2X^3T(a_1 + b_3) - 2XT^3(a_3 + b_1) + XT(a_2 - b_2 - 1) + X^3T^3(-a_2 + b_2 + 1) = 0.$$

This equation is satisfied by three conditions

$$b_1 = -a_3$$
$$b_2 = a_2 - 1$$
$$b_3 = -a_1$$

for the six parameters $a_1, \ldots, b_3$; thus three arbitrary constants remain: a_1, a_2, a_3. The factors $X(x)$ and $T(t)$ are governed by Equation (3.71) and

$$(T')^2 = -a_3 T^4 + (a_2 - 1)T^2 - a_1, \tag{3.75}$$

respectively, implying that they can be expressed as the Jacobi elliptic functions introduced in Appendix C [11, 40].

For appropriate choices of a_1, a_2, and a_3, three different solutions are obtained for the boundary value problem posed in Equations (3.68) [19]. These solutions are presented below, together with limiting approximations that show their qualitative behaviors. In constructing the approximations, the identities in Table C.1 for certain elliptic functions are used.

Plasma oscillations

In this case
$$u(x,t) = 4\arctan[A\operatorname{cn}(\beta x; k_x)\operatorname{cn}(\omega t; k_t)], \tag{3.76}$$
where
$$k_x^2 = \frac{A^2[\beta^2(1+A^2)+1]}{\beta^2(1+A^2)^2}$$
$$k_t^2 = \frac{A^2[\omega^2(1+A^2)-1]}{\omega^2(1+A^2)^2}$$

and β, ω, and A are related by the nonlinear dispersion equation
$$\omega^2 - \beta^2 = \frac{1-A^2}{1+A^2}. \tag{3.77}$$

Allowed values of β in these formulas are determined by the boundary conditions as
$$\beta_n = \frac{2n}{l}K(k_x),$$
where n is the number of antinodes (loops) in the standing wave and $K(\cdot)$ is the complete elliptic integral of the first kind.

As $A \to 0$, $k_x \to 0$, and $k_t \to 0$,
$$u(x,t) \to 4A\cos\beta x\cos\omega t,$$
with
$$\omega^2 - \beta^2 \doteq 1.$$

Josephson junction mavens call this a "plasma oscillation" because—in the limit $\beta \to 0$—the dispersion relation reduces to $\omega^2 = 1$, implying that the system is everywhere oscillating at its intrinsic frequency
$$\frac{c}{\lambda_J} = \sqrt{\frac{2eI_0}{\hbar C}},$$
like a bowl of jelly.

In this low amplitude limit, sinusoidal traveling-waves are propagating in both directions and interfering in a way that satisfies the boundary conditions at $x = 0$ and $x = l$.

Breathers

A distinctly different sort of solution involves the bound oscillation of a kink and an antikink, which can be located either at the center or at the ends of the system. Thus

$$u(x,t) = 4\arctan\{A\,\mathrm{dn}[\beta(x-x_0); k_x]\,\mathrm{sn}(\omega t; k_t)\}, \qquad (3.78)$$

where

$$k_x^2 = 1 - \left[\frac{1-\beta^2(1+A^2)/A^2}{\beta^2(1+A^2)}\right]$$

$$k_t^2 = \frac{A^2[1-\omega^2(1+A^2)]}{\omega^2(1+A^2)},$$

and the nonlinear dispersion equation is

$$\beta = \omega A.$$

The boundary conditions require

$$\beta_n = \frac{n}{l}K(k_x)$$

with three possible values for x_0: 0, $l/2$, and l.

In contrast to the plasma oscillations, there is a minimum value of A below which these breather oscillations do not exist. To see this, note that the expression for k_x^2 would become greater than unity for $A^2 < A_c^2 \equiv \beta_1^2/(1-\beta_1^2)$. As $l \to \infty$, however, $\beta_1 \to 0$, forcing A_c to zero. In this long wavelength limit, $k_x^2 = 1$. Also $k_t^2 = 0$ for $\omega^2(1+A^2) = 1$, implying the dispersion equation

$$\omega^2 + \beta^2 = 1;$$

thus Equation (3.78) reduces to

$$u(x,t) = 4\arctan\left[\frac{\beta\sin\omega t}{\omega\cosh\beta(x-x_0)}\right], \qquad (3.79)$$

called a breather. The oscillations described by Equation (3.78) correspond to breathers that are constrained to live in a box of length l.

Kink-antikink oscillations

A third solution to the nonlinear boundary value problem is

$$u(x,t) = 4\arctan[A\,\mathrm{dn}(\beta x; k_x)\mathrm{sc}(\omega t; k_t)], \qquad (3.80)$$

(where sc ≡ sn/cn) and

$$k_x^2 = 1 - \left[\frac{\omega^2(A^2-1)-1}{\beta^2(A^2-1)}\right]$$

$$k_t^2 = 1 - \left[\frac{A^2[\omega^2(A^2-1)-1]}{\omega^2(A^2-1)}\right].$$

The nonlinear dispersion relation is again

$$\beta = \omega A$$

with β limited to the values

$$\beta_n = \frac{n}{l} K(k_x)$$

by the boundary conditions. Setting $\omega^2(A^2-1) = 1$ implies $k_x^2 = k_t^2 = 1$, so

$$u(x,t) = 4\arctan\left[\frac{\beta}{\omega}\frac{\sinh \omega t}{\cosh \beta x}\right], \quad (3.81)$$

with the dispersion equation

$$\beta^2 - \omega^2 = 1.$$

This equation represents an oscillation that can be described as a kink traveling in the $+x$-direction that is reflected at $x = l$ as an antikink. The antikink propagates in the $-x$-direction, being reflected at $x = 0$ as a kink, and so on. In the context of the Josephson strip-line, this nonlinear mode can be driven by a steady source of electric current, providing the basis for generators of electromagnetic waves at hundreds of gigahertz.[4]

3.2.5 Two-soliton solutions

Returning to the infinite domain, the SG equation has an exact solution of the form

$$u_{kk}(x,t) = 4\arctan\left[\frac{v\sinh(x/\sqrt{1-v^2})}{\cosh(vt/\sqrt{1-v^2})}\right], \quad (3.82)$$

representing a kink of velocity $+v$ colliding at the origin with a kink of velocity $-v$. A plot of this solution with $v = \frac{1}{2}$ is presented in Figure 3.6.

Similarly, in Figure 3.7 is plotted a kink-antikink collision described by

$$u_{kak}(x,t) = 4\arctan\left[\frac{\sinh(vt/\sqrt{1-v^2})}{v\cosh(x/\sqrt{1-v^2})}\right] \quad (3.83)$$

with $v = \frac{1}{2}$.

---

[4]The nonlinear standing waves described in this section can be viewed on http://www.math.h.kyoto-u.ac.jp/~takasaki/soliton-lab/gallery/solitons/sg-e.html.

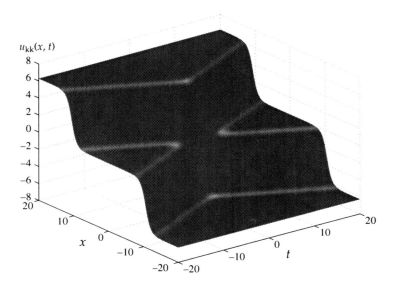

FIG. 3.6. *Surface plot of a kink-kink solution of the SG equation from Equation (3.82) with $v = \frac{1}{2}$.*

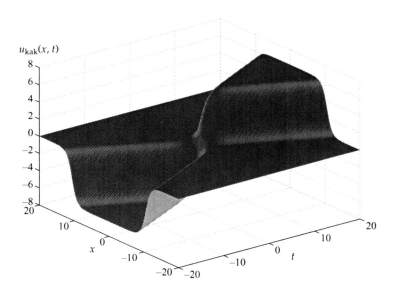

FIG. 3.7. *Surface plot of a kink-antikink solution plotted from Equation (3.83) with $v = \frac{1}{2}$.*

In the 1950s, these equations were proposed by Seeger et al. [63] to describe interactions of crystal dislocations. In the context of nonlinear field theories for elementary particles, they were rediscovered by Perring and Skyrme in 1962 [52].

We could derive these solutions using the nonlinear separation of variables discussed in the previous section (for example, with $v \equiv \omega/\beta$, Equations (3.83) and (3.81) are identical), but it is more instructive to use a Bäcklund transformation (BT). To see how this goes, recall from Chapter 1 that under the independent variable transformation

$$u(x,t) \to \tilde{u}(\xi, \tau) = u(x,t)$$
$$x \to \xi = (x+t)/2$$
$$t \to \tau = (x-t)/2,$$

the SG equation becomes

$$\frac{\partial^2 \tilde{u}}{\partial \xi \, \partial \tau} = \sin \tilde{u},$$

which was known in the nineteenth century to have the BT

$$\frac{\partial \tilde{u}_1}{\partial \xi} = 2a \sin\left(\frac{\tilde{u}_1 + \tilde{u}_0}{2}\right) + \frac{\partial \tilde{u}_0}{\partial \xi}$$

$$\frac{\partial \tilde{u}_1}{\partial \tau} = \frac{2}{a} \sin\left(\frac{\tilde{u}_1 - \tilde{u}_0}{2}\right) - \frac{\partial \tilde{u}_0}{\partial \tau}.$$

Just as with KdV, the permutative property of this transformation is indicated by the diagrams

$$\tilde{u}_0 \xrightarrow{a_1} \tilde{u}_1 \xrightarrow{a_2} \tilde{u}_{12} = \tilde{u}_{21},$$

and

$$\tilde{u}_0 \xrightarrow{a_2} \tilde{u}_2 \xrightarrow{a_1} \tilde{u}_{21} = \tilde{u}_{12},$$

the first of which represents the equations

$$\frac{\partial(\tilde{u}_1 - \tilde{u}_0)}{\partial \xi} = 2a_1 \sin\left(\frac{\tilde{u}_1 + \tilde{u}_0}{2}\right)$$

$$\frac{\partial(\tilde{u}_{12} - \tilde{u}_1)}{\partial \xi} = 2a_2 \sin\left(\frac{\tilde{u}_{12} + \tilde{u}_1}{2}\right),$$

while the second diagram represents

$$\frac{\partial(\tilde{u}_2 - \tilde{u}_0)}{\partial \xi} = 2a_2 \sin\left(\frac{\tilde{u}_2 + \tilde{u}_0}{2}\right)$$

$$\frac{\partial(\tilde{u}_{21} - \tilde{u}_2)}{\partial \xi} = 2a_1 \sin\left(\frac{\tilde{u}_{21} + \tilde{u}_2}{2}\right).$$

Because the BTs permute,
$$\tilde{u}_{21} = \tilde{u}_{12},$$
allowing the derivatives with respect to ξ to be eliminated. Thus

$$\tan\left(\frac{u_{12} - u_0}{4}\right) = \left(\frac{a_1 + a_2}{a_1 - a_2}\right) \tan\left(\frac{u_2 - u_1}{4}\right), \quad (3.84)$$

which holds both for $u(x,t)$ and for $\tilde{u}(\xi,\tau)$.

Assuming that u_2 is not identically equal to u_1, and setting $a_1 = a_2$, implies that $u_{12} = u_0 \pmod{2\pi}$; thus one cannot create two different kinks with the same speed, just as was found previously for KdV from Equation (3.32). Formulas for kink-antikink collisions can be constructed by setting $u_0 = 0$ in Equation (3.84) and letting

$$u_1 = 4\arctan\left[\exp\left(+\frac{x - v_1 t}{\sqrt{1 - v_1^2}}\right)\right]$$

$$u_2 = 4\arctan\left[\exp\left(+\frac{x - v_2 t}{\sqrt{1 - v_2^2}}\right)\right],$$

where
$$v_i = \frac{1 - a_i^2}{1 + a_i^2}, \quad i = 1, 2$$

or
$$a_i = \pm\sqrt{(1 + v_i)/(1 - v_i)}.$$

In particular, setting $v_1 = -v_2 \equiv v$ with both a_1 and a_2 positive implies that

$$\frac{a_1 + a_2}{a_1 - a_2} = \frac{1}{v},$$

leading directly to the kink-antikink collision of Equation (3.83). Equation (3.82) for a kink-kink collision can be generated in a similar manner.

The symmetric kink-antikink equation takes an interesting form if the velocity parameter (v) is allowed to be imaginary. For example, setting

$$v = i\omega/\sqrt{1 - \omega^2}, \quad \omega < 1,$$

Equation (3.83) becomes the stationary breather

$$u_{\text{sbr}}(x,t) = 4\arctan\left[\frac{\sqrt{1 - \omega^2}}{\omega} \frac{\sin \omega t}{\cosh\sqrt{1 - \omega^2}\,x}\right], \quad (3.85)$$

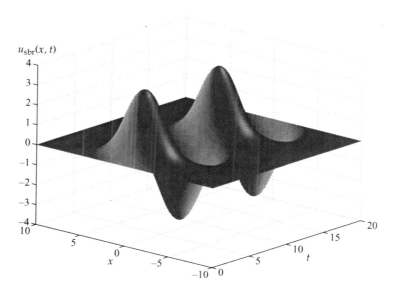

FIG. 3.8. *Surface plot of a stationary breather from Equation (3.85) with $\omega = \pi/5$.*

a localized, oscillating solution that we met before in Equation (3.79) and is plotted in Figure 3.8. The stationary breather can be made to move by introducing the Lorentz transformation of Equations (3.52). Thus

$$u_{\mathrm{mbr}}(x,t) = 4\arctan\left\{\frac{\sqrt{1-\omega^2}}{\omega}\sin\left[\frac{\omega(t-v_e x)}{\sqrt{1-v_e^2}}\right]\operatorname{sech}\left[\frac{\sqrt{1-\omega^2}(x-v_e t)}{\sqrt{1-v_e^2}}\right]\right\}$$
(3.86)

is an exact solution of the SG equation with an envelope velocity (v_e) equal to the reciprocal of its carrier velocity.

Multi-soliton solutions of the SG equation are given in Equations (B.4) and (B.5) of Appendix B.[5]

3.2.6 *More spatial dimensions*

Josephson junctions with both transverse dimensions larger than the characteristic length defined in Equation (3.48) are essentially two-dimensional structures. Defining a surface current vector as [58]

$$\mathbf{I} = I_x \mathbf{i} + I_y \mathbf{j},$$

---

[5]The kink-kink and kink-antikink collisions described in this section can be viewed on http://homepages.tversu.ru/ s000154/collision/main.html.

where **i** and **j** are respectively unit vectors in the x- and y-directions, Equations (3.43) can be generalized to

$$\operatorname{grad} V = -L\frac{\partial \mathbf{I}}{\partial t}$$
$$\operatorname{div} \mathbf{I} = -C\frac{\partial V}{\partial t} - I_0 \sin \phi. \qquad (3.87)$$

In these equations, the units may be chosen as: $V \sim$ volts, $L \sim$ henries/square, $\mathbf{I} \sim$ amps/meter, $C \sim$ farads/meter^2, $\phi \sim$ radians, and $I_0 \sim$ amps/meter^2. Also Equation (3.45) becomes

$$\operatorname{grad} \phi = -(2eL\hbar)\mathbf{I}, \qquad (3.88)$$

so the unnormalized SG equation describing ϕ (the change in phase of the superconducting wave function across the barrier) assumes the two-dimensional form

$$\frac{\partial^2 \phi}{\partial x^2} + \frac{\partial^2 \phi}{\partial y^2} - \frac{1}{c^2}\frac{\partial^2 \phi}{\partial t^2} = \frac{1}{\lambda_J^2}\sin \phi, \qquad (3.89)$$

where c and λ_J are defined by expressions that are formally identical to Equations (3.47) and (3.48). Assuming that v is the velocity of an observer moving in the x-direction, these equations are invariant under the Lorentz transformation

$$x \to x' = \gamma(x - vt)$$
$$t \to t' = \gamma(t - vx/c^2) \qquad (3.90)$$

under which

$$V \to V' = \gamma(V - LvI_x)$$
$$I_x \to I_x' = \gamma(I_x - CvV)$$
$$I_y \to I_y' = I_y$$
$$\phi \to \phi' = \phi,$$

where

$$\gamma \equiv 1/\sqrt{1 - v^2/c^2}. \qquad (3.91)$$

If a stationary observer measures a surface current (I_x) and transverse voltage (V), a moving observer's measurements of V' and I_x' will depend upon both I_x and V. This is analogous to the way that observations of Maxwell's electric and magnetic field intensities in three-dimensional space depend upon the motion of an observer.

Can the two-dimensional SG equation support solutions that are stable and localized in space? If there were such solutions, one could impart motion through the Lorentz transformation—just as for the stationary breather of

Equation (3.86)—discovering thereby two-dimensional "elementary particles" on a large Josephson junction. One way to begin this search is to assume time independent solutions with radial symmetry. In this case, however, numerical studies show that a boundary condition at the origin of the form

$$\operatorname{grad}\phi \sim -\alpha \mathbf{i_r}/r \qquad (3.92)$$

must be satisfied, where α is a positive constant and $\mathbf{i_r}$ is a unit vector in the outward radial direction [2]. Such a boundary condition could be satisfied near a small short across the junction—not difficult for the experimenter to imagine— but the translational symmetry necessary for a Lorentz transformation would then be destroyed.

Nonetheless, several research problems related to the SG with more than one space dimension come to mind, including the following.

1. *Bias currents.* Two-dimensional current flows in large Josephson junctions are becoming of practical interest [12–14,62]. Such studies are necessary to predict and control the bias currents in structures that are more intricate than the one-dimensional strip-line of Figure 3.4. Among these structures are exponentially tapered flux-flow oscillators, in which a train of fluxons is forced to move toward the load by the variations in transverse geometry [4,5].

2. *Internal flux loops.* Assuming a two-dimensional Josephson junction with independent variables extending over the domain

$$-\infty < x < +\infty$$
$$-\infty < y < +\infty$$
$$0 \le t < +\infty,$$

one can impose initial conditions by specifying the locus and velocity of a closed curve, across which ϕ changes by 2π. In a direction normal to the curve this is a kink soliton, representing a fluxon loop. If a circular fluxon loop is launched with an outward velocity, it will reach a maximum radius where the initial kinetic energy has been entirely transformed into rest energy of the longer loop [18]. After reaching this maximum radius, the ring is drawn back into the origin by forces derived from the reduction of loop energy with decreasing radius, in a manner that was suggested for elementary particles by Derrick in 1964 [20]. Interestingly, the collapsing loop is reflected from the origin to become a metastable "pulson" [8,9,17]. Recent numerical studies of elliptic rings show that pulson lifetimes can reach several cycles of oscillation [16].

3. *Three-dimensional dynamics.* The above two-dimensional studies have been extended to the three-dimensional SG equation

$$\frac{\partial^2 \phi}{\partial x^2} + \frac{\partial^2 \phi}{\partial y^2} + \frac{\partial^2 \phi}{\partial z^2} - \frac{\partial^2 \phi}{\partial t^2} = \sin\phi, \qquad (3.93)$$

where the initial conditions are imposed across a spherical surface, and again a metastable pulsating behavior is observed [10]. Referring to the discussions in Sections 1.2 and 1.3.5, Equation (3.93) may play a role in the description of elementary particles. As in the two-dimensional case, a singularity is expected at the origin of a spherically symmetric solution, but this might be avoided in more complex structures—for example, a closed three dimensional curve (or knot), along which there is directed a singularity of the sort generated by Equation (3.92). In the numerical search for such solutions, one might compare the dynamics of surfaces that are topologically equivalent to spheres with those that emerge from initial conditions imposed across various tori and closed surfaces of higher order.

As listed, these problems are increasingly computer intensive, but numerical power is becoming ever more readily available. It is hoped that they will be continued by nonlinear enthusiasts of the present century.

3.3 The nonlinear Schrödinger (NLS) equation

3.3.1 *Nonlinear wave packets*

As we have seen in Chapter 2, it is often convenient to write the solution of a linear wave problem in the form

$$u(x,t) = \frac{1}{2\pi} \int_{-\infty}^{\infty} F(k)e^{\mathrm{i}(kx-\omega t)}dk, \qquad (3.94)$$

where $F(k)$ is the Fourier transform of $u(x,0)$ and ω is related to k through the dispersion relation $\omega = \omega(k)$. Unless $\omega = k$, each component in Equation (3.94) travels at a different speed (ω/k), and the wave disperses—hence the name.

A wave packet is a special form of Equation (3.94) with the Fourier components lying close to some propagation number (k_0) and the corresponding frequency (ω_0). In other words, $F(k)$ has its maximum value at $k = k_0$, falling rapidly as $|k - k_0|$ increases and allowing the dispersion relation to be expanded as a power series about k_0.

With the notation

$$\omega = \omega_0 + b_1(k - k_0) + b_2(k - k_0)^2 + \cdots,$$

Equation (3.94) becomes

$$u(x,t) \doteq e^{\mathrm{i}(k_0 x - \omega_0 t)} \frac{1}{2\pi} \int_{-\infty}^{\infty} F(k) e^{\mathrm{i}[(k-k_0)x - b_1(k-k_0)t - b_2(k-k_0)^2 t]} \, dk,$$

where the factor

$$e^{\mathrm{i}(k_0 x - \omega_0 t)}$$

is a carrier wave with velocity

$$v_c = \omega_0/k_0.$$

Riding over the carrier is an envelope wave

$$\phi(x,t) = \frac{1}{2\pi} \int_{-\infty}^{\infty} F(\kappa + k_0) e^{i(\kappa x - b_1 \kappa t - b_2 \kappa^2 t)} d\kappa, \qquad (3.95)$$

where the variable of integration has been changed from k to $\kappa \equiv k - k_0$.

Taking the time derivative of Equation (3.95), one finds

$$\frac{\partial \phi}{\partial t} = \frac{1}{2\pi} \int_{-\infty}^{\infty} -i(b_1 \kappa + b_2 \kappa^2) F(\kappa + k_0) e^{i(\kappa x - b_1 \kappa t - b_2 \kappa^2 t)} d\kappa$$

$$= -b_1 \frac{\partial \phi}{\partial x} + i b_2 \frac{\partial^2 \phi}{\partial x^2};$$

thus

$$i\left(\frac{\partial \phi}{\partial t} + b_1 \frac{\partial \phi}{\partial x}\right) + b_2 \frac{\partial^2 \phi}{\partial x^2} = 0 \qquad (3.96)$$

is a PDE that governs time evolution of the envelope.

Up to this point, the discussion has remained within the realm of linear theory, but we now consider how a small amount of nonlinearity will alter Equation (3.96). If a is the local amplitude of the wave envelope, the lowest order contribution to the dispersion relation will be proportional to a^2. First-order terms in a do not appear because this would imply dependence of envelope dynamics on the sign of the carrier wave, which changes each half cycle of the carrier frequency.[6]

From Equation (3.96), the dispersion relation for the envelope of a linear wave packet is

$$\omega = b_1 \kappa + b_2 \kappa^2.$$

Noting the square of the amplitude $a^2 = |\phi|^2$, it follows that if Equation (3.96) is augmented to the nonlinear PDE

$$i\left(\frac{\partial \phi}{\partial t} + b_1 \frac{\partial \phi}{\partial x}\right) + b_2 \frac{\partial^2 \phi}{\partial x^2} + \alpha |\phi|^2 \phi = 0, \qquad (3.97)$$

where α is a real number, the nonlinear dispersion relation becomes

$$\omega = b_1 \kappa + b_2 \kappa^2 - \alpha a^2.$$

---

[6] For small amplitude plasma oscillations of the SG equation, we have seen in Equation (3.77)—and also from Problem 32—that the dispersion relation becomes

$$\omega^2 = 1 + k^2 - \tilde{\alpha} a^2,$$

where $\tilde{\alpha}$ is a real constant.

Equation (3.97) is the nonlinear Schrödinger (NLS) equation, which has the following salient properties.

1. NLS describes the propagation of an envelope wave (or modulation), riding over a carrier.
2. If the envelope varies sufficiently slowly with x and is of small enough amplitude, then the last two terms on the left-hand side can be neglected, and the modulation moves with the group velocity

$$v_g = b_1 = \left.\frac{\partial \omega}{\partial k}\right|_{k=k_0}.$$

3. The term $b_2(\partial^2 \phi/\partial x^2)$ introduces wave dispersion into the problem at the lowest level of approximation. Similarly, the term $\alpha|\phi|^2\phi$ introduces nonlinearity at the lowest level of approximation. Thus the NLS equation is generic, arising whenever one wishes to consider the lowest order effects of dispersion and nonlinearity on a wave packet.

Under an appropriate transformation of the variables, Equation (3.97) can be put into the standard form of Equation (3.3), but in making such a transformation, the relative algebraic signs of α and b_2 are of importance, with the condition $\alpha b_2 > 0$ leading to a positive sign in the last term of Equation (3.3), and the condition $\alpha b_2 < 0$ leading to a negative sign. Thus a more general version of Equation (3.3) is

$$\mathrm{i}\frac{\partial u}{\partial t} \pm \frac{\partial^2 u}{\partial x^2} + 2|u|^2 u = 0, \tag{3.98}$$

where the two versions of the equation are referred to as NLS(+) and NLS(−). Sometimes, these two equations are referred to respectively as the focusing and defocusing versions of the NLS equation.

Constant amplitude wave solutions of NLS(+) (plane waves) show Benjamin–Feir instability, noted in Section 3.2.3 in connection with periodic solutions of the SG equation. This instability, in turn, leads to the formation of stable traveling-waves of modulation.

3.3.2 Modulated traveling-wave solutions of NLS(+)

Let us look for bounded solutions of NLS(+) on the infinite domain $-\infty < x < +\infty$ and $-\infty < t < +\infty$. Evidently, the standard traveling-wave assumption, $u(x,t) = \tilde{u}(x-vt)$, is a nonstarter because it requires the velocity to be imaginary. Thus it is interesting to consider solutions of the more general form

$$u(x,t) = \phi(x,t) e^{\mathrm{i}\theta(x,t)}, \tag{3.99}$$

where both ϕ and θ are assumed to be real.

Substituting Equation (3.99) into NLS(+) and equating real and imaginary parts leads to

$$\phi_{xx} - \phi\theta_x^2 - \phi\theta_t + 2\phi^3 = 0$$
$$\phi\theta_{xx} + 2\phi_x\theta_x + \phi_t = 0, \tag{3.100}$$

suggesting modulated traveling-wave solutions of the NLS equation in which

$$\theta(x,t) = \tilde{\theta}(x - v_c t)$$
$$\phi(x,t) = \tilde{\phi}(x - v_e t).$$

Under these assumptions, Equations (3.100) become

$$\tilde{\phi}_{xx} - \tilde{\phi}\tilde{\theta}_x^2 + v_c\tilde{\phi}\tilde{\theta}_x + 2\tilde{\phi}^3 = 0$$
$$\tilde{\phi}\tilde{\theta}_{xx} + 2\tilde{\phi}_x\tilde{\theta}_x - v_e\tilde{\phi}_x = 0. \tag{3.101}$$

Upon multiplying the second of these equations by $2\tilde{\phi}$, it integrates to

$$\tilde{\phi}^2(2\tilde{\theta}_x - v_e) = \text{constant}, \tag{3.102}$$

which can be solved for $\tilde{\theta}_x$ and substituted into the first of Equations (3.101).

To simplify the subsequent integration, it is convenient to choose the constant of integration in Equation (3.102) to be zero. Then $\tilde{\theta}_x = v_e/2$, and the first of Equations (3.101) integrates to

$$\int^{\tilde{\phi}} \frac{dy}{\sqrt{P(y)}} = \pm(x - v_e t), \tag{3.103}$$

where

$$P(y) \equiv C + \tfrac{1}{4}(v_e^2 - 2v_e v_c)y^2 - y^4. \tag{3.104}$$

In general, Equation (3.103) is an elliptic integral, but if $C = 0$, then

$$\tilde{\phi} = a \,\text{sech}[a(x - v_e t)],$$

where

$$4a^2 = v_e^2 - 2v_e v_c. \tag{3.105}$$

Thus, any two of the three wave parameters (a, v_e, and v_c) can be independently chosen. Because the most significant parameters are the wave amplitude (a) and the envelope velocity (v_e), it is convenient to write

$$u(x,t) = a \exp\left[i\frac{v_e}{2}x + i\left(a^2 - \frac{v_e^2}{4}\right)t\right] \text{sech}[a(x - v_e t - x_0)], \tag{3.106}$$

as an exact solution of the NLS equation. Setting $v_e = 0$, this reduces to the stationary breather
$$u(x,t) = ae^{ia^2 t}\text{sech}[a(x-x_0)], \quad (3.107)$$
which is localized about $x = x_0$ and oscillates at a frequency equal to a^2.

One might be concerned that setting $v_e = 0$ in Equation (3.105) would force a^2 to be zero, but this is not so if we allow the magnitude of the carrier velocity (v_c) to become arbitrarily large. This is not as strange as it might seem. In linear dispersive wave systems, it is not uncommon for the phase velocity to go to infinity as the group velocity of a wave packet goes to zero. From a physical perspective, an infinite carrier (or phase) velocity merely implies that all parts of a system are oscillating in phase, as is the case for a stationary breather.

Choosing the integration constant in Equation (3.104) to be negative, allows $P(y)$ to be written as
$$P(y) = (a^2 - y^2)(y^2 - b^2),$$
which is real for $a \geq y \geq b$. The integral
$$\int^{\tilde{\phi}} \frac{dy}{\sqrt{P(y)}} = \pm(x - v_e t)$$
then implies that
$$\tilde{\phi} = a\,\text{dn}[a(x - v_e t); k],$$
a Jacobi elliptic function defined in Appendix C with $k^2 = 1 - b^2/a^2$. As a function of x, $\tilde{\phi}$ oscillates between a maximum value of a and a minimum of b with a period of $\lambda = (2/a)K(k)$, where
$$4(a^2 + b^2) = v_e^2 - 2v_e v_c.$$

Thus the corresponding solution of Equation (3.3) is
$$u(x,t) = a\exp\left[i\frac{v_e}{2}x + i\left(a^2 + b^2 - \frac{v_e^2}{4}\right)t\right]\text{dn}[a(x - v_e t - x_0); k]. \quad (3.108)$$

3.3.3 Dark soliton solutions of NLS(−)

Let us briefly return to Equation (3.98) with the '−' sign, which is called the defocusing version of NLS or NLS(−). At first glance, this choice of sign seems uninteresting because constant amplitude waves (plane waves) are stable, so the usual solitons—as in Equation (3.106)—do not form. Surprisingly, the NLS(−) equation does support regions of low amplitude (dark spots in an optical context) that propagate on the stable plane-wave background with constant shape and speed. Called dark solitons, a family of these low amplitude solutions is given by the expression [34]
$$u(x,t) = a\exp\left(2ia^2 t\right)\{(\cos\theta)\tanh\left[a(\cos\theta)(x - vt)\right] - i\sin\theta\}, \quad (3.109)$$

where a is the amplitude of the plane-wave background, $-\pi/2 \leq \theta \leq \pi/2$ determines the minimum value of the field, and $v = a\sin\theta$. Such dark spots can carry bits of information as well as bright spots (ordinary NLS solitons); thus they may also provide a useful basis for the design of optical computers.

3.3.4 A BT for NLS(+)

Using ideas related to the inverse scattering transform method, Chen has obtained the following BT for in the normalization of Equation (3.3) [15]

$$(u_n + u_{n-1})_x = -2i\xi(u_n + u_{n-1})$$
$$+ (u_n - u_{n-1})\sqrt{4\eta^2 - |u_n + u_{n-1}|^2}$$
$$(u_n + u_{n-1})_t = +2\xi(u_n + u_{n-1})_x$$
$$+ i(u_n - u_{n-1})_x\sqrt{4\eta^2 - |u_n + u_{n-1}|^2}$$
$$+ i(u_n + u_{n-1})(|u_n|^2 + |u_{n-1}|^2). \tag{3.110}$$

In this these equations $\lambda = \xi + i\eta$ is the locus of an upper half plane pole of an associated scattering operator, indicating the presence of an additional soliton in an N-soliton solution. A corresponding result, derived in the context of the AKNS formulation of the inverse scattering method (see Chapter 6), has been obtained by Konno and Wadati [35]. This approach leads to BTs for the KdV, SG, and NLS equations, among others.

Setting $u_n = u_1$ and $u_{n-1} = u_0 = 0$, and also $\xi = 0$, Equations (3.110) reduce to

$$u_{1,x} = u_1\sqrt{4\eta^2 - |u_1|^2}$$
$$u_{1,t} = 4i\eta^2 u_1, \tag{3.111}$$

which integrate to the stationary breather of Equation (3.107), with $\eta = a/2$. The moving breather of Equation (3.106) can be found by integrating Equations (3.110) with $u_{n-1} = 0$ and $\xi = -v_e/4$. It can also be obtained from the stationary breather by noting that if $u(x,t)$ is a solution of NLS, then

$$\exp\left[i\frac{v_e}{2}x - i\frac{v_e^2}{4}t\right] u(x - v_e t, t) \tag{3.112}$$

is another solution.

Interest in multisoliton solutions of the NLS equation was first motivated by dramatic experimental observations of pulse narrowing at high power levels on optical fibers by Mollenauer, et al. [47]. Clearly a nonlinear effect, these observations can be understood in the context of analytical studies of NLS with the initial conditions [57]

$$u(x,0) = N\operatorname{sech} x.$$

For $N = 1$, Equation (3.107) becomes

$$u_{\text{nls1}}(x,t) = e^{it}\operatorname{sech} x.$$

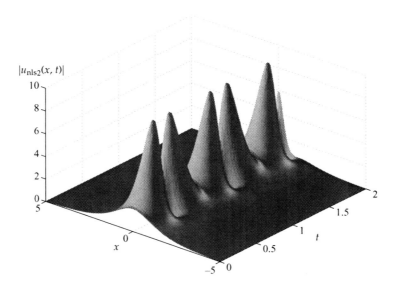

FIG. 3.9. *A plot of* $|u_{\text{nls}2}((x,t)|$ *from Equation (3.113)*.

For $N = 2$, Satsuma and Yajima used the inverse scattering method to show that

$$u_{\text{nls}2}(x,t) = 4\,e^{it}\,\frac{\cosh 3x + 3\,e^{8it}\cosh x}{\cosh 4x + 4\cosh 2x + 3\cos 8t}, \quad (3.113)$$

the magnitude of which is plotted in Figure 3.9. As was observed experimentally on optical fibers, this function becomes more narrow as t increases from zero.

To generate Equation (3.113) from a succession of BTs, one can use the permutative property to obtain a formula—analogous to Equation (3.32) for KdV and Equation (3.84) for SG—that does not require integration. Defining transformations by the diagrams

$$u_0 \xrightarrow{a_1} u_1 \xrightarrow{a_2} u_{12} = u_{21}$$

and

$$u_0 \xrightarrow{a_2} u_2 \xrightarrow{a_1} u_{21} = u_{12},$$

where it is assumed that $\xi = 0$ and $4\eta^2 = a^2$, implies

$$(u_{12} - u_2)\sqrt{a_1^2 - |u_{12} + u_2|^2} - (u_{12} - u_1)\sqrt{a_2^2 - |u_{12} + u_1|^2}$$
$$= (u_2 - u_0)\sqrt{a_2^2 - |u_2 + u_0|^2} - (u_1 - u_0)\sqrt{a_1^2 - |u_1 + u_0|^2}. \quad (3.114)$$

With $a_1 = 1$ and $a_2 = -3$, and

$$u_0 = 0 \quad u_1 = e^{it}\operatorname{sech} x \quad u_2 = -3\,e^{9it}\operatorname{sech} 3x,$$

$u_{1,2}(x,t)$ is equal to $u_{\text{nls2}}(x,t)$ defined in Equation (3.113).

The foregoing results suggest that the NLS equation has exact N-soliton solutions, and these are presented in Equation (B.6) of Appendix B.

3.3.5 Transverse phenomena

Among the several applications of the NLS equation, the most important are in the area of nonlinear optics [49]. In this context, Equation (3.97) models a plane electromagnetic wave propagating in the x-direction, with

$$\phi = |\mathbf{E}|,$$

an electric field amplitude that remains constant on surfaces perpendicular to the direction of propagation. While such a situation can be experimentally arranged, it is not always the case of interest. In this section, some generalizations of NLS are noted that describe variations in the transverse plane.

The vector NLS equation

If a plane electromagnetic wave is propagating in the x-direction, the envelope of the electric field will be a transverse vector of the form

$$\mathbf{E} = E_y(x,t)\mathbf{j} + E_z(x,t)\mathbf{k},$$

where $\mathbf{j}$ and $\mathbf{k}$ are respectively unit vectors in the y- and z-directions. In general, the wave has two orthogonal polarization components (E_y and E_z), and Equation (3.97) becomes

$$i\left(\frac{\partial \mathbf{E}}{\partial t} + b_1 \frac{\partial \mathbf{E}}{\partial x}\right) + b_2 \frac{\partial^2 \mathbf{E}}{\partial x^2} + \alpha |\mathbf{E}|^2 \mathbf{E} = 0,$$

which can be normalized as in Equation (3.3) and written as a pair of coupled NLS equations

$$\begin{aligned} i\frac{\partial u_y}{\partial t} + \frac{\partial^2 u_y}{\partial x^2} + 2(|u_y|^2 + |u_z|^2)u_y = 0 \\ i\frac{\partial u_z}{\partial t} + \frac{\partial^2 u_z}{\partial x^2} + 2(|u_y|^2 + |u_z|^2)u_z = 0, \end{aligned} \qquad (3.115)$$

with u_y (u_z) representing y (z) polarization.

Manakov has shown that these coupled NLS equations are a true soliton system, with associated linear operators upon which the inverse scattering calculations described in Chapter 6 can be based [44]. From our previous studies of NLS solitons, it is clear that Equations (3.115) have solutions of the form

$$\mathbf{u} = u_y \mathbf{j} + u_z \mathbf{k}$$
$$= a\mathbf{p} \exp\left[i\frac{v_e}{2}x + i\left(a^2 - \frac{v_e^2}{4}\right)t\right] \text{sech}[a(x - v_e t - x_0)],$$

where **p** is a unit vector in the direction of the wave polarization. Interactions between such vector solitons, Manakov demonstrates, are more interesting than those between scalar solitons for the following reason. Under a collision between vector solitons $\mathbf{u_1}$ and $\mathbf{u_2}$, the corresponding polarization vectors ($\mathbf{p_1}$ and $\mathbf{p_2}$) will change (in addition to the expected displacements of the locating parameters x_{01} and x_{02}), an effect having implications for interactions between NLS solitons that carry bits of information on optical fibers.

The system of Equations (3.115) is readily generalized to the vector NLS equation

$$i\frac{\partial \mathbf{u}}{\partial t} + \frac{\partial^2 \mathbf{u}}{\partial x^2} + 2(\mathbf{u}, \mathbf{u})\mathbf{u} = 0, \qquad (3.116)$$

where $\mathbf{u} \equiv (u_1, u_2, \ldots, u_N)$ and $(\mathbf{u}, \mathbf{u}) \equiv |u_1|^2 + |u_2|^2 + \cdots + |u_N|^2$.

Under certain simplifying assumptions, this formulation (with $N = 3$) can be used to study the launching and propagation of solitons along the three spines of an alpha-helix in protein (see Figure 8.6) [60].

Filamentation

The phenomenon of filamentation is widely observed, including such diverse examples as the formation of: streams and rivers, atmospheric discharges of electricity (lightning), hydrodynamic jets, and animal paths in the wilderness, among others. In each of these examples, there is a nonlinear effect acting transversely to the direction of motion, drawing the disturbance together into a relatively narrow channel.

Although laser engineers would like to design a device that emits a uniform plane wave, filamentation often emerges, changing the output beam into an untidy collection of bright spots. To control and eliminate this undesired effect, one should understand it; thus the motivation for studying filamentation in nonlinear optics. The simplest model is obtained by assuming slowly varying propagation in the x-direction, independence of time (steady state), and a transverse nonlinearity acting only in the y-direction. Under these assumptions, the envelope of the electric field is governed by an equation that can be written in the normalized form

$$i\frac{\partial u}{\partial x} + \frac{\partial^2 u}{\partial y^2} + 2|u|^2 u = 0 \qquad (3.117)$$

with $u \propto |\mathbf{E}|$, which is formally identical to Equation (3.3). This application motivated study of the NLS equation in the classic paper of Zakharov and Shabat [77]. Evidently, one solution is the steady (ribbon shaped) beam

$$u(x,y) = a e^{ia^2 x} \text{sech}[a(y - y_0)].$$

While Equation (3.117) is relatively easy to solve, it is unrealistic because it does not allow self-focusing in the z-direction. Thus it is also interesting to

consider
$$i\frac{\partial u}{\partial x} + \frac{\partial^2 u}{\partial y^2} + \frac{\partial^2 u}{\partial z^2} + 2|u|^2 u = 0, \tag{3.118}$$
describing steady state self-focusing of a plane wave into a circular beam.

Letting x represent time ($x \to t$) and assuming circular symmetry in the yz-plane, solutions of Equation (3.118) are found to collapse into the origin at a finite time ($t_0 < \infty$), a behavior that differs from that of the oscillating SG [43]. Although the amplitude at the origin goes to infinity at $t = t_0 < \infty$, an exact description of this phenomenon seems difficult to obtain, both analytically and numerically [45].

Blow-up and collapse

The above considerations motivate study of the nonlinear system
$$i\frac{\partial u}{\partial t} + \nabla^2 u + |u|^{2\sigma} u = 0, \tag{3.119}$$
where the space-derivative operator ∇^2 acts in $D = 1, 2$, or 3 space dimensions. This work has been reviewed by Rasmussen and Rypdal [53], who show that necessary conditions for the existence of localized states are
$$\sigma < 2/(D-2) \text{ for } D > 2, \text{ and } \sigma < \infty \text{ for } D \leq 2.$$
Given existence, a sufficient condition for stability of the localized state is
$$0 < \sigma < 2/D.$$
The case $D = 3$ and $\sigma = 2$ is of particular interest as it describes the dynamics of Langmuir waves in plasmas[7] [76].

If the stability condition for a localized state is not satisfied, an instability develops in which the solution becomes unbounded at some point in space (say $\mathbf{r_0}$) at a finite time (say t_0). Often the terms "blow-up" and "collapse" are used interchangeably for this phenomenon, but a distinction can be made in terms of the "mass" or "number"
$$N(u) = \int |u|^2 \, d\tau, \tag{3.120}$$
where $d\tau$ is a differential volume element of D dimensions. Because
$$\dot{N} = 0,$$
collapse can be said to occur when all of the mass is located at a single point, the term blow-up then implies the less restrictive requirement that $|u(\mathbf{r_0})| \to \infty$ as $t \to t_0 < \infty$. In either case, an unbounded solution is unphysical, indicating that certain assumptions in the derivation of Equation (3.119) have been violated.

---

[7]Studied by US chemist Irving Langmuir (1881–1957).

More general NLS-like equations

A rather general nonlinear system that reduces to the NLS equation with an appropriate choice of parameters is the complex Swift–Hohenberg equation

$$\frac{\partial u}{\partial t} = au + b\nabla^2 u + c|u|^2 u + d(1+\nabla^2)^2 u \qquad (3.121)$$

where a, b, c, and d are complex parameters and

$$\nabla^2 \equiv \frac{\partial^2}{\partial x^2} + \frac{\partial^2}{\partial y^2}.$$

This equation is of particular interest for those who would understand and control transverse instabilities in large aperture semiconductor lasers [41]. Recent numerical studies demonstrate a variety of localized standing wave solutions that interact in unexpected ways [56].

With $d = 0$, Equation (3.121) reduces to the complex Ginzburg–Landau equation, which has been widely studied as a nonlinear system that exhibits either filament formation or two-dimensional spatio-temporal chaos, depending upon the values chosen for the complex parameters. Opportunities to explore the dynamics of such solutions have been made available on the internet at http://www.cmp.caltech.edu/~mcc/Patterns/index.html by Michael Cross in the physics department of the California Institute of Technology. It is suggested that the reader devote a few hours to such studies. With $d = 0$ and a, b, and c real, for example, Equation (3.121) reduces to a two-dimensional nonlinear diffusion system that is discussed in the following chapter. Thus Equation (3.121) establishes a connection between NLS solitons and nonlinear diffusion at the frontiers of modern research.

3.4 Summary

This chapter primarily describes the three fundamental soliton equations (KdV, SG, and NLS), each of which arises in several areas of applied science. Salient physical applications are discussed, and means for solving the three equations are presented, including traveling-wave analysis, averaged Lagrangian analysis of modulated periodic waves, nonlinear separation of variables, Hirota's method, and Bäcklund transformations. As extensions of these systems to include more dependent variables and space dimensions provide opportunities for challenging future work, a few such developments are briefly sketched.

3.5 Problems

1. (a) Use Equation (1.1) and the quotations on page 2 to estimate the depth of the Union Canal where Russell discovered the hydrodynamic soliton.

   (b) Estimate error bars for your answer.

2. (a) For the KdV system defined in Equation (3.1), describe all bounded solutions in the linear (low amplitude) approximation.

   (b) Repeat for the SG system defined in Equation (3.2).

   (c) Repeat for the NLS system defined in Equation (3.3).

3. Find an appropriate set of units for $\tilde{u}$, ζ, and τ in Equation (3.5) so the KdV equation reduces to the form of Equation (3.1).

4. Integrate Equation (3.8) to obtain the KdV traveling-pulse solution of Equation (3.9).

5. (a) Find the limiting forms (for $k = 0$ and $k = 1$) of all twelve Jacobi elliptic functions that are defined in Appendix C. What are the periods of each of these functions?

   (b) Using Figures C.2 and C.3, sketch the twelve elliptic functions for $k = 0.9$ and $0 \leq \zeta < K$.

6. (a) From the trigonometric formulas

   $$\sin^2 y + \cos^2 y = 1, \quad d\sin y = \cos y \, dy, \quad \text{and} \quad d\cos y = -\sin y \, dy,$$

   show that $\sin y$ and $\cos y$ satisfy the differential equation

   $$(\psi')^2 = 1 - \psi^2.$$

   (b) From the elliptic function identities [11]

   $$\text{sn}^2 y + \text{cn}^2 y = 1, \quad k^2 \text{sn}^2 y + \text{dn}^2 y = 1, \quad \text{and} \quad d\text{cn}\, y = -\text{sn}\, y \, d\text{n}\, y \, dy,$$

   show that $\text{cn}(y; k)$ satisfies

   $$(\psi')^2 = (1 - \psi^2)(1 - k^2 + k^2 \psi^2),$$

   and $\text{sn}(y; k)$ satisfies

   $$(\psi')^2 = (1 - \psi^2)(1 - k^2 \psi^2).$$

   (c) Discuss how your results in (b) reduce to those in (a) as $k \to 0$.

7. (a) Make plots of $\tilde{u}(\xi)$ from Equation (3.12) with $\xi_1 = 0$, $a_1 = 0$, $a_2 = 1$ and $a_3 = 1, 4,$ and 25.

   (b) What are the traveling-wave velocities (v) in each of these cases?

   (c) What are the values of the traveling-wave period (λ) in each case?

8. Construct a general solution of the KdV equation for the periodic boundary condition $u(x, t) = u(x + l, t)$.

9. Consider a Lagrangian density for a wave process $w(x,t)$, with the functional dependence
$$\mathcal{L} = \mathcal{L}(w_{xx}, w_x, w_t),$$
and a variation (δw) of the true solution with the constraints that $\delta w(x,t) = 0$ for $x = x_1, x_2$ and $t = t_1, t_2$.

(a) Show that the condition
$$\delta \int_{x_1}^{x_2} \int_{t_1}^{t_2} \mathcal{L}\, dx\, dt = 0,$$
for such path variations, implies that
$$\frac{\partial^2}{\partial x^2}\frac{\partial \mathcal{L}}{\partial w_{xx}} - \frac{\partial}{\partial x}\frac{\partial \mathcal{L}}{\partial w_x} - \frac{\partial}{\partial t}\frac{\partial \mathcal{L}}{\partial w_t} = 0.$$

[Hint: refer to Appendix A.]

(b) If
$$\mathcal{L} = -\tfrac{1}{2} w_x w_t + w_x^3 + \tfrac{1}{2} w_{xx}^2,$$
find the pde for $w(x,t)$. How is the result of (b) related to the KdV equation?

10. In the PDE of the previous problem, let
$$w = \psi + \Phi(\theta),$$
where $\Phi(\theta)$ is a periodic function (with zero mean) of the fast variable $\theta = kx - \omega t$, and $\psi = \beta x - \gamma t$ is a slow variation of the solution. With this formulation, $u = w_x$ takes the form
$$u = \beta + k\Phi_\theta,$$
where β is the average value of u [71].

(a) Show that the average of $\mathcal{L}$ over a period of the solution
$$\langle \mathcal{L} \rangle \equiv \frac{1}{2\pi} \int_0^{2\pi} \mathcal{L}\, d\theta$$
$$= kW(A, B, v) + \tfrac{1}{2}[B + (A + \gamma)\beta - v\beta^2],$$
where
$$W(A, B, v) \equiv \frac{1}{2\pi} \oint \sqrt{2y^3 + vy^2 - Ay - B}\, dy,$$
A and B are integration constants defined as in Equation (3.11), and
$$v = \omega/k$$
is the local value of the traveling-wave speed.

(b) From variation of $\langle \mathcal{L} \rangle$ with respect to A, show that
$$\beta = -2kW_A.$$

(c) Show that variation of $\langle \mathcal{L} \rangle$ with respect to ψ implies

$$\frac{\partial}{\partial t}\left(\frac{\beta}{2}\right) + \frac{\partial}{\partial x}\left(v\beta - \frac{A+\gamma}{2}\right) = 0.$$

(d) For ψ to be defined as a function of x and t, show that

$$\frac{\partial \beta}{\partial t} + \frac{\partial \gamma}{\partial x} = 0.$$

(e) From parts (c) and (d), show that one may assume

$$\gamma = v\beta - \frac{A}{2},$$

implying

$$\langle \mathcal{L} \rangle = kW(A, B, v) + \frac{1}{2}\left(B + \frac{A}{2}\beta\right).$$

(f) Show that the expression for $\langle \mathcal{L} \rangle$ obtained in part (e) is a function only of a_1, a_2, and a_3, as defined in Equation (3.11) and shown in Figure 3.2.

11. Discuss the conditions on P and Q for the differential form

$$du = P(x,t)\,dx + Q(x,t)\,dt$$

to be an exact differential. [Hint: assume u is a well defined function of x and t.]

12. (a) For the linear PDE

$$u_{xx} - u_{tt} = u$$

with $-\infty < x < +\infty$, and u bounded as $x \to \pm\infty$, derive a nontrivial BT.

(b) Show that your BT introduces a new eigenfunction into the solution with each "turn of the Bäcklund crank."

13. Consider the class of nonlinear equations that have traveling-wave solutions. Show that there is always a BT that generates a traveling-wave from the vacuum solution.

14. (a) Integrate the KdV BT defined in Equations (3.22) and (3.23) with $w_1 = 0$ to obtain the singular solution

$$w_2(x,t) = -2\kappa \coth[\kappa(x - 4\kappa^2 t - x_0)],$$

where $\lambda \equiv -\kappa^2$.

(b) Use Equation (3.8) to show that $w_{2,x}$ is a solution of Equation (3.6).

15. (a) Demonstrate that the 2-soliton solution for KdV given in Equation (3.28) has the asymptotic forms given in Equations (3.29) and (3.30).

(b) Show this numerically.

16. Write the KdV BT of Equations (3.22) and (3.23) in the form of Equations (3.14).

17. Show that the x-derivative of Equation (3.33) gives the 2-soliton KdV solution of Equation (3.28).

18. Motivated by Equation (3.28), derive a general 2-soliton solution for the KdV equation.

19. Show that
    (i) $D_t D_x (a \cdot b) = a_{xt} b + a b_{xt} - a_t b_x - a_x b_t$,
    (ii) $D_t D_x (a \cdot a) = 2(a a_{xt} - a_x a_t)$,
    (iii) $D_x^m (a \cdot a) = 0$ for m odd,
    (iv) $D_x^2 (a \cdot a) = 2(a a_{xx} - a_x^2)$,
    (v) $D_x^4 (a \cdot a) = 2(a a_{xxxx} - 4 a_x a_{xxx} + 3 a_{xx}^2)$,
    (vi) $D_t^m D_x^n (\exp \theta_1 \cdot \exp \theta_2) = (\omega_2 - \omega_1)^m (k_1 - k_2)^n \exp(\theta_1 + \theta_2)$ where $\theta_i = k_i x - \omega_i t + \alpha_i, i = 1, 2$.

20. Show that the right-hand side of Hirota's equation (3.42) is zero.

21. Use Equations (B.1), (B.2), and (B.3) to generate solutions of KdV for $N = 1, 2,$ and 3.

22. (a) Use Maxwell's equations to show that the voltage variable (V) in Figure 3.4 is related to the electric field intensity (E_y) of a TEM wave by
    $$V = b E_y.$$
    (b) Similarly show that the current (I) is related to the magnetic field intensity (H_z) by
    $$I = a H_z.$$
    (c) What is the power flowing in the x-direction?

23. The derivation of Equation (3.45) gives
    $$\frac{\partial \phi}{\partial x} = -\frac{2eL}{\hbar} I + \text{constant},$$
    What is the physical significance of this constant of integration?

24. The parameter I_0 in Equation (3.43) indicates the strength of Josephson tunneling between two superconducting metals.
    (a) How does I_0 depend upon the width b of the insulating barrier? [Hint: consider the decay of a quantum wave function in a potential barrier.]
    (b) As $I_0 \to 0$, what happens to the normalization of Equation (3.46)?

(c) Use Maxwell's equations and the fact that magnetic fields enter into the surfaces of superconductors by a distance of λ_L (called the "London penetration depth") to show that the limiting speed of a fluxon on a Josephson strip-line is

$$c = \sqrt{\frac{b}{\mu_0 \epsilon (b + 2\lambda_L)}},$$

where μ_0 is the magnetic permeability of space and ϵ is the dielectric permittivity of the insulator region. (This limiting speed is called the Swihart velocity [65].)

25. (a) Show that the normalized SG equation is invariant under the independent variable transformation $(x, t) \to (\xi, \tau)$, where

$$\xi = \frac{x - vt}{\sqrt{1 - v^2}} \quad \text{and} \quad \tau = \frac{t - vx}{\sqrt{1 - v^2}}.$$

(b) Describe what happens to the dependent variables in Equation (3.43) under this Lorentz transformation?

26. (a) Write Equation (3.49) as the first-order system

$$d\tilde{u}/d\xi = w \qquad dw/d\xi = \sin \tilde{u}/(1 - v^2),$$

and discuss solution trajectories on the $(\tilde{u}, w)$ phase plane.

(b) Find the solution trajectories that correspond to the kinks and antikinks in Equation (3.51).

(c) Repeat (b) for the tachyons in Equation (3.53).

(d) Find solution trajectories corresponding to the four general traveling-waves indicated in Section 3.2.3.

27. (a) Find an elliptic function for the SG kink solution of case 1 in Section 3.2.3.

(b) Repeat for the solution of case 3.

28. (a) From an elastic band and a few dozen dressmaker pins, construct the mechanical model of the SG equation shown in Figure 3.5.

(b) Use Newton's second law in the continuum limit to show that your model approximates the SG equation.

(c) Demonstrate waves corresponding to cases 1 and 3 in Section 3.2.3.

(d) Describe what happens when a kink collides with an antikink on your model.

(e) Can you observe the Lorentz contraction of a kink?

29. (a) Find an elliptic function expression for the propagation of a kink with $|v| < 1$ on a system with the periodic boundary conditions $u(x, t) = u(x + l, t)$.

(b) Describe this solution in the context of the mechanical model of the previous problem.

30. Show that Equation (3.57) is a Lagrangian density for the SG equation.

31. (a) For periodic solutions of type 3 defined in Equation (3.56), show that
$$\tilde{u}(\theta) = \arccos\left[1 - (1 - A)\operatorname{sn}^2\left(2\theta K(k)/\pi\,;k\right)\right],$$
where $k^2 = (1 - A)/2$. [Hint: see Appendix C.]

   (b) Show that the nonlinear dispersion relation, given in Equation (3.60), is
$$\omega^2 - \beta^2 = [\pi/2K(k)]^2,$$
where $K(k)$ is the complete elliptic integral of the first kind, defined in Equation (C.8).

   (c) Determine the algebraic sign of $I(A)d^2I/dA^2$, where $I(A)$ is defined in Equation (3.63).

32. Consider small amplitude solutions of the previous problem in which
$$1 - A \equiv \varepsilon \to 0.$$

   (a) Show that
$$\tilde{u}(\theta) \sim \sqrt{2\varepsilon}\sin\theta.$$

   (b) Show that the nonlinear dispersion equation is
$$\omega^2 - \beta^2 = 1 - \varepsilon/4 + O(\varepsilon^2).$$

   (c) Discuss the dynamics of a wave system with this nonlinear dispersion equation to $O(\varepsilon)$. (See reference [50] for additional insight.)

33. (a) Find values of a_1, a_2, and a_3 in Equations (3.71) and (3.75) that lead to the plasma wave solution of Equation (3.76).

   (b) Repeat for the breather solution of Equation (3.78).

   (c) Repeat for the oscillating kink-antikink solution of Equation (3.80).

34. (a) Show that the SG BT of Equations (1.9) implies that if $u_0(\xi,\tau)$ satisfies Equation (1.8), then $u_1(\xi,\tau)$ will also.

   (b) Derive Equation (1.10) from Equations (1.9) with $u_0 = 0$.

   (c) Show that Equation (1.8) transforms to Equation (1.11) under Equations (1.12).

   (d) Check that Equation (1.13) is a solution of Equation (1.11).

35. Derive Equation (3.82) using the SG BT.

36. (a) Find an expression for the phase shift experienced during the kink-kink collision described by Equation (3.82)

   (b) Repeat for the kink-antikink solution of Equation (3.83).

37. Use the nonlinear separation of variables method for the SG equation presented in Section 3.2.4 to derive Equations (3.82) and (3.83).

38. Describe in detail a succession of BTs leading to the stationary breather of Equation (3.85).

39. (a) Assuming that $\alpha b_2 > 0$, find a transformation that carries Equation (3.97) into (3.3).

   (b) Assuming that $\alpha b_2 < 0$, find a transformation that carries Equation (3.97) into the NLS(−) equation
   $$iu_t - u_{xx} + 2|u|^2 u = 0.$$

40. Consider the linear wave system
    $$iu_t + u_{xx} + u = 0.$$

   (a) Derive the dispersion relation of this system for solutions of the form
   $$u(x,t) \sim e^{i(kx-\omega t)}.$$

   (b) Show that the phase velocity of a wave packet approaches infinity as its group velocity approaches zero.

   (c) Relate the result of (a) to Equation (3.105).

41. (a) Derive Equation (3.108) for a periodic solution of the NLS(+) equation and show that it reduces to Equation (3.106) as $b \to 0$.

   (b) Discuss the restrictions on selecting the carrier and envelope velocities in Equation (3.108).

42. Using the techniques outlined in Section 3.3.2, show that Equation (3.109) is a solution of the NLS(−) equation
    $$iu_t - u_{xx} + 2|u|^2 u = 0.$$

43. Show that if $u(x,t)$ is a solution of the NLS(+) equation, then another solution is
    $$\exp\left[i\frac{v_e}{2}x - i\frac{v_e^2}{4}t\right] u(x - v_e t, t).$$

44. Integrate the Bäcklund transformation of Equations (3.110) with $u_{n-1} = 0$, $4\eta^2 = a^2$, and $4\xi = -v_e$ to obtain the moving breather of Equation (3.106).

45. (a) Show that $u(x,t)$, given in Equation (3.113), equals $2\operatorname{sech} x$ at $t = 0$.

   (b) Plot $|u(x,t)|$ as a function of x at times $t = 0$, 0.2, and 0.4.

   (c) Find a solution of NLS that equals $2\operatorname{sech} x$ at $t = 0$ and moves with an envelope velocity v_e.

46. Check that if $u_0 = 0$, $u_1 = e^{it}\operatorname{sech} x$, and $u_2 = -3\,e^{9it}\operatorname{sech} 3x$ in Equation (3.114), then u_{12} is given by Equation (3.113).

47. (a) Find a solution of Equation (3.117) that satisfies the boundary condition
$$u(0, y) = 2\operatorname{sech} y.$$

   (b) Discuss the behavior of your solution for various values of $x > 0$.

48. Show that the quantity defined in Equation (3.120) is independent of time.

49. Discuss physical considerations that would limit the phenomenon of blow-up in solutions of Equation (3.119).

REFERENCES

1. T Barnard. $2N\pi$ ultrashort pulses. *Phys. Rev. A* 7 (1973) 373–376.
2. A Barone, F Esposito, C J Magee, and A C Scott. Theory and applications of the SG equation. *Riv. Nuovo Cimento* 1 (1971) 227–267.
3. A Barone and G Paternó. *Physics and Applications of the Josephson Effect*. John Wiley, New York, 1982.
4. A Benabdallah, J G Caputo, and A C Scott. Exponentially tapered Josephson flux flow oscillator. *Phys. Rev. B* 54 (1996) 16139–16146.
5. A Benabdallah, J G Caputo, and A C Scott. Laminar phase flow for an exponentially tapered Josephson oscillator. *J. Appl. Phys.* (6) 88 (2000) 3527–3540.
6. T Brooke Benjamin and J E Feir. The disintegration of wave trains in deep water. *J. Fluid Mech.* 27 (1967) 417–430.
7. D J Benney and A C Newell. The propagation of nonlinear wave envelopes. *J. Math. Phys.* 46 (1967) 133–139.
8. I L Bogolyubskii. Oscillating particle-like solutions of the non-linear Klein–Gordon equation. *JETP Lett.* 24 (1976) 535–538.
9. I L Bogolyubskii and V G Makhankov. Lifetime of pulsating solitons in certain classical models. *JETP Lett.* 24 (1976) 12–14.
10. I L Bogolyubskii and V G Makhankov. Dynamics of spherically symmetrical pulson of large amplitude. *JETP Lett.* 25 (1977) 107–110.
11. P F Byrd and M D Friedman. *Handbook of Elliptic Integrals for Engineers and Scientists*. Springer-Verlag, New York, 1954.

REFERENCES

12. J G Caputo, N Flytzanis, Y Gaididei, and M Vavalis. Two-dimensional effects in Josephson junctions: I. Static properties. *Phys. Rev. E* 54 (1996) 2092–2101.
13. J G Caputo, N Flytzanis, and M Vavalis. A semi-linear elliptic pde model for the static solution of Josephson junctions. *Int. J. Mod. Phys. C* 6 (1995) 241–262.
14. J G Caputo, N Flytzanis, and M Vavalis. Effect of geometry on fluxon width in a Josephson junction. *Int. J. Mod. Phys. C* 7 (1996) 191–216.
15. H H Chen. General derivation of Bäcklund transformations from inverse scattering problem. *Phys. Rev. Lett.* 33 (1974) 925–928.
16. P L Christiansen, N G Jensen, P S Lomdahl, and B A Malomed. Oscillations of eccentric pulsons. *Phys. Scr.* 55 (1997) 131–134.
17. P L Christiansen and P S Lomdahl. Numerical study of $2+1$ dimensional SG solitons. *Physica D* 2 (1981) 482–494.
18. P L Christiansen and O H Olsen. Return effect for rotationally symmetric solitary wave solutions of the SG equation. *Phys. Lett. A* 68 (1978) 185–188.
19. G Costabile, R D Parmentier, B Savo, D W McLaughlin, and A C Scott. Exact solutions of the SG equation describing oscillations in a long (but finite) Josephson junction. *Appl. Phys. Lett.* 32 (1978) 587–589.
20. G H Derrick. Comments on nonlinear wave equations as models for elementary particles. *J. Math. Phys.* 5 (1964) 1252–1254.
21. P G Drazin and R S Johnson. *Solitons: An Introduction*. Cambridge University Press, Cambridge, 1989.
22. W Döring. Über die Trägheit der Wände zwischen weisschen Bezirken. *Z. Naturforsch.* 31 (1948) 373–379.
23. L P Eisenhart. *A Treatise on the Differential Geometry of Curves and Surfaces*. Dover, New York, 1909/1960.
24. U Enz. Discrete mass, elementary length, and a topological invariant as a consequence of a relativistic invariant variational principle. *Phys. Rev.* 131 (1963) 1392–1394.
25. J L Fergason and G H Brown. Liquid crystals and living systems. *J. Am. Oil Chem. Soc.* 45 (1968) 120–127.
26. H Flaschka and D W McLaughlin. Some comments on Bäcklund transformations and inverse scattering problems. In *Bäcklund Transformations*, R M Miura, ed., Springer-Verlag, Berlin, 1976.
27. J Frenkel and T Kontorova. On the theory of plastic deformation and twinning. *J. Phys. (USSR)* 1 (1939) 137–149.
28. B D Fried and Y H Ichikawa. On the nonlinear Schrödinger equation for Langmuir waves. *J. Phys. Soc. Jpn.* 34 (1973) 1073–1082.
29. A Hasegawa and F Tappert. Transmission of stationary nonlinear optical pulses in dispersive dielectric fibers. *Appl. Phys. Lett.* 23 (1973) 142–144.
30. R Hirota. Exact solutions of the Korteweg–de Vries equation for multiple collisions of solitons. *Phys. Rev. Lett.* 27 (1971) 1192–1194.
31. Y H Ichikawa, T Imamura, and T Taniuti. Nonlinear wave modulation in collisionless plasma. *J. Phys. Soc. Jpn.* 34 (1972) 189–197.
32. V I Karpman and E M Kruskal. Modulated waves in nonlinear dispersive medium. *Sov. Phys. JETP* 28 (1969) 277–281.
33. P L Kelley. Self-focusing of optic beams. *Phys. Rev. Lett.* 15 (1965) 1005–1008.

34. Y S Kivshar and B Luther-Davies. Dark optical solitons: physics and applications. *Physics Reports* 298 (1998) 81-197.
35. K Konno and M Wadati. Simple derivation of Bäcklund transformation from Riccati form of the inverse method. *Prog. Theor. Phys.* 53 (1975) 1652–1656.
36. D J Korteweg and H de Vries. On the change of form of long waves advancing in a rectangular canal, and on a new type of long stationary waves. *Philos. Mag.* 39 (1895) 422–443.
37. G L Lamb, Jr. Propagation of ultrashort optical pulses. *Phys. Lett. A* 25 (1967) 181–182.
38. G L Lamb, Jr. Analytical descriptions of ultrashort optical pulse propagation in a resonant medium. *Rev. Mod. Phys.* 43 (1971) 99–124.
39. G L Lamb, Jr. *Elements of Soliton Theory*. John Wiley, New York, 1980.
40. D F Lawden. *Elliptic Functions and Applications*. Springer-Verlag, New York, 1989.
41. J Lega, J V Moloney, and A C Newell. Swift–Hohenberg equation for lasers. *Phys. Rev. Lett.* 73 (1994) 2978–2981.
42. S Leibovich. Weakly nonlinear waves in rotating fluids. *J. Fluid Mech.* 42 (1970) 803–822.
43. P S Lomdahl, O H Olsen, and P L Christiansen. Return and collapse of solutions to the nonlinear Schrödinger equation in cylindrical symmetry. *Phys. Lett. A* 78 (1980) 125–128.
44. S V Manakov. On the theory of two-dimensional stationary self-focusing of electromagnetic waves. *Sov. Phys. JETP* 38 (1974) 248–253.
45. D W McLaughlin, G C Papanicolaou, C Sulem, and P L Sulem. Focusing singularity of the cubic Schrödinger equation. *Phys. Rev. A* 34 (1986) 1200–1210.
46. R M Miura. Korteweg–de Vries equation and generalizations. I. A remarkable explicit nonlinear transformation. *J. Math. Phys.* 9 (1968) 1202–1204.
47. L F Mollenauer, R H Stolen, and J P Gordon. Experimental observation of picosecond pulse narrowing and solitons in optical fibers. *Phys. Rev. Lett.* 45 (1980) 1095–1098.
48. K Naugolnykh and L Ostrovsky. *Nonlinear Wave Processes in Acoustics*. Cambridge University Press, Cambridge, 1998.
49. A C Newell and J V Moloney. *Nonlinear Optics*. Addison-Wesley, Reading, MA, 1992.
50. L A Ostrovsky. Propagation of wave packets and space-time self-focusing in a nonlinear medium. *Sov. Phys. JETP* 24 (1967) 797–800.
51. N F Pedersen. Solitons on Josephson transmission lines. In *Solitons*, S E Trullinger, V E Zakharov, and V L Pokrovsky, eds., Elsevier, Amsterdam, 1986.
52. J K Perring and T H R Skyrme. A model unified field equation. *Nucl. Phys.* 31 (1962) 550–555.
53. J Juul Rasmussen and K Rypdal. Blow-up in nonlinear Schroedinger Equations—I. A general review. *Phys. Scr.* 33 (1986) 481–497.
54. M Remoissenet. *Waves Called Solitons: Concepts and Experiments*. 2nd edition. Springer-Verlag, Berlin, 1996.
55. N Rosen and H B Rostenstock. The force between particles in a nonlinear field theory. *Phys. Rev.* 85 (1952) 257–259.

REFERENCES

56. H Sakaguchi and H R Brand. Localized patterns for the quintic complex Swift–Hohenberg equation. *Physica D* 117 (1998) 95–105.
57. J Satsuma and N Yajima. Initial value problems of one-dimensional self-modulation of nonlinear waves in dispersive media. *Supp. Prog. Theor. Phys.* 55 (1974) 284–306.
58. A C Scott. A nonlinear Klein–Gordon equation. *Am. J. Phys.* 37 (1969) 52–61.
59. A C Scott. *Active and Nonlinear Wave Propagation in Electronics*. John Wiley, New York, 1970.
60. A C Scott. Launching a Davydov soliton: I. Soliton analysis. *Phys. Scr.* 29 (1984) 279–283.
61. A C Scott, F Y F Chu, and D W McLaughlin. The soliton: A new concept in applied science. *Proc. IEEE* 61 (1973) 1443–1483.
62. A C Scott and W J Johnson. Internal flux motion in large Josephson junctions. *Appl. Phys. Lett.* 14 (1969) 316–318.
63. A Seeger, H Donth, and A Kochendörfer. Theorie der Versetzungen in eindimensionalen Atomreihen. *Z. Phys.* 134 (1953) 173–193.
64. T H R Skyrme. A nonlinear theory of strong interactions. *Proc. R. Soc. (London)* A247 (1958) 260–278.
65. J C Swihart. Field solutions for a thin-film superconducting strip transmission line. *J. Appl. Phys.* 32 (1961) 461–469.
66. V I Talanov. Self-focusing of wave beams in nonlinear media. *JETP Lett.* 2 (1965) 138–141.
67. T Taniuti and H Washimi. Self-trapping and instability of hydromagnetic waves along the magnetic field in a cold plasma. *Phys. Rev. Lett.* 21 (1968) 209–212.
68. F Tappert and C M Varma. Asymptotic theory of self trapping of heat pulses in solids. *Phys. Rev. Lett.* 25 (1970) 1108–1111.
69. H D Wahlquist and F B Estabrook. Bäcklund transformation for solutions of the Korteweg–de Vries equation. *Phys. Rev. Lett.* 31 (1973) 1386–1390.
70. H Washimi and T Taniuti. Propagation of ion-acoustic solitary waves of small amplitude. *Phys. Rev. Lett.* 17 (1966) 399–403.
71. G B Whitham. *Linear and Nonlinear Waves*. John Wiley & Sons, New York, 1974.
72. L van Wijngaarden. On the equation of motion for mixtures of liquid and gas bubbles. *J. Fluid Mech.* 33 (1968) 465–474.
73. N J Zabusky. Solitons and bound states of the time independent Schrödinger equation. *Phys Rev.* 168 (1968) 124–128.
74. N J Zabusky. Solitons and energy transport in nonlinear lattices. *Comput. Phys. Commun.* 5 (1973) 1–10.
75. N J Zabusky and M D Kruskal. Interaction of solitons in a collisionless plasma and the recurrence of initial states. *Phys. Rev. Lett.* 15 (1965) 240–243.
76. V E Zakharov. Collapse of Langmuir waves. *Sov. Phys. JETP* 35 (1972) 908–914.
77. V E Zakharov and A B Shabat. Exact theory of two-dimensional self-modulation of waves in nonlinear media. *Sov. Phys. JETP* 34 (1972) 62–69.

4

REACTION-DIFFUSION SYSTEMS

At about the time that John Scott Russell was systematically studying hydrodynamic solitons on a Scottish canal, Michael Faraday, the self-taught English experimental physicist and physical chemist, began to present his annual "Christmas Lectures" on selected aspects of natural philosophy to young people at London's Royal Institution. Among the most popular of these were a series of six talks entitled "The Chemical History of a Candle," and beginning with a bold claim [20]:

There is no better, there is no more open door by which you can enter into the study of natural philosophy than by considering the physical phenomena of a candle.

Although this assertion may have startled some of his listeners, Faraday went on to support it with a series of simple and elegant experiments that demonstrate the structure and composition of a flame and lead to an analogy with the process of respiration in biological organisms. From this perspective, as we have noted in Chapter 1, the flame proceeds down the candle at a velocity (v) that is given by the ratio of the power being dissipated by the flame (P) to the chemical energy stored in the wax (E) per unit length of the candle. That is to say

$$P = vE, \qquad (4.1)$$

implying that the flame digests energy at the same rate it is eaten, as does a living creature.

Why be concerned about a candle in a discussion of nonlinear science? Because it is an example of a reaction-diffusion process, which is the subject of this chapter. In a lighted candle, heat generated by the flame diffuses into the solid wax, causing the release of a vapor that carries chemical energy upward into the flame. Combustion of the vaporized wax provides the heat. This entire structure—the flame, its thermal precursor, and the rising column of vaporized fuel—is a traveling-wave, moving down the candle at a fixed speed.

From Equation (4.1), this speed will be equal to P/E, where P (related to the size of the flame) remains about the same for candles of different diameters. The parameter E, on the other hand, increases with the square of the candle diameter (d), because it is proportional to the cross-sectional area. Thus the speed of a candle flame is expected to depend on diameter as $1/d^2$. This expectation is roughly confirmed in the log-log plot of Figure 4.1, where the dashed line with a slope of -2 indicates a $1/d^2$ dependence. (The occurrence of slower than

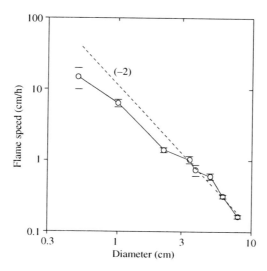

FIG. 4.1. *The dependence of the velocity of flame propagation (v) upon the diameter (d) of a candle. (Data courtesy of Lela Scott MacNeil [72].)*

expected flame speeds for the smaller candles is probably because these have smaller flames, reducing P in Equation (4.1).)

In the present chapter, we begin by analyzing simple models of nonlinear diffusion—observing that the power balance of Equation (4.1) plays more than a metaphoric role in describing the process—and then consider more realistic models of a nerve impulse that include geometrical variations.

Both the flame of a candle and the nerve impulse are examples of nonlinear diffusion in one space dimension. Generalizing to more space dimensions in the final section leads into a garden of unexpectedly beautiful phenomena waiting to be explored, including some of interest in theoretical biology and medical science.

4.1 Simple reaction-diffusion equations

4.1.1 *The Zeldovich–Frank-Kamenetsky (Z–F) equation*

As was noted in Chapter 1, one of the simplest nonlinear diffusion equations,

$$\frac{\partial^2 u}{\partial x^2} - \frac{\partial u}{\partial t} = f(u), \tag{4.2}$$

was proposed in the mid-1930s as a model for genetic diffusion [21, 39]. With $f(u)$ a cubic polynomial, this equation was studied in 1938 by Yakov Zeldovich and David Frank-Kamenetsky in the Soviet Union to represent flame front propagation [89]. Without reference to these previous studies, Equation (4.2) reappeared in the 1940s as a model of impulse conduction along an active nerve fiber, with $u(x,t)$ representing the voltage across a cell membrane [58].

Solutions of Equation (4.2) can be found by looking for traveling-waves of the form $u(x,t) = \tilde{u}(x - vt) = \tilde{u}(\xi)$, where $\xi \equiv x - vt$ is a traveling-wave variable, so

$$\frac{\partial \tilde{u}}{\partial x} = \frac{d\tilde{u}}{d\xi} \quad \text{and} \quad \frac{\partial \tilde{u}}{\partial t} = -v\frac{d\tilde{u}}{d\xi}. \tag{4.3}$$

Assumption of a traveling-wave form of the solution thus converts the PDE of Equation (4.2) into the ODE

$$\frac{d^2\tilde{u}}{d\xi^2} + v\frac{d\tilde{u}}{d\xi} = f(\tilde{u}),$$

which can be written as the first-order system

$$\begin{aligned} \frac{d\tilde{u}}{d\xi} &= w \\ \frac{dw}{d\xi} &= f(\tilde{u}) - vw. \end{aligned} \tag{4.4}$$

While progress has been made by reducing a nonlinear PDE to a nonlinear ODE, this reduction did not come for free. In the ODE system, the traveling-wave speed (v) is an unknown parameter that must be determined from the analysis. To fix ideas, it is convenient to choose the nonlinear function to be [89]

$$f(u) = u(u - a)(u - 1),$$

as shown in Figure 4.2(a). Then Equations (4.4) become

$$\begin{aligned} \frac{du}{d\xi} &= w \\ \frac{dw}{d\xi} &= u(u-a)(u-1) - vw, \end{aligned} \tag{4.5}$$

where the "tilde" on u is now dropped for typographical convenience. Equations (4.5) can be studied in the (u, w) phase plane of Figure 4.2(b) where:

- The "velocity" vector $(du/d\xi, dw/d\xi)$ of a solution trajectory is a unique function of the position vector (u, w).
- The system has three singular points (SPs), at which the velocity vector $(du/d\xi, dw/d\xi) = (0, 0)$. As shown on Figure 4.2(b), these SPs are at $(u, w) = (0, 0)$, $(a, 0)$, and $(1, 0)$.
- The SPs at $(0, 0)$ and $(1, 0)$ are saddle points, while the SP at $(a, 0)$ is a node or spiral.
- In general, solution trajectories of Equations (4.5) will grow without bound as $\xi \to \infty$.

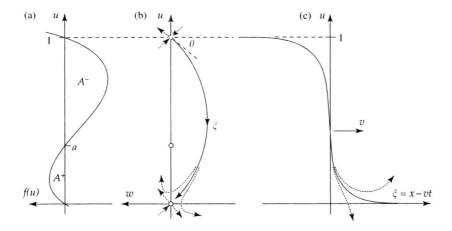

FIG. 4.2. *Finding a traveling-wave solution of Equation (4.2). (a) A cubic-shaped function $f(u)$. (b) Trajectories in the traveling-wave phase plane of Equations (4.4). (c) The traveling-wave solution.*

To find a trajectory that corresponds to a physically meaningful solution of the original PDE system of Equation (4.2), it is necessary that (u, w) remains bounded as $\xi \to \infty$. This requirement can be satisfied by adjusting the traveling-wave velocity v so that a trajectory leaving one of the saddle points at $\xi = -\infty$ lies on a trajectory approaching the other as $\xi \to +\infty$. Thus the traveling-wave speed is fixed by the condition that the solution and its derivative must remain finite.

In the particular example of Equations (4.5), a trial trajectory that intersects the two saddle points is

$$w = Ku(u-1),$$

implying that

$$\frac{dw}{du} = K(2u - 1).$$

But from dividing the left- and right-hand sides of Equations (4.5), another expression for this derivative is

$$\frac{dw}{du} = \frac{u-a}{K} - v.$$

To make these two expressions equal, it is necessary to choose $K = \pm 1/\sqrt{2}$ so

$$v = \pm(1 - 2a)/\sqrt{2}. \tag{4.6}$$

Finally, from the first of Equations (4.5)

$$\frac{du}{d\xi} = \pm \frac{u(u-1)}{\sqrt{2}},$$

which can be integrated to obtain monotonically increasing or decreasing traveling-wave solutions of the original PDE as

$$u(x,t) = \frac{1}{1 + \exp[\pm(x-vt)/\sqrt{2}]}, \qquad (4.7)$$

where the monotone decreasing case (+sign) is shown in Figure 4.2.

In thinking more generally about the solutions of Equation (4.5) represented by Equation (4.7), one must consider both $a > \frac{1}{2}$ and $a < \frac{1}{2}$ for both choices of the sign in Equation (4.6), leading to four distinct cases. These possibilities can be described as follows:

CASE 1: $a < \frac{1}{2}$ and $K > 0$. In this case, $v > 0$ and the solution trajectory goes from $(1,0)$ to $(0,0)$ as ξ goes from $-\infty$ to $+\infty$. This is the example shown in Figure 4.2.

CASE 2: $a > \frac{1}{2}$ and $K > 0$. In this case, $v < 0$ and the solution trajectory also goes from $(1,0)$ to $(0,0)$ as ξ goes from $-\infty$ to $+\infty$.

CASE 3: $a > \frac{1}{2}$ and $K < 0$. In this case, $v > 0$ and the solution trajectory goes from $(0,0)$ to $(1,0)$ as ξ goes from $-\infty$ to $+\infty$.

CASE 4: $a < \frac{1}{2}$ and $K < 0$. In this case, $v < 0$ and the solution trajectory goes from $(0,0)$ to $(1,0)$ as ξ goes from $-\infty$ to $+\infty$.

The differences between cases 1 and 4 and between cases 2 and 3 merely indicate that the original system has reflection symmetry in x; thus waves traveling in the $+x$-direction are physically equivalent to those traveling in the $-x$-direction. The differences between cases 3 and 4 (and between cases 1 and 2) are related to the constraints of the power balance condition expressed in Equation (4.1). Although formulated for the flame of a candle, this condition applies equally well for a variety of nonlinear diffusive processes, including active nerve fibers.

In case 3, for example, $a > \frac{1}{2}$ so the negative area under the curve

$$f(u) = u(u-a)(u-1)$$

is smaller than the positive area. Thus the overall effect of $f(u)$ is dissipative, corresponding to P being positive in Equation (4.1). As time goes from $-\infty$ to $+\infty$ in the original PDE system, the solution goes from $u = 1$ to $u = 0$, corresponding to E also being positive in Equation (4.1). Because both P and E are positive, v is also positive.

In case 1, on the other hand, the positive area under $f(u)$ is smaller than the negative area so the system acts as a source of energy, corresponding to P being negative (negative dissipation) in Equation (4.1). As time goes from $-\infty$ to $+\infty$ in the original PDE system, the solution goes from $u = 0$ to $u = 1$, corresponding to E also being negative in Equation (4.1). Because both P and E are negative, v is again positive.

Cases 1 and 2 listed above correspond to trajectories in the (u, w) phase plane that go as

$$(1, 0) \longrightarrow (u, w) \longrightarrow (0, 0) \quad \text{as} \quad -\infty \to \xi \to +\infty,$$

and cases 3 and 4 correspond to trajectories where

$$(0, 0) \longrightarrow (u, w) \longrightarrow (1, 0) \quad \text{as} \quad -\infty \to \xi \to +\infty.$$

Such trajectories are called heteroclinic, implying that they start at one SP and end at another.

For the special case $a = \frac{1}{2}$, these heteroclinic trajectories correspond to traveling-wave solutions of zero velocity, but there are additional stationary solutions. Assuming that $v = 0$, Equations (4.5) can be written as

$$\frac{d^2 u}{d\xi^2} = u(u - a)(u - 1), \tag{4.8}$$

with a solution trajectory defined by

$$w\,dw = u(u - a)(u - 1)\,du,$$

implying

$$w(u) = \pm \left[2 \int_0^u \alpha(\alpha - a)(\alpha - 1)\,d\alpha \right]^{1/2} \tag{4.9}$$

for $a < \frac{1}{2}$, with a corresponding expression for $a > \frac{1}{2}$. Thus Equation (4.8) has the homoclinic solution trajectories

$$\begin{cases} (0,0) \longrightarrow (0,0) \\ (1,0) \longrightarrow (1,0) \end{cases} \quad \text{for} \quad \begin{cases} a < 1/2 \\ a > 1/2, \end{cases}$$

the first of which is illustrated in Figure 4.3.

As functions of $\xi = x - vt$, the corresponding solutions of the original PDE are pulse like, with a maximum amplitude (u_m) near $\xi = 0$ and falling exponentially to zero as $x \to \pm\infty$. From Equation (4.9) it can be seen that u_m is determined by the condition that the shaded portion of Figure 4.3(a) has an area equal to A^+.

If A^+ is equal to A^-, one can construct heteroclinic trajectories, as in Figure 4.2, between the outer zeros of $f(u)$. These stationary solutions correspond to the transitions at $(a = \frac{1}{2})$ between Case 1 and Case 2 or between Case 3 and Case 4 in the discussion below Equation (4.7). Thus the qualitative

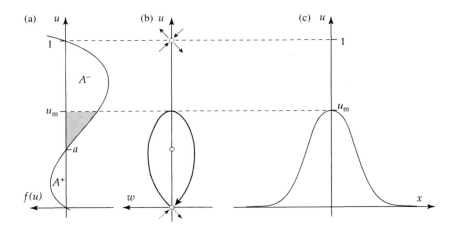

FIG. 4.3. *Zero-velocity, pulse-like solutions of Equation (4.2). (a) For $u = u_m$, the shaded area is equal to A^+. (b) A homoclinic trajectory in the (u, w) pahse plane. (c) The corresponding pulse-like solution.*

conclusions drawn in this discussion do not require the $f(u) = u(u - a)(u - a)$; a function of "cubic shape" is sufficient.

To this point, we have learned nothing about the stability of these traveling-wave solutions. In Section 4.4, it is shown that the monotonically increasing or decreasing traveling-wave solutions, corresponding to heteroclinic trajectories shown in Figure 4.2(b), are stable, while pulse-like solutions, corresponding to homoclinic trajectories in Figure 4.3(b), are unstable.

While the stable solutions (representing traveling waves that appear in the laboratory) are of greater interest to the experimentalist, the unstable, pulse-like solutions are not unimportant. These unstable solutions lead to the determination of threshold conditions for the emergence of stable traveling-waves.

4.1.2 The Burgers equation

Another simple nonlinear diffusion equation, proposed by Burgers in 1948 as a model for turbulence [9], can be written as

$$\frac{\partial u}{\partial t} + u \frac{\partial u}{\partial x} = D \frac{\partial^2 u}{\partial x^2}, \tag{4.10}$$

where D is a diffusion constant. In the limit $D \to 0$, this equation has solutions for which the higher amplitude parts of the wave travel faster than those of lower amplitude. Thus a pulse like initial condition will produce a wave that develops a shock on its leading edge, a phenomenon observed in nonlinear acoustics [55].

With $D > 0$, Equation (4.10) can be linearized through the Cole–Hopf transformation [11, 29]
$$u = -2D\phi_x/\phi.$$
Following the discussion by Whitham [79], this transformation is performed in two steps. First write
$$u = \psi_x$$
so upon integration (with the constant set to zero), Equation (4.10) becomes
$$\psi_t + \psi_x^2/2 = D\psi_{xx}.$$
Finally the transformation
$$\psi = -2D\log\phi$$
implies the linear diffusion equation
$$\phi_t = D\phi_{xx}, \tag{4.11}$$
the general solution of which was discussed in Chapter 2.

For all its appealing simplicity, the equation of Burgers is difficult to classify. Although not Hamiltonian, its initial value problem can be solved through the linearizing transformation
$$u(x,0) \to \phi(x,0) \to \phi(x,t) \to u(x,t),$$
just as for the inverse scattering transforms of Chapter 6.

By seeming to break the rules of mathematical etiquette, the Burgers equation teaches us to be wary of generalizations in the realm of nonlinear dynamics.

4.2 The Hodgkin–Huxley (H–H) system

The nerve cells (or neurons) of animal organisms have been crafted by evolution to carry impulses of electric voltage from one to another along interconnecting fibers called axons. The giant axon of the squid, for example, is a long cylinder or tube of membrane (as sketched in Figure 4.4), which is semipermeable to ions of sodium (Na^+) and potassium (K^+).

Consider the axis of a fiber to be in the x-direction, and let $V(x,t)$ be the transverse voltage across the cell membrane. Also let $i(x,t)$ be the electric current flowing through the tube in the axial direction. Applying Kirchhoff's voltage and current laws to the differential equivalent circuit in Figure 4.4(b) implies that

$$\begin{aligned}\frac{\partial V}{\partial x} &= -ri \\ \frac{\partial i}{\partial x} &= -c\frac{\partial V}{\partial t} - j_i,\end{aligned} \tag{4.12}$$

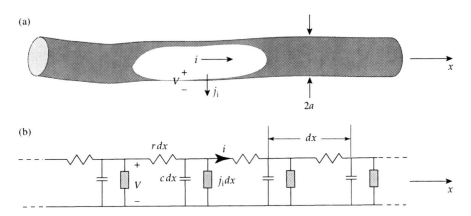

FIG. 4.4. *(a) Sketch of a squid nerve axon. (b) A differential equivalent circuit of the axon.*

where r is the longitudinal resistance per unit length of the fiber, c is the membrane capacitance per unit length, and j_i is the total ionic current flowing across the membrane per unit length. The first of these equations is merely a differential form of Ohm's law, but the second is more interesting. The term $c\,\partial V/\partial t$ represents a displacement current flowing through the membrane capacitance which adds to the effect of j_i in decreasing the flow of longitudinal current (i). The ionic current (j_i) is carried across the membrane through protein pores or channels [70].

Combining Equations (4.12) leads to a diffusion equation for $V(x,t)$ of the form

$$\frac{\partial^2 V}{\partial x^2} - rc\frac{\partial V}{\partial t} = rj_i, \qquad (4.13)$$

which is able to transmit impulses of voltage in the x-direction without attenuation because of nonlinearities in the ionic current. At this point, we face two problems. The first is to understand the dynamic nature of the membrane ionic current, and the second is to study traveling-wave solutions of Equation (4.13).

4.2.1 *Space-clamped squid membrane dynamics*

In the early 1950s, Hodgkin and Huxley made careful measurements of the sodium and potassium components of membrane ionic current for a dozen squid axons under "space clamped" conditions, which implies that membrane voltage is not permitted to vary along the axon [28]. (Experimentally, this is achieved by inserting a conducting wire longitudinally through the axon.) From these data they constructed the following phenomenological expressions for the ion current flowing out of the cell per unit area of membrane:

$$J_\text{i} = G_\text{K} n^4 (V - V_\text{K}) + G_\text{Na} m^3 h (V - V_\text{Na}) + G_\text{L}(V - V_\text{L}). \tag{4.14}$$

In the first term of this expression, $G_\text{K} n^4 V$ gives the conduction component of ionic current (forced through channels by the transmembrane electric field) and $-G_\text{K} n^4 V_\text{K}$ accounts for the diffusion component (driven by the difference in ionic concentrations across the membrane), and similarly for the other two components [70]. For an axon of circular cross section with radius a, the ionic current per unit length is

$$j_\text{i} = 2\pi a J_\text{i}.$$

Values obtained by H–H for the membrane capacitance per unit area (C) and the parameters in Equation (4.14) are given in Table 4.1. This table gives both the mean parameter values and their observed range of variation, in addition to values for a "standard" axon that will be discussed in the following section.

Dynamics enter the H–H formulation through the switching variables n, m, and h, which take values between zero and unity and are governed by the rate equations

$$\begin{aligned}\frac{dn}{dt} &= \alpha_n(1-n) - \beta_n n, \\ \frac{dm}{dt} &= \alpha_m(1-m) - \beta_m m, \\ \frac{dh}{dt} &= \alpha_h(1-h) - \beta_h h.\end{aligned} \tag{4.15}$$

The variables n and m are called, respectively, potassium and sodium "turn-on" variables because they rest near zero and approach unity as the voltage rises. Similarly, h is called a sodium "turn-off" variable because it rests near unity and approaches zero as the voltage rises.

Table 4.1. *H–H membrane parameters for squid axons. (Note: the value of V_L for the "standard axon" is chosen to make $J_\text{i} = 0$ at $V = 0$.)*

Parameter	Mean	Range	Standard	Units
C	0.91	(0.8–1.5)	1	$\mu\text{F}/\text{cm}^2$
G_Na	120	(65–260)	120	mmhos/cm^2
G_K	34	(26–49)	36	mmhos/cm^2
G_L	0.26	(0.13–0.5)	0.3	mmhos/cm^2
V_Na	109	(95–119)	+115	mV
V_K	−11	(9–14)	−12	mV
V_L	+11	(4–22)	+10.5995	mV

The parameters in Equations (4.15) have the following voltage dependence at a temperature of 6.3°C:

$$\alpha_n = \frac{0.01(10-V)}{\exp[(10-V)/10]-1},$$
$$\beta_n = 0.125\exp(-V/80),$$
$$\alpha_m = \frac{0.1(25-V)}{\exp[(25-V)/10]-1}, \qquad (4.16)$$
$$\beta_m = 4\exp(-V/18),$$
$$\alpha_h = 0.07\exp(-V/20),$$
$$\beta_h = \frac{1}{\exp[(30-V)/10]+1},$$

in units of milliseconds^{-1} with V in millivolts (mV).

At other temperatures, these rates are multiplied by the factor κ, where

$$\kappa = 3^{(T-6.3)/10}, \qquad (4.17)$$

and T is the Celsius temperature. To appreciate this equation, note that physiologists often express temperature dependence as a "Q_{10}" which indicates the factor by which a measurement increases with a temperature increase of ten degrees. In this language, Equation (4.17) states that "the Q_{10} of the H–H gating rates at 6.3°C is 3." Subsequent observations give values between 2 and 4 [27].

The membrane voltage V is measured with respect to the resting potential (about -65 mV for the squid axon), and an increase in the potential inside the axon is taken to be positive [70]. In Equation (4.14), the first (G_K) term accounts for potassium ion current, and the second (G_{Na}) accounts for sodium ion current, while the last term (G_L) accounts for all other ions under the heading of "leakage."

Hodgkin and Huxley came up with this formulation through the following process of reasoning. For clarity of thought and ease of computation, first of all, they wanted their switching variables (n, m, and h) to obey first order kinetics as in Equations (4.15). The first of these, for example, corresponds to the following kinetic diagram for hypothetical states of a protein that allows potassium ions to pass through the membrane.

$$\boxed{\text{OPEN }(n)} \;\underset{\alpha_n}{\overset{\beta_n}{\rightleftarrows}}\; \boxed{\text{CLOSED }(1-n)}$$

If the membrane's potassium permeability were proportional to n, potassium current would begin to rise linearly with a step (instantaneous) change in α_n and β_n resulting from a step change in membrane voltage (V). However, potassium current does not rise linearly with a step in V, but approximately as the fourth power of time, suggesting that n should appear to the fourth power in Equation (4.14). This is called a fourth-order process, which presumes that four independent substeps comprise the process of potassium turn on [27].

The h and m dependencies were treated similarly. The fall of sodium permeability (mediated by h) appears to be a first-order process, so this variable enters to the first power in Equation (4.14). Also, m was observed to rise approximately as the cube of time in response to a step change in V, leading to the cubic dependence in Equation (4.14).

Finally, the exponential dependencies on V in Equations (4.16) are consistent with voltage dependent energy barriers to ionic diffusion, as is the strong temperature dependence indicated in Equation (4.17).

However, one should not attach too much significance to such reasoning. Kenneth Cole, for example, has noted that the H–H data can be organized about higher powers of n and m in Equation (4.14) [12], and the functional forms used in Equations (4.16) to represent the voltage dependence of the rate parameters (αs and βs) are also not fixed by the data [27]. Indeed, several other formulations were suggested in the 1960s [23, 30, 75]. Thus, the H–H system can be viewed as a set of phenomenological equations that predict how membrane ion currents depend on voltage and time, without reference to underlying theory.

To understand how the H–H system switches, note that Equations (4.15) can also be written in the form:

$$\frac{dn}{dt} = -[n - n_0(V)]/\tau_n(V),$$
$$\frac{dm}{dt} = -[m - m_0(V)]/\tau_m(V), \quad (4.18)$$
$$\frac{dh}{dt} = -[h - h_0(V)]/\tau_h(V),$$

where, from Equations (4.16),

$$n_0(V) = \alpha_n/(\alpha_n + \beta_n) \quad \text{and} \quad \tau_n(V) = 1/(\alpha_n + \beta_n),$$
$$m_0(V) = \alpha_m/(\alpha_m + \beta_m) \quad \text{and} \quad \tau_m(V) = 1/(\alpha_m + \beta_m), \quad (4.19)$$
$$h_0(V) = \alpha_h/(\alpha_h + \beta_h) \quad \text{and} \quad \tau_h(V) = 1/(\alpha_h + \beta_h).$$

The voltage dependence of the stationary values of n, m and h and their corresponding switching times are plotted in Figure 4.5. From this figure, it is seen that the sodium turn-on time (τ_m) is about an order of magnitude smaller than potassium turn-on and sodium turn-off times (τ_n and τ_h). Thus the switching cycle of a standard space-clamped membrane can be described as follows:

1. At the resting potential ($V = 0$), the sodium conductance is almost zero. This is because, from Equations (4.19), $m_0(0) = 0.05293$ and $h_0(0) = 0.59612$ so sodium ion permeability

$$G_{\text{Na}} m_0^3(0) h_0(0) = 0.000\,089 G_{\text{Na}}.$$

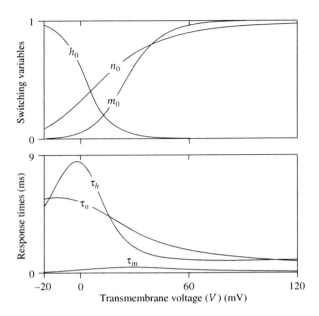

FIG. 4.5. *From Equations (4.19), the stationary values of the switching variables: $n_0(V)$, $m_0(V)$, and $h_0(V)$ (upper panel), and the corresponding switching times: τ_n, τ_m, and τ_h (lower panel).*

The potassium ion current is also small at rest because $n_0(0) = 0.31768$ so potassium ion permeability

$$G_K n_0^4(0) = 0.0102 G_K.$$

2. As the membrane voltage is increased from its resting value, sodium channels open ($m \to 1$) on a time scale of $\tau_m \sim 0.2\,\text{ms}$.
3. This influx of sodium ions brings the membrane voltage to a level approaching $+115$ mV with respect to its resting value. Thus $(V - V_{\text{Na}}) \sim 0$ so from Equation (4.14) the sodium ion current becomes small.
4. At this voltage, potassium ion permeability turns on ($n \to 1$) as sodium ion permeability turns off ($h \to 0$) on time scales of $\tau_n \sim 4\,\text{ms}$ and $\tau_h \sim 5\,\text{ms}$.
5. Because $(V - V_K) \sim +100$ mV, potassium ions flow out of the axon. This efflux of potassium ions carries the membrane voltage back to its resting value on a time scale of a few milliseconds.

In a space-clamped switching measurement through a membrane of area A, a source current (I_s) is injected into the cell by the experimenter which must equal the sum of the displacement current (through the membrane capacitance) and the ion current. Thus

$$CA\frac{dV}{dt} + J_i A = I_s, \qquad (4.20)$$

which is an ODE for the transmembrane voltage (V) that can readily be integrated numerically to compare predictions of the H–H formulation with empirical observations. Hodgkin and Huxley made such a comparison for several examples, obtaining convincing evidence that their equations do indeed represent the dynamics of a space clamped membrane [28]. Several convenient codes have recently become available that allow one to formulate and numerically solve versions of Equation (4.20) [19, 80].

Before leaving the subject of transmembrane ionic dynamics, three caveats should be mentioned. First, introductory discussions of nerve membrane dynamics sometimes leave the impression that sodium current ceases to flow at the peak of an action potential because the inward flow sodium ions alters the concentration ratio. In a squid giant axon, however, ionic ratios change by a negligible amount during the passage of a single nerve impulse. It is only after repeated stimulations of smaller fibers that ionic concentrations (and therefore the ionic batteries V_{Na} and V_K) are observed to change.[1]

Second, the H–H equations describe ionic currents that are averaged over many thousands of channels. It is now possible to observe currents flowing through individual protein channels using the "patch clamp" technique, in which a single channel is fixed on the end of a very fine glass microelectrode. Such measurements show that individual channels switch rapidly (in a small fraction of a millisecond) from being fully closed to being fully open, carrying currents on the order of a picoampere for times on the order of a millisecond. Thus, the H–H equations give probabilities of individual membrane channels being open as functions of time, and it is the temporal variation of these probabilities that are described by the dynamics of m, h, and n in Equations (4.18) and (4.19) [19, 27].

Finally, there are currently known to be several dozen types of ionic channels in neurons of the mammalian central nervous systems, exhibiting a variety of dynamic behaviors [27, 45]. Although two ionic species (sodium activating and potassium deactivating) may be satisfactory for qualitative studies, quantitative analyses often require formulations for the ionic current that include more or different terms than in Equation (4.14).

This last item is especially apt for mammalian cell bodies and dendrites, where calcium (Ca^{++}) ion depolarization replaces that of sodium ions in the

---

[1] Note, however, that Adelman and FitzHugh have augmented the Hodgkin–Huxley formulation to study build-up of external potassium ion concentration in a localized "periaxonal space" (30–40 nm across) surrounding a squid giant axon [1]. Widely observed in measurements on squid nerves, this increase of potassium concentration is caused by repeated stimulation of the fiber, which alters the equilibrium potential for potassium ions (V_K in Table 4.1) on time scales of the order of 100 ms.

emergence of action potentials. Because it is more difficult to make measurements on neocortical dendrites than on the giant axon of the squid, models based on calcium currents are less reliable than the H–H equations, but one example—the Morris–Lecar (M–L) model—is described below.

4.2.2 The H–H impulse

Let us return now to the reaction-diffusion system of Equation (4.13) and consider how to obtain a traveling-wave solution, corresponding to a nerve impulse or action potential. In this development, our previous study of the Z–F equation in Section (4.1.1) can provide a conceptual guide.

When the dependent variables are allowed to depend on x as well as t, the derivatives in Equations (4.15) or (4.18) become partial (rather than ordinary) derivatives with respect to time. Assuming the traveling-wave variable $\xi \equiv x - vt$, the H–H equations reduce to the following first-order ODE system

$$\frac{dV}{d\xi} \equiv W,$$
$$\frac{dW}{d\xi} = rj_i(V, m, h, n) - rcvw,$$
$$\frac{dn}{d\xi} = [n - n_0(V)]/v\tau_n(V), \qquad (4.21)$$
$$\frac{dm}{d\xi} = [m - m_0(V)]/v\tau_m(V),$$
$$\frac{dh}{d\xi} = [h - h_0(V)]/v\tau_h(V),$$

where the traveling-wave velocity v is an unknown parameter just as in the foregoing analysis of the Z–F equation. In the present case, the phase space of the above equations is five dimensional—one dimension for each dependent variable—and there is only one SP, where all variables are at their resting values. This SP is at

$$(V, W, n, m, h) = (0, 0, n_r, m_r, h_r),$$

with $n_r = n_0(0) = 0.31768$, $m_r = m_0(0) = 0.05293$, and $h_r = h_0(0) = 0.59612$.

A solution trajectory that corresponds to a physical impulse of the original PDE system must be homoclinic, as indicated in Figure 4.6, starting and ending at this SP as ξ advances from $-\infty$ to $+\infty$.

In the early 1950s, Hodgkin and Huxley did not have a digital computer, so they integrated Equations (4.21) on a mechanical hand calculator. This was a laborious task, the more so because to find a homoclinic trajectory it is necessary to perform the following iterative procedure: (i) assume a value for v, (ii) numerically extrapolate a trajectory emerging from the SP at $\xi = -\infty$, (iii) note how closely this trajectory returns to the SP as $\xi \to +\infty$, (iv) choose a new value of v,

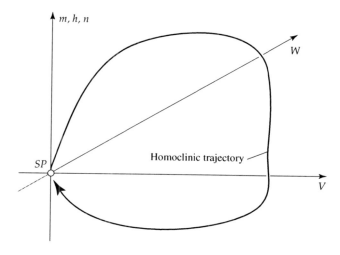

FIG. 4.6. *Schematic representation of a homoclinic trajectory in the (V, W, m, h, n) traveling-wave phase space defined in Equations (4.21). The assumed traveling-wave speed (v) must be adjusted so that a trajectory emerging from the SP at $\xi = -\infty$ approaches it again as $\xi \to +\infty$.*

Table 4.2. *Standard H–H parameters for the giant axon of the squid.*

Parameter	Value	Units
ρ	35.4	ohm-cm
a	238	μm
r	2.0×10^4	ohms/cm
c	1.5×10^{-7}	F/cm

and then (v) iterate steps (ii)–(iv) enough times that a sufficiently exact value for v and the corresponding homoclinic trajectory is obtained. Faced with this daunting task, H–H selected a particular axon (indicated as the "standard axon" in Table 4.1) for detailed parameter measurements, numerical integrations, and comparison of numerical results with observations of the corresponding nerve impulse.

The particular axon they chose had a radius of $a = 238$ microns (μm) and an axoplasmic resistivity of $\rho = 35.4$ ohm-cm, so the corresponding reaction-diffusion parameters (series resistance per unit length $r = \rho/\pi a^2$ and shunt capacitance per unit length $c = 2\pi a C$ in Equation (4.13) and Figure 4.4) are as given in Table 4.2. Using the above iterative procedure, H–H computed the large amplitude impulse shown in Figure 4.7 with a velocity of 18.8 m/s. The shape of this computed solution compares well with that shown in Figure 1.2, and the computed speed is close to the measured value of 21.2 m/s [28]. (Measurements on about two dozen squid axons made by the present author in 1980 indicate that at a radius of $a = 238$ μm and a temperature of 18.5° C the impulse speed is 20.3 m/s $\pm 5\%$ [70].) Thus Hodgkin and Huxley demonstrated that traveling-wave analysis of a carefully formulated reaction-diffusion equation can explain both the shape and the speed of a nerve impulse.

It turns out that there is a second homoclinic solution of Equations (4.21) which Huxley found (using a digital computer) in 1959 [31]. This solution corresponds to the smaller (dashed) curve in Figure 4.7 with a speed of 5.66 m/s. Although this lower amplitude solution is unstable, we will see that it is physically important because it determines threshold conditions for igniting the stable impulse.

The large amplitude impulse shown in Figure 4.7 is a stable solitary wave (moving with fixed shape and speed), but it is qualitatively different from the solitons of the previous chapter in several respects. First, the speed of a nerve impulse is fixed by system parameters, not by initial conditions as for a soliton. Second, a collision between two nerve impulses usually destroys both of them.

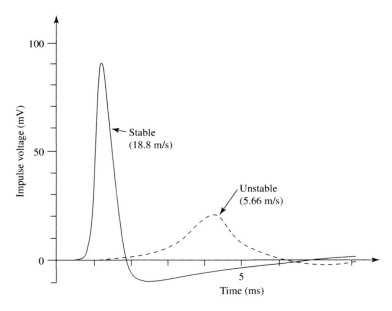

FIG. 4.7. *A full-sized action potential (at $v = 18.8$ m/s) and an unstable threshold impulse (at 5.66 m/s) for the H–H axon at 18.5°C. (Redrawn from [28] and [31].)*

Third, there are threshold conditions for launching a nerve impulse. Finally, the H–H system is not Hamiltonian; thus energy is not conserved, and time is unidirectional.

4.3 Simplified nerve models

Although the H–H equations provide a good representation of nerve impulse propagation on axons for which sodium and potassium ions govern the processes of activation and recovery, the large number of traveling-wave variables (five) make it difficult to obtain intuitive insight into the underlying dynamic processes. In this section, we describe three simpler nerve models, which have appeared in the neuroscience literature.

4.3.1 The Markin–Chizmadzhev (M–C) model

One of the simplest representations of a propagating nerve impulse was introduced by Kompaneyets and Gurovich in the mid-1960s [40] and developed in detail by Markin and Chizmadzhev in 1967 [48]. The M–C model assumes a reaction-diffusion equation of the form

$$\frac{\partial^2 V}{\partial x^2} - rc\frac{\partial V}{\partial t} = r j_{\mathrm{mc}}(x,t). \tag{4.22}$$

In this model, the ionic membrane current is not represented as a voltage-dependent variable, as in Equation (4.13), but by a prescribed function of space and time $j_{\mathrm{mc}}(x,t)$. (A yet simpler version of the M–C concept is the "integrate and fire" model of a neuron, in which the entire cell is approximated as a single switch in parallel with a capacitor [8, 37, 38]. The capacitor integrates incoming charge until a threshold voltage is reached, whereupon the switch closes briefly, discharging the capacitor and restarting the process.)

The prescribed membrane current is defined for two distinct cases. First, if the membrane voltage does not reach a threshold level (V_θ), then $j_{\mathrm{mc}}(x,t) = 0$, and there is no impulse. Second, if a point on the fiber does reach threshold, then the following current flows occur. (i) An inward current $j_{\mathrm{mc}}(x,t) = -j_1$ for a time τ_1, which represents the sodium current in the H–H model. (ii) An outward current $j_{\mathrm{mc}}(x,t) = +j_2$ for a time τ_2, representing potassium current. (iii) For $t > \tau_1 + \tau_2$, the membrane current is zero. In addition, it is assumed that $j_1\tau_1 = j_2\tau_2$ so that the total ionic charge crossing the membrane is zero.

Assuming an impulse propagating with velocity v and a traveling-wave variable $\xi = x - vt$, the fiber can be divided into four regions, as indicated in Figure 4.8. First, a precursor region in which $j_{\mathrm{mc}}(x,t) = 0$. It is convenient to define the precursor region to be where $\xi > 0$. Second, a sodium current region where $j_{\mathrm{mc}}(x,t) = -j_1$ and $-v\tau_1 < \xi < 0$. Third, a potassium current region where $j_{\mathrm{mc}}(x,t) = +j_2$ and $-v(\tau_1 + \tau_2) < \xi < -v\tau_1$. Fourth, a relaxation region where $j_{\mathrm{mc}}(x,t) = 0$ and $\xi < -v(\tau_1 + \tau_2)$.

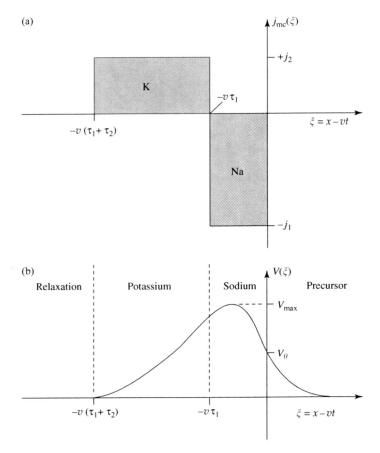

FIG. 4.8. (a) Ionic currents in the M–C model as a function of the traveling-wave variable (ξ). (b) Structure of the associated nerve impulse.

Within each of these four regions, Equation (4.22) is linear, and the voltage can be expressed as a sum of exponential terms. Doing so and requiring continuity of V and $dV/d\xi$ at the boundaries between regions implies that [70]

$$V(\xi) = C(v) \exp(-vrc\,\xi)) \tag{4.23}$$

in the precursor region ($\xi > 0$), where

$$C(v) = \frac{1}{v^2 rc^2} \left[j_1 + j_2 e^{-v^2 rc(\tau_1 + \tau_2)} - (j_1 + j_2) e^{-v^2 rc\tau_1} \right]. \tag{4.24}$$

Using values for j_1, j_1, τ_1, and τ_2 that correspond to the standard H–H squid axon, $C(v)$ has a maximum of 28.3 mV at $v \approx 10$ m/s, as indicated on

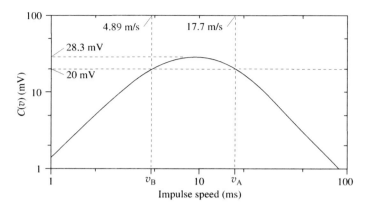

FIG. 4.9. *The function $C(v)$ defined Equation (4.24) is plotted for parameter values corresponding to the standard H–H squid axon [70].*

Figure 4.9. If the threshold voltage is larger than this maximum value, there is no transmembrane current therefore no impulse. If, on the other hand, $V_\theta <$ 28.3 mV, there are two values of velocity for which $C(v) = V_\theta$. Assuming a threshold voltage $V_\theta = 20$ mV, implies that these two impulse speeds are 17.7 m/s and 4.89 m/s, in fair agreement with the H–H values of 18.8 m/s and 5.66 m/s in Figure 4.7. (See [70] for further details.)

That the exponential precursor is a real neural phenomenon can be seen from Figure 1.2, where membrane voltage begins to rise before the membrane conductance has changed. From the M–C perspective, the precursor guides the impulse, setting its velocity, while the body of the impulse (the ionic currents) in turn generate the precursor.

4.3.2 The FitzHugh–Nagumo (F–N) model

In the previous section, the recovery effect in a nerve fiber was represented by assuming that transmembrane ionic current follows a prescribed function of time after the transmembrane voltage (V) reaches a threshold value. Although this M–C model offers physical insight, it has a certain ad hoc quality, lacking the dynamic character of the H–H system.

Here we consider a simple PDE system that was proposed by Jin-ichi Nagumo and his colleagues at the University of Tokyo in 1962 to represent a nerve impulse with recovery [54]. Because Nagumo had returned from an extended visit at the National Institutes of Health where he had interacted with Richard FitzHugh and studied his recently developed description of membrane dynamics [22], this system is usually called the FitzHugh–Nagumo (F–N) model.

In constructing this model, recall first from Figure 4.5 that the turn-on time for sodium (τ_m) is an order of magnitude shorter than the sodium turn-off time (τ_h) and the potassium turn-on time (τ_n). To take advantage of this difference in time scales, assume that $\tau_m = 0$, while $\tau_h = \infty$ and $\tau_n = \infty$ on the leading edge of the nerve impulse. In other words, along the initial rise of the transmembrane voltage assume that the sodium turn-on variable m follows instantly the stationary value $m_0(V)$, whereas h and n are frozen at their resting values of $h_0(0)$ and $n_0(0)$. Then the transmembrane ionic current in Equation (4.14) reduces to

$$G_K n_0^4(0)(V - V_K) + G_{Na} m_0^3(V) h_0(0)(V - V_{Na}) + G_L(V - V_L),$$

which is a function only of V with a cubic shape as in Figure 4.2(a). To this approximation, the H–H system reduces to a Z–F equation for the transmembrane voltage.

Next, normalize this Z–F equation by choosing units for x and t such that the diffusion constant is unity and add a single recovery variable (R) to obtain

$$\frac{\partial^2 V}{\partial x^2} - \frac{\partial V}{\partial t} = F(V) + R. \tag{4.25}$$

Finally, let the recovery variable be governed by the first-order rate equation

$$\frac{\partial R}{\partial t} = \varepsilon(V + c - bR). \tag{4.26}$$

Equations (4.25) and (4.26) comprise the F–N model, which represents a nerve fiber with the least number of dynamic variables. (A similar system was proposed independently as an electronic nerve or "neuristor" based on the Esaki diode by the present author [67].)

In the F–N system, ε, b, and c are constant parameters, $F(V)$ is a function with a cubic shape, as in Figure 4.2(a), providing a threshold for onset of sodium ion current, and R is a slowly changing recovery variable that represents potassium ion current. With reference to the H–H system, there is a rough correspondence between R and n because both variables allow the nerve to relax back to its resting state. Furthermore, there is an approximate correspondence between ε and κ/τ_n, where κ is the temperature factor defined in Equation (4.17), because both are rate constants for the recovery variable. Thus ε is sometimes called a "temperature parameter."

With $\varepsilon = 0$, R is constant, and Equation (4.25) reduces to the Z–F equation with the added feature that the level of the recovery variable in a region into which a pulse advances can be arbitrarily adjusted by changing R. Under a traveling-wave analysis, the F–N PDE system becomes the ODE system

$$\frac{dV}{d\xi} \equiv W,$$

$$\frac{dW}{d\xi} = F(V) + R - vW, \tag{4.27}$$

$$\frac{dR}{d\xi} = \frac{\varepsilon}{v}(bR - V - c),$$

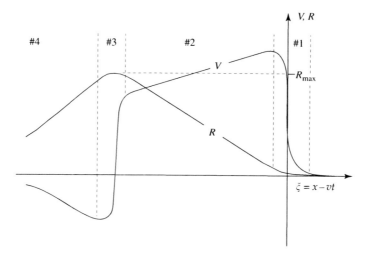

FIG. 4.10. *In the F–N system with $0 < \varepsilon \ll 1$, V and R are shown as functions of the traveling-wave variable ξ. (Region numbers correspond to trajectory branches in Figure 4.11.)*

defining solution trajectories in the three-dimensional phase space (V, W, R). With $F(V)$ as in Figure 4.2(a), there is a singular point at

$$(V, W, R) = (0, 0, 0),$$

and homoclinic trajectories from this point correspond to traveling-wave solutions.

A qualitative understanding of F–N impulse dynamics is obtained by assuming $0 < \varepsilon \ll 1$ with $b = 0$, $c = 0$, $F(V) = V(V - a)(V - 1)$, and $a < 1/2$. Then an impulse solution of F–N appears as in Figure 4.10, which can be understood by comparing with the corresponding homoclinic trajectory in Figure 4.11.

The leading edge of this impulse appears in region #1 of Figure 4.10, which corresponds to branch #1 of Figure 4.11. This is the rapid (order 1) upward transition (in time) of the Z–F equation that is shown in Figure 4.2(c) and described as Case 1 in Section 4.1.1. From Equation (4.6), the leading edge moves at $v \approx (1 - 2a)/\sqrt{2}$, which approaches an exact value as $\varepsilon \to 0$. Because $a < 1/2$, the positive area (A^+) in Figure 4.2(a) is less than the negative area (A^-), a necessary condition for Case 1.

The leading edge is followed by a rapidly varying (order 1) trailing edge, indicated by region #3 and branch #3 in Figures 4.10 and 4.11, respectively and corresponding to Case 3 in Section 4.1.1. Because the trailing edge must move

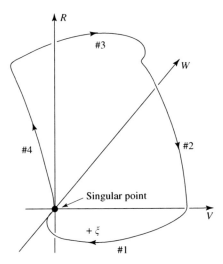

FIG. 4.11. *The homoclinic solution trajectory of Equations (4.27) corresponding to the traveling-wave impulse in Figure 4.10. (Note that branches #1 and #3 are parallel to the (V, W) plane, whereas branches #2 and #4 lie in the (V, R) plane.)*

at the same speed as the leading edge for the impulse to be a traveling-wave, the trailing edge occurs at a value of R (indicated as $R_{\max}$ in Figure 4.10) at which the positive area a (A^+) under $F(V) + R_{\max}$ is sufficiently large compared with the negative area (A^-). In other words, the value of $R_{\max}$ is determined by this area condition.

Between the two edges, in region #2, R and V vary slowly (order ε) with ξ, subject to the constraints

$$\frac{dR}{d\xi} = -\frac{\varepsilon V}{v}$$
$$F(V) + R \doteq 0$$
$$W \doteq 0.$$

Finally, in region #4, R and V relax slowly (order ε) back to zero.

To see what happens as ε increases, FitzHugh did a numerical study of Equations (4.27) from which Figure 4.12 is presented here [24]. Just as for the H–H system, he found two traveling-wave solutions for each value of ε, and these two solutions merge together at a critical value (ε_c), beyond which no traveling-waves exist. In this figure, the upper branch corresponds to the stable higher amplitude H–H impulse of Figure 4.7 and the lower branch to the unstable lower amplitude impulse. Thus the F–N model suggests that traveling-wave solutions

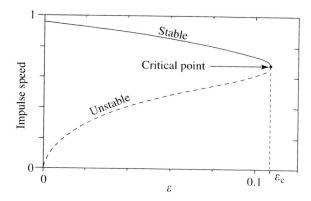

FIG. 4.12. *Propagation speeds for impulse solutions of a FitzHugh–Nagumo (F–N) system plotted against the temperature parameter ε. (Redrawn from [24] with $F(V) = V^3/3 - V$, $b = 0.8$, and $c = 0.7$.)*

are no longer possible at a sufficiently high temperature, in qualitative agreement with experimental observations on real nerves.

Connection with the M–C model of the previous section can be established by considering an increase of ε in F–N to be qualitatively equivalent to an increase of threshold voltage (V_θ) in M–C. In both cases, the greater and lesser wave speeds draw closer together, eventually merging at a critical point beyond which traveling-wave solutions do not exist. These are but two examples of several such parameters (temperature, fatigue, narcotic concentrations, etc.) that tend to inhibit nerve impulse propagation [70].

Yet another example of a merging together of fast and slow solutions is provided by closed trajectories (or cycles) in the phase space of Equations (4.27), which correspond to periodic traveling-wave solutions of the original PDE system. Periodic solutions of F–N have been studied numerically by Rinzel and Keller for the piecewise-linear function

$$F(V) = \begin{cases} V & \text{for } V < a, \\ (V-1) & \text{for } V > a, \end{cases} \quad (4.28)$$

and a typical result is sketched in Figure 4.13 [65]. These calculations show a critical value of wavelength (λ_c) below which periodic traveling-waves are not found. As $\lambda \to \infty$, the two traveling-wave speeds approach those of isolated impulses.

From a mathematical perspective, Figures 4.12 and 4.13 provide examples of bifurcations arising in parameter space [56]. Thus ε and λ are control parameters with critical values at which bounded trajectories in the traveling-wave phase space appear. In the parameter spaces, (v, ε) or (v, λ), these trajectories are loci at which it is possible to find solutions satisfying Equation (4.1).

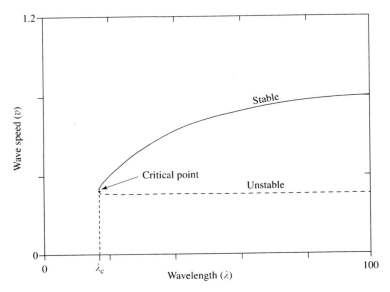

FIG. 4.13. *Propagation speeds for periodic traveling-wave solutions of a F–N system plotted against the wavelength λ. (Redrawn from [65] with $b = 0$, $c = 0$, $\varepsilon = 0.05$, and $F(V)$ the piecewise-linear function of Equation (4.28) with $a = 0.3$.)*

4.3.3 Morris–Lecar (M–L) models

Membrane switching based on activation via calcium ions (rather than sodium ions) play a role in many active fibers, including those of the mammalian central nervous system [45]. Although detailed measurements of membrane properties are difficult to perform on such small nerves, a study of calcium mediated membrane switching on the giant muscle fiber of the barnacle was reported by Catherine Morris and Harold Lecar in 1981, leading to a simple model for the transmembrane ionic current [52]. In this work, it was found that the initial inward flow of calcium ions did not experience deactivation; thus their model for membrane ionic current does not include the "turn-off" (h) component of the H–H formulation, and Equation (4.14) is replaced with the expression

$$J_{\mathrm{ml}} = G_{\mathrm{K}} n \left(V - V_{\mathrm{K}}\right) + G_{\mathrm{Ca}} m \left(V - V_{\mathrm{Ca}}\right) + G_{\mathrm{L}}(V - V_{\mathrm{L}}), \qquad (4.29)$$

where

$$\frac{dm}{dt} = -[m - m_0(V)]/\tau_m(V),$$
$$\frac{dn}{dt} = -[n - n_0(V)]/\tau_n(V).$$

In these equations

$$m_0(V) = [1 + \tanh((V - V_1)/V_2)]/2,$$
$$n_0(V) = [1 + \tanh((V - V_3)/V_4)]/2, \quad (4.30)$$

and

$$\tau_m(V) = \tau_{m0}\text{sech}[(V - V_1)/2V_2],$$
$$\tau_n(V) = \tau_{n0}\text{sech}[(V - V_3)/2V_4]. \quad (4.31)$$

To appreciate Equations (4.30) and (4.31), refer to Figure 4.5 where the behaviors of $m_0(V)$, $n_0(V)$, $\tau_m(V)$, and $\tau_n(V)$ for the H–H system are displayed. Under M–L, evidently, m and n have the same off-to-on switching behavior with increasing transmembrane voltage (V), and their switching times rise to maxima where $m_0(V) = \frac{1}{2}$ and $n_0(V) = \frac{1}{2}$ as do the corresponding H–H variables. Several sets of parameter values that give space-clamped oscillatory behaviors in accord with those observed on various biological preparations are presented in Table 4.3.

Just as with H–H, the turn-on time for the activating variable (τ_m) is much shorter than that for the recovery variable (τ_n). For times long compared with τ_m but short compared with τ_n, $m(t) \approx m_0(V)$ and $n(t) \approx n(V_R)$. In this limit, the right-hand side of Equation (4.29) becomes

$$J_1 \approx G_K n(V_R)(V - V_K) + G_{Ca} m_0(V)(V - V_{Ca}) + G_L(V - V_L),$$

Table 4.3. *Suggested parameter values for the M–L model of membrane current in barnacle muscle fiber at room temperature (ca. 22° C) [19, 52]. (The voltages given in this table are total values, rather than with respect to the resting potential (V_R), as in Section 4.2.)*

Item	(1) [52]	(2) [52]	(1) [19]	(2) [19]	Units
C	20	20	20	20	μF/cm^2
G_{Ca}	6	4	4.4	4	mmhos/cm^2
G_K	12	8	8	8	mmhos/cm^2
G_L	2	2	2	2	mmhos/cm^2
V_{Ca}	100	100	120	120	mV
V_K	-70	-70	-84	-84	mV
V_L	-50	-50	-60	-60	mV
V_1	0	10	-1.2	-1.2	mV
V_2	15	15	18	18	mV
V_3	10	-1	2	12	mV
V_4	10	14.5	30	17.4	mV
τ_{m0}	1	0	0	0	ms
τ_{n0}	10	15	25	15.15	ms
$V_R \approx$	-50	-50	-60	-60	mV

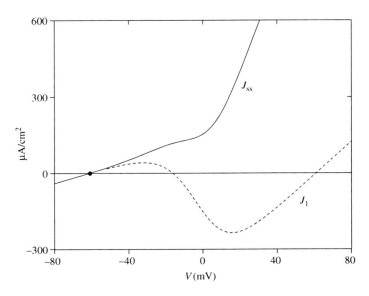

FIG. 4.14. *Plots of the initial ionic current (J_1) and the steady state current (J_{ss}) under the M–L model. (Parameters are from the third column of Table 4.3.)*

which is the cubic shaped function plotted as a dashed line in Figure 4.14. This is the proper approximation for describing the initial inflow of calcium ions on the leading edge of a propagating impulse. Thus, the reaction-diffusion equation

$$\frac{\partial^2 V}{\partial x^2} - rc\frac{\partial V}{\partial t} = rj_{ml}, \qquad (4.32)$$

(where $j_{ml} = 2\pi a J_1$ and a is the fiber radius) reduces to the Z–F equation of Section 4.1.1.

At times long compared with τ_n, $m(t)$ remains equal to $m_0(V)$ and $n(t) \to n_0(V)$, so

$$J_{ml} \to J_{ss} = G_K n_0(V)(V - V_K) + G_{Ca} m_0(V)(V - V_{Ca}) + G_L(V - V_L).$$

This is the steady state membrane current (plotted as the solid line in Figure 4.14), which forces the system to return to its resting state at

$$[V, m, n] = [V_R, m_0(V_R), n_0(V_R)].$$

For sufficiently large values of τ_{n0}, therefore, the Equation (4.32) will support a nerve impulse with recovery.

Presently, the M–L model of nerve membrane dynamics is becoming of increasing interest to computational neuroscientists for several reasons. First,

M–L is simpler than H–H. Second, dynamics generated by calcium ions are important in many neurons, especially dendrites [70]. Finally, M–L is based on biological measurements. In using this model, however, the following caveats should be kept in mind.

1. The membrane capacitance (C) of 20 μF/cm^2 seems unrealistically high compared with the H–H value of about 1 μF/cm^2 [70]. Is the membrane capacitance of a barnacle muscle really this large? Or are neglected features of the dynamics missed by the M–L formulation and compensated for by increasing the value of C? Noting that $c = 2\pi a C$ in Equation (4.32), one expects a significant lowering of impulse propagation speed from such a large value of C. Future experimental research should strive to resolve this discrepancy.
2. As with H–H, the original M–L formulation proposed a turn-on time for potassium ions (τ_{n0}) that was an order of magnitude greater than that for the calcium ions (τ_{m0}) [52]. Subsequent analyses often choose (τ_{m0}) to be zero, thereby decreasing the order of the system to that of the F–N model [19]. Although convenient, this assumption may be no more justified under M–L than H–H, where it approximately doubles estimates of impulse speed [60, 61, 70].
3. Under M–L, dependencies upon m and n in Equation (4.29) are as the first power rather than the third and fourth powers respectively of H–H in Equation (4.14). To explain this assumption, Morris and Lecar state [52]: "In general we have used linear relations for the instantaneous current-voltage curves through open channels. Although our own experiments on instantaneous Ca^{++} current in barnacle muscle (unpublished data) show departures from linearity, under situations of high permeable-ion gradient, in all but one instance we stick for the sake of simplicity to the linear driving force approximation." While this perspective is in accord with the previously noted observation by Cole that the powers of m and n selected by Hodgkin and Huxley are not particularly critical [12], the assumption of linearity may introduce noticeable errors under some circumstances.
4. Assuming first order kinetics, the forms of Equations (4.30) and (4.31) can be understood through the following simple argument. As in Equations (4.15), dynamics of (say) m are given by

$$\frac{dm}{dt} = \alpha_m(1-m) - \beta_m m = (\alpha_m + \beta_m)\left(\frac{\alpha_m}{\alpha_m + \beta_m} - m\right).$$

If one takes $\alpha_m \propto \exp[(V-V_1)/V_2]$ and $\beta_m \propto \exp[-(V-V_1)/V_2]$, then

$$m_0(v) = \frac{\alpha_m}{\alpha_m + \beta_m} = \frac{1}{2}\left[1 + \tanh\left(\frac{V-V_1}{V_2}\right)\right]$$

is as in Equation (4.30). However

$$\tau_m \propto \operatorname{sech}[(V-V_1)/V_2],$$

which misses the factor of 2 in the denominator of the argument in Equations (4.31). Although this discrepancy is carried along in current M–L formulations [19], it probably should be eliminated in future calculations by changing Equations (4.31) to

$$\tau_m(V) = \tau_{m0}\operatorname{sech}[(V - V_1)/V_2] \quad \text{and} \quad \tau_n(V) = \tau_{n0}\operatorname{sech}[(V - V_3)/V_4],$$

and revising values for the voltage parameters: V_1, V_2, V_3, and V_4.

4.4 Stability analyses

In the previous sections of this chapter, we have considered a variety of models for nerve impulse propagation, finding solutions in the traveling-wave approximation. Although such studies are important for developing the theory of a nerve impulse, they tell nothing about the stability of an impulse. If a traveling-wave solution is slightly perturbed, does it recover to its original shape and speed (like a candle flame briefly disturbed by a passing breeze)? Or does the disturbance grow until the original impulse is lost?

There are three ways to consider questions of stability. First come qualitative arguments, which provide intuitive insight while lacking numerical precision and mathematical rigor. Second, there are numerical studies, providing reliable results for particular sets of parameter values. Finally, there have been some mathematical analyses that provide theorems for stability and instability in the context of well defined models. In this section, are presented currently available stability results for the models that have previously been introduced.

4.4.1 The Z–F equation

A linear stability analysis of Equation (4.2) was published by Zeldovich and Barenblatt in 1959 in the context of flame-front propagation [90], the details of which are presented in Appendix D.

Briefly, this analysis proceeds as follows. A coordinate (independent variable) transformation is first introduced into a frame of reference that moves with the same speed as the traveling-wave. This transformation yields a nonlinear PDE describing the dynamics of deviations from the traveling-wave solution. Assuming that deformations of the traveling wave are small, the nonlinear PDE is then approximated by a linear PDE, which is obtained by linearizing the nonlinear PDE about the traveling-wave solution. Finally, the linear PDE is studied to learn whether small deviations wax or wane with time, thereby determining whether the traveling-wave is unstable or stable. A traveling-wave solution is said to be asymptotically stable if all small changes relax back to zero with increasing time, unstable if any small deviation grows without bound in the linear approximation, and stable otherwise.

From the discussion in Appendix D, we have the following result of Zeldovich and Barenblatt [90].

- *Theorem:* Consider a traveling-wave solution $u(\xi)$ of Equation (4.2), where $f(u)$ has the cubic character indicated in Figure 4.2(a). If $du(\xi)/d\xi$ has no zero crossings, then $u(\xi)$ is stable but not asymptotically stable. If $du(\xi)/d\xi$ has one or more zero crossings, then $u(\xi)$ is unstable.

If $u(\xi)$ is a monotone transition waveform as indicated in Figure 4.2(c), then $du(\xi)/d\xi$ has no zero crossings. Thus, all monotone increasing or decreasing transition waveforms (Cases 1–4 in Section 4.1.1) are stable but not asymptotically stable.

There is a physical reason behind the fact that transition waveforms are not asymptotically stable. Upon experiencing a small change, such a wave may relax to $u(\xi + \xi_0)$, where ξ_0 is a real constant. Thus, as time increases,

$$u(\xi) + \text{deformation} \to u(\xi + \xi_0), \qquad (4.33)$$

which is also a traveling-wave solution of the Z–F equation.

In other words, altering a stable traveling-wave solution may cause a displacement of that traveling wave in the ξ-direction with respect to where it would have been without the influence of the alteration. Because a displaced traveling-wave is still a solution of Equation (4.2), residual effects of the disturbance persist in time without growing or decaying.

Consider, on the other hand, the bell-shaped solution shown in Figure 4.3(c), where $u(\xi)$ rises to a maximum value and then returns to zero. In this case, $du(\xi)/d\xi$ has a zero crossing (at the maximum value of the solution), so $u(\xi)$ is unstable. Small changes in $u(\xi)$ therefore grow with time, eventually distorting it beyond recognition.

To get an intuitive feeling for this instability, refer to Figure 4.3(a), which determines the amplitude of the impulse-shaped solution (u_m) by the requirement that the total integral of $f(u)$ over the waveform is zero, with the positive area just canceled by the (shaded) negative area. If u_m is slightly decreased, the losses from the positive area exceed energy input from the negative area, and the solution relaxes to zero. If u_m is slightly increased, on the other hand, the increased energy input from the negative-going area overcomes the losses, and the wave grows.

4.4.2 The M–C model

Under the M–C model of Section 4.3.1, the impulse is guided by an exponential precursor, which is seen on Figure 1.2. The form of this precursor is given in Equation (4.23), and its amplitude is plotted as a function of the impulse velocity for parameters corresponding to a standard H–H squid axon in Figure 4.9 [70]. As shown on this figure, traveling-wave speeds are found at the two values of impulse speed (v_A and v_B) for which

$$C(v) = V_\theta.$$

The stability properties of impulses at these two wave speeds can be understood through the following qualitative argument which is based on the assumption that the impulse speed will increase for $C(v) > V_\theta$ and decrease for $C(v) < V_\theta$.

Suppose first that the impulse is traveling at the lesser of these two speeds, $v_{\rm B}$, where $C(v)$ increases with increasing v. Now let the impulse speed be increased slightly so that $v > v_{\rm B}$ and therefore that $C(v) > V_\theta$. Because the amplitude of the precursor is greater than threshold, the impulse will increase its speed, making $C(v)$ even greater. If v is made slightly smaller than $v_{\rm B}$, on the other hand, $C(v) < V_\theta$, which will further decrease v. Thus $v_{\rm B}$ is an unstable value of impulse speed.

Suppose next that the impulse is traveling at the greater of these two speeds, $v_{\rm A}$, where $C(v)$ decreases with increasing v. Now let the impulse speed be increased slightly so that $v > v_{\rm A}$ and therefore that $C(v) < V_\theta$. Because the amplitude of the precursor is less than threshold, the impulse will decrease its speed, bringing v back to $v_{\rm A}$. If v is made smaller than $v_{\rm A}$, on the other hand, $C(v) > V_\theta$, causing v to increase back toward $v_{\rm A}$. Thus $v_{\rm A}$ is a stable value of impulse speed. This impulse is not asymptotically stable because any change in wave speed will result in a net displacement of the impulse as indicated in Equation (4.33).

4.4.3 The F–N model

In studies of F–N impulses, all three approaches to stability analysis have been considered, yielding complementary results.

Qualitative studies

The above mentioned numerical observations are in accord with the following qualitative arguments. First, as we have seen, mathematical studies of leading-edge dynamics show the upper curve to be stable and the lower curve unstable in the limit $\varepsilon \to 0$ of Figure 4.12.

Second, for the F–N impulse with $0 < \varepsilon \ll 1$ (as shown in Figure 4.10), we can ask what happens if an impulse gets a little too long or too short. Along the upper branch of Figure 4.12, the answer to this question is that an excessively long impulse will become shorter, and an impulse that is too short will become longer, with both converging on the traveling-wave shown in Figure 4.10. To see this, consider the following argument.

From the discussion in Section 4.3.2, the correct value of $R_{\max}$ is determined by the condition that the speed of the trailing edge (region #3) must equal that of the leading edge (region #1). Moving in the $-\xi$-direction (the $+t$-direction), the leading edge is an upward transition between the two outer zeros of $F(V)$, whereas the trailing edge is a downward transition between the two outer zeros of

$$\tilde{F}(V) = F(V) + R_{\max}.$$

Now, consider the following two cases. First, $R_{\max}$ is larger than the correct value, which indicates that the impulse is too long. Then the positive-going area (A^+) under $\tilde{F}(V)$ is increased, making it more dissipative. Because the trailing edge is a downward transition, requiring the dissipation of energy, this change in the shape of $\tilde{F}(V)$ accelerates the trailing edge, which shortens the impulse and returns $R_{\max}$ to its correct value. Second, $R_{\max}$ is smaller than the correct value, which indicates that the impulse is too short. Then the positive-going area under $\tilde{F}(V)$ is decreased, which slows down the trailing edge, thereby lengthening the impulse and again returning $R_{\max}$ to its correct value.

Thus the nonlinearities in the system allow an impulse to automatically adjust itself for the correct length, as shown in Figure 4.10. These qualitative considerations suggest that the upper curve of Figure 4.12 is stable for some range $0 < \varepsilon \ll \varepsilon_c$.

Numerical results

One way to assess the stability of a traveling-wave solution is to use it as the initial condition for numerical computations that are based on the original PDE system given in Equation (4.27). Assuming that the numerical procedure is reliable, divergence of the numerical solution from traveling-wave initial conditions implies instability; otherwise, the impulse is stable. Such computations have now been carried out many times for the F–N system for a wide range of parameters, and all results have been found to be consistent with the following statements: traveling-wave impulses corresponding to the upper curve in Figure 4.12 are stable; and impulses corresponding to the lower curve are unstable [70].

Because there may be some unexplored combinations of parameters giving contradictory results, such numerical studies do not provide mathematical proofs but they offer circumstantial evidence.

Mathematical analysis

Because the initial growth of an instability is governed by a PDE obtained by linearizing the nonlinear PDE about the nerve impulse, a general stability study can be based on a mode analysis of that linearized PDE. This is the approach presented in Appendix D. The key idea is to look at individual modes (eigenfunctions) of the linearized PDE and ask whether they grow or decay with time. A sufficient condition for instability is that at least one mode of the linearized PDE grows with time. A necessary condition for asymptotic stability, on the other hand, is that all modes decay with time.

The first step is to express the original nonlinear PDE in a moving coordinate system, for which the independent variables become $\xi = x - v_0 t$ and $\tau = t$, where v_0 is the speed of the undisturbed traveling-wave. Thus, ξ and τ are, respectively, the distance and time in the moving system, where τ is measured by a clock in the stationary system. It is in the moving coordinate system that the original nonlinear PDE is linearized about the traveling-wave.

Because the linearized PDE does not depend explicitly on time in the moving system, the method of separation of variables implies that the temporal behavior of a mode will be as $\exp(\lambda\tau)$ where λ is an eigenvalue of the corresponding eigenfunction (or mode). Eigenvalues may be of two types: discrete eigenvalues, occurring at isolated points in the complex λ-plane, corresponding to localized dynamics (internal oscillations, for example) of an impulse, and continuous eigenvalues, which are dense on lines in the λ-plane and correspond to modes of radiation from the traveling-wave. If the real part of λ is positive for any mode, that mode will grow with time, implying instability. If, on the other hand, the real parts of the eigenvalues of all modes are less than zero, then the corresponding traveling-wave will be asymptotically stable.

There is always one mode for which $\lambda = 0$ is a discrete eigenvalue: the derivative of the traveling wave with respect to the traveling-wave variable ξ. Adding this mode to a traveling-wave merely translates that traveling-wave in the ξ direction; thus, it is called a translation mode. Because this translation mode neither grows nor decays with time, a nerve impulse is at most stable but not asymptotically stable.

The mathematical basis for stability analysis of nerve axon models was established during the 1970s in an important series of papers by Evans. In the first of these, he formulated a general set of nonlinear PDEs, for which the F–N system is a special case, showing that the pulse is stable relative to the nonlinear PDE if and only if it is stable in the linearized PDE [15]. The second paper studied stability of the resting state, from which it is seen that the resting state of the F–N system is stable [16]. In the third paper of the series, Evans showed that disturbed impulses relax exponentially back into (possibly displaced) traveling waves if and only if [17]:

- There are no eigenvalues for which $\text{Re}[\lambda] > 0$.
- The translation mode is the only mode for which $\lambda = 0$.

Finally, in the fourth paper, Evans showed how to construct an analytic function that allows the preceding stability conditions to be deduced [18]. Thus, zeros of this "Evans function" are eigenvalues, and multiple zeros indicate that the corresponding eigenvalues are not simple.

The work of Evans gave criteria for the stability of pulses but left open whether any traveling-wave representing a nerve impulse in the F–N system is actually stable. The first full result of impulse stability is due to Jones [33, 34] for the fast F–N pulse. He verified stability in the limit $0 < \varepsilon \ll 1$, for which the leading edge is separated from the trailing edge by a long intermediate phase. As we just have seen, the front and back are individually stable for their relevant reduced systems, with the recovery variable fixed. The proof of stability then involves showing that the front and back lock to each other under the dynamics of the PDE rather than drift apart. Discerning which of these effects actually occurs, amounts to showing that the eigenvalue associated with relative motion of the front and back lies in the left half-plane, where $\text{Re}[\lambda] < 0$. Jones used the Evans function to show that the corresponding zero lies in the left half of the complex λ-plane. Yanagida [87] took a similar approach to give a somewhat more qualitative proof of fast pulse stability.

There is no known stability proof away from the singular limit because the decomposition of the pulse into front and back is then lost. A continuation argument can be used as ε is increased to show that no bifurcations of other pulses take place up to the parameter values at which the fast and slow pulses coalesce. Internal oscillations, however, may destabilize the impulse, and it is an analytic challenge to find techniques that rule this out, as is suggested by numerical evidence.

4.4.4 The H–H and M–L systems

As described in Section 4.2, the H–H system is distinguished by being based on experimental measurements that were made on real active membranes. With reference to Figure 4.7, much numerical and empirical evidence has accumulated over the past half century indicating that the impulse of greater speed and amplitude is stable, whereas the slower and smaller traveling-wave solution is unstable.

In the limit
$$\tau_m \to 0, \tau_n \to \infty, \quad \text{and} \quad \tau_h \to \infty,$$
the H–H system reduces to the Z–F equation [70]. Excepting the translation mode (at $\lambda = 0$), all discrete eigenvalues of the Z–F stability equation that corresponds to the leading edge of the large amplitude impulse in this limit—Equation (D.4) in Appendix D—have negative real parts of finite magnitude. From a continuity argument, therefore, one expects some range of values $0 < \tau_m \ll 1$, $0 < 1/\tau_n \ll 1$, and $0 < 1/\tau_m \ll 1$ over which the large amplitude wave is stable. The formulation of Evans, which was discussed in the previous section for F–N [15–18], applies also to H–H, but an Evans function has not yet been constructed for the H–H system.

The M–L system is also based on empirical evidence, and it is a reduced version of the H–H system; thus the above statements for H–H are expected to hold also for M–L. From a mathematical perspective, it is an interesting challenge to construct an Evans function for M–L, corresponding to the work of Jones and Yanagida for F–N [33, 34, 87].

4.5 Decremental conduction

The growth of scientific knowledge is not as steady as some would have us believe, and our collective picture of natural phenomena has occasionally been trapped in cul-de-sacs of misunderstanding. There is perhaps no better example of such a misconception than the over emphasis that many neurophysiologists have placed on the concept of "all-or-nothing" propagation in nerve. For much of the twentieth century, it was widely held that nerve fibers either conduct fully developed action potentials or passively diffuse, with no intermediate phenomena allowed.

Even in a lightly narcotized region of nerve, it was believed, no vestige of the impulse, no "decremental conduction," would remain. This misunderstanding goes back to a Stockholm meeting of the International Physiological Congress in 1926, where a flawed demonstration of the total loss of conduction in a narcotized region of nerve was dramatically presented [46]. From a mathematical perspective, however, it is naive to suppose that a nerve fiber must either carry a fully developed action potential or passively diffuse, with no intermediate behavior.

On the H–H axon, for example, decreasing the maximum values of sodium and potassium conductances (G_{Na} and G_K) by a narcotizing factor (η) lowers (raises) the amplitude and speed of the stable (unstable) traveling wave as is shown in Figure 4.15 [13]. At the critical value $\eta_c = 0.261$, the stable and unstable traveling-waves merge, and decremental conduction can be observed for

$$\eta < \eta_c.$$

In Figure 4.16, for example, are shown calculations on an H–H nerve with $\eta = 0.25$ [13].

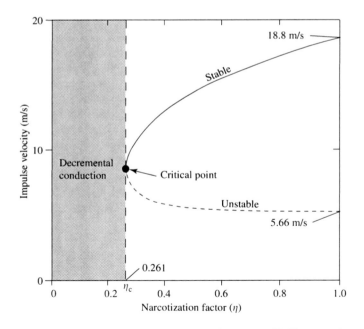

FIG. 4.15. *Amplitude of a traveling-wave impulse on an H–H axon plotted against a narcotization factor—η—reducing the maximum sodium and potassium conductances. (Redrawn from [13].)*

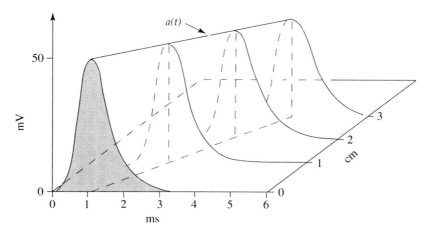

FIG. 4.16. *Propagation of a decremental impulse on an H–H axon narcotized by the factor $\eta = 0.25$. (Redrawn from [13] and [41].)*

There are several parameters of nerve fibers that, upon being altered, can lead to decremental conduction, including the following:

1. A decrease in one or both of the maximum sodium and potassium conductances, as noted above. Such a decrease might be experimentally induced by the action of a narcotic on the nerve, or it could arise in a natural way through a decrease in the number of ion channels per unit area of the nerve membrane.
2. Changes in the external or internal concentrations of sodium or potassium ions will alter the sodium or potassium potentials (V_{Na} or V_K in Table 4.1), leading to decremental conduction.
3. As shown in Figure 4.12, an increase in temperature can induce decremental conduction.
4. Increasing the leakage conduction—G_L—of a H–H axon from the value given in Table 4.1 to about $9\,\text{mmhos/cm}^2$ leads to decremental conduction [12].
5. For a periodic traveling-wave, decremental conduction should be observed when the wavelength is reduced below its minimum value (λ_c), as is indicated in Figure 4.13.

Nerve, therefore, has several avenues for evolving toward a state of decremental conduction. From a theoretical perspective, this behavior can be studied by writing the F–N system of Equations (4.25) and (4.26) (with $b = 0$ and $c = 0$) in the form

$$\frac{\partial^2 V}{\partial t^2} + \varepsilon V = \frac{\partial^3 V}{\partial x^2 \partial t} - F'(V)\frac{\partial V}{\partial t},$$

where conservative terms are collected on the left-hand side. The conservative terms are recognized because they can be derived by substituting the Lagrangian density $(\varepsilon V^2 - V_t^2)/2$ into the Lagrange–Euler equations (Equations (A.2) in Appendix A), implying that the total energy of the traveling-wave is

$$\mathcal{E} = \frac{1}{2}\int_{-\infty}^{\infty}\left[\left(\frac{\partial V}{\partial t}\right)^2 + \varepsilon V^2\right] dx. \tag{4.34}$$

This expression represents the interaction of electrostatic energy (stored in the capacitance of the cell membrane) with a phenomenological inductance of the membrane that has been studied by Mauro and his colleagues [49]. (It is important to note that this inductance is unrelated to the storage of magnetic field energy, which is negligible in nerve [68].) Phenomenological inductance arises from time delays associated with changing the sodium turn-off and potassium turn-on parameters, h and n in Equations (4.15). Roughly speaking, a patch of squid membrane acts as a (parallel LC) resonant circuit with a frequency of $100\,\mathrm{Hz}$ at $10°C$ and $250\,\mathrm{Hz}$ at $20°C$ [70].

Differentiating Equation (4.34) with time and using the F–N equation implies that

$$\frac{d\mathcal{E}}{dt} = -\int_{-\infty}^{\infty}\left[\left(\frac{\partial^2 V}{\partial x \partial t}\right)^2 + \frac{dF(V)}{dV}\left(\frac{\partial V}{\partial t}\right)^2\right] dx. \tag{4.35}$$

For a traveling-wave solution with

$$V(x,t) = \tilde{V}(x - vt) \equiv \tilde{V}(\xi),$$

the right-hand side of Equation (4.35) must be zero; thus a necessary condition for a traveling-wave is the existence of a range of voltage over which

$$\frac{dF(V)}{dV} < 0.$$

The right-hand side of Equation (4.35) being zero is a condition of power balance ($P = vE$) that was described in connection with Equation (4.1). Therefore

$$P = \int_{-\infty}^{\infty}\left(\frac{\partial^2 V}{\partial x \partial t}\right)^2 dx$$

and

$$vE = -\int_{-\infty}^{\infty}\frac{dF(V)}{dV}\left(\frac{\partial V}{\partial t}\right)^2 dx.$$

At the critical point, the right hand side of Equation (4.35) will be zero for a solution of the form $V(x,t) = V_c(x - v_c t) = V_c(\xi)$. Suppose a decremental impulse is attenuated as

$$V(x,t) \doteq a(t)\, V_c(x - v_c t),$$

where the amplitude factor—$a(t)$—is approximately equal to unity. Because $\mathcal{E} \propto a^2(t)$, the amplitude factor should vary with time as

$$\frac{da}{dt} \approx -av_c^2 \frac{\int \left[(d^2V_c/d\xi^2)^2 + (dV_c/d\xi)^2 \, dF(V)/dV|_{V=aV_c} \right] d\xi}{\int \left[v_c^2 \, (dV_c/d\xi)^2 + \varepsilon V_c^2 \right] d\xi}. \tag{4.36}$$

Considering the importance of decremental conduction for the function of nerves, especially on the dendrites (or incoming fibers) of a nerve cell, this problem deserves a careful theoretical, numerical, and experimental investigation.

4.6 Nonuniform fibers

We have seen how an understanding of nonlinear wave dynamics has unraveled the mystery of nerve impulse propagation that puzzled Helmholtz and Nernst during the nineteenth century. It is now understood that the basic mechanism involves a release of electrostatic energy stored in the membrane capacitance and its subsequent dissipation by circulating ionic currents. By itself this explanation is important, but there is more to the story. Like any good theory, the H–H equations not only account for a variety of previously known phenomena but have allowed electrophysiologists to predict novel aspects of neural behavior that may play roles in the functioning of biological nervous systems. As experimental techniques in electrophysiology become more sophisticated, the possibilities for constructive interactions between such measurements and the theory of nonlinear diffusion are ever more interesting. The aim of this section is to introduce a few of the salient research opportunities with the hope that the reader might consider taking part in this activity [70].

4.6.1 Tapered fibers

In the 1960s, Lindgren and Buratti suggested making electronic nerve models (called "neuristors"), having parameters varying exponentially with distance along the fiber as [44]

$$r = r_0 e^{\gamma x}, \quad c = c_0 e^{-\gamma x}, \quad \text{and} \quad j_\mathrm{i} = j_0 e^{-\gamma x}, \tag{4.37}$$

where r_0, c_0, and j_0 are independent of x. The first-order PDEs of the system

$$\frac{\partial V}{\partial x} = -ri$$

$$\frac{\partial i}{\partial x} = -c\frac{\partial V}{\partial t} - j_\mathrm{i}$$

then imply a nonlinear diffusion equation of the form (where $j_0 = j_0(V)$)

$$\frac{\partial^2 V}{\partial x^2} - \gamma \frac{\partial V}{\partial x} - r_0 c_0 \frac{\partial V}{\partial t} = r_0 j_0.$$

For traveling-wave solutions, this reduces to the ODE

$$\frac{d^2V}{d\xi^2} + r_0c_0\left(v - \frac{\gamma}{r_0c_0}\right)\frac{dV}{d\xi} = r_0j_0,$$

implying that the traveling-wave speed in the $+x$-direction increases by γ/r_0c_0 as a result of the exponential taper.

For biological nerve fibers, the parameter dependence of Equations (4.37) is unrealistic because the parameters vary as $r \propto 1/d^2$, $c \propto d$, and $j_i \propto d$, where d is the diameter of the fiber. Thus for an exponential tapering of the fiber diameter,

$$r = r_0 e^{2\gamma x}, \quad c = c_0 e^{-\gamma x}, \quad \text{and} \quad j_i = j_0 e^{-\gamma x}, \tag{4.38}$$

leading to the PDE

$$\frac{\partial^2 V}{\partial x^2} - 2\gamma\frac{\partial V}{\partial x} - r_0 c_0 e^{\gamma x}\frac{\partial V}{\partial t} = r_0 j_0 e^{\gamma x}. \tag{4.39}$$

If the size of a nerve impulse is small compared with $1/\gamma$, the exponential factors in Equation (4.39) remain approximately constant, and at $x = 0$ the traveling-wave speed is increased by $2\gamma/r_0c_0$.

If the fiber diameter changes greatly over the length of a nerve impulse, the above considerations are no longer valid. With a sufficiently abrupt increase of diameter, an impulse can actually be blocked as is indicated by the numerical computations on linearly tapered H–H axons recorded in Table 4.4 [3]. A taper length of 0.088 cm is very close to an abrupt widening, which barely passes an isolated impulse at a ratio of 5:1 with a time delay of about 0.8 milliseconds [36].

The physical reason for this delay can be understood as the time required for the capacitance of the enlarged membrane area to become charged to a voltage that exceeds the threshold [70]. Such delays occur whenever a nerve impulse encounters a varicosity (local enlargement) as is observed for the dendrites (incoming signal pathways) of a variety of cochlear (auditory) neurons [6].

Do such dendritic varicosities adjust impulse timings in connection with dendritic information processing? To answer this question, we must consider more carefully the threshold conditions for launching an impulse on an active nerve fiber.

Table 4.4. *Diameter ratio for blocking and passage as functions of taper length for H–H axons [3].*

Taper length (cm)	Blocking	Passage
0.088	5.5 : 1	5 : 1
0.785	6 : 1	5.5 : 1
1.76	8 : 1	7 : 1
3.81	>10 : 1	10 : 1

4.6.2 Leading-edge charge and impulse ignition

How does one initiate an impulse on a nerve axon? From both experimental observations [12, 35, 57] and integration of the full H–H system [13], it appears that a certain threshold of electric charge (Q_θ) is required. In other words, if a short pulse of current is injected into a nerve fiber, the integral over time of this current pulse must exceed Q_θ for an impulse to emerge. In the standard H–H axon at a temperature of 18.5°C, the threshold charge in coulombs (C) is about

$$Q_\theta = 1.33 \times 10^{-9} \text{C}.$$

We can understand this theoretically confirmed experimental observation by considering the leading edge of an H–H impulse, which obeys the first-order PDEs

$$\frac{\partial V}{\partial x} = -ri, \tag{4.40}$$

$$\frac{\partial i}{\partial x} = -c\frac{\partial V}{\partial t} - j_\text{i}. \tag{4.41}$$

Note from Figure 1.2 that the action potential has reached almost its full amplitude before the membrane permeability is fully on. In other words,

$$|j_\text{i}| \ll c|\partial V/\partial t|$$

over much of the leading edge of a typical impulse.

Neglecting ionic current on the leading edge of the propagating impulse, Equation (4.41) becomes the approximate conservation law

$$\frac{\partial i}{\partial x} + \frac{\partial (cV)}{\partial t} \doteq 0. \tag{4.42}$$

With reference to the discussion of conservation laws in Appendix A, it is seen that i is the flow of the conserved quantity past a fixed point per unit of time, and cV is the corresponding density of the conserved quantity per unit of distance along the axon. From the units, the conserved quantity is evidently an electric charge, which is given by the integral expression

$$Q_0 = \int_\text{le} i\, dt,$$

where the subscript indicates that the integration is across the leading edge of an H–H impulse. Substituting Equation (4.40) and noting that V is a traveling-wave of speed v, this expression for the conserved charge becomes

$$Q_0 = -\frac{1}{r}\int_\text{le} \frac{\partial V}{\partial x} dt = +\frac{1}{rv}\int_\text{le} \frac{\partial V}{\partial t} dt,$$

which reduces to the simple expression

$$Q_0 = V_{\max}/rv, \qquad (4.43)$$

where $V_{\max}$ is the total amplitude of the nerve impulse.

Upon inserting the values $r = 2\times 10^4$ ohm/cm and (from Figure 4.7) $v = 1880$ cm/s and $V_{\max} = 90.5 \times 10^{-3}$ V gives a leading edge charge of [69]

$$Q_0 = 2.4 \times 10^{-9} \mathrm{C},$$

which is substantially larger than the experimental value for Q_θ recorded previously.

Should we be disappointed that the leading-edge charge that we have just calculated is almost twice the amount of charge needed to launch an impulse? Not at all. The fact that

$$Q_0 = 1.8 Q_\theta$$

tells us that impulse propagation on a squid fiber has a safety factor of about 80%. In other words, the leading-edge charge can be reduced by a factor of almost one-half before propagation fails.

Finally, we can use Equation (4.43) and parameters for the small-amplitude unstable impulse in Figure 4.7 to calculate Q_θ. Keeping r the same but using a velocity of 566 cm/s and a voltage amplitude of 18×10^{-3} V gives a threshold charge of

$$Q_\theta = 1.6 \times 10^{-9} \mathrm{C},$$

which is close to the experimental value quoted at the beginning of this section.

4.6.3 Dendritic logic

As an example of a real neuron, consider the Purkinje cell of the human cerebellum, the dendritic field of which is displayed in Figure 4.17 [62]. Like an espaliered fruit tree, this dendritic field is essentially planar, receiving some 80 to 100 thousand input signals. Does the output signal pathway (or axon) merely record whether a weighted linear sum of these inputs exceeds some threshold? Or do the dendritic trees process information on the incoming signals before they reach the cell body?

For several decades of the previous century, electrophysiologists have believed that the nerve fibers of dendritic trees are entirely passive in their electrical behavior, obeying the laws of linear diffusion. Although there is now much experimental evidence to the contrary [70], the classical notion persists. The reason for this oversight may be related to an overemphasis upon the concept of all-or-nothing propagation for an active fiber. If one supposes (consciously or otherwise) that active impulses are never extinguished, then a firing event occurring anywhere in the dendrites shown in Figure 4.17 would cause the entire tree structure to fire, and the dendrites could play no integrative role in neural

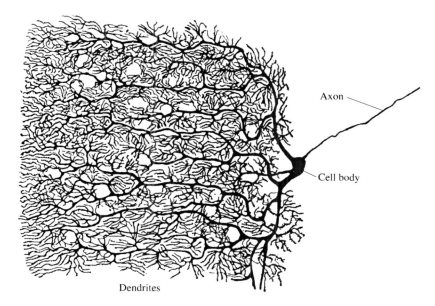

FIG. 4.17. *A Purkinje cell of the human cerebellum. The dendrites of this cell spread over an area of about 0.25 mm × 0.25 mm. (Redrawn from [62].)*

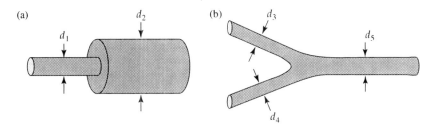

FIG. 4.18. *(a) Abrupt widening of a nerve fiber. (b) Branching region.*

functioning. Thus a classical view of nerve impulse propagation leads to the erroneous conclusion that dendritic dynamics must be passive.

It is now known that active nerve impulses can be blocked at varicosities, especially at high impulse rates. In a dendritic tree, the signal path always becomes enlarged at branchings, where two or more daughter fibers merge into a parent. Could such branchings be the locus of dendritic information processing? How should one describe a branch?

To answer such questions, it is helpful to consider Figure 4.18, which compares an abrupt widening of a nerve fiber with a branching region [70]. The characteristic admittance Y of a uniform fiber is given by[2]

$$Y = \sqrt{\frac{\text{shunt admittance per unit length}}{\text{series resistance per unit length}}} \propto \sqrt{\frac{d}{1/d^2}} = d^{3/2}.$$

This implies that linear propagation through the discontinuity (from d_1 to d_2) in Figure 4.18(a) will depend on the "geometric ratio" (or GR) of $(d_2/d_1)^{3/2}$ [26]. To have the same linear propagation from branch d_3 into branches d_4 and d_5 of Figure 4.18(b) requires

$$\text{GR} = \frac{d_2^{3/2}}{d_1^{3/2}} = \frac{d_4^{3/2} + d_5^{3/2}}{d_3^{3/2}}. \tag{4.44}$$

The same conclusion follows from the ignition analysis of the previous section. From Equation (4.43), the leading-edge charge carried by an active fiber is $Q_0 = V_{\max}/rv$, where $V_{\max}$ is the amplitude of the impulse and v is its traveling-wave speed. $V_{\max}$ is an intrinsic property of the nerve membrane, while $v \sim \sqrt{d}$ and $r \sim 1/d^2$. This implies $Q_0 \propto d^{3/2}$, and from a similar argument $Q_\theta \propto d^{3/2}$. Thus Equation (4.44) also follows from the assumption that the ratio of incoming leading-edge charge to the total outgoing threshold charge is the same in the abrupt widening of Figure 4.18(a) as in the branching structure of Figure 4.18(b).

In his book *The Problem of Excitation*, Boris Khodorov has assembled the results of many numerical computations of impulse delay and blockage [35]. For abrupt widenings of H–H fibers (at 20°C), as shown in Figure 4.18(a), he reports impulse blockage at widening ratios greater than 5.5:1 and passage at ratios of 5:1 or less. Recently, Altenberger et al. have carried through more refined calculations and found the critical widening ratio for blockage of a standard H–H impulse (at 18.5°C) to be [2]

$$\frac{d_2}{d_1} > 5.43,$$

with an error of ±0.05%. Assuming that the logic behind Equation (4.44) is correct, we can calculate a condition on the branch diameters in Figure 4.18(b) for an incoming impulse on d_3 to become blocked as

$$\frac{d_4^{3/2} + d_5^{3/2}}{d_3^{3/2}} > 5.43^{3/2} = 12.7.$$

Similarly, an incoming impulse on d_4 will be blocked if

$$\frac{d_3^{3/2} + d_5^{3/2}}{d_4^{3/2}} > 12.7.$$

---

[2]In electrical engineering jargon, admittance is a generalization of the concept of conductance, and the characteristic admittance of a transmission system is the admittance seen looking into one end of a semi-infinite length.

If these inequalities are satisfied, incoming impulses on both of the daughter fibers (d_3 and d_4) are required to ignite the parent fiber d_5. In this case, the branching region can be described in the jargon of computer engineering as an AND junction.

If, on the other hand,

$$\frac{d_4^{3/2} + d_5^{3/2}}{d_3^{3/2}} < 12.7 \quad \text{and} \quad \frac{d_3^{3/2} + d_5^{3/2}}{d_4^{3/2}} < 12.7,$$

then incoming impulses on either daughter can ignite the parent, and the branching region is an OR junction.

Although examination of the dendritic branchings of Figure 4.17 reveals few if any branches with GRs as large as 12.7, there are several reasons for suspecting that this condition may be too severe for real dendrites [70].

First, action potentials on dendrites are not well described by the standard H–H equations, because calcium channels play an important role. Using a version of the M–L system, for example, Altenberger et al. have computed a critical GR of 3.4 for the abrupt widening of Figure 4.18(a) [2].

Second, although sodium channels are often present in addition to calcium channels, they may have lower density (number of channels per unit area of membrane). Also, much of the dendritic membrane is covered by synapses, which could lessen the widening ratio necessary for blockage.

Third, changes in ionic concentrations and temperature can also lower the safety factor for impulse propagation, thereby raising threshold conditions and easing the geometrical requirements for an AND junction. Body temperatures of mammals, for example, are typically larger than the value of 18.5°C used in H–H calculations of critical widening.

Fourth, the fiber length required for an impulse to grow from threshold to its full amplitude is greater than the lengths of some dendritic segments in Figure 4.17, implying that voltage amplitudes of impulses arriving at a branch may be less than their full values. This effect also lowers the geometric ratio needed for blockage.

Fifth, inspection of Figure 4.17 reveals several delta-shaped enlargements at dendritic bifurcations, which increase the total membrane capacitance and impede impulse transmission.

Finally, the incoming impulses may not be isolated but spaced with intervals as small as a few milliseconds. Khodorov reports numerical calculations for H–H impulses (at 20°C) separated by an interval of 2.5 ms, in which the second impulse is blocked at a widening ratio that is less than 3:1 and greater than 1.5:1, implying a critical GR within the range $1.8 < \text{GR}_{\text{crit}} < 5.2$ [35]. Correspondingly effects have been observed in experiments on branching squid nerves by the present author [70].

For all of these reasons, it is expected that studies of neural information processing in dendritic trees will be of increasing interest in coming years.

4.7 More space dimensions

Just as the candle provides a vivid example of reaction-diffusion in one space dimension, fairy rings of mushrooms offer a memorable image of this phenomenon in two dimensions. For those who have not yet had the pleasure of seeing such rings on a summer morning, the following description by Parker-Rhodes may be of interest [59].

A large grass field carrying a system of fairy rings can be visualized as like a pond in a light shower of rain. Each ring grows outwards at a constant (or irregularly fluctuating) rate, and new rings are added to the system with approximately constant frequency. When rings of the same species meet, then, unlike ring waves on water, their intersected portions are obliterated; but as between different species either or both may survive the intersection. The important thing is that no ring ever remains still, but only stops growing when it dies. However, all these processes take place with extreme slowness; in the rain-drop analogy the rings grow at a few feet per second, and new ones appear at the order of ten per square foot per second, whereas fairy rings grow at a few feet per year and appear less often than ten per hundred acres per year.

Other examples of such nonlinear diffusion effects in more than one spatial dimension abound. The periodic prairie fires that prevented forestation of the Great Plains of North America provide a striking example in two dimensions, as do lichens, social amoebae (slime molds), and chemical oscillators that are based on interacting ionic species (see Figure 1.4) [81, 83, 88]. Let us consider such phenomena in somewhat greater detail.

4.7.1 Two-dimensional nonlinear diffusion

Consider nonlinear diffusion in the (x, y)-plane by writing the F–N equations from Section 4.3.2 in two-dimensional form as [50, 77]

$$D\left(\frac{\partial^2 u}{\partial x^2} + \frac{\partial^2 u}{\partial y^2}\right) - \frac{\partial u}{\partial t} = f(u) + w,$$
$$\frac{\partial w}{\partial t} = \varepsilon(u + c - bw),$$
(4.45)

where D is the linear diffusion constant for the propagator (or activation) variable (u), and the recovery variable (w) does not diffuse. What solutions can this system support?

One possibility is obvious. From Equations (4.45) can emerge plane waves that propagate in any direction on the (x, y)-surface. As this phenomenon was discussed in Section 4.3.2, no more needs to be said here.

A second possibility is the ring wave that was observed by Zaikin and Zhabotinsky in chemical solutions [88] (shown in Figure 1.4) and by Parker-Rhodes in fields of mushrooms. In this case, the solutions do not depend upon the azimuthal angle; thus it is convenient to write Equations (4.45) as

$$D\frac{\partial^2 u}{\partial r^2} + \frac{D}{r}\frac{\partial u}{\partial r} - \frac{\partial u}{\partial t} = f(u) + w,$$
$$\frac{\partial w}{\partial t} = \varepsilon(u + c - bw), \tag{4.46}$$

where r is the radial distance from the point of initiation. In this formulation, the number of independent variables has been reduced from three (x, y, and t) to two (r and t), but a price has been paid—the coefficient D/r depends on the dependent variable r. This prevents us from making the traveling-wave assumption (that u and w are functions of $\xi \equiv r - vt$) unless the radius is large enough that the term $(D/r)(\partial u/\partial r)$ can be neglected, in which case the solution becomes a plane wave moving at a constant speed.

If the term $(D/r)(\partial u/\partial r)$ cannot be neglected, comparison with the results on tapered nerve fibers in Section 4.6.1 implies that the outward speed of the ring wave is slowed by an amount that is proportional to the local curvature $(1/r)$ without otherwise greatly altering its qualitative behavior. Thus for $r \gg D/v_0$, a point of constant amplitude on the wave front moves outward with a radial velocity

$$v \doteq v_0 - D/r, \tag{4.47}$$

where v_0 is the speed of a plane wave.

Another interesting feature of ring waves was noted by Parker-Rhodes in the above quotation: "When rings... meet, their intersected portions are obliterated." Obliteration occurs because the recovery variable w requires a time of order $1/\varepsilon$ to relax back to its resting value, thereby inhibiting the propagation of a new ring. Thus the rings do not pass through each other—as with water waves on the surface of a pond—but instead form cusps in the wave fronts as shown in Figure 1.4. Such cusps are a characteristic feature of intersecting rings of mushrooms.

Figure 4.19 shows a pair of spiral waves, each emerging from a rotor (or vortex) in a system of excitable chemical reagents, a phenomenon that was discovered by Winfree in 1972 [81]. Self-sustaining spirals are observed in a variety of two-dimensional, excitable media.

Spiral waves can be generated, both numerically and experimentally, by forcing a plane wave to have a free end that is allowed to propagate into an active region. Both numerical studies and experiments with chemical systems show that such a free end curls up into an organizing center that generates periodic solutions of

$$D\frac{\partial^2 u}{\partial r^2} + \frac{D}{r}\frac{\partial u}{\partial r} + \frac{D}{r^2}\frac{\partial^2 u}{\partial \theta^2} - \frac{\partial u}{\partial t} = f(u) + w,$$
$$\frac{\partial w}{\partial t} = \varepsilon(u + c - bw), \tag{4.48}$$

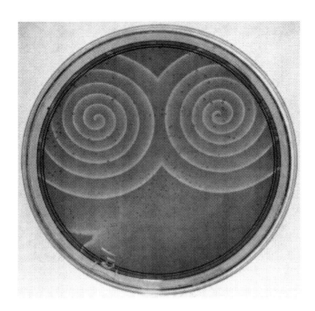

FIG. 4.19. *Spiral waves in an excitable chemical reagent. (Courtesy of A.T. Winfree.)*

which is equivalent to Equations (4.45). For $r \gg D/v_0$, points of constant wave amplitude are Archimedean spirals with the form

$$r = \pm \lambda \theta / 2\pi,$$

where λ is the radial spacing between successive wave fronts. These spirals rotate at radian frequency

$$\omega = \pm 2\pi v \lambda,$$

where v is given by Equation (4.47).

For $r \sim D/v_0$, the dynamics are more intricate. Winfree has cataloged numerical solutions of Equations (4.48), showing how the center of the rotor meanders randomly or traces out a variety of minute geometrical figures that change as the parameters ε, b, and c are varied [51, 85].

4.7.2 Nonlinear diffusion in three dimensions

Proceeding to the study of nonlinear diffusion phenomena in three space dimensions, let us generalize the system of interest to [76, 82, 84, 85, 91, 92]

$$\begin{aligned} D_u \nabla^2 u - \frac{\partial u}{\partial t} &= f(u, w) \\ D_w \nabla^2 w - \frac{\partial w}{\partial t} &= g(u, w), \end{aligned} \quad (4.49)$$

where the operator

$$\nabla^2 \equiv \frac{\partial^2}{\partial x^2} + \frac{\partial^2}{\partial y^2} + \frac{\partial^2}{\partial z^2}$$

is in a rectangular coordinate system. Equations (4.49) can represent the nonlinear diffusion of oscillating chemical reagents (u and w), proposed by Luther in 1906 [47], rediscovered by Bray in 1921 [7], and rediscovered yet again by Belousov in the Soviet Union in the 1950s.

Evidently a three-dimensional F–N system is a special case of Equations (4.49) with $D_w = 0$. Just as two-dimensional nonlinear diffusion offers more interesting phenomena than is found on the one-dimensional nerve axon, three-dimensional systems provide an even greater variety of dynamic phenomena that are not yet entirely cataloged. In addition to a plane wave (or 3D version of the standard nerve impulse) there is a 3D version of the ring wave, called a spherical wave, with $u = \tilde{u}(r - vt)$.

To calculate the speed of a spherical wave of large radius, recall that for spherical symmetry

$$\nabla^2 = \frac{\partial^2}{\partial r^2} + 2/r \frac{\partial}{\partial r}.$$

Thus for $r \gg D/v_0$, a point of constant amplitude travels outward at velocity

$$v \doteq v_0 - 2D/r, \tag{4.50}$$

suggesting that in three dimensions the natural speed of a curved wave front is reduced by the diffusion constant times the sum of the two curvatures (of $1/r$ each) in perpendicular directions.

In three dimensions there are a great many self-exciting configurations because the organizing center of a spiral wave can be extended into a vortex line, and this line can be tied into a variety of linked rings and knots. The simplest of these is the straight line of a scroll wave, from which the propagating surface unrolls—like a piece of ancient Chinese art—at a speed given by Equation (4.47). Next is the scroll ring shown in Figure 4.20, which has been experimentally observed in the chemical medium of Belousov and Zhabotinsky [32].

If the vortex line of a scroll ring is a circle of radius R, its dynamics can be studied in the following way. The outer surfaces at radius $R+\rho$ should propagate outward at speed

$$v_{\text{outer}} = v_0 - D\left(\frac{1}{\rho} + \frac{1}{R+\rho}\right),$$

where ρ is the distance from the vortex ring to the propagating surface. Likewise, an inner surface at radius $R - \rho$ propagates inward at speed

$$v_{\text{inner}} = -v_0 + D\left(\frac{1}{\rho} - \frac{1}{R-\rho}\right)$$

because the curvature around the inside of the ring is negative, while that across the ring is positive.

The average of v_{outer} and v_{inner} is

$$\langle v \rangle = (v_{\text{outer}} + v_{\text{inner}})/2$$
$$= -\frac{D}{2}\left(\frac{1}{R+\rho} + \frac{1}{R-\rho}\right) = -\frac{DR}{R^2 - \rho^2}.$$

As $\rho/R \to 0$, the average velocity approaches that of the vortex line of radius R. In other words,

$$\langle v \rangle \to dR/dt = -D/R.$$

Upon integration, this equation implies that the radius of a vortex ring shrinks with time according to

$$(R_0^2 - R^2) = 2D(t - t_0), \tag{4.51}$$

in accord with experiments on excitable chemical systems [32].

Myriad variations on the theme of the vortex ring can be generated by thinking of a vortex line as a piece of string and imagining how it might be arranged in 3D space [86].

The 3D scroll ring that is displayed in Figure 4.20 is more than an experimental or numerical curiosity. The human heart is (among other things) an active medium that pumps blood by contracting as a segment of a spherical wave passes through, upon stimulation by the autonomous nervous system. Under certain

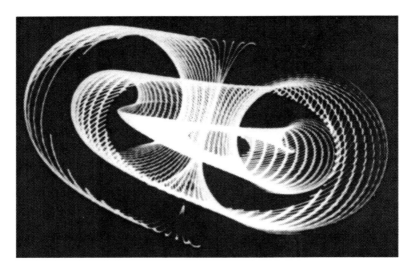

FIG. 4.20. *A computer generated plot of a scroll ring. (Courtesy of A.T. Winfree.)*

unhappy circumstances, however, a self-perpetuating scroll ring becomes established, whereupon normal pumping activity is interrupted by a phenomenon called "fibrillation."[3] If this behavior is not immediately suppressed, the owner of the heart succumbs to "sudden cardiac death" [84, 85].

How often does this happen? In his book *When Time Breaks Down*, Winfree tells us [83]: "Sudden cardiac death can occur at any time, unpredictably. If you spend half an hour reading this chapter, you can be confident that hundreds of individuals will meanwhile add to the worldwide statistics of this mishap. There are a thousand cases each day in the United States alone."

4.7.3 Turing patterns

Among the most important and least understood of natural phenomena is the development of a living organism from a single cell through its embryonic stages and infancy to adult form. Although a theory of such intricate dynamics lies beyond the present scope of scientific knowledge, there may be some contributions to be made by the applied mathematician. In his classic work *On Growth and Form*, for example, the Scottish polymath D'Arcy Thompson showed how considerations of surface energy can account for many features of the shapes of cells and their aggregates, thereby contributing to the science of biological morphogenesis [74]. But Thompson's discussion was almost entirely of static forms, and the concept of morpho*genesis* includes dynamics. Can the study of nonlinear diffusion lead to an understanding of embryonic development?

Interestingly, just such a proposal was made in 1952—the same year that the H–H paper appeared—by an English mathematician named Alan Turing [76]. To appreciate his ideas, let us return to Equations (4.49), describing the nonlinear diffusion of a propagator (u) and its recovery variable (w) in 3D space, and consider various relationships between the corresponding diffusion constants D_u and D_w.

1. *Nerve and muscle.* From the perspective of the H–H (or F–N) equations, the propagator is a transmembrane voltage, and the recovery variable is related to conformational states of protein molecules that are embedded in the cell membrane, mediating ionic currents. Thus for such systems

$$D_w = 0.$$

2. *Chemical diffusion.* In the Belousov–Zhabotinsky reaction, both the propagator and the recovery variables are chemical species in aqueous solution so

$$D_w \sim D_u.$$

---

[3]According to Winfree [83], the term was coined in France during the nineteenth century to suggest that individual fibers of the heart muscle are acting independently.

3. *Turing patterns.* For Equations (4.49) to develop stationary patterns, it is necessary that

$$D_w \gg D_u.$$

Turing patterns are of interest as models for biological morphogenesis because they don't move. To see how this comes about, consider Equations (4.49), and imagine the following sequence of events. (i) Stimulation at a point initiates growth of the excitatory variable (u). (ii) Increase of u stimulates growth of the inhibitory variable (w). (iii) Because w diffuses more rapidly than u, it quickly leaves the vicinity of the point of initial stimulation. (iv) Thus the excitatory variable finds itself surrounded by a spherical halo of inhibition, impeding outward motion.

Although Turing patterns are readily produced in numerical simulations [53], they are difficult to find experimentally. The reason for this difficulty is that in a chemical solution all ions diffuse at about the same rate, so $D_w \sim D_u$. In recent experiments that have demonstrated Turing patterns, D_u has been artificially reduced by fixing the excitatory component in a gel and allowing the inhibitor to diffuse at a normal rate beyond a porous barrier [10, 42]. Such experimental problems suggest that the dynamic aspects of biological morphogenesis are more complicated than was originally anticipated by Turing.

4.7.4 Hypercycles

As scroll waves are known to play roles in the dynamics of the slime mold *Dictyostelium discoideum* [73], there may be more general applications of dynamic spatial structures to the emergence of multicelled organisms during the course of evolution [4, 5]. In developing this idea, Boerlijst and his colleagues have built upon the concept of a hypercycle, introduced in 1979 by Eigen and Schuster as a fundamental dynamic principle of life [14].

A hypercycle is a cycle of cycles of cycles. One example of a basic catalytic cycle is the Krebs (or citric acid) cycle,[4] which uses a molecule of oxaloacetate over and over again to extract the energy from acetic acid, driving a variety of higher level cycles that require the consumption of energy. At a higher level of organization, several such basic catalytic cycles can comprise an "autocatalytic cycle" that is able to instruct its own reproduction—an example being a protein-DNA structure that can reproduce itself. Several of these autocatalytic cycles can be organized at yet a higher level into a "catalytic hypercycle"—such as a virus—standing at the threshold of life because it can evolve into more efficient forms. Also called "complex adaptive systems," these structures are among the most challenging for modern mathematics [43]. According to Eigen and Schuster,

---

[4] Named for Hans Adolf Krebs (1900–1981), a German-born British biochemist. It was the Krebs cycle that inspired Belousov to invent his autocatalytic reaction in the early 1950s.

the dynamic structure of an organism is as follows:

catalytic hypercycle

↑ ↓

autocatalytic cycle

↑ ↓

basic catalytic cycle,

each level being modeled by an independent set of nonlinear differential equations. Life, it seems, emerges from all of the functionally significant levels.

In its original formulation, the hypercycle was considered to be described by ODEs in time, with uniformity in space. A difficulty arises with the appearance of parasites, which compete effectively with the biomolecules of the basic cycles because they don't need to contribute energy to catalysis. Put in other words, the basic molecules tend to lose out because they are guilty of the Darwinian sin of altruism (feeding the parasites for free), leading to the destruction of the hypercycle.

Boerlijst and his colleagues have constructed simple numerical models of such processes that are based on PDEs, thereby allowing the development of spiral waves of activity. In such systems, they have shown, parasites are swept outward, leaving the organizing center undisturbed. Thus there is evolutionary selection at the level of the spiral, encouraging the altruistic behavior of the molecular components and providing protection against parasites. This analysis, of course, employs a vastly oversimplified model of evolution, but it does show that the emergence of dynamic spatial patterns in multicelled organisms can mitigate some of the gloomier aspects of Darwinian selection.

4.8 Summary

It is hoped that the reader concludes this chapter with an appreciation for the wisdom behind Michael Faraday's introduction to the study of a candle. Through his "open door," science enters a domain describing the propagation and processing of information on nerves, normal and pathological contractions of the heart, and aspects of morphogenesis, among other biological applications.

In these applications, energy is not conserved by the emergent structures but dissipated, implying that time is directed away from the past and toward the future. Such coherent structures are characterized by thresholds of stimulation, below which they fail to emerge, and a collision between two structures can result in their mutual annihilation.

Thus the phenomena of nonlinear diffusion in excitable media are fundamentally different from those of energy conserving (Hamiltonian) processes considered in Chapter 3. Although different, nonlinear diffusion is equally important from a general scientific perspective and more so in the realm of biology.

4.9 Problems

1. Measurements of the flame speeds (v) for candles of different diameter (d) are recorded in the following table [72].

d (cm)	0.5	1.0	2.2	3.44	3.8	4.9	6.1	7.9
v (cm/h)	14.9	6.36	1.40	1.03	0.74	0.60	0.32	0.165
RMS errors	5	0.8	0.11	0.12	0.13	0.06	0.02	0.008

   (a) Find a formula for v as a function of d.

   (b) Discuss the physical implications of your formula.

2. (a) For a one-dimensional diffusion process with the diffusivity depending upon the concentration of the diffusing substance (i.e. $D = D(u)$), construct an appropriate nonlinear partial differential equation.

   (b) Find a function $D(u)$ for which $u(x,t)$ is a traveling-wave.

3. Consider an arbitrary partial differential equation with x and t as the independent variables. Consider and discuss the relationship between the traveling-wave assumption that
   $$u(x,t) = \tilde{u}(x - vt)$$
   and the construction of a Bäcklund transformation that generates a solution $u_1(x,t)$ from the vacuum state $u_0 = 0$.

4. Use the results of the previous problem to find solitary wave solutions of Equation (4.2) with $f(u) = u(u-a)(u-1)$.

5. Consider Equation (4.2) with $f(u) = u(1-u)$. Show that bounded traveling-wave solutions are unstable. [Hint: examine linear stability at the stationary points.]

6. (a) Show that the singular points at $(0,0)$ and at $(1,0)$ are saddle points in the (u,w) phase plane for solutions of Equations (4.5).

   (b) Sketch trajectories in the phase plane.

   (c) Study how the solution trajectories change as one varies the assumed traveling-wave speed v.

7. (a) Find all the homoclinic trajectories of Equation (4.8).

   (b) Does the system of Equation (4.5) have a homoclinic trajectory for $v \neq 0$? [Hint: use an "energy" argument.]

8. Assume the Z–F equation to be
   $$D\frac{\partial^2 u}{\partial x^2} - \frac{\partial u}{\partial t} = \tau^{-1} u(u-a)(u-1),$$
   where D is a diffusion constant, τ is a time constant, and u is unitless.

(a) If u is independent of x, how does it vary with t?

(b) If $u(x,t) = \tilde{u}(x - vt)$, find both $\tilde{u}(\cdot)$ and v.

(c) How are the results of (b) related to Equations (1.6) and (1.7)?

9. For the Z–F equation as defined in Equation (4.4), let
$$f(u) = \begin{cases} u & \text{for } u < a, \text{ and} \\ u - 1 & \text{for } u > a, \end{cases}$$
where $0 < a < 1$.

  (a) For traveling-wave solutions corresponding to heteroclinic trajectories, show that
  $$v = \pm(1 - 2a)/\sqrt{a(1-a)}.$$

  (b) Find the corresponding solutions of the form $u(x - vt)$. [Hint: assume exponential behavior for $u > a$ and $u < a$.]

10. Repeat the previous problem for
$$f(u) = K_1 \sin \pi u + K_2 \sin 2\pi u,$$
(where $K_2 > 0$ and $0 < u < 1$), showing that the traveling-wave speed [66]
$$v = \pm \frac{K_1}{\sqrt{2K_2/\pi}}.$$

11. Repeat the previous problem for
$$f(u) = \begin{cases} g_1 u & \text{for } u < a, \text{ and} \\ g_2(u - 1) & \text{for } u > a, \end{cases}$$
(where $0 < a < 1$ and $g_1, g_2 > 0$), showing that the traveling-wave speed [78]
$$v = \pm \frac{\left[\left(\frac{1-a}{a}\right)^2 g_2 - g_1\right]}{\sqrt{\left[\frac{1-a}{a^2}\left(\frac{1-a}{a}g_2 + g_1\right)\right]}}.$$

12. Derive and sketch the traveling-wave solutions related to Equation (4.8).

13. Consider the equation
$$\frac{\partial^2 u}{\partial x^2} - \frac{1}{c^2}\frac{\partial^2 u}{\partial t^2} - \frac{\partial u}{\partial t} = u(u - a)(u - 1).$$

(a) Find traveling-wave solutions corresponding to those in Equations (4.6) and (4.7) for $v < c$.

(b) Describe the behavior of your solutions as $v \to c$ from below.

(c) Consider this system to be a "relativistic correction" to Equation (4.2) for a nerve model. Assuming that $v = 20\,\text{m/s}$ without the correction and c is the speed of light, estimate the magnitude of the error made in neglecting the correction.

14. Consider Equation (4.10) with $-\infty < x < +\infty$, $t > 0$, and the initial condition [79]
$$u(x,0) = f(x).$$
Use the Cole–Hopf transformation and results from Chapter 2 to show that the solution is
$$u(x,t) = \frac{\int_{-\infty}^{\infty} [(x-y)/t]\exp(-G/2D)\,dy}{\int_{-\infty}^{\infty}\exp(-G/2D)\,dy},$$
where
$$G(x,y,t) \equiv \int_0^y f(y')\,dy' + \frac{(x-y)^2}{2t}.$$

15. (a) Assuming that $D \to 0$ in the previous problem, show that the dominant contributions to the integrals in the expression for $u(x,t)$ are where $\partial G/\partial y = 0$ or
$$f(y) - \frac{x-y}{t} = 0.$$

(b) Show that this implies
$$u(x,t) \to f(\xi),$$
where $\xi = x - ut$.

(c) For $f(x)$ a pulse like function, sketch $u(x,t)$ in the limit $D \to 0$.

16. Show that the Cole–Hopf transformation of Section 4.1.2 can be viewed as the Bäcklund transform
$$\partial \phi/\partial x = -u\phi/2D$$
$$\partial \phi/\partial t = u^2\phi/4D - (\phi/2)\partial u/\partial x,$$
carrying solutions of Equation (4.10) into those of Equation (4.11).

17. Discuss the qualitative differences between the Burgers equation and the Korteweg–de Vries equation from both physical and mathematical perspectives.

18. Use Kirchhoff's voltage and current laws to derive the first-order "cable equations" (4.12) for a nerve fiber from the equivalent circuit in Figure 4.4(b).

19. The sodium and potassium voltages (V_{Na} and V_{K}) for an H–H nerve are related to the outside and inside ionic concentrations by

$$V_{\text{Na}} = \frac{kT}{e} \log\left(\frac{[\text{Na}^+]_o}{[\text{Na}^+]_i}\right) + 65\,\text{mV}$$

$$V_{\text{K}} = \frac{kT}{e} \log\left(\frac{[\text{K}^+]_o}{[\text{K}^+]_i}\right) + 65\,\text{mV},$$

where values for k and e are given Sections 8.1 and 8.2, and T is the absolute (Kelvin) temperature.

(a) From these relations, compute the ratios of outside to inside concentrations for the sodium and potassium ions at body temperature.

(b) From the results of part (a), discuss in qualitative terms the dynamics of membrane switching.

(c) Find the resting potential of an H–H axon.

20. (a) Show that the H–H ODE system of Equations (4.21) has only one SP.

(b) Find the values of V, W, n, m, and h at this SP?

(c) Use a small amplitude analysis to investigate the behavior of solution trajectories in the vicinity of this SP.

21. In Equations (4.21), assume that $\tau_m = 0$ and $\tau_h = \tau_n = \infty$.

(a) Show that $j_i = j_i[V, m(V), h_0(0), n_0(0)]$.

(b) Plot j_i as a function of V for the H–H parameters from Table 4.1.

(c) Use the velocity formula obtained in Section 4.1.1 to estimate the speed of a nerve impulse on the H–H axon.

(d) Discuss how this estimate might be improved.

22. (a) Using results of the previous problem, show that the traveling-wave speed of a nerve impulse on the H–H axon varies with the fiber diameter as $v \propto \sqrt{d}$.

(b) Estimate the proportionality constant at a temperature of $6.3°C$.

23. Measurements on axons of the common squid *Loligo vulgaris* over a temperature range from 15 to 22°C show the impulse velocity to be given by

$$v = 20.3\sqrt{d/476}\,[1 + 0.038(T - 18.5)]\,\text{m/s}$$

(to an experimental accuracy of about ±5%), where d is the axon diameter in microns and T is the temperature in degrees Celsius [70]. Explain why the F–N system shows a decrease in impulse speed with increasing values of the temperature parameter (ε), whereas impulse speed increases with temperature on real nerves.

24. Consider the F–N system given in Equations (4.25) and (4.26) with
$$F(V) = V(V-a)(V-1),$$
$\varepsilon = 0$, and assume that $R = R_0$.

   (a) Find and plot the speed of a stable traveling-wave solution as a function of R_0. [Hint: express the cubic function $F(V) + R_0$ in the original form.]

   (b) Find and plot the shape of the traveling-wave solution as a function of R_0.

25. Consider the F–N system given in Equations (4.25) and (4.26) with $b = 0$ $c = 0$ and
$$F(V) = V^3/3 - V.$$

   (a) Show that trajectories near the origin of the (V, W, R) phase space are governed by the linear system
$$\frac{d}{d\xi}\begin{bmatrix} V \\ W \\ R \end{bmatrix} \doteq \begin{bmatrix} 0 & 1 & 0 \\ -1 & -v & 1 \\ -\varepsilon/v & 0 & 0 \end{bmatrix} \begin{bmatrix} V \\ W \\ R \end{bmatrix}.$$

   (b) Discuss in detail the behavior of the trajectories near $(0,0,0)$.

26. In the previous problem with
$$F(V) = V(V-a)(V-1),$$
construct two homoclinic trajectories in the (V, W, R) phase space for the limiting case that $\varepsilon \to 0$.

27. (a) Compute values for the resting potentials of the M–L parameters given in Table 4.3 that are accurate to .001 mV.

   (b) Explain why the resting potentials (V_R) in Table 4.3 are close to the leakage potentials (V_L).

28. Discuss the similarities and differences between the F–N and M–L models of active nerve membrane.

29. Assume a cell of membrane area A, into which an electrophysiologist injects a steady (d.c.) current of I_{app}. If the membrane is described by the M–L model of Equations (4.29), (4.30), and (4.31), the system is governed by the ODE
$$CA\frac{dV}{dt} + J_{\text{ml}}A = I_{\text{app}}.$$

   (a) Find necessary conditions on (4.29) and the parameters of J_{ml} for the system to oscillate.

(b) What are sufficient conditions for oscillation?

30. Consider the larger amplitude traveling-wave solution of a F–N system with $0 < \varepsilon \ll 1$. Show that the distance between the leading and trailing edges is stable. [Hint: compare speeds of the two edges.]

31. A physically realistic description of a Josephson junction transmission line is

$$\beta \frac{\partial^3 u}{\partial x^2 \partial t} + \frac{\partial^2 u}{\partial x^2} - \frac{\partial^2 u}{\partial t^2} - \alpha \frac{\partial u}{\partial t} + \gamma - \sin u = 0,$$

where γ represents a current bias and dissipative effects enter through the terms proportional to α and β [71].

(a) Write the F–N system of Equations (4.25) and (4.26) as a single, third-order PDE, and compare with the above equation.

(b) For the Josephson junction system, discuss the time scales and range of parameters for which the Hamiltonian

$$H(u) \equiv \int_{-\infty}^{\infty} \left[\frac{1}{2} \left(\frac{\partial u}{\partial x} \right)^2 + \frac{1}{2} \left(\frac{\partial u}{\partial t} \right)^2 + 1 - \cos u \right] dx$$

remains approximately constant.

32. (a) From stability analysis of the traveling-wave solution given in Equation (4.7), find the eigenfunction with zero eigenvalue.

(b) Graphically add a small amount of this eigenfunction to the original traveling-wave and show that it introduces a translation.

33. Construct an argument to show that traveling-wave solutions of nonlinear wave equations are never asymptotically stable. [Hint: consider the linear eigenfunction with zero eigenvalue.]

34. Use the concepts of linear stability analysis developed in Appendix D to investigate the stability of a SG kink.

35. (a) Assuming that a periodic solution of a F–N system has the qualitative behavior shown in Figure 4.13, use the relation $v = f\lambda$ to sketch the nonlinear dispersion relation giving f as a function of $1/\lambda$.

(b) In the context of a conservation law for nerve impulses (see Appendix A), show that $D_t + F_x = 0$, where $F = f$ and $D = 1/\lambda$.

(c) Use the conservation law obtained in part (b) to show that a condition for the instability of an almost periodic burst of impulses is [63, 64]

$$\frac{df}{d(1/\lambda)} < 0.$$

(d) Have you experienced a related phenomenon that occasionally occurs in automobile traffic? If so, describe the dynamics. [Hint: consider how a driver's speed depends upon the distance between vehicles.]

36. Consider the Z–F equation as defined in Section 4.1.1, where the energy stored by the wave is taken to be
$$\mathcal{E} = \frac{1}{2} \int_{-\infty}^{\infty} u_t^2 \, dx.$$

(a) Show that
$$\frac{d\mathcal{E}}{dt} = - \int_{-\infty}^{\infty} [u_{xt}^2 + f'(u) u_t^2] \, dx.$$

(b) For $f(u) = u(u-a)(u-1)$ and $u(x,t)$ given by the traveling-wave of Equation (4.7), show that $d\mathcal{E}/dt = 0$.

(c) Interpret your result in the context of Equation (4.1).

(d) How does this interpretation differ from those of Section 4.1.1?

37. Repeat the previous problem for an energy that is taken to be
$$\mathcal{E} = \frac{1}{2} \int_{-\infty}^{\infty} u_x^2 \, dx.$$

38. Derive Equation (4.36) for the amplitude of a decremental nerve impulse.

39. Explain why one should expect $r \propto 1/d^2$, $c \propto d$, and $j_i \propto d$ for a biological nerve fiber.

40. Consider impulse solutions of the tapered nerve equation (4.39), assuming that $1/\gamma$ is much larger than the size of the impulse. Discuss how the impulse speed depends upon γ and x.

41. Use the parameters of the H–H axon given in Tables 4.1 and 4.2 to calculate how the traveling-wave speed changes with the exponential tapering of the fiber cross-section that is indicated in Equations (4.38). Over what range of γ are your results valid?

42. (a) How does the leading edge charge, Q_0, of a H–H nerve impulse depend upon the axon diameter?

(b) How does it depend upon temperature?

43. Assume a two dimensional version of the Z–F equation to be
$$D \left(\frac{\partial^2 V}{\partial x^2} + \frac{\partial^2 V}{\partial y^2} \right) - \frac{\partial V}{\partial t} = \tau^{-1} V(V-a)(V-1),$$

where V is unitless.

(a) Transform this equation to a circular coordinate system.

(b) How does the propagation speed of a circularly symmetric wave front depend upon D, τ, a, and its radius?

(c) Use the result of (b) to estimate the threshold radius of a circularly symmetric initial condition.

44. If the velocity of a periodic plane wave of two-dimensional diffusion is $v_0(\lambda)$ as indicated in Figure 4.13, derive a formula for the rotation frequency (ω) of a spiral wave as a function of λ and r. (For a moving plot of this formula, visit http://www.gbar.dtu.dk/~c958213/integralinvariant/)

45. Using the Z–F equation in the two-dimensional form

$$\lambda^2 \left(\frac{\partial u}{\partial x^2} + \frac{\partial u}{\partial y^2} \right) - \tau \frac{\partial u}{\partial t} = u(u - 0.25)(u - 1)$$

to describe the dynamics of a fairy ring of mushrooms that is 1 m wide and propagates outward at a speed of 1 m/year:

(a) Estimate values for the parameters λ and τ.

(b) What is the linear diffusion coefficient for this system?

46. Show that Equation (4.46) for ring waves follows from Equation (4.45) under the assumption that the propagator and recovery variables are functions only of r and t.

47. A lichen comprises a fungus growing together with an alga on a rock or tree trunk. In Figure 4.21 are shown two photographs of a lichen growing on a brick wall in Scotland. Taken 12 years apart, these observations suggest that the lichen is (among other things) a spiral wave.

(a) Assuming this to be so, estimate v_0 in Equation (4.47).

(b) What is the range of error in your estimate?

48. Given

$$\nabla^2 \equiv \frac{\partial^2}{\partial x^2} + \frac{\partial^2}{\partial y^2} + \frac{\partial^2}{\partial z^2}$$

in a rectangular coordinate system, explain how to derive the form of this operator for a spherical coordinate system.

49. Consider the nonlinear Schrödinger equation

$$i\frac{\partial \psi}{\partial t} + \nabla^2 \psi + 2|\psi|^2 \psi = 0,$$

where $\psi = u + iw$. Compare the equations for u and w with those in Equations (4.49).

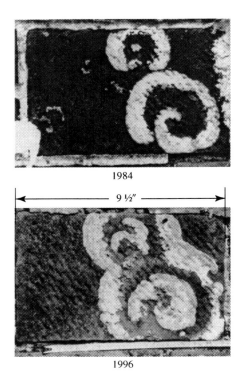

FIG. 4.21. *Lichens growing on a wall in Scotland may be a biological example of spiral waves. (Photographs by A. Burgess and A.T. Winfree.)*

50. Repeat the previous problem for the complex Ginzburg–Landau equation

$$i\frac{\partial \psi}{\partial t} + (1 + ic_0)\psi + (1 + ic_1)\nabla^2\psi + (2 + ic_2)|\psi|^2\psi = 0,$$

where c_0, c_1, and c_2 are real parameters.

51. (a) Repeat the previous problem for the quintic complex Swift–Hohenberg equation of nonlinear optics

$$\frac{\partial \psi}{\partial t} = a\psi + b|\psi|^2\psi - c|\psi|^4\psi - d(1 + \nabla^2)^2\psi + If\nabla^2\psi,$$

where a, b, c, d, and f are complex parameters.

(b) For what parameter values does the Swift–Hohenberg equation reduce to the nonlinear Schrödinger equation?

52. Suppose the vortex line of a 3D ring wave solution of Equations (4.49) is an ellipse. How do you expect the eccentricity of the ellipse to vary with time? Can you express your result quantitatively?

53. (a) Discuss why the development of stationary (Turing) patterns of nonlinear diffusion requires $D_w \gg D_u$, as defined in Equations (4.49).

(b) For a Turing system of one space dimension as defined in Equations (4.49) with $D_u = 1$, $D_w = 12$, and

$$f(u, w) = u - 30 + 4uw/(1 + u^2)$$
$$g(u, w) = -12[u - uw/(1 + u^2)],$$

analyze the spatially independent oscillation in the (u, w) phase plane. (See reference [10] for corresponding experimental details.)

REFERENCES

1. W J Adelman, Jr. and R FitzHugh. Solutions of the Hodgkin–Huxley equations modified for potassium accumulation in a periaxonal space. *Fed. Proc. (Fed. Am. Soc. Exp. Biol.)* 34 (1975) 1322–1329.
2. R Altenberger, K A Lindsay, J M Ogden, and J R Rosenberg. The interaction between membrane kinetics and membrane geometry in the transmission of action potentials in non-uniform excitable fibres: A finite element approach. *J. Neurosci. Meth.* 112 (2001) 101–117.
3. M B Berkinblit, N D Vvedenskaya, L S Gnedenko, S A Kovalev, A V Kholopov, S V Fomin, and L M Chailakhyan. Computer investigation of the features of conduction of a nerve impulse along fibers with different degrees of widening. *Biophysics* 15 (1970) 1121–1130.
4. M C Boerlijst and P Hogeweg. Spiral wave structure in prebiotic evolution: Hypercycles stable against parasites. *Physica D* 48 (1991) 17–28.
5. M C Boerlijst, M E Lamers, and P Hogeweg. Evolutionary consequences of spiral waves in a host-parasitoid system. *Proc. R. Soc. (London)* B 253 (1993) 15–18.
6. L S Bogoslovskaya, I A Lyubinskii, N V Pozin, Ye V Putsillo, L A Shmelev, and T M Shura-Bura. Spread of excitation along a fiber with local inhomogeneities (results of modelling). *Biophysics* 18 (1973) 944–948.
7. W C Bray. A periodic reaction in homogeneous solution and its relation to catalysis. *J. Am. Chem. Soc.* 43 (1921) 1262–1267.
8. P C Bressloff. A dynamical theory of spike train transitions in networks of integrate-and-fire oscillators. *SIAM J. Appl. Math.* 60 (2000) 820–841.
9. J M Burgers. A mathematical model illustrating the theory of turbulence. *Adv. Appl. Mech.* 1 (1948) 171–199.
10. V Castets, E Dulos, J Boissonade, and P De Kepper. Experimental evidence of a sustained standing Turing-type nonequilibrium chemical pattern. *Phys. Rev. Lett.* 64 (1990) 2953–2956.
11. J D Cole. On a quasilinear parabolic equation occurring in aerodynamics. *Q. Appl. Math.* 9 (1951) 225–236.
12. K S Cole. *Membranes, Ions and Impulses.* University of California Press, Berkeley, 1968.

13. J W Cooley and F A Dodge. Digital computer solutions for excitation and propagation of the nerve impulse. *Biophys. J.* 6 (1966) 583–599.
14. M Eigen and P Schuster. *The Hypercycle: A Principle of Natural Self-organization.* Springer-Verlag, Berlin, 1979.
15. J W Evans. Nerve axon equations: I. Linear approximations, *Indiana Univ. Math. J.* 21 (1972) 877–885.
16. J W Evans. Nerve axon equations: II. Stability at rest. *Indiana Univ. Math. J.* 22 (1972) 75–90.
17. J W Evans. Nerve impulse equations: III. Stablity of the nerve impulse. *Indiana Univ. Math. J.* 22 (1972) 577–593.
18. J W Evans. Nerve impluse equations: IV. The stable and unstable impluse. *Indiana Univ. Math. J.* 24 (1975) 1169–1190.
19. C P Fall, E S Marland, J M Wagner, and J J Tyson. *Computational Cell Biology.* Springer-Verlag, New York, 2002.
20. M Faraday. *Faraday's Chemical History of a Candle.* Chicago Review Press, Chicago, 1988. (A republication of *A Course of Six Lectures on the Chemical History of a Candle*, which first appeared in 1861.)
21. R A Fisher. The wave of advance of advantageous genes. *Ann. Eugen.* (now *Ann. Hum. Gen.*) 7 (1937) 355–369.
22. R FitzHugh. Impulses and physiological states in theoretical models of nerve membrane. *Biophys. J.* 1 (1961) 445–466.
23. R FitzHugh. A kinetic model for the conductance changes in nerve membranes. *J. Cell. Compar. Physiol.* 66 (1965) 111–117.
24. R FitzHugh. Mathematical models of excitation and propagation in nerve. In *Biological Engineering*, H P Schwann, ed., McGraw-Hill, New York, 1969, pp. 1–85.
25. F Gabbiani, W Metzner, R Wessel, and C Koch. From stimulus encoding to feature extraction in weakly electric fish. *Nature* 384 (1996) 564–567.
26. S S Goldstein and W Rall. Changes of action potential, shape and velocity for changing core conductor geometry. *Biophys. J.* 14 (1974) 731–757.
27. B Hille. *Ion Channels of Excitable Membranes.* 3rd edition. Sinauer Association, Sunderland, Massachusetts, 2001.
28. A L Hodgkin and A F Huxley. A quantitative description of membrane current and its application to conduction and excitation in nerve. *J. Physiol.* 117 (1952) 500–544.
29. E Hopf. The partial differential equation $u_t + uu_x = \mu u_{xx}$. *Comm. Pure Appl. Math.* 3 (1950) 201–230.
30. R C Hoyt. The squid giant axon: Mathematical models. *Biophys. J.* 3 (1963) 399–431.
31. A F Huxley. Can a nerve propagate a subthreshold disturbance? *J. Physiol. (London)* 148 (1959) 80P–81P.
32. W Jahnke, C Henze, and A T Winfree. Chemical vortex dynamics in three-dimensional excitable media. *Nature* 336 (1988) 662–665.
33. C K R T Jones. Some ideas in the proof that the FitzHugh–Nagumo pulse is stable. In *Nonlinear Partial Differential Equations*, J Smoller ed., Contemporary Mathematics 17, American Mathematical Society, Providence, 1984, pp. 287–292.

34. C K R T Jones, Stability of the travelling wave solution of the FitzHugh–Nagumo system. *Trans. Am. Math. Soc.* 286 (1984) 431–469.
35. B I Khodorov. *The Problem of Excitability*. Plenum, New York, 1974.
36. B I Khodorov, Ye N Timin, S Ya Vilenkin, and F B Gul'ko. Theoretical analysis of the mechanisms of conduction of a nerve impulse over an inhomogeneous axon. I. Conduction through a portion with increased diameter. *Biophysics* 14 (1969) 323–335.
37. B W Knight. Dynamics of encoding in a population of neurons. *J. Gen. Physiol.* 59 (1972) 734–766.
38. B W Knight. The relationship between the firing rate of a single neuron and the level of activity in a population of neurons. *J. Gen. Physiol.* 59 (1972) 767–778.
39. A Kolmogoroff, I Petrovsky, and N Piscounoff. Étude de l'équation de la diffusion avec croissance de la quantité de matière et son application a un problème biologique. *Bull. Univ. Moscow, Sér. Int.* A1 (1937) 137–149.
40. A S Kompaneyets and V T Gurovich. Propagation of an impulse in a nerve fiber. *Biophysics* 11 (1966) 1049–1052.
41. K N Leibovic. *Nervous System Theory: An Introductory Study*. Academic Press, New York, 1972.
42. I Lengyel, S Kádár, and I R Epstein. Quasi-two-dimensional Turing patterns in an imposed gradient. *Phys. Rev. Lett.* 69 (1992) 2729–2732.
43. S A Levin. Complex adaptive systems: Exploring the known, the unknown and the unknowable. *Bull. Am. Math. Soc.* 40 (2002) 3–19.
44. A G Lindgren and R J Buratti. Stability of waveforms on active non-linear transmission lines. *Trans. IEEE Circuit Theory* 16 (1969) 274–279.
45. R R Llinás. The intrinsic electrophysiological properties of mammalian neurons: Insights into central nervous system function. *Science* 242 (1988) 1654–1664.
46. R Lorente de Nó and G A Condouris. Decremental conduction in peripheral nerve: Integration of stimuli in the neuron. *Proc. Natl. Acad. Sci. USA* 45 (1959) 593–617.
47. R Luther. Räumliche Fortpflanzung chemischer Reaktionen. *Z. Elektrochem.* 12(32) (1906) 596–600. [English translation in *J. Chem. Ed.* 64 (1987) 740–742.]
48. V S Markin and Y A Chizmadzhev. On the propagation of an excitation for one model of a nerve fiber. *Biophysics* 12 (1967) 1032–1040.
49. A Mauro, F Conti, F Dodge, and R Schor. Subthreshold behavior and phenomenological impedance of the squid giant axon. *J. Gen. Physiol.* 55 (1970) 497–523.
50. E Mehron. Pattern formation in excitable media. *Phys. Rep.* 218 (1992) 1–66.
51. A S Mikhailov, V A Davydov, and V S Zykov. Complex dynamics of spiral waves and motions of curves. *Physica D* 70 (1994) 1–39.
52. C Morris and H Lecar. Voltage oscillations in the barnacle giant muscle. *Biophys. J.* 71 (1981) 193–213.
53. E Mosekilde. *Topics in Nonlinear Biology*. World Scientific, Singapore, 1996.
54. J Nagumo, S Arimoto, and S Yoshizawa. An active impulse transmission line simulating nerve axon. *Proc. IRE* 50 (1962) 2061–2070.
55. K Naugolnykh and L Ostrovsky. *Nonlinear Wave Processes in Acoustics*. Cambridge University Press, Cambridge, 1998.

56. A H Nayfeh and B Balachandran. *Applied Nonlinear Dynamics*. John Wiley, New York, 1995.
57. D Noble and R B Stein. The threshold conditions for initiation of action potentials by excitable cells. *J. Physiol. (London)* 187 (1966) 129–142.
58. F Offner, A Weinberg, and C Young. Nerve conduction theory: Some mathematical consequences of Bernstein's model. *Bull. Math. Biophys.* 2 (1940) 89–103.
59. A F Parker-Rhodes. Fairy ring kinetics. *Trans. Br. Mycol. Soc.* 38 (1955) 59–72.
60. V F Pastushenko, Y A Chizmadzhev, and V S Markin. Speed of excitation in the reduced Hodgkin–Huxley model: I. Rapid relaxation of sodium current. *Biophysics* 20 (1975) 685–692.
61. V F Pastushenko, Y A Chizmadzhev, and V S Markin. Speed of excitation in the reduced Hodgkin–Huxley model: II. Slow relaxation of sodium current. *Biophysics* 20 (1975) 894–901.
62. S Ramón y Cajal. *Histologie du Système Nerveux de l'Homme et des Vertébrés*. Malaine, Paris, 1909.
63. J Rinzel. Spatial stability of traveling-wave solutions of a nerve conduction equation. *Biophys. J.* 15 (1975) 975–988.
64. J Rinzel. Neutrally stable traveling-wave solutions of nerve conduction equations. *J. Math. Biol.* 2 (1975) 205–217.
65. J Rinzel and J Keller. Traveling-wave solutions of a nerve conduction equation. *Biophys. J.* 13 (1973) 1313–1337.
66. P Rissman. The leading edge approximation to the nerve axon equation. *Bull. Math. Biol.* 39 (1977) 43–58.
67. A C Scott. Neuristor propagation on a tunnel diode loaded transmission line. *Proc. IEEE* 51 (1963) 240.
68. A C Scott. Effect of the series inductance of a nerve axon upon its conduction velocity. *Math. Biosci.* 11 (1971) 277–290.
69. A C Scott. Strength duration curves for threshold excitation of nerves. *Math. Biosci.* 18 (1973) 137–152.
70. A C Scott. *Neuroscience: A Mathematical Primer*. Springer-Verlag, New York, 2002.
71. A C Scott, F Y F Chu, and S A Reible. Magnetic-flux propagation on a Josephson transmission line. *J. Appl. Phys.* 47 (1976) 3272–3286.
72. L Scott MacNeil. Report to the 7th Grade Science Class at Green Fields Country Day School, Tucson, Arizona, March 1997.
73. F Siegert and C J Weijer. Three dimensional scroll waves organize *Dictyostelium* slugs. *Proc. Natl. Acad. Sci. USA* 89 (1992) 6433–6437.
74. D W Thompson. *On Growth and Form*. Abridged edition. Cambridge University Press, Cambridge, 1961.
75. J Tille. A new interpretation of the dynamic changes of the potassium conductance in the squid giant axon. *Biophys. J.* 5 (1965) 163–171.
76. A Turing. The chemical basis of morphogenesis. *Philos. Trans. R. Soc. (London)* B 237 (1952) 37–72.
77. J J Tyson and J P Keener. Singular perturbation theory of traveling waves in excitable media. *Physica D* 32 (1988) 327–361.

78. Yu L Vorontsov, M I Kozhevnikova, and I V Polyakov. Wave processes in active RC–lines. *Rad. Eng. Elec. Phys.* 11 (1967) 1449–1456.
79. G B Whitham. *Linear and Nonlinear Waves*. John Wiley, New York, 1974.
80. H R Wilson. *Spikes, Decisions, and Actions: The Dynamical Foundations of Neuroscience*. Oxford University Press, Oxford, 1999.
81. A T Winfree. Spiral waves of chemical activity. *Science* 175 (1972) 634–636.
82. A T Winfree. Wavefront geometry in excitable media. *Physica D* 12 (1984) 321–332.
83. A T Winfree. *When Time Breaks Down: The Three-Dimensional Dynamics of Electrochemical Waves and Cardiac Arrhythmias*. Princeton University Press, 1987.
84. A T Winfree. Stable particle-like solutions to the nonlinear wave equations of three-dimensional excitable media. *SIAM Rev.* 32 (1990) 1–53.
85. A T Winfree. The geometry of excitability. In *1992 Lectures in Complex Systems*, L Nadel and D Stein, eds., Addison-Wesley, Reading, MA, 1992, pp. 207–298.
86. A T Winfree. Persistent tangles of vortex rings in excitable media. *Physica D* 84 (1995) 126–147.
87. E Yanagida. Stability of fast travelling pulse solutions of the FitzHugh–Nagumo equations. *J. Math. Biol.* 22 (1985) 81–104.
88. A N Zaikin and A M Zhabotinsky. Concentration wave propagation in two-dimensional liquid-phase self-oscillating systems. *Nature* 225 (1970) 535–537.
89. Ya B Zeldovich and D A Frank-Kamenetsky. K teorii avnomernogo rasprostranenia plameni. *Dokl. Akad. Nauk SSSR* 19 (1938) 693–697.
90. Y B Zeldovich and G I Barenblatt. Theory of flame propagation. *Combust. Flame* 3 (1959) 61–74.
91. A M Zhabotinsky, M Dolnik, and I R Epstein. Pattern formation arising from wave instability in a simple reaction-diffusion system. *J. Chem. Phys.* 103 (1995) 10306–10314.
92. A M Zhabotinsky and A N Zaikin. Autowave processes in a distributed chemical system. *J. Theor. Biol.* 40 (1973) 45–61.

5
NONLINEAR LATTICES

Hitherto we have considered nonlinear partial differential equations, in which both the time and space coordinates are continuous independent variables. In the present chapter, we take the spatial variable to be a lattice of points, so the objects of study are sets of nonlinear ordinary differential equations, with time remaining a continuous independent variable.

Such problems have a venerable place in the history of nonlinear wave studies, going back at least to the 1939 investigation by Frenkel and Kontorova of plastic deformations in crystals that led them to a spatially discrete version of the sine–Gordon equation [60]. Fourteen years later, the famous Fermi–Pasta–Ulam problem—spawning modern soliton research—was defined on a lattice so it could be numerically integrated on the recently inaugurated MANIAC computer at Los Alamos [49]. Finally it is appropriate to mention the system of masses interconnected by exponential interaction potentials that was introduced in 1967 by Toda [128]. He noted that such an "exponential lattice" is nonergodic, possessing nonlinear normal modes of oscillation, a result that was obtained independently of the work by Kruskal and his colleagues at about the same time [64].

Expanding upon such considerations, there are several reasons for interest in and studies of nonlinear lattices:

1. Certain nonlinear lattices are integrable in the sense defined in the following chapter, and therefore of enhanced theoretical interest.
2. Small molecules (benzene, methane, ammonia, etc.) exhibit nonlinear local modes, as was known by physical chemists of the 1920s.
3. Crystals, molecular crystals, and polymers are nonlinear lattices, which can support the formation of localized oscillating regions.
4. Classical nonlinear lattices provide models for studies of biochemical molecules such as proteins and DNA.
5. Interconnected one-, two-, and three-dimensional arrays of nonlinear oscillators are model systems with different modes competing for a common source of energy.
6. The myelinated structure of vertebrate motor nerves—which forces an impulse to jump from one active node to the next and leads to increased impulse velocity—can be viewed as a nonlinear biological lattice.
7. Cellular automata are nonlinear lattices.
8. The most intricate dynamic structure known to exist in the universe, the human brain, is sometimes described as a lattice of neurons.

Books have been written on each of these items; thus the aim here is merely to describe a few of the key ideas, with the hope that the reader will consider entering some phase of this research.

5.1   Spring-mass lattices

The basic idea of a spring-mass lattice, shown in the upper part of Figure 5.1, is an infinite chain of identical masses (m) interconnected with nonlinear springs. The springs are assumed to have potential energy $U(r)$, where r is the increase in distance between adjacent masses from the resting value at which the spring energy is a minimum and its force

$$F = -dU/dr$$

is zero. If y_n is the longitudinal displacement of the nth mass from its equilibrium position, Newton's second law implies that

$$m\frac{d^2 y_n}{dt^2} = U'(y_{n+1} - y_n) - U'(y_n - y_{n-1}).$$

With $r_n = y_{n+1} - y_n$, this gives the infinite set of ODEs

$$m\frac{d^2 r_n}{dt^2} = \left(\frac{dU(r_{n+1})}{dr_{n+1}} - \frac{dU(r_n)}{dr_n}\right) - \left(\frac{dU(r_n)}{dr_n} - \frac{dU(r_{n-1})}{dr_{n-1}}\right), \qquad (5.1)$$

where $-\infty < n < +\infty$.

In this section, we consider some of the theoretical results that have been obtained for this system. For clarity, we call these systems "spring-mass lattices," noting that they are also called the "FPU lattices" in honor of the seminal studies of Fermi et al. [49].

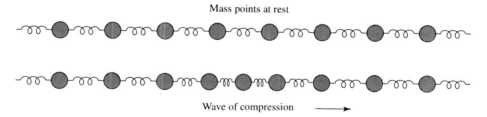

FIG. 5.1. *The upper figure shows a spring-mass lattice at rest with all values of y_n and $r_n = y_{n+1} - y_n$ equal to zero. The lower figure shows a compressive wave, corresponding to a Toda-lattice soliton (TLS) given in Equation (5.5). (For a TLS, all values of r_n are less than zero.) (Note that there is one more mass on the lower part of the diagram.)*

5.1.1 The Toda-lattice soliton

In a seminal paper that appeared in 1967, Toda studied a spring-mass lattice with anharmonic potentials of the form

$$U(r) = \frac{a}{b}(e^{-br} + br - 1), \tag{5.2}$$

where $a, b > 0$. Evidently the force derived from this potential is zero at $r = 0$. With unit masses ($m = 1$), Equation (5.1) takes the form

$$\frac{d^2 r_n}{dt^2} = a(2e^{-br_n} - e^{-br_{n+1}} - e^{-br_{n-1}}), \tag{5.3}$$

which are called the Toda-lattice equations.

In the limit $b \to 0$ with ab finite, these equations reduce to the linear difference-differential equations

$$\frac{d^2 r_n}{dt^2} = ab(r_{n+1} - 2r_n + r_{n-1}), \tag{5.4}$$

with solutions having a long wavelength speed of $\sqrt{ab}$ lattice points per unit time.

If the parameter b is not small, Equations (5.1) have exact solitary wave solutions of the form

$$r_n = -\frac{1}{b} \log[1 + \sinh^2 \kappa \operatorname{sech}^2(\kappa n \pm t\sqrt{ab} \sinh \kappa)], \tag{5.5}$$

which are derived in Appendix E. The velocity of this localized lattice wave is given in terms of the amplitude parameter (κ) as

$$v = \sqrt{ab} \, \frac{\sinh \kappa}{\kappa}. \tag{5.6}$$

Results for such solutions of the Toda lattice can be summarized as follows.

1. Solitary wave solutions of the Toda lattice can also be generated via a Bäcklund transform (BT) (see Problem 6), so Equation (5.5) describes a lattice soliton [129, 136]. Collisions of these solitons are exactly described by the N-soliton solutions given in Equations (B.7) and (B.8) of Appendix B.
2. The minus sign in Equation (5.5) indicates that the Toda-lattice soliton (TLS) is a compression wave, where the masses come closer together as shown in Figure 5.1.
3. As the amplitude of a TLS is reduced to zero (by allowing $\sinh \kappa$ to approach zero), it reduces to a solution of the linear equation (5.4) moving at $v = \sqrt{ab}$.
4. The speed of a TLS is greater than that of small amplitude waves.

5. The Toda-lattice formulation is related to a family of self-dual lattices, from which it emerged in the mid-1960s [28, 126, 127, 130]. Self-dual lattices have been proposed as a means of coding [123].

Interestingly, these results hold for all forms of the Toda potential, from the linear limit of Equation (5.4) to the "hard sphere" limit, where $b \to \infty$ while ab remains finite.

5.1.2 Lattice solitary waves

From the perspective of the early 1970s, it is remarkable that a spring-mass lattice is completely integrable (in the sense of Chapter 6) for the special form of the interaction potential introduced by Toda in Equation (5.2). With any other potential, exact integrability is lost, but it is still possible to find solitary wave solutions. Starting with Equations (5.1) and assuming a traveling-wave solution of the form

$$r_n(t) = R(n - vt) \equiv R(z) \tag{5.7}$$

with $R(z) \to 0$ as $z \to \pm\infty$ implies the equation

$$v^2 R''(z) = U'[R(z+1)] - 2U'[R(z)] + U'[R(z-1)]. \tag{5.8}$$

Eilbeck and Flesch have shown how to obtain numerical solutions for such equations, using pseudospectral techniques and path following methods to trace out a family of solutions for different values of v [39]. Under this procedure, $R(z)$ is approximated as a finite Fourier cosine series (of n terms) over a finite interval, and curves showing the relationship between pulse speed and amplitude can be calculated handily on a standard work station. The convergence of these calculations is observed to be "superalgebraic" in the number of Fourier terms, meaning that errors ultimately fall off faster than any finite power of $1/n$.

Several techniques for finding corresponding analytic approximations for lattice solitary waves (LSWs) have been proposed. For a potential of the form

$$U(r) = r^2/2 + ar^3/3, \tag{5.9}$$

Collins has suggested an approximate inversion of the difference operator on the right-hand side of Equation (5.8) [25], and Roseneau [111] and Hochstrasser et al. [73] have obtained an approximation as the Boussinesq equation. Wattis has developed a rather general approach based on Fourier transform analysis, which is sketched in Appendix F [37]. (A discussion of these approximation methods has been presented by Remoissenet [110].)

With the potential as in Equation (5.9), Equation (F.3) can be written as

$$R'' = \left(\frac{12(v^2-1)}{v^2}\right) R - \left(\frac{12a}{v^2}\right) R^2.$$

Under an analysis identical to that for the KdV equation in Section 3.1.2, this equation has the solitary-wave solution

$$R(z) = r_n(t) = (3/2a)(v^2 - 1)\operatorname{sech}^2\left[(n - vt)\sqrt{3(v^2 - 1)/v^2}\right]. \qquad (5.10)$$

From studies of these equations, the following comments can be made about LSWs.

1. With the intermass potential given in Equation (5.9) LSWs can be either compressive (for $a < 0$) or expansive (for $a > 0$). (A wave of compression is shown in Figure 5.1.)
2. In both cases ($a > 0$ and $a < 0$), the speed of the LSW is greater than unity, the wave speed in the low-amplitude limit. Thus these are dynamic objects with no zero-velocity limits.
3. From Figure 5.1, a wave of compression ($a < 0$ and $r_n < 0$) transports one or more mass points in the direction of propagation. Similarly, an expansive wave moves mass in the opposite direction.
4. As v^2 approaches unity from above with $a < 0$, the solution given in Equation (5.10) approaches the corresponding expression for a TLS in Equation (5.5).
5. As v^2 approaches unity from above, the amplitude of the LSW approaches zero, its width increases without bound and the accuracy of the approximation improves. Thus Equation (5.10) describes "large" solitary waves that extend over many lattice spacings.

For systems more intricate than the simple spring-mass lattice of Figure 5.2, there is a rich array of phenomena to be explored. Wattis has used a quasi-continuum approximation to study the behavior of lattices with second neighbor interactions, which allow waves with oscillating (non-monotone) behavior at subsonic speeds [140]. More recently diatomic lattices have been studied in the quasi-continuum approximation, extending the results of previous analyses to allow propagation over a range of speeds [125, 142] rather than at a single speed [105].

Although a mechanical ladder with linear springs and nonlinear masses is physically unreasonable, a corresponding electrical system is readily realized using shunt nonlinear capacitors interconnected with series linear inductors [99, 110]. This "discrete nonlinear telegraph equation" is a generalization of Toda's nonlinear electric filter (discussed in Appendix B4), which has recently been studied in a quasi-continuum approximation and shown to have both localized and periodic solutions [141].

5.1.3 Existence of lattice solitary waves

Toda's results demonstrate the existence of solitary waves on spring-mass lattices, but only for the particular value of inter-mass potential given in Equation (5.2). The numerical results cited in the previous section, however, suggest the existence of solitary waves on spring-mass lattices for a wider class of potential functions. It was in this context that Friesecke and Wattis showed existence of lattice solitary waves if $U(r)$ is superquadratic (grows faster than r^2) for either positive or negative values of r [63]. By viewing solutions as minimizers of an associated variational problem, they proved the following.

- *Theorem:* Assume that $U(r)$ has a second derivative, with $U(0) = 0$ and $U(r) \geq 0$ in some neighborhood of the origin. A sufficient condition for solitary wave solutions of Equation (5.1) to exist is that $U(r)$ be superquadratic on at least one side. (In other words, $U(r)/r^2$ increases strictly with $|r|$ either for r between zero and some positive value R_p, or between zero and some negative value $-R_n$, where R_p and R_n could be finite or infinite.) Furthermore, these solitary waves are supersonic, with arbitrarily large and small amplitudes, and they are either all positive or all negative (entirely expansive or compressive).

Subsequent results on the existence of traveling-waves include these. First, constructive proofs for small-amplitude traveling-waves by Deift et al. [34] and by Iooss [75]. Second, a proof of solitary-wave existence for a prescribed speed by Smets and Willem [122]. Third, theorems by Filip and Venakides that demonstrate existence of periodic waves for all amplitudes while requiring $U(r)$ to be convex ($U'' > 0$ for all r) [50]. Filip and Venakides also describe numerical methods for constructing periodic solutions and formulate Whitham's averaged-Lagrangian method (see Section 3.2.3 and reference [144]) for evolution of the periodic-wave parameters. Finally, the stability of traveling-waves on spring-mass lattices is under investigation by Friesecke and Pego [61, 62].

To fix ideas, we can test the Friesecke–Wattis (FW) theorem against exact solutions of the Toda lattice. With $U(r)$ as in Equation (5.2), $U(0) = 0$, $U''(0) = 0$, and with $r < 0$,

$$U(r)/r^2 = (ab/2) + (ab^2/6)|r| + (ab^3/24)|r|^2 + \cdots.$$

Thus $U(r)$ is superquadratic for negative r, and solitary waves are expected, in agreement with Equation (5.5). These Toda lattice solitons are supersonic, with amplitudes in the range $(-\infty, 0)$, and entirely compressive, in accord with the theorem. For $r > 0$, $U(r)$ is not superquadratic, and solitary wave solutions do not exist.

For the cubic potential of Equation (5.9), this theorem provides new information. With positive a and $r > 0$,

$$U(r)/r^2 = 1/2 + a|r|/3,$$

while for negative a and $r < 0$

$$U(r)/r^2 = 1/2 + |a||r|/3;$$

thus solitary waves are anticipated in both of these cases. These solitary waves are not limited to the small amplitudes for which Equation (5.10) is valid, in accord with numerical results [39]. For $a = 0$, $U(r)$ is not superquadratic, and there are only nonlocalized linear waves.

Another example is the quartic potential

$$U(r) = r^2/2 + br^4/4, \qquad (5.11)$$

for which

$$U(r)/r^2 = 1/2 + b|r|^2/4.$$

This potential is superquadratic for $b > 0$ but not for $b < 0$, implying solitary waves for positive values of b but not for negative values, again in agreement with numerical studies [39].

Thus the FW theorem is useful in offering an easily applied test for existence of solitary wave solutions on spring-mass lattices. This theorem is also scientifically important for two reasons. First, FW demonstrates that compression and expansion waves in spring-mass lattice are generic, to be expected for most physically reasonable intermass potentials. Second, FW predicts supersonic solitary waves, providing a theoretical explanation for Scott Russell's observation that "the sound of a cannon firing travels faster than the command to fire it."

5.1.4 *Intrinsic localized modes and intrinsic gap modes*

During the 1970s, many physical chemists assumed that perfect crystals could only have solutions sharing the lattice symmetry, thereby precluding the possibility of anharmonic localization. In the late 1980s, however, Dolgov [35], Sievers and Takeno [120], and Page [100] showed that one should also expect to find "small" localized modes which occupy only a few lattice sites and oscillate.

The basic idea of such a mode can be understood by assuming a uniform spring-mass lattice with the quartic inter-mass potential given in Equation (5.11), leading to the difference-differential system

$$\frac{d^2 r_n}{dt^2} = r_{n+1} - 2r_n + r_{n-1} + b(r_{n+1}^3 - 2r_n^3 + r_{n-1}^3), \qquad (5.12)$$

where all masses are assumed unity, $r_n = y_{n+1} - y_n$, and y_n is the displacement of the nth mass.

With $b = 0$ (no anharmonicity), L lattice sites, and periodic boundary conditions ($r_{n+L} = r_n$), these equations have the solutions

$$r_n(t) = A\cos(kn - \omega t), \qquad (5.13)$$

with A an arbitrary amplitude and

$$\omega = 2|\sin(k/2)| \quad \text{and} \quad k = 2\pi\nu/L. \qquad (5.14)$$

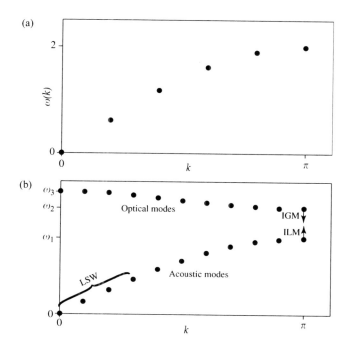

FIG. 5.2. (a) A plot of the acoustic mode dispersion relation for a uniform lattice given in Equation (5.14) with 10 masses and periodic boundary conditions. (b) A corresponding plot for two atoms in each of 20 unit cells, leading to a band of optical modes.

For L even, the allowed values of k are for $\nu = 0, \pm 1, \ldots, \pm(L/2 - 1), L/2$, and for L odd, $\nu = 0, \pm 1, \ldots, \pm(L-1)/2$.

Values of ω vs. k for these solutions are plotted in Figure 5.2(a) for a system of $L = 10$ mass points, and it is helpful to consider this figure carefully. Note first that there are 10 modes indicated (for $k = 0$, $\pm\pi/5$, $\pm 2\pi/5$, $\pm 3\pi/5$, $\pm 4\pi/5$, and π), corresponding to the 10 masses of the system. Second, the mode with $k = 0$ and $\omega = 0$ corresponds to a uniform translation of the chain (i.e. no dynamics). Third, from Equation (5.13) it is evident that k indicates the phase shift of a mode between adjacent lattice sites. This phase shift is zero for the translation mode ($k = 0$) and a maximum for $k = \pi$, where two adjacent mass points are moving 180° out of phase; thus anharmonic effects are maximum where $k = \pi$. In the language of solid-state physics, the plot of ω vs. k shown in Figure 5.2(a) is called an "acoustic mode" band.

With more than one atom in each unit cell, the spectrum becomes more interesting, as indicated in Figure 5.2(b). Here it is assumed that there are two different masses in each unit cell and 20 unit cells, whereupon an "optical mode" band additionally appears.[1] (The names are suggested by the experimental fact that optical modes appear in the infra-red range, whereas acoustic mode frequencies can fall to zero.)

With $b > 0$, Equation (5.12) shows that the frequencies in Figure 5.2(a) will increase by an amount that depends upon bA^2. To estimate this increase, write

$$\cos^3(kn - \omega t) = \tfrac{3}{4}\cos(kn - \omega t) + \tfrac{1}{4}\cos(3kn - 3\omega t)$$
$$\approx \tfrac{3}{4}\cos(kn - \omega t),$$

where the $\cos(3kn - 3\omega t)$ component is neglected because it oscillates at three times the fundamental frequency and is therefore not resonant with the fundamental components. This approximation is called the "rotating-wave approximation" (RWA) in physics [91], "harmonic balance" in engineering [5], and a "truncated Birkhoff normal form" in mathematics [17]. Under the RWA, Equation (5.14) becomes

$$\omega \approx 2\sqrt{1 + 3bA^2/4}|\sin(k/2)|,$$

which indicates that mode frequencies increase with amplitude.

As mode amplitudes are increased, the mode of highest frequency—at $k = \pi$ in Figure 5.2(a)—evolves into an intrinsic localized mode (ILM) that oscillates and is nonzero over only a few lattice sites [120]. Thus ILMs are qualitatively different from the LSWs which were discussed in the previous section. An excellent introduction to ILMs in one- and three-dimensional crystal models is available at http://www.lassp.cornell.edu/%7Esievers/ilm/index.html, from which some of the following items are extracted.

1. ILMs are becoming of interest in solid state physics for their structural stability, which appears when the oscillation frequency and its overtones lie within stop bands of the crystal lattice, where small amplitude phonons cannot propagate. Because stop bands stem from discreteness of the lattice, this means that discreteness encourages the stability of ILMs.
2. If $b < 0$ in Equation (5.11), mode frequencies are depressed (into the acoustic mode band) with increasing amplitude and no ILM appears. Using a model with two different masses in each unit cell, however, the dispersion relation contains two branches, describing optical modes in addition to acoustic modes as shown in Figure 5.2(b). In this case, a negative value for the parameter b results in the emergence of an "intrinsic gap mode" (IGM) from the lower edge of the optical mode band.

---

[1] In the optical mode at frequency (ω_2), the two atoms of a unit cell are in phase with each other, whereas adjacent unit cells are out of phase. At the other end of the band (ω_3), the atoms within a unit cell are out of phase, while adjacent unit cells are in phase. Notice that the number of modes (40) is equal to the number of masses (2×20).

3. The quartic potential of Equation (5.11) is not a realistic interatomic potential, which would include a cubic term among several others [82]. When more realistic nearest neighbor potentials are examined, lowering of the vibrational frequency with mode amplitude may preclude the existence of ILMs, but IGMs may still appear [82]. Three dimensional lattices present interesting variations on these basic themes.
4. Empirically, ILMs and IGMs are expected to be energized at finite temperatures; thus they may play a role in determining the thermal properties of crystals. In their original publications, Sievers and Takeno proposed that the existence of such states can rectify inconsistencies between calculations and measurements of specific heat in solid helium [120] and in glass [121]. Evidence of ILMs has also been found from observations of ionic mobilities and the spectra of ionic crystals [83, 120, 124].

Mathematical existence of ILMs on spring-mass lattices has recently been established by James [77]. He shows that a sufficient condition on the inter-site potential $U(r)$ is

$$U''(0)U''''(0) > 2[U'''(0)]^2. \tag{5.15}$$

This condition is satisfied for the quartic potential of Equation (5.11); thus ILMs are expected in this case, in accord with the numerical observations of Dolgov [35] and of Sievers and Takeno [120]. On the other hand, James's existence condition fails for the cubic potential of Equation (5.9) and for the Toda potential.

5.2 Lattices with nonlinear on-site potentials

In the above discussions, we have considered chains of equal masses interconnected by nonlinear springs. Here we look at lattices of nonlinear oscillators. The basic equation is

$$\frac{d^2 u_n}{dt^2} = \varepsilon(u_{n+1} + u_{n-1}) - \frac{dV(u_n)}{du_n}, \tag{5.16}$$

where n indicates lattice sites and $V(u_n)$ is the nonlinear potential of an "on-site" oscillator. When this equation is used to model the dynamics of molecular crystals, the on-site oscillation is an interatomic vibration, such as the CH stretching oscillation shown in Figure 1.3. In this context, the nearest neighbor coupling strength (ε) stems from electromagnetic dipole–dipole interactions between on-site oscillations. Thus the physical interpretation of Equation (5.16) is quite different from that of the spring-mass lattice in Equation (5.1).

If $V(u_n)$ is constant, Equation (5.16) reduces to a linear lattice equation, whereas if ε is zero it represents a collection of uncoupled nonlinear oscillators, each governed by the on-site equation

$$\frac{d^2 u_n}{dt^2} = -\frac{dV(u_n)}{du_n}. \tag{5.17}$$

In the language of physical chemistry, these two limits are called the "normal-mode limit" and the "local-mode limit" respectively.

In 1994, this system was studied with mathematical rigor by MacKay and Aubrey, who proved the existence of exponentially localized "discrete breather" solutions of Equation (5.16) in one or more spatial dimensions, under the following conditions [92]. (i) Equation (5.17) is a nonlinear oscillator that does not resonate with the lattice, and (ii) ε is sufficiently small (which they called the "anti-continuum limit").[2] The dynamics of Equation (5.16) are discussed in detail in the review by Flach and Willis [57].

As physical chemists of the 1920s had gathered convincing empirical evidence for nonlinear local modes in benzene, one should not be surprised to find them among the solutions of Equation (5.16). Interestingly, there have been several phases in the development of local-mode theory, the first extending from the beginning of infrared spectroscopy to the formulation of quantum mechanics in the late 1920s [107]. During this period, it was recognized that vibrational energy is localized in molecules, providing straightforward explanations for observed properties of stretching-mode overtones and measurements of specific heat. In the second phase, lasting from 1930 to 1950, overtone spectra were carefully measured and quantitatively analyzed by Reinhard Mecke and his colleagues in Germany, using concepts similar to those of modern local-mode theory. From 1950 to 1970 the linear perspective of normal-mode theory was ascendant, and the early work was ignored. Around 1970, local modes were rediscovered, seemingly without knowledge of the earlier studies [72, 86].

At the onset of this rediscovery phase, manuscripts on local-mode experiments were sometimes rejected on the basis that they must be at variance with quantum theory. Under this view, the fact that the symmetry of quantum eigenfunctions must conform to that of the system Hamiltonian, was believed to exclude the possibility of local modes. As is discussed in Chapter 8, however, there are three ways of seeing that this theoretical rejection of local modes is incorrect.

1. For several quantum units of oscillator energy, exact wave functions are close to those obtained in a Hartree approximation that is constructed from solutions of Equation (5.16) in the RWA.
2. Exact quantum analyses in the RWA show that the lowest energy levels are quasidegenerate, meaning that they have dispersion times of order $(N-1)!$, where N is the fundamental quantum number. Thus a local-mode wave packet of a few quanta can remain localized over an experimentally significant period of time.

---

[2]It should be noted that some use the term "discrete breather" for all localized oscillatory states on a lattice, including the ILMs and IGMs of spring-mass ladders, which were discussed in the previous section.

3. Although sharing the symmetry of the system Hamiltonian, the eigenfunctions of local-mode states imply a greater probability of quanta being located near each other than do eigenfunctions of nonlocalized normal modes.

From the current perspective, therefore, the local-mode picture and normal-mode theory are not antagonistic formulations [118]. As the nonlinear parameters approach zero, vibrational energy becomes spread out over the system, and a linear normal-mode analysis is more efficient. In the local-mode limit, on the other hand, energy becomes localized, and one solves the quantum version of Equation (5.17). It is in the range between these two limits that quantum studies are difficult.

5.2.1 The discrete sine–Gordon equation

Defining the sine–Gordon (SG) equation of Chapter 3 on a spatial lattice implies the difference–differential system

$$\frac{d^2 u_n}{dt^2} = u_{n+1} - 2u_n + u_{n-1} - \Gamma^2 \sin u_n, \qquad (5.18)$$

which is called the discrete sine–Gordon (DSG) equation. As was noted above, this system is of historical interest, having been proposed in 1939 to model the motions of crystal dislocations [60], but it is currently being investigated for three additional applications.

1. A ladder network of shunt Josephson junctions that are interconnected by inductors is described by Equation (5.18) and can be viewed as a discrete version of the differential equivalent circuit shown in Figure 3.3(b). With appropriate values of the parameter Γ, this system serves as the basis for a variety of high speed information processing systems [102, 146], including analog to digital converters [30], millimeter wave generators [96], digital processors [36, 80], discrete vortex flow transistors [31], and arrays for pattern recognition [74]. A further motivation for these applications is the recent discovery of materials that remain superconducting at temperatures up to about 100 K, because Josephson junctions are more easily fabricated as lattices from these materials.
2. Molecular crystals of 4-methyl-pyridine (with the chemical formula C_6H_7N), which demonstrate resonances that involve partially hindered rotations of the methyl (CH_3) group [51]. Measurements on partially deuterated crystals suggest that these are collective resonances that can be modeled by the DSG equation with $\Gamma = 0.82$ [51–54].
3. Biological molecules of deoxyribonucleic acid (DNA), which carry the genetic codes of living organisms [43, 145]. In this application, base pairs are attracted inward (from rotations about the double helix backbones) by hydrogen-bond potentials, and the DSG equation models opening of the base pairs.

Discrete sine–Gordon kinks

A stationary DSG kink is a solution of the time independent system

$$u_{n+1} - 2u_n + u_{n-1} = \Gamma^2 \sin u_n \tag{5.19}$$

where $u_n \to 0$ as $n \to -\infty$ and $u_n \to 2\pi$ as $n \to +\infty$. There are two such solutions, one of which is centered at a lattice site and another that is centered between lattice sites. Numerical computations of these solutions can be obtained by constructing recursion relations and "shooting" at appropriate boundary conditions.

The energy of a static solution is given by

$$E_s = \sum_n \left[\tfrac{1}{2}(u_n - u_{n-1})^2 + \Gamma^2(1 - \cos u_n) \right],$$

and the static solution with higher energy is unstable to collapse into the lower energy solution. The difference between these two energies—sometimes called the "Peierls barrier"[3]—is the energy that must be overcome in order for the center of a kink to move from one lattice site to the next.

As $\Gamma \to 0$, slowly varying solutions of Equation (5.18) approach those of the continuum SG equation, supporting moving kinks of the form

$$u_n(t) = 4 \arctan\left[\exp\left(\Gamma \frac{n - vt}{\sqrt{1 - v^2}} \right) \right]. \tag{5.20}$$

Thus it was interesting for Eilbeck to seek numerical solutions of

$$v^2 U''(z) = U(z+1) - 2U(z) + U(z-1) - \Gamma^2 \sin U(z),$$

where $U(z) = U(n - vt) = u_n(t)$, with the boundary conditions

$$U(z) \to 2\pi \quad \text{as } z \to +\infty, \quad \text{and}$$
$$U(z) \to 0 \quad \text{as } z \to -\infty.$$

Interestingly, he was unable to find such solutions [37]. The closest numerical result involved periodic tails of small amplitude, suggesting that the nonlinear medium must be primed with background radiation for the kink to move without loss. This observation is in accord with numerical calculations by Peyrard and Kruskal on the full ODE system, showing that a DSG kink loses energy through radiation as it propagates [104].

---

[3] For Rudolph E. Peierls (1907–1995), a German condensed matter physicist.

Discrete Josephson-junction breathers

Shunt Josephson-junction (JJ) ladder structures are described by the DSG equation, on which breather-like solutions might be expected to take the form

$$u_n(t) = 4 \arctan f(n,t).$$

Imposing periodic boundary conditions so

$$f(n,t) = f(n+L,t) \tag{5.21}$$

and assuming that f is an even function of both space and time, f can be represented as the double Fourier series

$$f(n,t) = \sum_i \sum_j a_{ij} \cos \frac{2\pi i n}{L} \cos j\omega t.$$

A degenerate solution of this sort was found by Parmentier to be [101]

$$f(n,t) = A\,\mathrm{cd}(\omega t, k), \tag{5.22}$$

where $\mathrm{cd}(\omega t, k)$ is a Jacobi elliptic function of modulus $k = A^4$ (see Table C.1 of Appendix C), and $\omega = \Gamma/(1+A^2)$. In this solution, the oscillation amplitude remains constant along the lattice. Numerical studies by Feddersen reveal that a discrete breather bifurcates from Parmentier's solution as Γ is made sufficiently large [37]. Subsequent numerical computations using an accurate and efficient symplectic integration scheme suggest that the discrete breather is quasi-stable [38], in agreement with related studies by Boesch and Peyrard [18]. These numerical results for stationary discrete breathers are supported by mathematical analyses that proceed from the anti-continuum limit ($\Gamma^2 \gg 1$), where existence and stability have been demonstrated [92, 93].

For the case of moving DSG breathers, numerical results near this limit are available [22], and Wattis has used a combination of variational methods and perturbation theory to develop analytic approximations to the true solutions [138, 139]. These latter results show that small amplitude breathers move easily through the lattice and are of particular interest as they may provide a basis for reinterpreting the above mentioned neutron scattering spectra from pyridines.

Breathers have been seen experimentally on ladder structures that include JJs as both series and shunt elements [15, 16, 55, 58, 131]. Observed with scanning laser microscopy, these states involve oscillations of a few nearby loops and are of symmetric, asymmetric, and mixed symmetry. (Under scanning laser microscopy, a focused laser scans the superconducting substrate as the corresponding junction resistance is being measured. If a particular JJ is oscillating and being scanned, its resistance will increase, leading to an observation of voltage.) For further details and animated views of these breathers, visit the websites at sites http://www.pi3.physik.uni-erlangen.de/ustinov/ind_lectures.html and http://www.mpipks-dresden.mpg.de/flach/animation.DIR/A11.html.

5.2.2 Nonlinear Schrödinger lattices

Investigation of Equation (5.16) in the RWA leads to several models that are useful for describing local modes in small molecules, molecular crystals, and chain-like biomolecules. These systems can also be viewed as lattice versions of the nonlinear Schrödinger (NLS) equation which was introduced in Chapter 3.

The discrete NLS equation

Perhaps the most widely applicable form of a lattice NLS equation is

$$i\frac{du_n}{dt} - \omega_0 u_n + \gamma |u_n|^2 u_n + \varepsilon(u_{n+1} + u_{n-1}) = 0, \tag{5.23}$$

where the index n ranges over the lattice. The lattice may be infinite ($n = 0, \pm 1, \pm 2, \ldots$) or finite ($n = 1, 2, \ldots, L$), and in the latter case one often assumes periodic boundary conditions. The parameters γ and ε respectively introduce anharmonicity and dispersion, and the discrete dependent variables—the u_n—are complex functions of time.

As is discussed in Section 8.1.3, a simple linear oscillator (say a mass connected to a spring) can be described in two ways: in terms of the position and momentum of the mass, and as a complex rotating wave amplitude, having a real part that is proportional to momentum and an imaginary part proportional to position. This rotating wave picture is exact for a linear oscillator, and it remains a useful approximation for weakly nonlinear oscillators.

From this perspective, Equation (5.23) can be interpreted as follows:

1. An oscillator at the nth lattice site is described by its complex rotating wave amplitude u_n.
2. With both γ and ε set to zero,

$$u_n(t) = u_n(0)e^{-i\omega_0 t},$$

   indicating a linear (sinusoidal) oscillation at the site frequency ω_0. In this limit, the rotating wave picture is exact.
3. With $\gamma = 0$ and $\varepsilon > 0$, Equation (5.23) describes a linear dispersive lattice that can be completely solved. (This is the normal-mode limit of physical chemistry.)
4. With $\gamma > 0$ and $\varepsilon = 0$, Equation (5.23) describes a set of uncoupled, anharmonic oscillators that can also be completely solved. (This is the local-mode or anti-continuum limit [7].)
5. With both $\gamma > 0$ and $\varepsilon > 0$, there is competition between the effects of nonlinearity and those of dispersion. This struggle leads either to the emergence of stable coherent structures or chaos, depending on details of the initial conditions.

The ω_0 term in Equation (5.23) is readily removed through the transformation

$$u_n(t) = \phi_n(t)e^{-i\omega_0 t},$$

whereupon Equation (5.23) reduces to the normalized form

$$i\frac{d\phi_n}{dt} + \gamma|\phi_n|^2\phi_n + \varepsilon(\phi_{n+1} + \phi_{n-1}) = 0, \tag{5.24}$$

solutions of which are constrained by the conserved number

$$N(\phi_n) = \sum_n |\phi_n|^2 = \text{constant}. \tag{5.25}$$

We shall refer to Equation (5.24) as the discrete nonlinear Schrödinger (DNLS) equation.[4]

Because Equation (5.23)—and therefore Equation (5.24)—is constructed to represent a lattice of coupled, undamped oscillators, its solutions should conserve energy. This property is expressed by the existence of a Hamiltonian

$$H(\phi_n) = -\sum_n \left[\frac{\gamma}{2}|\phi_n|^4 + \varepsilon\phi_n^*(\phi_{n+1} + \phi_{n-1})\right], \tag{5.26}$$

which remains constant under the evolution of Equation (5.24). In terms of a Poisson bracket[5] that is defined as

$$\{U, V\} \equiv \sum_n \left[\frac{\partial U}{\partial \phi_n}\frac{\partial V}{\partial \phi_n^*} - \frac{\partial V}{\partial \phi_n}\frac{\partial U}{\partial \phi_n^*}\right], \tag{5.27}$$

Equation (5.24) can be compactly written as

$$i\frac{d\phi_n(t)}{dt} = \{\phi_n, H\}. \tag{5.28}$$

For two lattice sites ($L = 2$), the DNLS system is exactly integrable by the Liouville–Arnold theorem, because the number of freedoms is equal to the number of conserved quantities H and N [6, 27, 42, 117]. Although DNLS is not exactly integrable for systems with more than two lattice sites, there is numerical evidence that it supports solitary wave solutions that move through the lattice [37]. These solutions have been computed in the following manner. First, ϕ_n is written as

$$\phi_n(t) = \phi(n - vt)e^{i(kn - \omega t)}$$
$$= \phi(z)e^{i(kn - \omega t)}, \tag{5.29}$$

---

[4]Some properties of DNLS systems with higher degrees of nonlinearity are discussed in reference [10]. In these models, $|\phi_n|^2$ is replaced by $|\phi_n|^{2\sigma}$, where $\sigma > 1$. With $\sigma > 2$, a strong tendency to concentrate vibrational energy on a single freedom is observed, a phenomenon that is related to the blow-up in continuum systems that was mentioned in Chapter 3 [108].

[5]After French mathematician Simeon Poisson (1781–1840).

where v is the speed of the traveling-wave envelope. Imposing the periodic boundary conditions $\phi_n(t) = \phi_{n+L}(t)$ requires that

$$\phi(z) = \phi(z+L) \quad \text{and} \quad k = 2\pi m/L,$$

where m is an integer. From substitution of Equation (5.29) into (5.24), $\phi(z)$ must satisfy

$$iv\phi'(z) = \omega\phi(z) + \gamma|\phi(z)|^2\phi(z) + \varepsilon[e^{ik}\phi(z+1) + e^{-ik}\phi(z-1)]. \tag{5.30}$$

This complex nonlinear differential–advance–delay equation has stationary solutions of the form $\phi(z) = $ constant, which can be normalized by the condition

$$\sum_{n=1}^{L} |\phi|^2 = 1,$$

leading to the dispersion relation

$$\omega = -\frac{\gamma}{L} - 2\varepsilon \cos k.$$

At sufficiently large values of γ/ε, localized traveling-wave solutions are found to bifurcate from these stationary solutions.

From these numerical studies, there appears to be a window of anharmonicity where traveling-waves are observed. Thus γ must be large enough for a solitary wave to bifurcate from the corresponding non-localized, stationary solution but not so large that this solution becomes pinned to the lattice. From a mathematical perspective, existence of these moving solutions remains an open question [40].

The DNLS equation has additional solutions, some of which can be found from writing the solution to Equations (5.24) as

$$\phi_n(t) = \psi_n(t)e^{i\omega t},$$

where $\psi(t)$ is a solution of

$$i\frac{d\psi_n}{dt} - \omega\psi_n + \gamma|\psi_n|^2\psi_n + \varepsilon(\psi_{n+1} + u_{\psi-1}) = 0. \tag{5.31}$$

In contrast to Equation (5.23), ω is an adjustable parameter in this family of solutions.

With the ψ_n independent of time, Equation (5.31) reduces to

$$-\omega\psi_n + \gamma|\psi_n|^2\psi_n + \varepsilon(\psi_{n+1} + u_{\psi-1}) = 0,$$

which for real ψ_n can be viewed as a recurrence equation. Then by choosing different values of ω in a series of shooting calculations, localized solutions can be constructed that have their maximum values at $n = 0$ (say) and fall exponentially as $n \to \pm\infty$ [119]. These are examples of the discrete breathers that are guaranteed for sufficiently small ε by the existence theorem of MacKay and Aubry [92].

If the ψ_n in Equation (5.31) are allowed to be functions of time, yet more interesting solutions appear. Among these are localized periodic solutions for which

$$\psi_n(t) = \psi(t + T),$$

where both T and ω are continuously adjustable parameters in solutions of Equation (5.24) [40]. If ωT is not a rational number, these are examples of the quasiharmonic breathers that were suggested by MacKay and Aubry [92], demonstrated for the DNLS equation by Johansson and Aubry [78, 81], and established for a general on-site potential by Bambusi and Vella [9].

DNLS equations in more spatial dimensions

Generalizing the DNLS equation to two space dimensions, there have been several studies of the system

$$\mathrm{i}\phi_{m,n} = \gamma|\phi_{m,n}|^2\phi_{m,n} + \varepsilon[\phi_{m+1,n} + \phi_{m-1,n} + \phi_{m,n+1} + \phi_{m,n-1} - 4\phi_{m,n}], \quad (5.32)$$

where $m = 0, \pm 1, \pm 2, \ldots$ and $n = 0, \pm 1, \pm 2, \ldots$, and the bracketed terms are a discrete version of the two-dimensional Laplacian [81]. Importantly, the MacKay and Aubry theorem also guarantees localized solutions of this equation in the anti-continuum limit [92].

The most simple of these localized solution has a maximum amplitude at $(m, n) = (0, 0)$, falling off exponentially as one moves away from the origin of the lattice [90, 97]. As the ratio of γ to ε decreases, this solution decreases more gradually, eventually becoming unstable to a blow-up solution of the sort described in Section 3.3.5 [143]. Not surprisingly, the same phenomenon is found on a three-dimensional lattice [56]. It has been suggested that this instability can be observed experimentally as a delocalizing transition of Bose-Einstein condensates in optical lattices [79].

The Ablowitz–Ladik equation

In 1976, Ablowitz and Ladik presented and discussed the lattice system [1, 2]

$$\mathrm{i}\frac{d\phi_n}{dt} + (\phi_{n+1} + \phi_{n-1})\left(1 + \frac{\gamma}{2}|\phi_n|^2\right) = 0. \quad (5.33)$$

This system is theoretically interesting because—like the Toda lattice—it is exactly integrable in the sense defined by Kruskal and his colleagues [45, 64].[6] From the perspective of applications to molecular crystal dynamics, the system is

---

[6]Yet another completely integrable lattice NLS equation was introduced by Izergin and Korepin in 1981 [45, 76].

interesting because it allows for nonlinearity in the couplings between adjacent oscillators on a chain. We will call Equation (5.33) the Ablowitz–Ladik (AL) equation and note its relationship to the DNLS equation, described in the previous section and the NLS equation of Chapter 3.

In particular, solutions of AL approach those of DNLS in the continuum limit where $\gamma \to 0$ and $(\phi_n - \phi_{n-1})/\phi_n \to 0$. Also with the transformation

$$\phi_n = \psi_n e^{2\mathrm{i}t},$$

solutions of both AL and DNLS approach those of the NLS equation

$$\mathrm{i}\frac{\partial \psi}{\partial t} + \gamma |\psi|^2 \psi + \frac{\partial^2 \psi}{\partial x^2} = 0$$

as the ratio of anharmonicity to dispersion (γ) approaches zero. Thus AL and DNLS can be viewed respectively as integrable and nonintegrable approximations to NLS.

As the AL equation is integrable, there are exact analytical expressions for its traveling-wave solutions, including [113]

$$\phi_n(t) = A \operatorname{cn}[\beta(n - vt); k] e^{-\mathrm{i}(\omega t + \alpha n + \phi_0)}, \tag{5.34}$$

where $\operatorname{cn}[\,\cdot\,; k]$ is a Jacobi elliptic function of modulus k (see Appendix C). Choosing the units of ϕ_n so $\gamma/2 = 1$, the parameters in Equation (5.34) can be written as

$$A = \frac{k \operatorname{sn}(\beta; k)}{\operatorname{dn}(\beta; k)}$$

$$\omega = -\frac{2 \operatorname{cn}(\beta; k) \cos \alpha}{\operatorname{dn}^2(\beta; k)}$$

$$v = -\frac{2 \operatorname{sn}(\beta; k) \sin \alpha}{\beta \operatorname{dn}(\beta; k)},$$

where α, β, and k are free parameters with the ranges

$$-\pi \leq \alpha \leq \pi, \quad 0 < \beta < \infty, \quad \text{and} \quad 0 < k < 1.$$

As $k \to 0$ and $\beta \to 0$, Equation (5.34) approaches a solution of the linear ($\gamma = 0$) system. As $k \to 1$, on the other hand, one finds the lattice soliton

$$\phi_n(t) = \sinh \beta \operatorname{sech}[\beta(n - vt)] e^{-\mathrm{i}(\omega t + \alpha n + \phi_0)}, \tag{5.35}$$

with

$$\omega = -2 \cosh \beta \cos \alpha$$

$$v = -(2/\beta) \sinh \beta \sin \alpha,$$

which can move at any speed.

Among the conserved quantities possessed by Equation (5.33) is a number

$$N(\phi_n) = \frac{2}{\gamma} \sum_n \log\left(1 + \frac{\gamma}{2}|\phi_n|^2\right), \tag{5.36}$$

evidently reducing to the DNLS number of Equation (5.25) as $\gamma \to 0$. In addition, there is a Hamiltonian [45, 65, 113]

$$H(\phi_n) = -\sum_n \phi_n^*(\phi_{n+1} + \phi_{n-1}), \tag{5.37}$$

which is defined in relation to a nonstandard Poisson bracket as

$$\{U, V\} \equiv \sum_n \left[\frac{\partial U}{\partial \phi_n}\frac{\partial V}{\partial \phi_n^*} - \frac{\partial V}{\partial \phi_n}\frac{\partial U}{\partial \phi_n^*}\right]\left(1 + \frac{\gamma}{2}\phi_n\phi_n^*\right). \tag{5.38}$$

Thus the AL equation can be written as

$$i\frac{d\phi_n}{dt} = \{\phi_n, H\}.$$

In Chapter 8, we shall see that this formulation is convenient for a quantum analysis of the AL equation.

Salerno's equation

To recapitulate the above results, note that the DNLS equation can be written (for $\varepsilon = 1$) as

$$i\frac{du_n}{dt} + u_{n+1} - 2u_n + u_{n-1} + \gamma|u_n|^2 u_n = 0, \tag{5.39}$$

having a conserved number

$$N_1 = \sum_n |u_n|^2 \tag{5.40}$$

and a conserved energy

$$H_1 = -\sum_n \left[u_n^*(u_{n+1} + u_{n-1}) - 2|u_n|^2 + \frac{\gamma}{2}|u_n|^4\right]. \tag{5.41}$$

Similarly, the AL equation can be written as

$$i\frac{du_n}{dt} + u_{n+1} - 2u_n + u_{n-1} + \frac{\gamma}{2}|u_n|^2(u_{n+1} + u_{n-1}) = 0, \tag{5.42}$$

with the conserved number

$$N_2 = \frac{2}{\gamma} \sum_n \log\left(1 + \frac{\gamma}{2}|u_n|^2\right) \tag{5.43}$$

and the conserved energy

$$H_2 = -\sum_n \left[u_n^*(u_{n+1} + u_{n-1}) - \frac{4}{\gamma} \log\left(1 + \frac{\gamma}{2}|u_n|^2\right) \right]. \tag{5.44}$$

In 1992, Salerno proposed a Hamiltonian of the form [71, 112]

$$H_3 = -\sum_n \left[u_n^*(u_{n+1} + u_{n-1}) + \eta|u_n|^2 - (2+\eta)\frac{\eta}{\epsilon} \log\left(1 + \frac{\epsilon}{\eta}|u_n|^2\right) \right], \tag{5.45}$$

where ϵ and η are parameters that are constrained by the relationship

$$\eta = 2\epsilon/(\gamma - \epsilon). \tag{5.46}$$

Two special cases are of particular interest.

- $\epsilon \to 0$: From Equation (5.46), we see that as $\epsilon \to 0$, $\eta \to 0$, and $\epsilon/\eta \to \gamma/2$. In this case,
$$H_3 \to H_2,$$
the AL Hamiltonian.
- $\epsilon \to \gamma$: In this case, $\epsilon/\eta \to 0$ and
$$H_3 \to H_1,$$
the DNLS Hamiltonian.

With a Poisson bracket defined as

$$\{U, V\} \equiv \sum_n \left[\frac{\partial U}{\partial u_n}\frac{\partial V}{\partial u_n^*} - \frac{\partial V}{\partial u_n}\frac{\partial U}{\partial u_n^*} \right] \left(1 + \frac{\epsilon}{\eta} u_n u_n^*\right), \tag{5.47}$$

reducing to Equation (5.38) as $\epsilon \to 0$ and to Equation (5.27) as $\epsilon \to \gamma$, the equation

$$i\frac{du_n}{dt} = \{u_n, H_3\}$$

becomes

$$i\frac{du_n}{dt} + (\epsilon|u_n|^2 - 2)u_n + \left(1 + \frac{\epsilon}{\eta}|u_n|^2\right)(u_{n+1} + u_{n-1}) = 0. \tag{5.48}$$

Evidently Salerno's equation reduces to AL for $\epsilon = 0$ and to DNLS for $\epsilon = \gamma$; thus it interpolates between these two lattice NLS equations. A conserved quantity is the number

$$N_3 = \frac{\eta}{\epsilon} \sum_n \log\left(1 + \frac{\epsilon}{\eta}|u_n|^2\right), \tag{5.49}$$

which—together with the Hamiltonian H_3—provides the basis for a quantum analysis. For applications to molecular crystals, Salerno's equation is interesting for two reasons. First, it allows the independent adjustment of nonlinear parameters describing both on-site potentials and nearest neighbor interactions. Second, the corresponding quantum formulation can be exactly solved.

5.2.3 The discrete self-trapping equation

Here we return to the physical motivation behind the DNLS equation, introduced in the previous section. The picture was of a set of nonlinear oscillators, each interacting with its nearest neighbors on a one-dimensional lattice. The individual oscillators are described by their complex, rotating wave amplitudes, the u_n. This context is now generalized to a lattice of arbitrary structure, allowing interactions of varying strengths between any pair of oscillators. Thus we study the system of f oscillators (freedoms) [42]

$$\left(\mathrm{i}\frac{d}{dt} - \omega_0\right)\mathbf{U} + \gamma D(|\mathbf{U}|^2)\mathbf{U} + \varepsilon M \mathbf{U} = \mathbf{0}, \tag{5.50}$$

where

$$\mathbf{U} = \mathrm{col}(u_1, u_2, \ldots, u_f)$$

is an f-component column vector, each component representing the complex rotating wave amplitude of a particular oscillator. The diagonal matrix

$$D(|\mathbf{U}|^2) \equiv \mathrm{diag}(|u_1|^2, |u_2|^2, \ldots, |u_f|^2),$$

appearing in the second term of Equation (5.50), represents the tendency of the system to localize energy through contributions from two sorts of anharmonicity:

1. Intrinsic anharmonicity, arising from nonlinearity in the interactive forces. (In the context of molecular oscillations, intrinsic anharmonicity is caused by a reconfiguration of the electronic cloud as an interatomic bond is stretched.)
2. Extrinsic anharmonicity, arising from polaron-like interactions between an oscillator and the nearby lattice. (In the context of a molecular crystal, stretching of an oscillator distorts the local crystal structure, altering the interactive force constant.)

The third term of Equation (5.50) involves a real, symmetric, $f \times f$ matrix

$$M = [m_{jk}],$$

where εm_{jk} represents the linear electromagnetic and mechanical interactions between the jth and kth oscillators. From the reciprocity theorem for passive, linear systems, it follows that $m_{jk} = m_{kj}$.

We refer to Equation (5.50) as the discrete self-trapping (DST) equation. Historically, the DST equation was formulated in the early 1980s to describe experimental observations of anomalous amide resonances (related to the CO–stretching oscillation) in molecular crystals of acetanilide [19, 41], and extrinsic local modes in a globular protein [42]. This motivation is close to that for subsequent studies of local modes on one-dimensional monatomic and diatomic lattices by Pnevmatikos et al. [59, 105].

In applications of the DST equation to globular proteins, the interaction matrix M is perhaps 200×200, making the task of numerical computation formidable. Thus, it is interesting to employ the DST equation as a model for resonant oscillations on small molecules, taking ammonia (NH_3) as a simple example. There are three NH stretching oscillators, arranged with equal angles between each pair of them. A moment's reflection suggests that the interaction matrix for the three NH oscillators is of the form

$$M = \begin{bmatrix} 0 & 1 & 1 \\ 1 & 0 & 1 \\ 1 & 1 & 0 \end{bmatrix}. \qquad (5.51)$$

This matrix has zeros on the diagonal because these elements are carried by the ω_0 terms in Equation (5.50), which can be eliminated through the transformation

$$\mathbf{U} \to \mathbf{U} e^{-i\omega_0 t}.$$

Thus the DST equation reduces to the standard form[7]

$$i\dot{\mathbf{U}} + \gamma D(|\mathbf{U}|^2)\mathbf{U} + \varepsilon M \mathbf{U} = \mathbf{0}. \qquad (5.52)$$

Stationary solutions

Stationary solutions of the DST equation are assumed to have the exponential time dependence

$$\mathbf{U}(t) = \boldsymbol{\phi} e^{i\omega t}, \qquad (5.53)$$

where $\boldsymbol{\phi} = \mathrm{col}(\phi_1, \phi_2, \ldots, \phi_f)$ satisfies the nonlinear, algebraic eigenvalue problem

$$-\omega \boldsymbol{\phi} + \gamma D(|\boldsymbol{\phi}|^2)\boldsymbol{\phi} + \varepsilon M \boldsymbol{\phi} = 0. \qquad (5.54)$$

With $\gamma = 0$, Equation (5.54) reduces to the linear problem

$$\varepsilon M \boldsymbol{\phi} = \omega \boldsymbol{\phi}, \qquad (5.55)$$

---

[7]This system has been studied in some detail since the mid-1980s, and many references to it can be found at the Web site:

http://www.ma.hw.ac.uk/~chris/dst/

with eigenvalues and eigenvectors that are readily calculated by the standard methods of normal mode theory.

For $\gamma \gg \varepsilon$ (the anti-continuum limit), $\omega\phi_j = \gamma|\phi_j|^2\phi_j$. Because ω is assumed to be constant throughout the lattice, each ϕ_j must satisfy either $\phi_j = 0$ or $|\phi_j|^2 = \omega/\gamma$. If the components of $\boldsymbol{\phi}$ are real, they must take one of the three values[8]

$$\phi_j = 0, \pm\sqrt{\omega/\gamma}.$$

Between these two cases, it is sometimes possible to find analytic expressions that connect the limiting solutions (for example, the complete graph model in which all $m_{jk} = 1$ with $j \neq k$). When analytic solutions are not known, numerical path following methods enable numerical solutions to be generated [42].

As a specific example, consider the interaction matrix of Equation (5.51), modeling the NH stretching oscillators in ammonia. Normalizing as

$$N = \sum_{j=1}^{3} |\phi_j|^2 = 1$$

and assuming $\varepsilon = 1$ (so γ is the ratio of anharmonicity to dispersion), the relations between γ and ω for different branches are displayed in Figure 5.3. In this figure, the symbols "↑" and "↓" indicate in-phase and out-of-phase oscillations that maintain equal amplitude as $\varepsilon/\gamma \to 0$ (the anti-continuum limit), and the symbol "·" implies an amplitude that goes to zero in this limit. Ignoring degeneracies, there are five distinct asymptotic solutions with the following relative amplitudes:

$$(\uparrow\uparrow\downarrow), (\uparrow\downarrow\, 0), (\uparrow\uparrow\uparrow), (\uparrow\downarrow\, \cdot), \text{ and } (\uparrow\, \cdot\, \cdot).$$

The three-fold degenerate $(\uparrow\, \cdot\, \cdot)$ branch is of particular interest. As the DST Hamiltonian is

$$H = -\frac{\gamma}{2}\sum_j |u_j|^4 - \varepsilon \sum_{j,k} m_{jk} u_j^* u_k, \qquad (5.56)$$

the $(\uparrow\, \cdot\, \cdot)$ mode has the lowest energy in the limit of large anharmonicity. It thus describe a classical solution where almost all of the vibrational energy is concentrated in a single oscillator—a local mode.

Stability of stationary solutions

The alert reader will have noticed that some of the functional relationships in Figure 5.3 are plotted as solid lines, while others are dotted lines. This is done to indicate stability (solid) or instability (dotted) of the corresponding stationary solutions, following a linear stability theory developed by Carr and Eilbeck [20].

---

[8] Note the scale invariance: if $\boldsymbol{\phi}$ is a solution of Equation (5.54), then so is $\boldsymbol{\phi}\exp(i\alpha)$ for any real α.

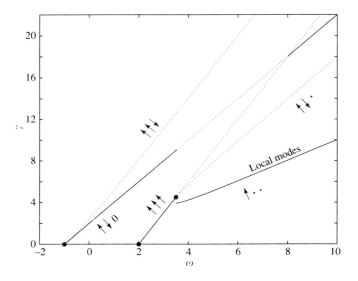

FIG. 5.3. *The relations between γ and ω for stationary solutions of DST with three lattice sites, M as in Equation (5.51), and $\varepsilon = 1$. Branch designations are explained in the text. (Courtesy of J.C. Eilbeck.)*

Under this analysis, a general solution of Equation (5.52) is written in the form

$$\mathbf{U} = [\boldsymbol{\phi} + \boldsymbol{\psi}(t)]e^{i\omega t}, \tag{5.57}$$

where ω and the vector $\boldsymbol{\phi}$ satisfy Equation (5.54), and $\boldsymbol{\psi}(t)$ is an arbitrary perturbation vector.

Considering first the case when $\boldsymbol{\phi}$ is real, substitute Equation (5.57) into the standard form of the DST equation to obtain

$$i\frac{d\boldsymbol{\psi}}{dt} - \omega\boldsymbol{\psi} + \gamma \operatorname{diag}(\phi_j^2)(2\boldsymbol{\psi} + \boldsymbol{\psi}^*) + \varepsilon M \boldsymbol{\psi} \doteq \mathbf{0}, \tag{5.58}$$

where "$\doteq$" indicates that the terms of quadratic and higher order in $\boldsymbol{\psi}$ have been neglected. Equation (5.58) thus provides the basis for a linear stability analysis. Because the perturbation is introduced in the rotating frame of the stationary solution, this equation is free of explicit dependence upon time, simplifying the analysis.

Writing

$$\boldsymbol{\psi} = \boldsymbol{\psi}_1 + i\boldsymbol{\psi}_2,$$

where $\boldsymbol{\psi}_1$ and $\boldsymbol{\psi}_2$ are real vectors, Equation (5.58) takes the form

$$\frac{d}{dt}\begin{bmatrix}\boldsymbol{\psi}_1\\\boldsymbol{\psi}_2\end{bmatrix} = \begin{bmatrix}0 & C\\B & 0\end{bmatrix}\begin{bmatrix}\boldsymbol{\psi}_1\\\boldsymbol{\psi}_2\end{bmatrix}, \tag{5.59}$$

with

$$C \equiv -\varepsilon M + \omega I - \gamma \operatorname{diag}(\phi_j^2)$$
$$B \equiv +\varepsilon M - \omega I + \gamma \operatorname{diag}(3\phi_j^2),$$

and I is the $f \times f$ unit matrix.

Eigenvalues of Equation (5.59) are given by the roots of

$$\det[BC - \lambda^2 I] = 0, \tag{5.60}$$

and a necessary condition for stability of a stationary solution is that all the roots (λ_j) have real parts ≤ 0. From Equation (5.60), this condition can only be satisfied if all the λ_j^2 are purely real and ≤ 0, so the λ_j are all purely imaginary.

A sufficient condition for instability is that at least one λ_j^2 is not purely real and ≤ 0, leading to a λ_j with a positive real part.

When ϕ is complex, so

$$\phi_j = \phi_{j1} + i\phi_{j2},$$

the problem can be treated in a similar manner. In this case, Equation (5.59) becomes

$$\frac{d}{dt}\begin{bmatrix}\psi_1\\\psi_2\end{bmatrix} = \begin{bmatrix}A & C\\B & D\end{bmatrix}\begin{bmatrix}\psi_1\\\psi_2\end{bmatrix}, \tag{5.61}$$

where

$$A \equiv -2\operatorname{diag}(\phi_{j1}\phi_{j2})$$
$$D \equiv +2\operatorname{diag}(\phi_{j1}\phi_{j2})$$
$$C \equiv -\varepsilon M + \omega I - \gamma \operatorname{diag}(\phi_{j1}^2 + 3\phi_{j2}^2)$$
$$B \equiv +\varepsilon M - \omega I + \gamma \operatorname{diag}(3\phi_{j1}^2 + \phi_{j2}^2),$$

and the eigenvalues are roots of

$$\det[BC - (A - \lambda I)(D - \lambda I)] = 0.$$

If all eigenvalues have real parts ≤ 0, a necessary condition for stability is fulfilled. A sufficient condition for instability is for at least one eigenvalue to have a real part > 0. (See also references [93–95].)

Chaos

Returning to Figure 5.3, recall that the dynamic behavior of a DST system with three or more freedoms ($f \geq 3$) is regular in two cases: the anti-continuum limit ($\varepsilon \to 0$) where energy is concentrated on nonlinear local modes that don't interact, and the normal-mode limit ($\gamma \to 0$) in which energy is distributed onto the linear extended eigenfunctions of the matrix M. Between these two limits,

Figure 5.3 suggests an interweaving of regular and chaotic trajectories that is most intricate when $\gamma N \sim \varepsilon$, and chaotic-looking solutions were observed in this range in early studies of the DST equation [42].

The chaotic character of phase space for the three freedom DST system with interaction matrix of Equation (5.51) has been studied numerically for various values of the parameters γ and ε [28]. The approach was to choose initial conditions randomly (subject to appropriate constraints on the number) and record the fraction—ρ—that led to chaotic trajectories. Using 50–100 trajectories examined for each datum, a maximum value of

$$\rho_{\max} = 0.6 \pm 10\%$$

was observed, with $\rho \to 0$ in the limiting cases $\gamma \to 0$ and $\gamma^{-1} \to 0$. From the perspective of modern nonlinear systems theory, these numerical results are calculations of the measure in phase space of chaotic behavior.

Another study of DST chaos with $f = 3$ has been carried out by Hennig et al., who assumed $m_{13} = m_{23} \ll m_{12} = m_{21}$ [69]. In the limit $m_{13} = m_{23} = 0$, the system reduces to $f = 2$, which is exactly integrable because the number of freedoms equals the number of conserved quantities (N and H) [6, 27, 42, 117]. Then for small $m_{13} = m_{23}$ a Melnikov calculation was used to show homoclinic chaos. Similar approaches were taken with $f = 4$ to show Arnold diffusion [48, 70].

5.3 Biological solitons

Biological organisms function on many levels of dynamic activity, the lowest being that of the biomolecules: polysaccharides, proteins, and nucleic acids. At this level, it is possible to determine atomic coordinates with some degree of accuracy and to ask how these structures might oscillate under the influence of their interatomic forces. Such studies are at the realm of energy conserving (Hamiltonian) systems, where biomolecular dynamics are characterized by three features of particular interest to practitioners of nonlinear science. First, the level of description is sufficiently low that averaging over microstates is not appropriate. Thus many dissipative and diffusive phenomena do not arise. Second, the level of description is sufficiently high that quantum corrections are largely negligible. Finally, biomolecules are often constructed in the form of one-dimensional lattices (polymers), which are candidates for the support of nonlinear wave processes.

5.3.1 Alpha-helix solitons in protein

At its most basic level of description, a protein molecule is a polymer (or chain) of amino acids, coupled by valence bonds. Because 20 distinct amino acids are available to biological organisms, there are some 20^{200} possible protein molecules that are composed of 200 amino acids each. This is a very large number; indeed, there is not enough matter in the entire universe to construct but a single molecule of each possible protein.

Throughout the eons of life's evolution on earth, a small fraction—yet a substantial number—of all possible proteins have been explored, and the corresponding amino acid sequences (the primary structures) of useful proteins have been recorded in the genetic codes of living organisms. These primary strands are organized into secondary structure, which can be alpha-helix, beta-sheet, or random. Finally the overall organization of a protein molecule is referred to as its tertiary structure.

Although alpha-helix is a major component of the structural proteins in hair and skin, it also plays a functional role in the transmembrane proteins that pump ions across the active membrane of a nerve cell and in muscular contraction [116]. An alpha-helical region is sketched in Figure 5.4, showing three chains with structure

$$\cdots \text{H-N-C=O} \cdots \text{H-N-C=O} \cdots \text{H-N-C=O} \cdots \text{H-N-C=O} \cdots$$

which extend along the helix. (Not shown are the residues, or side chains which distinguish individual amino acids.) Here we sketch the dynamics of CO stretching oscillations (also called Amide-I vibrations) on these three chains using the insights of nonlinear lattice theory.

In particular, the ability of alpha-helix to store and transport CO vibrational energy is considered, using a means that was suggested by Davydov in 1973 [32, 33]. Davydov's mechanism for energy transport is essentially that of a polaron, acting to localize energy of the CO stretching oscillations. Thus localized vibrational energy strains the protein structure in its neighborhood by stretching adjacent hydrogen bonds (see Figure 5.4). Retroactively, the local distortion of the lattice acts as a potential well, trapping the vibrational energy and preventing

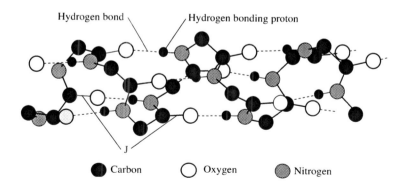

FIG. 5.4. *A short section of alpha helix. The dashed lines indicate relatively weak hydrogen bonds. (Residue groups are not shown.)*

its dispersion. The quantum theory of this phenomenon is discussed in Chapter 8, but from a classical perspective, governing equations can be written as

$$i\frac{d\phi_n}{dt} - \omega_0 \phi_n + J(\phi_{n+1} + \phi_{n-1}) = \chi(\beta_{n+1} - \beta_n)\phi_n,$$
$$3M\frac{d^2\beta_n}{dt^2} - 3w(\beta_{n+1} - 2\beta_n + \beta_{n-1}) = \chi(|\phi_n|^2 - |\phi_{n-1}|^2). \tag{5.62}$$

In the first of these equations, ϕ_n is the complex amplitude (in the rotating wave approximation) of a CO oscillator, ω_0 is the site frequency, and J is the electromagnetic dipole–dipole coupling between adjacent CO oscillators. In the second equation β_n is the longitudinal displacement of a CO oscillator from its equilibrium position, M is the mass of an amino acid, and w is the spring constant of a hydrogen bond. (The factors of 3 enter the second equation because the dynamics along three chains are considered together.) Each of these equations is linear, anharmonicity entering through the coupling parameter χ. With $\chi = 0$, the first of Equations (5.62) represents the linear dispersion of CO vibrational energy at frequency ω_0. Similarly, the second equation describes the propagation of longitudinal sound waves along the helical lattice.

In physical terms, χ indicates the change in the frequency (ω_0) of a CO oscillator with variations in the length ($\beta_{n+1} - \beta_n$) of an adjacent hydrogen bond. Thus

$$\chi = \frac{d\omega_0}{d(\beta_{n+1} - \beta_n)},$$

which can be determined experimentally by comparing spectral measurements of the CO stretching band at various temperatures with corresponding x-ray diffraction measurements of the hydrogen-bond length as a function of temperature [115]. With $\chi \neq 0$, localized vibrational energy acts as a source of longitudinal sound waves in the second of Equations (5.62), and the helical distortion produced by this sound reacts as a potential well in the first equation. As we have often seen, the overall phenomenon can be summarized in the feedback diagram:

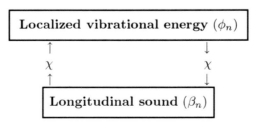

To see that solitary wave solutions are to be expected, assume a wave process that is moving slowly compared with the longitudinal sound speed. Then the kinetic energy term $3Md^2\beta_n/dt^2$ can be neglected in the second of Equations (5.62), leading to[9]

$$\beta_{n+1} - \beta_n \doteq -\frac{\chi}{3w}|\phi_n|^2.$$

Upon substitution into the first of Equations (5.62), one obtains the DNLS equation

$$i\frac{d\phi_n}{dt} - \omega_0\phi_n + \frac{\chi^2}{3w}|\phi_n|^2\phi_n + J(\phi_{n+1} + \phi_{n-1}) \doteq 0, \qquad (5.63)$$

which was discussed in Section 5.2.2. (It is to be expected that the anharmonic parameter $\gamma = \chi^2/3w$ depends upon the square of χ, because χ appears twice in the loop of the above feedback diagram.)

A feature of this formulation is that each parameter in Equations (5.62) can be independently defined and measured; thus no parameter adjustment is permitted in seeking a correspondence between numerical computations and experimental observations. Using established values of M, w, and J, Feddersen has shown numerically that Equations (5.62) support solitary waves that do not become pinned to the lattice (by the Peierls barrier) until χ reaches two or three times experimental estimates [47].

Thus one might expect that real alpha-helix can store and transport realistic amounts of biological energy. Before jumping to this conclusion, however, it is necessary to consider two difficulties that Davydov's soliton theory faces.

1. *Thermal effects.* Because biological organisms function at a temperature of about 310 K (37°C), the static aspects of Figure 5.4 are misleading. Realistically, one should consider this diagram as indicating average positions of atoms that move randomly under the influence of thermal forces. Such forces can be introduced into Equations (5.62) by adding a random homogeneous solution to $\beta_n(t)$, as has been done in a number of numerical studies during the 1980s.
2. *Additional interactions.* The simplified ball-and-stick diagram of Figure 5.4 may neglect interatomic interactions of biological significance. One effect of such interactions would be to limit the lifetime of a solitary wave through radiation into the surrounding protein structure. Whether this is a serious limitation depends upon both the strengths of the neglected interactions and the time scale of the functional process under consideration.

The evidence for evaluating these effects is reviewed in Section 8.5.3.

5.3.2 Self-trapping in globular proteins

A way to investigate the dynamics of a globular protein for which the structure has been determined (by x-ray diffraction studies) is to consider it as a set of masses interconnected by nonlinear springs. To model the springs, a molecular

---

[9]This is sometimes called the "adiabatic approximation."

dynamics code can be used, but these codes neglect polaronic energy terms of the form

$$\sum_n \chi(\beta_{n+1} - \beta_n)|\phi_n|^2,$$

which generate the right hand sides of Equations (5.62).

To study polaronic effects in a realistic geometry, one can generalize the formulation of Equations (5.62) from the secondary structure (alpha-helix) to the tertiary structure of an entire globular protein. It was this generalization that first suggested the DST equation, discussed above in Section 5.2.3.

Such a numerical study has been carried out by Feddersen for adenylate kinase, a structure determined enzyme that catalyzes energetic reactions in the cell [46]. Because it is composed of 194 amino acids ($f = 194$), there are

$$f(f-1)/2 = 18{,}721$$

distinct off-diagonal elements in the interaction matrix εM of Equation (5.50). Each of these elements indicates the strength of the dipole–dipole interaction between the CO stretching oscillations of a particular pair of amino acids which can be automatically computed from Maxwell's equations and the atomic coordinates. In the matrix εM, there are also f diagonal elements (the site frequencies) that could not be determined from the x-ray data; thus Feddersen selected them at random from an appropriate frequency interval [132]. In this manner the entire set of $f(f+1)/2 = 18{,}915$ distinct elements in the interaction matrix were determined and stored in a computer file.

Although the mapping of a complete bifurcation diagram (like that shown in Figure 5.3 for $f = 3$) is out of the question for 194 freedoms, some insights were gained from the following computational procedure. First, with γ much larger than all elements of εM, a particular amino acid (freedom of CO oscillation) was selected and all of the vibrational energy localized thereon. Second, using curve following techniques, the value of γ was reduced to physically reasonable values. Finally, as anharmonicity (γ) was changed, the stability of the corresponding stationary solution was computed, employing a curve following technique. Also localization was defined by the parameter

$$\alpha = \sum_{j=1}^{f} |\phi_j|^4,$$

where the single quantum normalization

$$\sum_{j=1}^{f} |\phi_j|^2 = 1$$

was imposed. Thus $\alpha = 1$ implies that the solution is localized at a particular lattice site, and $\alpha = 1/f$ indicates a solution that is spread over the entire lattice. Plots of α as a function of γ for various stationary solutions show how the phenomenon of localization is influenced by anharmonicity.

Two conclusions were drawn from this study:

1. At experimentally reasonable levels of anharmonicity (γ), stable localized stationary solutions were observed near some—but not all—of the amino acids. This numerical result suggests that local modes can support active sites in enzymes, which may trap and store biologically useful amounts of energy.
2. Unexpectedly, these observations of anharmonic localization were distinctly different from "Anderson localization," a property of randomly interacting linear systems [3, 4]. Thus none of the stationary states that were observed to be localized at large γ remained so as $\gamma \to 0$. Also, none of the states that were localized at $\gamma = 0$ (i.e. Anderson localized) remained so as γ was increased to a physically reasonable level. Rather than reinforcing each other, as one might expect, these two types of localization (Anderson and nonlinear) appear to be disjoint; where one is strong, the other is weak. Kopidakis and Aubry have recently published a more systematic study of the complex relationship between localization from disorder and nonlinear localization [84, 85].

5.3.3 *Solitons in DNA*

Codes for the primary sequences of amino acids in an organism's proteins are stored in the DNA of its chromosomes. The double-helix structure of DNA (sketched in Figure 5.5) consists of two helical backbones, composed of alternating sugar and phosphate groups and joined together by pairs of bases.

Thanks to their complementary shapes, these bases are paired as follows: adenine (A) is always joined (by two hydrogen bonds) to thymine (T), and cytosine (C) is always coupled (by three hydrogen bonds) to guanine (G). In other words, the series of bases

$$-A-C-T-C-G-A-G-T-C-A-T-$$

can only be paired with the complementary series

$$-T-G-A-G-C-T-C-A-G-T-A-$$

like the negative of a photographic image.

Thus, one strand of DNA acts as a template to create a complementary strand. And this complementary strand, at a later stage, letter by letter, itself serves as a template for stamping out protein molecules. Cells manufacture a wide variety of proteins after consulting their internal blueprints, the DNA, to find the appropriate primary structures.

The DNA structure of Figure 5.5 appears as the answer to an applied mathematician's dreams. Of central importance to the mystery of life, it is a one-dimensional, nonlinear lattice that is expected to support solitary waves of the

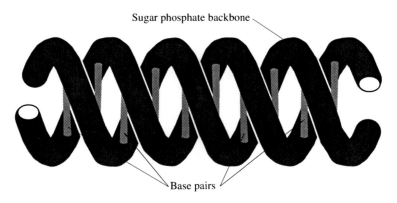

FIG. 5.5. *A cartoon of a short section of DNA double helix.*

sort considered in this chapter. Consequently, many studies have been devoted to the task of finding solitons on various approximations to the structure of DNA over the past two decades.

The task of sorting through this growing body of work is eased by a book by Yakushevich entitled *Nonlinear Physics of DNA*, in which she surveys the currently available publications in the area [145]. This work contains descriptions of measurements, tables of relevant physical and chemical constants, and a complete bibliography, in addition to an introduction to the chemistry of DNA and discussions of relations among the many mathematical models.

These models, constructed primarily by physicists and applied mathematicians, fall into two broad classes.

1. *Linear models.* Useful for interpreting spectral data, linear dynamic models represent DNA as: elastic rods and disks coupled by springs, displaying longitudinal, torsional, and bending motions; helical models, both continuous and discrete, and comprising one or two helices (see Figure 5.5); molecular models, accounting for the motions of phosphate, sugar, and base groups; and complete atomic models.
2. *Nonlinear models.* Inspired by the successes of nonlinear studies in other areas of applied science, anharmonic models of DNA include: the DSG equation, which represents the uncoupling of base pairs as they rotate about the backbone; coupled pairs of DSG equations (nonlinear rods and double rods; nonlinear helices and double helices; nonlinear molecular models; and nonlinear atomic models). The last of these lead to very challenging numerical problems at the present time, making less complicated models attractive for specific applications.

Although this activity is pursued vigorously in the communities of applied mathematics and condensed matter physics, it is not without criticism from biochemists. (Indeed, the term "hijacking" has been used in an esteemed biochemistry journal to describe applications of nonlinear dynamics to the study of DNA.) Arising in part from diverse perspectives of different scientific subcultures, concerns about the intrusions of nonlinear science into biochemistry stem also from objective considerations. The activity of model making can become ingrown, carried on with scant concern for its relevance to the real problems of biology, and presented in terms that are difficult for the biochemist to understand.

Concerns of this sort have led Krumhansl (who was among the first to suggest a DSG model for DNA) to comment on the state of biomathematics, offering the following suggestions as guides to future work [87, 88]:

Identify the phenomenon to be studied. Do so only after consulting with experimentalists to find out what they believe to be important. Initially, what *we* think is important may not be relevant; try to do their problem first. Reduction from 'first principles' hardly ever works.

Use physically realistic variables, i.e. 'order parameters'. The endemic difficulty here is that formal mathematical or theoretical scientists expect experimental scientists to pull out from general models the important results that operate to their need. This is unrealistic; the interaction must be a two way street. The nonlinear scientist must make the effort to learn enough about biology... to talk to the experimentalist and to use common sense in modeling. Clearly, it is important to find the minimum number of functionally important coordinates, but not more—a clever trick.

In an attempt to bridge the gap between the two cultures studying biomolecular dynamics, a tutorial workshop was organized at Les Houches in the French Alps in early June of 1994. The lectures from this meeting have been gathered together in a book, entitled *Nonlinear Excitations in Biomolecules* [103], which is also suggested as a first step for a novice biomathematician wishing to enter the field of DNA dynamics.

Of particular interest to physical scientists at this workshop were the lectures by biologists, emphasizing the intricacy of the transcription process, whereby genetic information is copied from the DNA double helix onto an auxiliary nucleic acid molecule: messenger ribonucleic acid (mRNA) [109]. A key component in this process is the enzyme (protein) RNA polymerase (RNAP) that crawls along the double helix, uncoupling 10–20 base pairs at an energetic expense of 50–100 kcal/mole. Interestingly, *in vitro* experiments with purified RNAP and DNA indicate that no energy need be supplied to this system, suggesting that the required energy is borrowed from the thermal reservoir and localized in the region of open base pairs. Can nonlinear lattice dynamics help to explain this observation?

Aiming to provide an answer, Dauxois and Peyrard have investigated a simple lattice model of DNA, where stretching of the hydrogen bonds connecting the base pairs—y_n—is the only dependent variable [29]. Their Hamiltonian is

$$H = \sum_n \left[\tfrac{1}{2} m \dot{y}_n^2 + V(y_n) + W(y_n, y_{n-1}) \right], \tag{5.64}$$

where

$$V(y_n) = D \left(e^{-a y_n} - 1 \right)^2$$

is an on-site potential, and

$$W(y_n, y_{n-1}) = \frac{K}{2} \left(1 + \rho e^{-\alpha(y_n + y_{n+1})} \right) (y_n - y_{n-1})^2$$

is a nonlinear interaction potential between nearest neighbors. Thus their model is strongly nonlinear in both the site dynamics and in the nearest-neighbor coupling. This model reproduces the growth of localized "bubbles" (regions of open base pairs) with increasing ambient temperature, confirming earlier studies of a similar nature [24,98]. The model also demonstrates that a lattice is more effective in gathering energy from a thermal reservoir than a nonlinear continuum model would be, because discreteness encourages the coagulation of solitary waves.

But Equation (5.64) is far from a realistic model of DNA, as it neglects the role of enzyme RNAP which is essential for the transcription of a genetic code. Thus it seems unlikely that anything as simple as a soliton dominates the dynamics of DNA transcription.

5.4 Nonconservative lattices

To this point in our survey of nonlinear lattices, we have considered only systems that conserve energy in the sense of Appendix A. Such a constraint is often inappropriate for models of biological systems, and also for engineering systems where resistors, amplifiers, and nonlinear conductive elements play decisive roles. In the present section, some dissipative lattices are described.

5.4.1 Quasiharmonic lattices

Two identical oscillators sharing a common source of energy have a stationary solution of equal amplitudes, but this solution is often unstable. One of the oscillators will inevitably gain a bit of amplitude and consume more than its share of the available power, leading to its further growth and the eventual extinction of the other oscillation [106]. This situation arises in diverse areas of engineering and science including:

1. Various "flip-flop" circuits, used by computer engineers to store information.
2. Multimode lasers where two or more optical modes compete for the energy available from an active medium (inverted population). Often several

modes manage to coexist—to the despair of laser engineers—by becoming dominant in different regions of the laser medium. This effect is especially noticeable when the modes have transverse structure.
3. Interacting biological species with a common food supply. Often several such species manage to coexist—to the delight of the naturalist—by finding different ecological niches: related birds, for example, that confine themselves to different levels of the forest.
4. The neocortex of a human brain, where many concepts or ideas struggle for the dominance of attention.

As a model for such phenomena, consider a two-dimensional array of identical oscillators, each interacting only with its nearest neighbors. The analytical tool to be used is called the method of harmonic balance, introduced by van der Pol in 1934 [106] and developed by applied mathematicians and engineers in the Soviet Union during the 1930s and 1940s [5, 89].

Shown in Figure 5.6, a unit cell of the array consists of a nonlinear conductance $I(v)$ in parallel with a capacitance C, connecting a lattice point to a ground plane. Each lattice point is attached to its four nearest neighbors through an inductance L. The nonlinear conductance is represented as

$$I(v) = -Gv\left(1 - 4v^2/3V_0^2\right), \tag{5.65}$$

so for $-V_0/2 < v < +V_0/2$ the differential conductance (dI/dv) is negative.

Assume a large, square lattice of these unit cells with zero voltage (short circuit) boundary conditions on the edges, as in Figure 5.7. In a zero-order approximation, the nonlinear conductance can be neglected, and in a continuum approximation, this linear, lossless system supports an arbitrary number (say n) modes, leading to a total voltage

$$v(x,y,t) = \mathcal{V}_1(x,y)\cos\theta_1 + \mathcal{V}_2(x,y)\cos\theta_2 + \cdots + \mathcal{V}_n(x,y)\cos\theta_n, \tag{5.66}$$

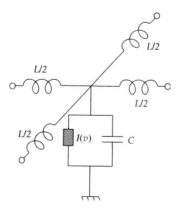

FIG. 5.6. *A unit cell of the two-dimensional tunnel diode array.*

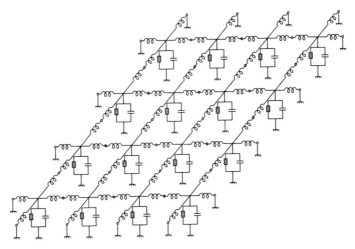

FIG. 5.7. *A 4×4 array ($N^2 = 16$) of the unit cells in Figure 5.6.*

where

$$\mathcal{V}_i(x, y) \equiv V_i \cos k_{xi} x \cos k_{yi} y$$

$$\theta_i \equiv \omega_i t + \phi_i.$$

At the next order of approximation, substitute Equation (5.66) into Equation (5.65) to obtain

$$i(v) = I_1 \cos \theta_1 + I_2 \cos \theta_2 + \cdots + I_n \cos \theta_n + \text{overtones},$$

where, for example,

$$I_1 = -G\mathcal{V}_1 \left[1 - \frac{\mathcal{V}_1^2 + 2(\mathcal{V}_2^2 + \cdots + \mathcal{V}_n^2)}{V_0^2} \right]$$

and similarly for the other I_j.

Neglecting overtones and resonant interactions between the modes, the average power into a lattice point for the ith mode is

$$p_i = -\lim_{T \to \infty} \frac{1}{T} \int_0^T i(v) \mathcal{V}_i \cos \theta_i \, dt = I_i \mathcal{V}_i / 2.$$

Assuming further that all the k_x and k_y are different, the total power into the first mode is

$$P_1 = \frac{N^2 G V_1^2}{8} \left[1 - \frac{(9/8) V_1^2 + V_2^2 + \cdots + V_n^2}{2 V_0^2} \right],$$

where N^2 is the total number of lattice points in the array. Because the energy in the ith mode is related to its amplitude by

$$U_i = (N^2C/8)V_i^2,$$

the rate of change of energy into the first mode is

$$\frac{dU_1}{d\tau} = U_1\left[1 - \alpha\left(\tfrac{9}{8}U_1 + U_2 + \cdots + U_n\right)\right], \tag{5.67}$$

with

$$\tau \equiv Gt/C$$
$$\alpha \equiv 4/(N^2CV_0^2).$$

For n excited modes, there is a set of n equations—similar to Equation (5.67) but with the indices appropriately altered—for the rates of change of the mode energies as functions of those energies. These nonlinear, autonomous equations have the same form as those introduced by Volterra to describe the interaction of biological species competing for the same food supply [134], and they are similar to models suggested by Greene [66] and by Cowan [26] to describe interactions between assemblies of neurons in the brain.

The autonomous system defined by Equation (5.67) evidently has stationary solutions with

$$U_1 = U_2 = \cdots = U_n = 1/(n + 1/8)\alpha. \tag{5.68}$$

From this it is seen that unexcited modes have linear growth rates given by

$$\frac{dU_j}{d\tau} = U_j/(8n + 1), \tag{5.69}$$

implying that a stationary solution of n modes is unstable to the emergence of additional oscillations. This conclusion neglects energy losses in the inductors, which can be expressed in terms of a quality factor

$$Q_j = \omega_j L/R,$$

where R is the series resistance of an inductor. (In other words, the quality factor is the ratio of the maximum energy stored in an inductor to the energy lost in a radian of the oscillatory period.) Thus the number of oscillatory modes (n) is expected to grow until

$$\frac{C/G}{8n+1} < \frac{L}{R},$$

implying

$$n > \frac{1}{8}\left(\frac{RC}{GL} - 1\right). \tag{5.70}$$

If Equation (5.70) is satisfied, linear analysis of Equation (5.67) near the stationary point defined by Equation (5.68) indicates stability.

In systems of d-space dimensions, equations of interacting mode energies become

$$\frac{dU_1}{d\tau} = U_1 \left[1 - \alpha \left(KU_1 + U_2 + \cdots + U_n\right)\right]$$

$$\vdots \qquad (5.71)$$

$$\frac{dU_n}{d\tau} = U_n \left[1 - \alpha \left(U_1 + U_2 + \cdots + KU_n\right)\right],$$

where

$$K = 3^d/2^{d+1}.$$

A linear analysis of Equations (5.71) indicates multimode stability for $K > 1$, implying $d > 1$.

Thus for systems of two or three space dimensions, it is possible for more than one mode to be stable. This phenomenon is important in the design of semiconductor lasers, where large transverse dimensions are introduced to increase the power output. Unfortunately, such a design allows several modes to oscillate together, thereby decreasing the spectral purity of the output beam of light.

A variety of such oscillator arrays have been realized using semiconductor (Esaki) tunnel diodes [114], superconductor tunnel diodes [74], and integrated circuits [8]. In these experiments, one observes a variety of stable multimode oscillations, some of which are quasiperiodic and others periodic, indicating mode locking. In a manner that is qualitatively similar to that proposed by Greene in his model of the brain's neocortex [66], the oscillator array can be induced to switch from one stable mode configuration to another.

At the level of subjective perception, similar switchings in the brain are observed as one stares at the Necker cube shown in Figure 5.8. Constructed by a Swiss geologist in the mid-1800s, this image seems to jump from one metastable orientation to another, like the flip-flop circuit of a computer engineer.

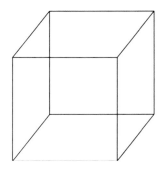

FIG. 5.8. *A Necker cube.*

Of course one should not carry the tunnel diode array beyond a simple metaphor for the brain's dynamics—a real brain is a far more intricate nonlinear lattice than any yet understood by applied mathematicians [116]—yet Haken takes a similar approach in his recent book *Principles of Brain Functioning* [67]. Defining order parameters (ξ_1 and ξ_2) to represent neural activities related to the two perceptions of the Necker cube, he suggests the equations

$$\frac{d\xi_1}{dt} = \xi_1[1 - C\xi_1^2 - (B+C)\xi_2^2]$$
$$\frac{d\xi_2}{dt} = \xi_2[1 - C\xi_2^2 - (B+C)\xi_1^2]$$
(5.72)

as an appropriate dynamic description. With $U_j = \xi_j^2$ ($j = 1, 2$), Equations (5.72) are identical to Equations (5.71), showing single mode stability for $B > 0$. Study of these systems in the (U_1, U_2)- or (ξ_1, ξ_2)-plane reveals behavior that is similar to one's subjective experience of Figure 5.8.

5.4.2 Myelinated nerves

Following the Hodgkin–Huxley formulation of nerve impulse dynamics for the giant axon of the squid, most mathematical studies have focused on smooth nerve fibers, as in the previous chapter. Although this picture is appropriate for the squid axon, many vertebrate nerves—including motor nerves and axons of mammalian brains—are bundles of discrete, periodic structures, comprising active nodes (also called "nodes of Ranvier") separated by relatively long fiber segments that are insulated by a fatty material called myelin. In such "myelinated" nerves, the wave of activity jumps from one node to the next, and should be modeled by nonlinear difference-differential equations rather than by PDEs. In an evolutionary context, myelinated nerve structures are useful because they allow an increase in the speed of a nerve impulse while decreasing the diameter of the nerve fiber [116].

The phenomenon of saltatory conduction on myelinated nerves introduces two qualitatively important features. On the up side is the above-mentioned increase in speed of conduction with reduced fiber diameter and energy dissipation. On the down side, however, is the possibility of failure when the distance—or electrical resistance—between successive active nodes becomes too large.

In Figure 5.9(a) is sketched a single myelinated nerve fiber showing active nodes separated by regions of the fiber that are insulated by myelin, and Figure 5.9(b) corresponds to the difference-differential equations

$$V_n - V_{n+1} = (R_\mathrm{i} + R_\mathrm{o})I_n \qquad (5.73)$$

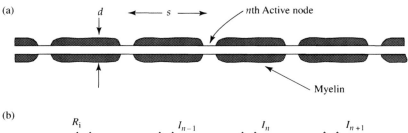

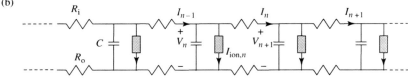

FIG. 5.9. *(a) A single myelinated nerve fiber (not to scale). (b) The corresponding electric circuit diagram.*

and

$$I_{n-1} - I_n = C\frac{dV_n}{dt} + I_{\text{ion},n}. \tag{5.74}$$

In these equations, the index n indicates successive active nodes, each characterized by a transverse voltage across the membrane (V_n). A second dynamic variable is the current (I_n) flowing longitudinally through the fiber from node n to node $n+1$. Thus Equation (5.73) is merely Ohm's law, which relates the voltage difference between two adjacent nodes of the current flowing between them times the sum of the inside and outside resistances, R_i and R_o. Equation (5.74) says that the current flowing into the nth node from the $(n-1)$th node (I_{n-1}) minus the current flowing out of it to the $(n+1)$th node (I_n) is equal to the following two components of transverse (inside to outside) current leaving the node: capacitive current, $C\,dV_n/dt$, and ionic current, $I_{\text{ion},n}$, comprising mainly a sodium component.

The time delay for the onset of sodium ion permeability is rather short (in the frog nerve it is about 0.1 ms), whereas the time delay for the onset of potassium ion permeability is several milliseconds. Thus, it is assumed that the sodium ion current begins without delay and the potassium permeability remains equal to its resting value over the leading edge of the impulse. From the Hodgkin–Huxley formulation of Chapter 4, these assumptions imply that the total ionic current is a function of the transmembrane voltage and can be represented by a nonlinear conductivity. It is further assumed that this ionic current is given by the cubic polynomial function

$$I_{\text{ion},n} = \left(\frac{G}{V_2(V_2 - V_1)}\right) V_n(V_n - V_1)(V_n - V_2). \tag{5.75}$$

Table 5.1. *Parameter values in the myelinated model for a standard frog motor axon [116].*

Parameter	Value	Units
s	2	mm
d	14	μm
C	3.7	pF
R_f	28	Mohms
G	0.57	μmhos
V_1	25	mV
V_2	122	mV

Measuring voltages in units of (V_2) to obtain the dimensionless voltage variables $\mathcal{V}_n \equiv V_n/V_2$, Equations (5.73) and (5.74) become the discrete reaction diffusion system

$$RC\frac{d\mathcal{V}_n}{dt} = (\mathcal{V}_{n+1} - 2\mathcal{V}_n + \mathcal{V}_{n-1}) - \left(\frac{RG}{1-a}\right)\mathcal{V}_n(\mathcal{V}_n - a)(\mathcal{V}_n - 1),$$

where $a \equiv V_1/V_2$ and $R = R_\mathrm{i} + R_\mathrm{o}$, and all parameters of the model are given in Table 5.1.

In order to mimic the effects of evolutionary pressures, the model is normalized so internode spacing is an independent parameter. Noting that $R \propto s$, a discreteness parameter is defined as

$$D \equiv \left(\frac{2\mathrm{mm}}{s}\right) = \frac{R_f}{R},$$

where $R_f = 28$ Mohms, the internode resistance of the standard frog axon. In other words, $1/D$ is the spacing between nodes in units of 2 mm, so $D = 1$ implies the discreteness of the standard frog axon. Under this formulation, the dynamic equation becomes

$$D(\mathcal{V}_{n+1} - 2\mathcal{V}_n + \mathcal{V}_{n-1}) = R_f C \frac{d\mathcal{V}_n}{dt} + \left(\frac{R_f G}{1-a}\right)\mathcal{V}_n(\mathcal{V}_n - a)(\mathcal{V}_n - 1). \quad (5.76)$$

The nature of the wave propagation on a discrete nerve model can be characterized by looking at the relative change in voltage between two adjacent nodes. If this relative change is small, then the voltages and currents are relatively smooth functions of distance and the system can be described by partial differential equations—the corresponding continuum system. If, on the other hand, the fractional change in voltage amplitude from one node to the next is large compared with unity, then the conduction process is saltatory. We refer to these two cases as the "continuum limit" and "saltatory limit," respectively.

Continuum limit. If the internode spacing s is small enough so that the continuum limit is reached, Equation (5.76) can be written as the partial differential equation

$$s^2 D \frac{\partial^2 \mathcal{V}}{\partial x^2} - R_{\mathrm{f}} C \frac{\partial \mathcal{V}}{\partial t} = \left(\frac{R_{\mathrm{f}} G}{1-a}\right) \mathcal{V}(\mathcal{V}-a)(\mathcal{V}-1),$$

where we have assumed $D \gg 1$ and let $ns \to x$. This is the Z–F equation which was discussed in Section 4.1.1, where time is measured in units of $(1-a)C/G$ and distance in units of $s\sqrt{(1-a)/RG}$. Thus a traveling-wave front (the leading edge of the impulse) has speed

$$v_{\mathrm{c}} = \sqrt{\frac{G}{RC^2}} \left(\frac{1-2a}{\sqrt{2(1-a)}}\right) \text{ nodes/s}. \tag{5.77}$$

To get the corresponding impulse speed in (say) meters/second, as is plotted in Figure 5.10, merely multiply this expression by the number of meters between adjacent nodes (s).

Saltatory limit. For D equal to or less than unity, the wave of excitation jumps from node to node in a discontinuous manner. An additional feature of the saltatory limit is the possibility that the switching of one node may be unable to ignite the adjacent node. In this situation, called "failure," the impulse ceases to propagate. Because failure of impulse conduction is an undesired property of a real nerve, we expect the node spacing for frog axons to lie comfortably beyond this range.

If the internode spacing s is increased so that D is reduced to a critical value D^*, failure of impulse propagation occurs because the fully developed voltage at one node is unable to bring the next node above threshold. This occurs when the internode distance, and therefore R, becomes too large with respect to the node resistance $1/G$.

With the cubic form of the sodium ion current in Equation (5.75), Erneux and Nicolis [44] have shown that the critical value of the discreteness parameter is given to lowest order in a ($\equiv V_1/V_2$) by

$$D^* \approx \frac{R_{\mathrm{f}} G a^2}{4(1-a)} = 0.21. \tag{5.78}$$

For D slightly larger than D^*, these same authors show that the impulse velocity $v \to v_{\mathrm{s}}$, where

$$v_{\mathrm{s}} = \frac{1}{\pi C} \sqrt{\frac{G(D-D^*)}{R_{\mathrm{f}}(1-a)}} \text{ nodes/s}. \tag{5.79}$$

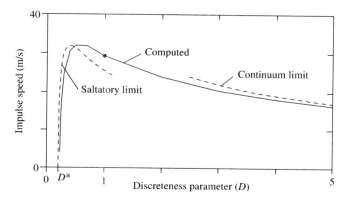

FIG. 5.10. *Leading-edge impulse velocity on a myelinated axon as a function of the discreteness parameter $D = 2\,\text{mm/s}$. The black dot at $D = 1$ indicates a standard frog axon of outside fiber diameter equal to $14\,\mu m$. (Numerical data courtesy of Stephane Binczak.)*

Again, the corresponding impulse speed in meters/second, as plotted in Figure 5.10, is obtained from multiplying this expression by s, the number of meters/node.

Equation (5.76) has been used to compute the wave-front velocity for a nerve impulse, which is plotted against the discreteness parameter ($D = 2\,\text{mm/s}$) in Figure 5.10 [12]. From these numerical data, one observes, first, that the continuum approximation of Equation (5.77) holds well for $D > 5$ or $s < 0.4$ mm. Second, failure in the model is accurately predicted by Equation (5.78) to occur at $D^* = 0.21$ or $s = 9.5$ cm, and near the failure point, impulse velocity is well-represented by Equation (5.79). Third, neither the continuum approximation formula nor Equation (5.79) give a satisfactory estimate of the impulse velocity at $D = 1$ (corresponding to the parameters of a real frog nerve). Finally, the computed velocity is in accord with measurements of impulse speeds on frog axons, suggesting how the parameters of myelinated nerves may have been optimized during the course of biological evolution [116].

5.4.3 *Emergence of form by replication*

At this point, the reader may be asking: Why are two such different subjects as "quasiharmonic lattices" and "myelinated nerves" treated in the same section of this book? Beyond the fact that these systems do not conserve energy, what have they in common? Interestingly, both systems are related to a means recently proposed for replication of form that may play a role in biological morphogenesis [133].

To see this relation, return to Figures 5.6 and 5.7, making each inductor an open circuit. Then the individual lattice points are isolated capacitors shunted by the nonlinear conductance indicated in Equation (5.65), with two stable voltage states at $v = \pm\sqrt{3}V_0/2$, and the system is a conventional computer memory, capable of storing

$$2^{N^2}$$

patterns or N^2 bits of information. Replacing the inductors with resistors (R) that are sufficiently large to preserve this storage capacity, leads to the system

$$\dot{v}_{j,k} = \kappa(v_{j+1,k} + v_{j-1,k} + v_{j,k+1} + v_{k,j-1} - 4v_{j,k}) - f(v_{j,k}),$$

where $\kappa \propto 1/R$ is sufficiently small. If $v_{j,k}$ is independent of the index k, this reduces to a myelinated nerve model of Equations (5.76).

Seeking a way to replicate form, Velarde et al. augmented the previous equation to a coupled system of two-dimensional reaction-diffusion equations; thus

$$\begin{aligned}
\dot{v}_{j,k} &= \kappa(v_{j+1,k} + v_{j-1,k} + v_{j,k+1} + v_{k,j-1} - 4v_{j,k}) - f(v_{j,k}) \\
&\quad - h(v_{j,k} - u_{j,k}) \\
\dot{u}_{j,k} &= \kappa(u_{j+1,k} + u_{j-1,k} + u_{j,k+1} + u_{k,j-1} - 4u_{j,k}) - f(u_{j,k}) \\
&\quad - h(u_{j,k} - v_{j,k}),
\end{aligned} \quad (5.80)$$

where

$$f(x) = x(x-a)(x-1). \quad (5.81)$$

With $a = 1/2$ and a rescaling, Equation (5.81) is equivalent to Equation (5.65).

Essentially a cellular automaton, Equations (5.80) are related to the two-dimensional reaction diffusion systems introduced in Section 4.6.1, with coupling provided through the factor h. If one subsystem (say v) carries a pattern and the other (say u) is a random array of black and white dots (and the coupling factor h is sufficiently large), then the u-system will become slaved to the v-system, allowing a transfer of the pattern from v to u. More particularly if $h > (1 - a - a^2)/6$, both systems will adjust themselves to the approximate the original pattern of the ordered system, an example of order emerging from chaos.

Velarde et al. present several striking demonstrations of pattern replication by Equations (5.80), and speculate that such processes may influence the development of biological form, perhaps even at the level of biomolecular structures.

5.5 Assemblies of neurons

As the most intricate dynamic system known to exist in the universe, the human brain has been an object of interest over the past 200 years, promising to play a central role in the sciences of coming decades. Comprising some ten to a hundred thousand million neurons, the brain's neocortex can be viewed as a very intricate nonlinear lattice, where any two points have the possibility of interacting. How does one explore the dynamic possibilities of such a system?

An answer to this question was proposed a half century ago by the Canadian psychologist Donald Hebb in a classic work entitled *The Organization of Behavior: A Neuropsychological Theory*, where he attempted to "bridge the long gap between the facts of psychology and those of neurophysiology" [68].[10] Central to Hebb's description of the brain's dynamics was the concept of a "cell assembly," which was introduced as follows:

> Any frequently repeated, particular stimulation will lead to the slow development of a "cell-assembly," a diffuse structure comprising cells...capable of acting briefly as a closed system, delivering facilitation to other such systems and usually having a specific motor facilitation. A series of such events constitutes a "phase sequence"—the thought process. Each assembly may be aroused by a preceding assembly, by a sensory event, or—normally—by both. The central facilitation from one of these activities on the next is the prototype of "attention."

To better understand what Hebb and others now mean by a cell assembly, one can compare the brain to a city and the individual neurons to its citizens. A particular person might be a member of several social assemblies such as a political association, a church, a bowling league, a parent-teacher's group, a hiking club, the Junior League, and so on, where the members of each assembly are interconnected. Often a pair of assemblies will have a few overlapping members, and lists of addresses and telephone numbers allow one organization to activate its own members—or even the members of a like-minded assembly—whenever an appropriate occasion arises.

Members of the hiking club, for instance, could encourage the 4-H club to resist the development of a theme park near their farms. Or the Junior League might enlist the support of teachers to close a pornographic bookstore. Just as an individual could be a member of both the hiking club and the League, a single

---

[10] Recent references to Hebb's work often focus attention on the postulate that the strength of a synapse (or neural interconnection) increases as it participates in firing a neuron, a property that some are in the habit of calling "Hebbian." Hebb was amused by this because the synaptic postulate was one of the few features of his theory that he did not consider to be original, Freud, among others, having previously suggested the possibility of learning by synaptic modifications.

nerve cell would—in Hebb's theory—participate in different assemblies of the brain. To understand the dynamics of a city, it is necessary to study how these groups interact with each other.

Thus, Hebb pointed out that to comprehend the brain's activity one should not necessarily focus attention on individual neurons. Just as Haken has emphasized the fundamental importance of assigning appropriate order parameters at each level of description [11, 67], Hebb's central idea is that the brain functions not merely by switching individual neurons on and off, but by activating assemblies of neurons. These assemblies thereby become the points of a new nonlinear lattice of assemblies that is embedded in the lattice of neurons.

If one considers the interactions of n assemblies, each described by its firing rate F_j (where $j = 1, 2, \ldots, n$), a simple model is [116]

$$\frac{dF_1}{dt} = +F_1(1 - F_1) - \alpha F_2 - \alpha F_3 - \cdots - \alpha F_n,$$

$$\frac{dF_2}{dt} = -\alpha F_1 + F_2(1 - F_2) - \alpha F_3 - \cdots - \alpha F_n,$$

$$\vdots$$

$$\frac{dF_n}{dt} = -\alpha F_1 - \alpha F_2 - \alpha F_3 - \cdots + F_n(1 - F_n),$$

where α is a parameter that is positive for inhibitory interactions among assemblies. This system is similar to those of Equations (5.71) and (5.72), which model perceptual dynamics of the Necker cube in Figure 5.8.

Cell assemblies share two salient properties of individual neurons.

1. *Threshold.* An assembly of randomly interconnected neurons has a threshold firing level for the onset of global activity. If this level is not attained, the assembly will not ignite, falling back to a quiescent state.
2. *All-or-nothing response.* If the threshold level is exceeded, firing activity of an assembly will rise rapidly to a maximum level.

These two properties are sufficient to ensure that assemblies of neurons can form assemblies of assemblies. Just as assemblies emerge from the nonlinear properties of individual neurons, assemblies of assemblies emerge from the nonlinear dynamic properties of assemblies. Repeated several times, this leads to the following picture of the brain as an emergent dynamic hierarchy.

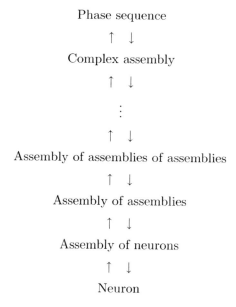

The currency passed up and down along the arrows is information, and from the perspective of Haken's synergetics, appropriate order parameters would be defined at each level of this hierarchy.

As we know from introspection, interactions among complex assemblies can be very complicated, and the means for such interactions are not well understood. Interestingly, a nonlinear traveling-wave of information was proposed by Beurle in 1956, soon after the Hodgkin and Huxley formulation for neural dynamics [13]. Beurle's localized waves can pass through each other—much like the solitons of Chapter 3—because they engage only a sparse subset of the total population of neurons, suggesting that cell assemblies are not necessarily static but may move through the neocortex, allowing yet more intricate interactions.

5.6 Summary

In this chapter, we have seen that nonlinear lattices arise as useful models for a wide variety of phenomena at many levels of description, ranging from the dynamics of molecules and crystals through interacting biological species and and myelinated nerve impulse propagation to the dynamics of the neocortex. Some of these models conserve energy, while others involve nonlinear interactions between sources and sinks of energy. Of the systems that conserve energy, some are completely integrable in the sense of soliton theory, but many—including those most important for applications—are not. In some examples, useful analytical tools are presently available, and others are waiting to be discovered. Prospects for research in the area of nonlinear lattices are indeed bright.

5.7 Problems

1. (a) Consider Equation (5.2) in the limit $a \to \infty$ and $b \to 0$ with ab finite, and show that it reduces to a linear spring.

   (b) Consider the same equation in the limit $a \to 0$ and $b \to \infty$ with ab finite, and show that it describes the interaction between hard spheres.

2. Plot the Toda potential from Equation (5.2) with $a = 1$, and $b = 0.1, 0.3, 1, 3,$ and 10.

3. Derive and discuss the significance of all solutions to the linear difference-differential equations defined by Equation (5.4) with $-\infty < n < \infty$.

4. Verify that the solitary wave in Equation (5.5) is a solution of Equations (5.1) with the anharmonic potential defined in Equation (5.2).

5. With reference to Appendix E,

   (a) Show that a periodic traveling-wave solution of Equations (5.1) with frequency ν and wavelength λ can be written as
   $$r_n = -\frac{1}{b} \log\left(1 + \frac{(2K\nu)^2}{ab}\left\{\mathrm{dn}^2[2K(\nu t \pm n/\lambda); k] - \frac{E}{K}\right\}\right),$$
   where $K(k)$ and $E(k)$ are respectively complete elliptic integrals of the first and second kind,
   $$2K\nu = \left[\frac{ab}{\mathrm{sn}^{-2}(2K/\lambda) - 1 + E/K}\right]^{1/2},$$
   and $\mathrm{sn}(\,\cdot\,;k)$ and $\mathrm{dn}(\,\cdot\,;k)$ are elliptic functions of modulus $0 \le k \le 1$.

   (b) As $k \to 0$, show that
   $$r_n \to -\frac{\omega^2 k^2}{8ab^2}\cos(\omega t \pm 2\pi n/\lambda) + O(k^4).$$

   (c) As $k \to 1$, show that r_n approaches the solitary wave given in Equation (5.5).

6. In 1975, Wadati and Toda proved that solutions of the Toda lattice can be created by a BT that can be formulated as follows [129, 136]. Let
   $$r_n = Q_n - Q_{n-1},$$

so Q_n is the displacement of the nth mass from its equilibrium position. With the potential normalized by setting $a = b = 1$ in Equation (5.2), a BT is then

$$\frac{d}{dt}(Q_{n+1} - \tilde{Q}_n) = a\left\{\exp[-(\tilde{Q}_{n+1} - Q_{n+1})] - \exp[-(\tilde{Q}_n - Q_n)]\right\}$$

$$\frac{d}{dt}(\tilde{Q}_n - Q_n) = \frac{1}{a}\left\{\exp[-(Q_{n+1} - \tilde{Q}_n)] - \exp[-(Q_n - \tilde{Q}_{n-1})]\right\},$$

where Q_n is a known solution and $\tilde{Q}_n$ is a new solution that is generated by the transformation.

(a) Starting with the vacuum solution

$$Q_n = 0,$$

show that this BT generates a single Toda-lattice soliton of the form

$$\exp(\tilde{Q}_n) = a\frac{\cosh(\kappa n + t\sinh\kappa)}{\cosh[\kappa(n+1) + t\sinh\kappa]},$$

where $a = e^{-\kappa}$.

(b) Check that this result is in agreement with Equation (5.5).

7. In the Toda-lattice BT, show that if $Q_n(t)$ is a solution of the Toda-lattice, then $\tilde{Q}_n(t)$ must also be such a solution.

8. Discuss in general (qualitative) terms how the Toda lattice BT can be used to derive N-soliton solutions without the need for integration.

9. (a) Show that the Toda-lattice equations are equivalent to those for the nonlinear electric filter defined by

$$\frac{d^2}{dt^2}\log[1 + V_n(t)] = V_{n+1}(t) - 2V_n(t) + V_{n-1}(t).$$

(b) Use this equivalence to construct a soliton for the nonlinear filter.

10. Derive the approximate lattice solitary-wave solution of Equation (5.10) from the preceding ODE.

11. (a) Use Equation (F.3) of Appendix F to construct an approximate solitary-wave solution for the quartic potential of Equation (5.11).

(b) Check that this solution is in accord with the Friesecke–Wattis theorem.

12. Show that the expression for a compressive LSW given in Equation (5.10) approaches that for a Toda-lattice soliton as v^2 approaches unity from above.

13. (a) Verify that a single pole and single zero approximation of $\Lambda(k)$ in Equation (F.2) is

$$\Lambda(k) \doteq \frac{1 - k^2/20}{1 + k^2/30}.$$

(b) Show that the corresponding differential equation in this case is

$$v^2(1 - D^2/30)R(z) \doteq (1 + D^2/20)\phi'[R(z)].$$

(c) Solve this equation for the potential defined in Equation (5.9) to find a traveling wave solution of amplitude

$$R(0) = \frac{1}{9a}\left[4v^2 - 9 + v\sqrt{5(14v^2 - 9)}\right].$$

(d) Compare your results from part (c) with Equation (5.10).

14. Show that the Toda-lattice potential fails to satisfy James's condition of Equation (5.15) (for ILMs to be found).

15. In Equation (5.19) for stationary DSG kinks, write $u_n = \phi_n + \pi$, and assume (i) $\phi_0 = 0$ and (ii) $\phi_n = -\phi_{-n}$.

    (a) Construct a recursion relation for ϕ_n in terms of ϕ_{n-1} and ϕ_{n-2}.

    (b) Assuming $\Gamma = 1$, solve this recursion relation for a sequence of ϕ_n that approach π as $n \to \infty$.

    (c) Compare your calculations with the continuum limit.

16. Repeat the previous problem assuming that the two central values of the ϕ_n are ϕ_1 and $\phi_{-1} = -\phi_1$.

17. (a) Calculate the Peierls barrier as the difference in energies of the static solutions obtained in the previous two problems.

    (b) Which of the two solutions do you expect to be unstable? Why?

18. For $\Gamma = 0.1, 0.3, 1, 3,$ and 10, compute and plot the values of u_n for the stable stationary kink solution of the DSG equation.

19. Derive Equation (5.22) for a DSG oscillation of constant amplitude..

20. (a) For $-\infty < n < +\infty$, find all solutions of the DNLS equation (5.23) with $\gamma = 0$, and $\varepsilon > 0$.

    (b) Find all solutions with $\gamma > 0$, and $\varepsilon = 0$.

21. (a) Assuming periodic boundary conditions in Equation (5.24), show that the Hamiltonian given in Equation (5.26) is independent of time.

    (b) If $-\infty < n < +\infty$ in Equation (5.24), what boundary conditions can be imposed as $n \to \pm\infty$ to ensure that the energy remains constant?

22. Repeat the previous problem for the number defined in Equation (5.25).

23. Show that the DNLS equation defined on two lattice sites is integrable. [Hint: use the facts that H and N are constants of the motion.]

24. Prove that Equation (5.28) for the DNLS equation is equivalent to Equation (5.24).

25. Show that as $k \to 0$ and $\beta \to 0$ in Equation (5.34) the general traveling-wave solution approaches that of

$$i\frac{d\phi_n}{dt} + \phi_{n+1} + \phi_{n-1} = 0.$$

26. (a) With the nonstandard Poisson bracket defined in Equation (5.38), find a Hamiltonian for the A–L equation

$$i\frac{d\phi_n}{dt} + \frac{\gamma}{2}|\phi_n|^2(\phi_{n+1} + \phi_{n-1}) + \phi_{n+1} - 2\phi_n + \phi_{n-1} = 0.$$

   (b) How are solutions of this equation related to those of Equation (5.33)?

27. (a) Assuming periodic boundary conditions in Equation (5.33), show that the Hamiltonian defined in Equation (5.37) is independent of time.

   (b) Repeat (a) for the number defined in Equation (5.36).

28. (a) For the numbers and Hamiltonians defined in Section 5.3.3, use appropriate definitions of the Poisson brackets to show that $\{N_1, H_1\} = 0$, $\{N_2, H_2\} = 0$, and $\{N_3, H_3\} = 0$.

   (b) Discuss the physical significance of these results.

29. (a) Show that the standard form of DST in Equation (5.52) can be derived from the Lagrangian

$$L = \frac{1}{2}\sum_j [i(u_j^* \dot{u}_j - u_j \dot{u}_j^*) + \gamma|u_j|^4] + \varepsilon \sum_{j,k} m_{jk} u_j^* u_k.$$

   (b) From this Lagrangian, derive the corresponding Hamiltonian.

   (c) Show that the DST number

$$N = \sum_j |u_j|^2$$

   and this Hamiltonian satisfy the condition

$$\{N, H\} = 0,$$

   where $\{\cdot, \cdot\}$ is the Poisson bracket defined in Equation (5.27).

228   NONLINEAR LATTICES

30. For the stationary DST solution in Equation (5.53), show that
$$\omega = -dH/dN,$$
where H and N are defined in the previous problem.

31. Find all the linear eigenvalues in Equation (5.55) with M as in Equation (5.51). Compare your results with Figure 5.3.

32. For the stationary solutions of the DST shown in Figure 5.3, find analytical expressions of the relationships between γ and ω for the modes: (a) ($\uparrow\uparrow\uparrow$), (b) ($\uparrow\uparrow\downarrow$), and (c) ($\uparrow\downarrow 0$).

33. (a) For a DST system with two freedoms, find all stationary solutions.

    (b) Determine the stability of each solution, and plot your results as in Figure 5.3.

34. Consider a DST system of f freedoms, with $3 \leq f < \infty$ and an interaction matrix $M = [m_{jk}]$, with
$$m_{jk} = \begin{cases} 1 & \text{for } j \neq k \\ 0 & \text{for } j = k. \end{cases}$$
Show that there is a local-mode solution of the form ($\uparrow \ldots$, etc., $\ldots$) with
$$\omega + \varepsilon = \frac{3\varepsilon\sqrt{f^2 - 3f + 3}\sin(\theta + \theta_0)}{\sin 3\theta}$$
$$\phi_1 = -\frac{2}{\sqrt{3\gamma}}\sqrt{\omega + \varepsilon}\,\sin(\theta - \pi/3)$$
$$\phi_2 = \phi_3 = \cdots = \phi_f = \frac{2}{\sqrt{3\gamma}}\sqrt{\omega + \varepsilon}\,\sin\theta$$
$$\theta_0 = \arctan\left[\frac{(f-1)\sqrt{3}}{f-3}\right] - \frac{\pi}{3}$$
$$0 < \theta < \frac{\pi}{6}.$$

35. (a) Construct a Hamiltonian for Davydov's system in Equations (5.62).

    (b) Give a physical interpretation for the term in this Hamiltonian that generates the right-hand sides of Equations (5.62).

36. A homogeneous solution—$h_n(t)$—of the second of Equations (5.62) satisfies the linear system
$$\tilde{M}\frac{d^2 h_n}{dt^2} - \tilde{w}(h_{n+1} - 2h_n + h_{n-1}) = 0.$$

    (a) Discuss how the addition of such a homogeneous solution to $\beta_n(t)$ influences the behavior of Davydov's system. [Hint: consider the influence of temperature.]

PROBLEMS

(b) Suggest how Equation (5.63) might be changed to include the effects discussed in part (a).

37. How many DNA base pairs are needed to encode the 20 amino acids of natural protein?

38. (a) Find the lattice corresponding to the Hamiltonian of Equation (5.64).

    (b) With $\alpha = 0$, construct the second-order continuum approximation for this lattice.

    (c) Find a solitary-wave solution of this continuum approximation.

    (d) Use this solitary-wave solution to construct an approximate solution to the original lattice.

39. (a) For the set of n autonomous equations implied by Equation (5.67), show that a stationary solution emerges when all active modes have the energy

    $$U_0 = 1/(n + \tfrac{1}{8})\alpha.$$

    (b) From a linear analysis near this stationary solution, show that it is stable if the unexcited modes are not permitted to grow.

40. Derive Equation (5.69) for the growth of unexcited modes from Equation (5.67).

41. (a) Repeat the calculations of Section 5.5.1 for a lattice of d spatial dimensions.

    (b) Show that stationary solutions are stable for $d \geq 2$ and unstable for $d < 2$.

42. (a) Consider a lossless $N \times N$ array (see Figure 5.7), and discuss the information storage capacity as a function of N. [Hint: Shannon information is defined as the $\log_2$ of the number of states.]

    (b) What is the information storage capacity of an $N \times N$ lattice when the mode energies are governed by Equation (5.67)?

43. Discuss phase plane trajectories of Haken's order parameters (ξ_1 and ξ_2) for perceptions of the Necker cube, given in Equations (5.72), with $B > 0$.

44. (a) Estimate the speed at which a row of dominoes will fall.

    (b) Compare your estimate with experimental observations.

45. Give a qualitative explanation of failure on a myelinated nerve, as shown in Figure 5.10.

46. Consider the pattern replication system defined in Equations (5.80), with $N = 4$, $a = 0.5$, $h = 0.4$, $\kappa = 0.006$, and periodic boundary conditions.

   (a) Let the initial pattern on the v-system be a black 2×2 square on a white background, while the u-pattern is random black and white dots. Show that the u-pattern converges to the v-pattern as time progresses.

   (b) Estimate the time scale for this convergence.

   (c) How many patterns will this system support?

   (d) How many of these patterns should be considered disordered?

   (e) Using ideas from Section 4.6.1, suggest augmentations of Equations (5.80) that would permit replication of a spiral wave.

REFERENCES

1. M J Ablowitz and J F Ladik. Nonlinear differential-difference equations and Fourier analysis. *J. Math. Phys.* 17 (1976) 1011–1018.
2. M J Ablowitz and J F Ladik. A nonlinear difference scheme and inverse scattering. *Stud. Appl. Math.* 55 (1976) 213–229.
3. P W Anderson. Absence of diffusion in certain random lattices. *Phys. Rev.* 109 (1958) 1492–1505.
4. P W Anderson. Local moments and localized states. *Rev. Mod. Phys.* 50 (1978) 191–201.
5. A A Andronov, A A Vitt, and S E Khaikin. *Theory of Oscillators.* Addison-Wesley, New York, 1966.
6. V I Arnold. *Mathematical Methods of Classical Mechanics.* 2nd edition. Springer-Verlag, New York, 1989.
7. S Aubry. Breathers in nonlinear lattices: Existence, linear stability and quantization. Reference [21], 201–250.
8. H M Aumann. Standing waves on a multimode ladder oscillator. *IEEE Trans. Circuits Syst.* CAS-21 (1974) 461–462.
9. D Bambusi and D Vella. Quasi periodic breathers in Hamiltonian lattices with symmetries. *Disc. Cont. Dyn. Syst.* 2 (2002) 389–399.
10. O Bang, J J Rasmussen, and P L Christiansen. Subcritical localization in the discrete nonlinear Schrödinger equation with arbitrary power nonlinearity. *Nonlinearity* 7 (1994) 205–218.
11. E Basar, H Flohr, H Haken, and A J Mandell (eds.). *Synergetics of the Brain.* Springer-Verlag, Berlin, 1983.
12. S Binczak, J C Eilbeck, and A C Scott. Ephaptic coupling of myelinated nerve fibers. *Physica D* 148 (2001) 159–174.
13. R L Beurle. Properties of a mass of cells capable of regerating pulses. *Philos. Trans. R. Soc.* (London) 240A (1956) 55–94.
14. S Binczak, J C Comte, B Michaux, P Marquié, and J M Bilbault. Experimental nonlinear electrical reaction–diffusion lattice. *Electron. Lett.* 34 (1998) 1061–1062.

15. P Binder, D Abraimov, and A V Ustinov. Diversity of discrete breathers observed in a Josephson ladder. *Phys. Rev. E* 62 (2000) 2858–2862.
16. P Binder and A V Ustinov. Breather "zoo"—large variety of breathers. *Phys. Rev. E* 66 (2002) 016603.
17. G D Birkhoff. *Dynamical Systems.* AMS Publications, Providence, 1927.
18. R Boesch and M Peyrard. Discreteness effects on a sine–Gordon breather. *Phys. Rev. B* 43 (1991) 8491–8508.
19. G Careri, U Buontempo, F Galluzzi, A C Scott, E Gratton, and E Shyamsunder. Spectroscopic evidence for Davydov-like solitons in acetanilide. *Phys. Rev. B* 30 (1984) 4689–4702.
20. J Carr and J C Eilbeck. Stability of stationary solutions of the discrete self-trapping equation. *Phys. Lett. A* 109 (1985) 201–204.
21. H Chaté and M Courbage (eds.). Lattice Dynamics. *Physica D* (1–4) 103 (1997) 1–612.
22. D Chen, S Aubry, and G P Tsironis. Breather mobility in discrete ϕ^4 lattices. *Phys. Rev. Lett.* 77 (1996) 4776–4779.
23. P L Christiansen, M F Jørgensen, and V B Kuznetsov. On integrable systems close to the Toda lattice. *Lett. Math. Phys.* 29 (1993) 165–173.
24. P L Christiansen, P S Lomdahl, and V Muto. On a Toda lattice model with a transverse degree of freedom. *Nonlinearity* 4 (1990) 477–501.
25. M A Collins. A quasi-continuum approach for solitons in an atomic chain. *Chem. Phys. Lett.* 77 (1981) 342–347.
26. J Cowan. Statistical mechanics of neural nets. In *Neural Networks*, E R Caianiello, ed., Springer-Verlag, New York, 1968.
27. L Cruzeiro-Hansson, P L Christiansen, and J Elgin. Comment on "self-trapping on a dimer: Time-dependent solutions of a discrete nonlinear Schrdinger equation." *Phys. Rev. B* 3 (1988) 7896–7897.
28. L Cruzeiro-Hansson, H Feddersen, R Flesch, P L Christiansen, M Salerno, and A C Scott. Classical and quantum analysis of chaos in the discrete self-trapping equation. *Phys. Rev. B* 42 (1990) 522–526.
29. T Dauxois and M Peyrard. A nonlinear model for DNA melting. Reference [103], 127–136.
30. A Davidson, S H Talisa, M G Forrester, J Talvacchio, J Gavaler, and M A Janocko. Progress in HTS electronics. Reference [102], 3–18.
31. B A Davidson, R D Redwing, Y Nguyen, and J E Nordman. Discrete vortex flow transistors using electron-beam scribed Josephson junctions in YBCO. Reference [102], 469–478.
32. A S Davydov. The theory of contraction of proteins under their excitation. *J. Theor. Biol.* 38 (1973) 559–569.
33. A S Davydov. *Solitons in Molecular Systems.* 2nd edition. Reidel, Dordrecht, 1991.
34. P Deift, T Kriecherbauer, and S Venakides. Forced lattice vibrations. I and II. *Comm. Pure Appl. Math.* 48 (1995) 1187–1249 and 1251–1298.
35. A S Dolgov. Localization of vibrations in a nonlinear crystalline structure. *Sov. Phys. Solid State* (6)28 (1986) 907–909.
36. N B Dubash, Y Zhang, and P-F Yuh. Linewidth measurements and phase locking of Josephson oscillators using RSFQ circuits. *Trans. IEEE on Appl. Superconductivity* 7 (1997) 3803–3811.

37. D B Duncan, J C Eilbeck, H Feddersen, and J A D Wattis. Solitons on lattices. *Physica D* 68 (1993) 1–11.
38. D B Duncan, C H Walshaw, and J A D Wattis. A symplectic solver for lattice equations. In *Nonlinear Coherent Structures in Physics and Biology*, M Remoissenet and M Peyrard, eds., Springer-Verlag, Berlin, 1991.
39. J C Eilbeck and R Flesch. Calculation of families of solitary waves on discrete lattices. *Phys. Lett. A* 149 (1990) 200–202.
40. J C Eilbeck and M Johansson. The discrete nonlinear Schrödinger equation— 20 years on. (Proceedings of the Conference on Localization and Energy Transfer in Nonlinear Systems, June 17–21, 2002, San Lorenzo de El Escorial, Madrid, Spain; to be published by World Scientific.)
41. J C Eilbeck, P S Lomdahl, and A C Scott. Soliton structure on crystalline acetanilide. *Phys. Rev. B* 30 (1984) 4703–4712.
42. J C Eilbeck, P S Lomdahl, and A C Scott. The discrete self-trapping equation. *Physica D* 16 (1985) 318–338.
43. S W Englander, N R Kallenbach, A J Heeger, J A Krumhansl, and A Litwin. Nature of the open state in long polynucleotide double helices: Possibility of soliton excitations. *Proc. Natl. Acad. Sci. USA* 77 (1980) 7222–7226.
44. T Erneux and G Nicolis. Propagating waves in discrete bistable reaction–diffusion systems. *Physica D* 67 (1993) 237–244.
45. L D Faddeev and L A Takhtajan. *Hamiltonian Methods in the Theory of Solitons*. Springer-Verlag, Berlin, 1987.
46. H Feddersen. Localization of vibrational energy in globular protein. *Phys. Lett. A* 154 (1991) 391–395.
47. H Feddersen. Numerical calculations of solitary waves in Davydov's equations. *Phys. Scr.* 47 (1993) 481–483.
48. H Feddersen, P L Christiansen, and M Salerno. Quantum chaology in the discrete self-trapping equation in the presence of Arnold diffusion. *Physica Scripta* 43 (1991) 353–355.
49. E Fermi, J R Pasta, and S M Ulam. Studies of nonlinear problems. Los Alamos Sci. Lab. Rep. LA–1940 (1955); also in *Collected Works of Enrico Fermi*, Vol. II. University Chicago Press, 1965.
50. A M Filip and S Venakides. Existence and modulation of traveling waves in particle chains. *Comm. Pure and Appl. Math.* 52 (1999) 693–735.
51. F Fillaux and C J Carlile. Collective rotation of methyl groups in isotopic mixtures of 4-methyl-pyridine at low temperature. Inelastic neutron scattering spectra. *Chem. Phys. Lett.* 162 (1989) 188–195.
52. F Fillaux and C J Carlile. Inelastic-neutron-scattering study of methyl tunneling and the quantum sine–Gordon breather in isotopic mixtures of 4-methyl-pyridine at low temperatures. *Phys. Rev. B* 42 (1990) 5990–6006.
53. F Fillaux, C J Carlile, and G J Kearly. Inelastic neutron-scattering study at low temperature of the quantum sine–Gordon breather in 4-methyl-pyridine with partially deuterated methyl groups. *Phys. Rev. B* 44 (1991) 12280–12293.
54. F Fillaux, C J Carlile, G J Kearly, and M Prager. Inelastic neutron-scattering study of methyl tunnelling and the quantum sine–Gordon breather mode is isotropic mixtures of 2,6-dimethyl-pyridine at low temperature. *Physica B* 202 (1994) 302–310.

REFERENCES

55. M V Fistul, A E Miroshnichenko, S Flach, M Schuster, and A V Ustinov. Incommensurate frequencies states. *Phys. Rev. B* 65 (2002) 174524.
56. S Flach, K Kladko, and R S MacKay. Energy thresholds for discrete breathers in one-, two-, and three-dimensional lattices. *Phys. Rev. Lett.* 78 (1997) 1207–1210.
57. S Flach and C R Willis. Discrete breathers. *Phys. Reps.* 295 (1998) 181–264.
58. L Floria. Superconducting ladders take breather. *Physics World.* 13 (2000) 23.
59. N Flytzanis, S Pnevmatikos, and M Remoissenet. Kink, breather and symmetric envelope or dark solitons in nonlinear chains: I. Monatomic chain. *J. Phys. C* 18 (1985) 4603–4629.
60. J Frenkel and T Kontorova. On the theory of plastic deformation and twinning. *J. Phys.* (USSR) 1 (1939) 137–149.
61. G Friesecke and R L Pego. Solitary waves on FPU lattices: I. Qualitative properties, renormalization and continuum limit. *Nonlinearity* 12 (1999) 1601–1628.
62. G Friesecke and R L Pego. Solitary waves on FPU lattices: II. Linear implies nonlinear stability. *Nonlinearity* 15 (2002) 1343–1359.
63. G Friesecke and J A D Wattis. Existence theorem for travelling waves on lattices. *Commun Math Phys.* 161 (1994) 391–418.
64. C S Gardner, J M Greene, M D Kruskal, and R M Miura. Method for solving the Korteweg–de Vries Equation. *Phys. Rev. Lett.* 19 (1967) 1095–1097.
65. V S Gerdjikov, M I Ivanov, and P P Kulish. Expansions over the 'squared' solutions and difference evolution equations. *J. Math. Phys.* 25 (1984) 25–34.
66. P H Greene. On looking for neural networks and "cell assemblies" that underlie behavior. *Bull. Math. Biophys.* 24 (1962) 247–275 and 395–411.
67. H Haken. *Principles of Brain Functioning: A Synergetic Approach to Brain Activity, Behavior and Cognition.* Springer-Verlag, Berlin, 1996.
68. D O Hebb. *The Organization of Behavior: A Neuropsychological Theory.* John Wiley, New York, 1949.
69. D Hennig, H Gabriel, M F Jørgensen, P L Christiansen, and C B Clausen. Homoclinic chaos in the discrete self-trapping trimer. *Phys. Rev. E* 51 (1995) 2870–2876.
70. D Hennig and H Gabriel. The 4-element discrete nonlinear Schrödinger equation—non-integrability and Arnold diffusion. *J. Phys. A* 28 (1995) 3749–3756.
71. D Hennig and G P Tsironis. Wave transmission in nonlinear lattices. *Phys. Rep.* 307 (1999) 334–432.
72. B R Henry. Local modes and their application to the analysis of polyatomic overtone spectra. *J. Phys. Chem.* 80 (1976) 2160–2164.
73. D Hochstrasser, F G Mertens, and H Büttner. An iterative method for the calculation of narrow solitary excitations on atomic chains. *Physica D* 35 (1988) 259–266.
74. L S Hoel, W H Keller, J E Nordman, and A C Scott. Niobium superconductive tunnel diode integrated circuit arrays. *Solid State Electron.* 15 (1972) 1167–1173.
75. G Iooss. Travelling waves in the Fermi–Pasta–Ulam lattice. *Nonlinearity* 13 (2000) 849–866.
76. A G Izergin and V E Korepin. A lattice model related to the nonlinear Schrödinger equation. *Sov. Phys. Dokl.* 26 (1981) 653–654.

77. G James. Existence of breathers on FPU lattices. *C. R. Acad. Sci. Paris* 332 (Series I) (2001) 581–586.
78. M Johansson and S Aubry. Existence and stability of quasiperiodic breathers in the discrete nonlinear Schr odinger equation. *Nonlinearity* 10 (1997) 1151–1178.
79. G Kalosakas, K Rasmussen, and A R Bishop. Delocalizing transition of Bose-Einstein condensates in optical lattices. *Phys. Rev. Lett.* 89 (2002) 030402-1-4.
80. V Kaplunenko, J Mygind, N F Pedersen, V Koshelets, and T Doderer. Nonlinear phenomena in RSFQ logic devices. Reference [102], 411–436.
81. P G Kevrekidis, K Ø Rasmussen, and A R Bishop. The discrete nonlinear Schrödinger equation: A survey of recent results. *Int. J. Mod. Phys. B* 15 (2001) 2833–2900.
82. S A Kiselev, S R Bickam, and A J Sievers. Properties of intrinsic localized modes in one-dimensional lattices. *Comments Condensed Matter Phys.* 17 (1995) 135–173.
83. S A Kiselev and A J Sievers. Generation of intrinsic vibrational gap modes in three-dimensional ionic crystals. *Phys. Rev. B* 55 (1997) 5755–5758.
84. G Kopidakis and S Aubry. Intraband discrete breathers in disordered non-linear systems. I. Delocalization. *Physica D* 130 (1999) 155–186.
85. G Kopidakis and S Aubry. Intraband discrete breathers in disordered non-linear systems. II. Localization. *Physica D* 139 (2000) 247–275.
86. A M Kosevich and A S Kovalev, Selflocalization of vibrations in a one-dimensional anharmonic chain. *Zh. Eksp. Teor. Fiz.* 67 (1974) 1793–1798 [*Sov. Phys. JETP* 40 (1974) 891–896].
87. J A Krumhansl. Nonlinear science: Toward the next frontiers. *Physica D* 68 (1993) 97–103.
88. J A Krumhansl. The intersection of nonlinear science, molecular biology, and condensed matter physics. Viewpoints. Reference [103], 1–9.
89. N Kryloff and N Bogoliuboff. *Introduction to Nonlinear Mechanics.* Princeton University Press, Princeton, 1947.
90. E W Laedke, K H Spatschek, V K Mezentsev, S L Musher, I V Ryzhenkova, and S K Turitsyn. Instability of 2-dimensional solitons in discrete-systems. *JETP Letters* 62 (1995) 677–684.
91. W H Louisell. *Coupled Mode and Parametric Electronics.* John Wiley & Sons, New York, 1960.
92. R S MacKay and S Aubry. Proof of existence of breathers for time-reversible or Hamiltonian networks of weakly coupled oscillators. *Nonlinearity* 7 (1994) 1623–1643.
93. R S MacKay and J-A Sepulchre. Stability of discrete breathers. *Physica D* 119 (1998) 148–162.
94. J L Marin and S Aubry. Finite size effects on instabilities of discrete breathers. *Physica D* 119 (1998) 163–174.
95. P J Martínez, L M Floría, J L Marín, and J J Mazo. Floquet instability of discrete breathers in anisotropic Josephson junction ladders. *Physica D* 119 (1998) 175–183.

96. S Matarazzo, S Pagano, G Filatrella, S Barbanera, F Murtas, C Romero, V Boffa, F Gatta, and U Gambardella. High T_c squid arrays as microwave generators. Reference [102], 227–234.
97. V K Mezentsev, S L Musher, I V Ryzhenkova, and S K Turitsyn. 2-dimensional solitons in discrete-systems. *JETP Letters* 60 (1994) 829–835.
98. V Muto, P S Lomdahl, and P L Christiansen. Two-dimensional models for DNA dynamics: Longitudinal wave propagation and denaturation. *Phys. Rev. A* 42 (1990) 7452–7458.
99. M S Ody, A K Common, and M I Sohby. Continuous symmetries of the discrete nonlinear telegraph equation. *Eur. J. Appl. Math.* 10 (1999) 265–284.
100. J B Page. Asymptotic solution for localized vibrational modes in strongly anharmonic periodic systems. *Phys. Rev. B* 41 (1990) 7835–7838.
101. R D Parmentier. Fluxons in long Josephson junctions. In *Solitons in Action*, K Lonngren and A Scott, eds., Academic Press, New York, 1978, 173–199.
102. R D Parmentier and N F Pedersen (eds.). *Nonlinear Superconducting Devices and High-T_c Materials*. World Scientific, Singapore, 1995.
103. M Peyrard (ed.). *Nonlinear Excitations in Biomolecules*. Springer-Verlag, Berlin and Les Editions de Physique, Les Ulis, 1995.
104. M Peyrard and M D Kruskal. Kink dynamics in the highly discrete sine–Gordon system. *Physica D* 14 (1984) 88–102.
105. S Pnevmatikos, N Flytzanis, and M Remoissenet. Soliton dynamics of nonlinear diatomic atoms. *Phys. Rev. B* 33 (1986) 2308–2321.
106. B van der Pol. The nonlinear theory of electric oscillations. *Proc. IRE* 22 (1934) 1051–1086.
107. M Quack. Spectra and dynamics of coupled vibrations in polyatomic molecules. *Ann. Rev. Phys. Chem.* 41 (1990) 839–874.
108. J Juul Rasmussen and K Rypdal. Blow-up in nonlinear Schroedinger equations—I. A general review. *Phys. Scr.* 33 (1986) 481–497.
109. C Reiss. Selected topics in molecular biology, in need of "hard" science. Reference [103], 29–55.
110. M Remoissenet. *Waves Called Solitons*. Springer-Verlag, Berlin, 1999.
111. P Rosenau. Dynamics of dense lattices. *Phys. Rev. A* 36 (1987) 5868–5876.
112. M Salerno. Quantum deformations of the discrete nonlinear Schrödinger equation. *Phys. Rev. A* 46 (1992) 6856–6859.
113. R Scharf and A R Bishop. Properties of the nonlinear Schrödinger equation on a lattice. *Phys. Rev. A* 43 (1991) 6535–6543.
114. A C Scott. Tunnel diode arrays for information processing and storage. *IEEE Trans. Syst., Man, Cybern.* SMC-1 (1971) 267–275.
115. A C Scott. Davydov's soliton. *Phys. Rep.* 217 (1992) 1–67.
116. A C Scott. *Neuroscience: A Mathematical Primer*. Springer-Verlag, New York, 2002.
117. A C Scott and P L Christiansen. A generalized discrete self-trapping equation. *Physica Scripta* 42 (1990) 257–262.
118. A C Scott, P S Lomdahl, and J C Eilbeck. Between the local mode and normal mode limits. *Chem. Phys. Lett.* 113 (1985) 29–36.

119. A C Scott and L MacNeil. Binding energy for a "small" stationary soliton. *Phys. Lett. A* 98 (1983) 87–88.
120. A J Sievers and S Takeno. Intrinsic localized modes in anharmonic crystals. *Phys. Rev. Lett.* 61 (1988) 970–973.
121. A J Sievers and S Takeno. Anharmonic resonant modes and the low-temperature specific heat of glass. *Phys. Rev. B* 39 (1989) 3374–3379.
122. D Smets and M Willem. Solitary waves with prescribed speed on infinite lattices. *J. Funct. Anal.* 149 (1997) 266–275.
123. K Suzuki, R Hirota, and K Yoshikawa. The properties of phase modulated soliton trains. *Jpn. J. Appl. Phys.* 12 (1973) 361–365.
124. B I Swanson, L A Brozik, S P Love, G F Strouse, A P Shreve, A R Bishop, W Z Wang, and M I Salkola. Observation of intrinsically localized modes in a discrete low dimensional material. *Phys. Rev. Lett.* 82 (1999) 3288–3291.
125. R B Tew and J A D Wattis. Quasi-continuum approximations for travelling kinks in diatomic lattices. *J. Phys. A: Math. Gen.* 34 (2001) 7163–7180.
126. M Toda. One dimensional dual transformation. *J. Phys. Soc. Jpn.* 20 (1965) 2095.
127. M Toda. One dimensional dual transformation. *Supp. Prog. Theor. Phys.* 36 (1966) 113–119.
128. M Toda. Vibration of a chain with nonlinear interaction. *J. Phys. Soc. Jpn.* 22 (1967) 431–436.
129. M Toda. *Theory of Nonlinear Lattices*. Springer-Verlag, Berlin, 1981.
130. M Toda, Y Okada, and S Watanabe. Nonlinear dual lattice. *J. Phys. Soc. Jpn.* 59 (1990) 4279–4285.
131. E Trias, J J Mazo, and T P Orlando. Discrete breathers in nonlinear lattices: Experimental detection in a Josephson array. *Phys. Rev. Lett.* 84 (2000) 741–744.
132. A T Tu. *Spectroscopy in Biology*. John Wiley, New York, 1982.
133. M G Velarde, V I Nekorkin, V B Kazantsev, and J Ross. The emergence of form by replication. *Proc. Natl. Acad. Sci. (USA)* 94 (1997) 5024–5027.
134. V Volterra. Principes de biologie mathématique. *Acta Biotheor.* 3 (1937) 1–36.
135. N K Voulgarakis, G Kalosakas, A R Bishop, and G P Tsironis. A multi-quanta breather model for PtCl. *Phys. Rev. B* 64 (2001) 020301-1.
136. M Wadati and M Toda. Bäcklund transformation for the exponential lattice. *J. Phys. Soc. Jpn.* 39 (1975) 1196–1203.
137. J A D Wattis. *Analytic Approximations to Solitary Waves on Lattices*. PhD thesis, Mathematics Department, Heriot-Watt University, 1993.
138. J A D Wattis. Variational approximations to breather modes in the discrete sine–Gordon equation. *Physica D* 82 (1995) 333–339.
139. J A D Wattis. Variational approximations to breathers in the discrete sine–Gordon equation II: Moving breathers and the Peierls-Nabarro energies. *Nonlinearity* 9 (1996) 1583–1598.
140. J A D Wattis. Approximations to solitary waves on lattices, III: Monatomic lattice with second neighbour interactions. *J. Phys. A* 29 (1996) 8139–8157.

141. J A D Wattis. Quasi-continuum approximations to lattice equations arising from the discrete nonlinear telegraph equation. *J. Phys. A: Math. Gen.* 33 (2000) 5925–5944.
142. J A D Wattis. Quasi-continuum approximations for diatomic lattices. *Phys Lett A* 284 (2001) 16–22.
143. M I Weinstein. Excitation thresholds for nonlinear localized modes on lattices. *Nonlinearity* 12 (1999) 673–691.
144. G B Whitham. *Linear and Nonlinear Waves*. John Wiley & Sons, New York, 1974.
145. L V Yakushevich. *Nonlinear Physics of DNA*. John Wiley, Chichester, 1998.
146. H S J van der Zant, M Barahona, A E Dewell, E Trías, T P Orlando, Shinya Watanabe, and Stephen Strogatz. Dynamics of one-dimensional Josephson-junction arrays. *Physica D* 119 (1998) 219–226.

6
INVERSE SCATTERING METHODS

In Chapter 3, we studied the soliton—that remarkable dynamic entity discovered experimentally by John Scott Russell in the nineteenth century and rediscovered in the course of numerical studies by Zabusky and Kruskal in the mid-1960s—and learned about Bäcklund transforms and N-soliton formulas, which will be used to expand the range of soliton perturbation theory in the next chapter. Throughout these discussions there floats an air of mystery. How is it that solutions of nonlinear partial differential equations are able to display such unexpected properties?

An answer to this question is that each soliton equation is related to a linear scattering operator in the following manner. Given an operator L that acts on the x-coordinate and is a function of $u(x,t)$ (a solution of the nonlinear PDE under investigation), the operator equation

$$L_t = [M, L] \equiv ML - LM \tag{6.1}$$

implies that $u(x,t)$ is a solution of the corresponding nonlinear PDE. This is a strong property not shared by most PDEs (including the nonlinear diffusion equations studied in Chapter 4), but when it is satisfied a wealth of related properties falls into the hands of the analyst.

To see the broad picture, suppose that we have such an operator, related to the nonlinear equation

$$u_t = N(u)$$

that is to be solved for the initial condition

$$u(x, 0) = f(x).$$

To this end, one can proceed as follows:

1. Calculate the scattering data for the operator L at $t = 0$ from $u(x, 0)$.
2. Use the operator M to determine the evolution in time of this scattering data.
3. From knowledge of the scattering data at $t > 0$, find $u(x, t)$, using an inverse scattering calculation for the operator L.

In this manner the solution of a nonlinear initial value problem is reduced to the three linear computations listed above.

The author thanks M.P. Sørensen for writing the first drafts of Sections 6.1 and 6.3

First described for the Korteweg–de Vries (KdV) equation in a classic 1967 paper by Gardner et al. [13], this "inverse scattering method" (ISM) involves a number of analytic details that should not obscure the underlying elegance of the approach. The aim here is to introduce the structure of the ISM without getting lost among the related equations.

This chapter begins with a study of the KdV equation

$$\frac{\partial u}{\partial t} = 6u\frac{\partial u}{\partial x} - \frac{\partial^3 u}{\partial x^3},$$

which is related in the above manner to the Schrödinger operator

$$L = -\frac{\partial^2}{\partial x^2} + u(x,t), \tag{6.2}$$

with

$$M = -4\frac{\partial^3}{\partial x^3} + 3u\frac{\partial}{\partial x} + 3u_x.$$

(At this point, the reader should pause to check that the operator equation $L_t = [M, L]$ implies that $u(x,t)$ must satisfy the KdV equation.)

In Chapter 2, plausibility arguments were presented for direct and inverse scattering calculations (steps 1 and 3), using the Schrödinger operator of Equation (6.2). In the first section of this chapter, these arguments are substantiated, using ideas from the theory of complex functions. Based upon comprehension of the Schrödinger scattering problem, the ISM for the KdV equation is then presented in Section 6.2.

At this point, it is expected that the reader will be familiar with the overall ISM strategy, facilitating its extension, in Section 6.3, to scattering operators that are also 2×2 matrices. As was shown in the classic 1972 paper of Zakharov and Shabat [31], this generalization allows one to use the ISM for the nonlinear Schrödinger (NLS) equation

$$\mathrm{i}\frac{\partial u}{\partial t} + \frac{\partial^2 u}{\partial x^2} + 2|u|^2 u = 0.$$

Shortly thereafter, Takhtajan and Faddeev [29] and Ablowitz et al. [2] independently obtained matrix scattering operators for the sine–Gordon (SG) equation

$$\frac{\partial^2 u}{\partial x^2} - \frac{\partial^2 u}{\partial t^2} = \sin u,$$

and other nonlinear wave equations of practical interest [3]. Following this matrix formulation of the linear scattering operator, applications of the ISM to the sine–Gordon and nonlinear Schrödinger equations are sketched in Sections 6.4 and 6.5. Finally, in Section 6.6, it is shown how an ISM formulation leads to the construction of a countably infinite number of independent conservation laws.

6.1 Linear scattering revisited

For an equation of the form
$$L\psi = \lambda\psi,$$

where L is a linear operator, there are two types of eigenfunctions: scattering solutions, which are bounded but nonzero as $x \to \pm\infty$, and bound states, which decay to zero as $x \to \pm\infty$.

For the Schrödinger operator defined in Equation (6.2), it was noted in Chapter 2 that these two types of eigenfunctions are related through analytic properties of transmission and reflection coefficients of the scattering solutions. In this section, the scattering problem is considered in greater detail, establishing a relationship between normalizing factors for the eigenfunctions and residues of the complex scattering functions. Based on this formulation, the Gel'fand–Levitan equation is rederived without appealing to the "pseudotime" that was introduced in Chapter 2.

6.1.1 Scattering solutions, bound states, and upper half plane poles

With L as in Equation (6.2) and
$$k^2 \equiv \lambda > 0,$$

the Schrödinger eigenvalue equation is

$$\frac{d^2\psi}{dx^2} + [k^2 - u(x)]\psi = 0, \tag{6.3}$$

where $\psi(x, k)$ is bounded as $x \to \pm\infty$ for all real values of k. If it is assumed that
$$u(x) \to 0 \quad \text{as } x \to \pm\infty,$$

the most general scattering solution of this second-order ODE can be written as

$$\psi(x, k) = A\psi^+(x, k) + B\psi^-(x, k) \quad \text{for } x \to \pm\infty, \tag{6.4}$$

where

$$\psi^+(x, k) \sim \begin{cases} e^{-ikx} + b(k)e^{ikx} & \text{for } x \to +\infty \\ a(k)e^{-ikx} & \text{for } x \to -\infty, \end{cases} \tag{6.5}$$

$$\psi^-(x, k) \sim \begin{cases} e^{ikx} + d(k)e^{-ikx} & \text{for } x \to -\infty \\ c(k)e^{ikx} & \text{for } x \to +\infty, \end{cases} \tag{6.6}$$

and $-\infty < k < +\infty$ is a real number.

As indicated in Figure 6.1, $b(k)$ and $a(k)$ are respectively the reflection and transmission coefficients for an incident wave from the right ($x = +\infty$), and

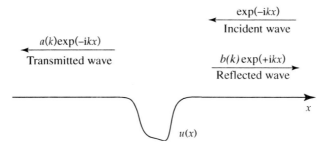

FIG. 6.1. *Amplitudes of scattering solutions.*

$d(k)$ and $c(k)$ play the same roles with respect to a wave incident from the left. In general the transmission and reflection coefficients are complex numbers.

Because the scattering is elastic or energy conserving, the intensity of the incoming wave must equal the sum of intensities of the reflected and the transmitted wave. In other words

$$|a|^2 + |b|^2 = 1, \quad |c|^2 + |d|^2 = 1. \tag{6.7}$$

We now extend $\psi(x, k)$ from the real k-axis into the upper half of the complex k-plane via analytic continuation, showing that bound state eigenfunctions correspond to poles of the reflection and transmission coefficient. To see this, let k approach the positive imaginary axis, $k \to i\kappa_n$ with $\kappa_n > 0$, so the exponentials in ψ^+ and ψ^- converge to $e^{\pm \kappa_n x}$. Is it possible to construct bound states for the k-values on the positive imaginary axis that equal $i\kappa_n$? In this case Equation (6.4) becomes

$$\psi(x, i\kappa_n) \sim \begin{cases} Ae^{\kappa_n x} + (Ab + Bc)e^{-\kappa_n x} & \text{for } x \to +\infty \\ (Aa + Bd)e^{\kappa_n x} + Be^{-\kappa_n x} & \text{for } x \to -\infty, \end{cases} \tag{6.8}$$

which diverges for $x \to \pm\infty$ unless appropriate coefficients of the exponentials are chosen. In order to obtain a bound state, A and B must go to zero as k approaches the positive imaginary axis at $i\kappa_n$, while the factors $Ab + Bc$ and $Aa + Bd$ stay finite. This implies that a, b, c, and d diverge as $k \to i\kappa_n$, a requirement that identifies the bound states.

From the reciprocity theorem for passive, linear systems (or see Problem 4), $a = c$. Thus $|b| = |d|$, and A must approach zero to the same power of $(k - i\kappa_n)$ as B, implying that A, B, a^{-1}, b^{-1}, c^{-1}, and d^{-1} all approach zero to the same power.

To be specific, let us choose a and b to have simple poles in the points $k = i\kappa_n$ and assume the following series representations:

$$a(k) = \frac{a_{-1,n}}{k - i\kappa_n} + a_{0,n} + a_{1,n}(k - i\kappa_n) + \cdots,$$

$$b(k) = \frac{b_{-1,n}}{k - i\kappa_n} + b_{0,n} + b_{1,n}(k - i\kappa_n) + \cdots, \tag{6.9}$$

$$A(k) = A_{1,n}(k - i\kappa_n) + A_{2,n}(k - i\kappa_n)^2 + \cdots,$$

$$B(k) = B_{1,n}(k - i\kappa_n) + B_{2,n}(k - i\kappa_n)^2 + \cdots.$$

Here $a(k)$ and $b(k)$ are Laurent series expansions with the desired property: $a, b \to \infty$ as $k \to i\kappa_n$. Using Equations (6.9) in (6.8), the bound state in the limit $k \to i\kappa_n$ goes as

$$\psi(x, i\kappa_n) \sim \begin{cases} (A_{1,n}b_{-1,n} + B_{1,n}c_{-1,n})e^{-\kappa_n x} & \text{for } x \to +\infty \\ (A_{1,n}a_{-1,n} + B_{1,n}d_{-1,n})e^{\kappa_n x} & \text{for } x \to -\infty. \end{cases} \tag{6.10}$$

Because $A(k)$ converges toward zero as k approaches $i\kappa_n$, the bound state can be interpreted as a situation where the incident waves vanish but the transmitted and reflected waves remain finite. In other words, the bound state is a normal mode (or eigenstate) of the system, not needing a cause to exist.

The residue $\text{Res}[b(k), i\kappa_n]$ of the function $b(k)$ at the point $k = i\kappa_n$ is $b_{-1,n}$. Straightforward application of the Laurent series for b in Equation (6.9) leads to $b'/b^2 \to -1/b_{-1}$ as $k \to i\kappa_n$ and hence

$$\text{Res}[b(k), i\kappa_n] = b_{-1,n} = -\frac{b(k)^2}{b'(k)}\bigg|_{k=i\kappa_n}, \tag{6.11}$$

the prime indicating differentiation with respect to k.

That the poles are simple is not yet proven. Perhaps one could choose poles of second order and let A and B converge toward zero as $(k - i\kappa_n)^2$. It will be shown in the next section that this is not possible and only poles of first order appear.

6.1.2 Why the upper half plane poles must be simple

Equations (6.4) and (6.10) give solutions of the Schrödinger equation in the limits $x \to \pm\infty$ where the potential $u(x)$ is assumed to vanish. In order to analyze Equation (6.2) in the region with nonvanishing potential, we are guided by the discussion of Section 2.6.2 to introduce the following two independent solutions:[1]

$$\phi_1(x,k) = e^{ikx} + \int_x^\infty K(x,z)e^{ikz}\,dz$$
$$\phi_2(x,k) = e^{-ikx} + \int_{-\infty}^x L(x,z)e^{-ikz}\,dz, \tag{6.12}$$

where the kernels $K(x,z)$ and $L(x,z)$ satisfy

$$\left.\begin{array}{l}K(x,z) \to 0\\ K_z(x,z) \to 0\end{array}\right\} \quad \text{for } z \to +\infty,$$

and

$$\left.\begin{array}{l}L(x,z) \to 0\\ L_z(x,z) \to 0\end{array}\right\} \quad \text{for } z \to -\infty,$$

the subscript z indicating partial differentiation with respect to z. At $k = i\kappa_n$, the function ϕ_1 is a bound state that can be normalized by multiplying with a constant c_n defined as

$$\frac{1}{c_n^2} = \int_{-\infty}^\infty \phi_1^2\,dx. \tag{6.13}$$

It was previously assumed that the transmission coefficient $a(k)$ and the reflection coefficient $b(k)$ have simple poles at bound state eigenvalues. To show that this is so, we are guided by Drazin and Johnson [10] and Vvedensky [30] to prove the following.

- *Theorem:* Let c_n be the normalization constant for $\phi_1(x, i\kappa_n)$ where $k = i\kappa_n$ is the eigenvalue for bound state number n. Then the relation

$$c_n^2 = i\frac{a(i\kappa_n)b(i\kappa_n)}{a'(i\kappa_n)}, \tag{6.14}$$

holds at $k = i\kappa_n$.

Proof: Because $\phi_1(x,k)$ and $\phi_1(x,-k)$ are independent solutions of the second-order ODE, $\phi_2(x,k)$ can be written as

$$\phi_2(x,k) = C(k)\phi_1(x,-k) + D(k)\phi_1(x,k).$$

After division by C, the limiting form of ϕ_2 at $x \to \pm\infty$ is

$$\frac{1}{C}\phi_2 = \begin{cases} e^{-ikx} + D/C e^{ikx} & \text{for } x \to +\infty\\ 1/C\,e^{-ikx} & \text{for } x \to -\infty, \end{cases} \tag{6.15}$$

---

[1]These are sometimes called "Jost functions" by soliton mavens, alluding to related equations that arise in the quantum theory for interacting particles. See reference [15] for details.

whereupon one can identify $a(k)\phi_2(x,k) = \psi^+(x,k)$, implying $a = 1/C$, $b = D/C$, and

$$\phi_2(x,k) = \frac{1}{a}\phi_1(x,-k) + \frac{b}{a}\phi_1(x,k). \tag{6.16}$$

Because $a(k)$ diverges for k approaching $i\kappa_n$, it is clear from the previous discussion on bound states that $b(k)$ must likewise diverge. Therefore the ratio $b(i\kappa_n)/a(i\kappa_n)$ remains finite for $k \to i\kappa_n$, implying that at a pole of $a(k)$ Equation (6.16) becomes

$$\phi_2(x, i\kappa_n) = \frac{b(i\kappa_n)}{a(i\kappa_n)}\phi_1(x, i\kappa_n). \tag{6.17}$$

Using Equation (6.15) to calculate the Wronskian[2] $W(\phi_1, \phi_2)$ at $x \to +\infty$, one obtains

$$W[\phi_1(x,k); \phi_2(x,k)] = -\frac{2ik}{a(k)}. \tag{6.18}$$

Differentiating the Schrödinger equation in (6.3) with respect to the propagation number k gives

$$\psi_{xxk} + 2k\psi + (k^2 - u)\psi_k = 0.$$

Multiplying by ψ and subtracting from the Schrödinger equation multiplied by ψ_k leads to

$$\psi_k\psi_{xx} - \psi\psi_{xxk} - 2k\psi^2 = 0.$$

This expression is equivalent to

$$\frac{dW(\psi_k; \psi)}{dx} = 2k\psi^2,$$

and by integration we find

$$2k\int_{-\infty}^{\infty}\psi^2 dx = W(\psi_k; \psi)|_{x \to +\infty} - W(\psi_k; \psi)|_{x \to -\infty}.$$

Choosing $\psi = \phi_1$ and letting $k \to i\kappa_n$, one gets

$$2i\kappa_n \int_{-\infty}^{\infty} \phi_1^2 dx = \frac{2i\kappa_n}{c_n^2} = W(\phi_{1,k}; \phi_1)|_{x \to +\infty} - W(\phi_{1,k}; \phi_1)|_{x \to -\infty}. \tag{6.19}$$

Differentiating the Wronskian in Equation (6.18) with respect to k gives

$$\frac{d}{dk}W(\phi_1; \phi_2) = W(\phi_{1,k}; \phi_2) + W(\phi_1; \phi_{2,k}) = -\frac{2i}{a(k)} + \frac{2ik}{a(k)^2}a'(k). \tag{6.20}$$

---

[2]The Wronskian of two independent solutions to a second-order ODE is defined as $W(v;w) \equiv vw' - wv'$. For the Schrödinger equation, the Wronskian is independent of x. Josef Maria Wronski (1778–1853) was a Polish born and German educated mathematician, who lived most of his life in France.

For $k = i\kappa_n$ Equation (6.20) can be written as

$$\frac{b(i\kappa_n)}{a(i\kappa_n)} W(\phi_{1,k}; \phi_1)|_{x\to-\infty} + \frac{a(i\kappa_n)}{b(i\kappa_n)} W(\phi_2; \phi_{2,k})|_{x\to-\infty} = -\frac{2\kappa_n}{a^2(i\kappa_n)} a'(i\kappa_n), \tag{6.21}$$

because $1/a(i\kappa_n)$ vanishes. Combining Equations (6.19) and (6.21) leads to

$$W(\phi_{1,k}; \phi_1)|_{x\to+\infty} + \frac{a(i\kappa_n)^2}{b(i\kappa_n)^2} W(\phi_2; \phi_{2,k})|_{x\to-\infty} = \frac{2i\kappa_n}{c_n^2} - \frac{2\kappa_n}{a(i\kappa_n)} \frac{a'(i\kappa_n)}{b(i\kappa_n)}.$$

By straightforward calculation the Wronskians are zero, implying Equation (6.14).
□

Assuming that the transmission and reflection coefficients have poles of order $m \geq 2$ at $k = i\kappa_n$, the right-hand side in Equation (6.14) diverges so the normalization constant c_n must also diverge (see Problem 9). Therefore no bound states exist with poles that are not simple. For simple poles, the limiting value of the right-hand side in (6.14) is

$$c_n^2 = -ib_{-1,n}, \tag{6.22}$$

as can be calculated from the Laurent series in Equation (6.9).

The fact that the upper half plane poles of the transmission and reflection coefficients for the Schrödinger equation are simple arises in several contexts, including the following:

1. A phase plane analysis of Equation (6.3) shows that different bound states have different values for $k^2 = \lambda$. This property was used in Chapter 4 in the course of a stability analysis of traveling-wave solutions for Fisher's equation. It will be invoked again in Chapter 8 in connection with a quantum analysis of the anharmonic oscillator.
2. Each upper half plane pole of the reflection (and transmission) coefficients corresponds to a soliton component of $u(x,t)$ in Equation (6.2). In Chapter 3, it was observed that the Bäcklund transform for KdV could not generate two solitons with the same speed, which is consistent with the fact that the UHP poles are simple.

6.1.3 The Gel'fand–Levitan equation again

To derive the Gel'fand–Levitan equation, consider ϕ_1, which satisfies the Schrödinger equation

$$\phi_{1,xx} + (k^2 - u)\phi_1 = 0. \tag{6.23}$$

Differentiating ϕ_1 in Equation (6.12) twice with respect to x gives

$$\phi_{1,xx} = -k^2 e^{ikx} - \frac{dK(x,x)}{dx} e^{ikx} - ikK(x,x)e^{ikx}$$
$$- \frac{\partial K(x,x)}{\partial x} e^{ikx} + \int_x^\infty K_{xx}(x,z)e^{ikz}\,dz, \tag{6.24}$$

where

$$\frac{\partial K(\underline{x}, x)}{\partial x} \equiv \frac{\partial K(x, y)}{\partial x}\bigg|_{y=x},$$

implying that the partial differentiation is with respect to the underlined variable.

Integrating in Equation (6.12) twice by parts yields

$$\phi_1(x,k) = \left[1 + \frac{i}{k}K(x,x) - \frac{1}{k^2}\frac{\partial K(x,\underline{x})}{\partial x}\right]e^{ikx} - \frac{1}{k^2}\int_x^\infty K_{zz}(x,z)e^{ikz}\,dz. \tag{6.25}$$

In the Schrödinger equation (6.23), substitute $\phi_{1,xx}$ by the expression in Equation (6.24) and ϕ_1 in the term $k^2\phi_1$ by the expression in Equation (6.25). For $u\phi_1$ use the definition of the wave function ϕ_1 in Equation (6.12). After these substitutions Equation (6.23) becomes

$$\left[-\frac{dK(x,x)}{dx} - \frac{\partial K(\underline{x},x)}{\partial x} - \frac{\partial K(x,\underline{x})}{\partial x} - u(x)\right]e^{ikx}$$
$$+ \int_x^\infty \left[K_{xx}(x,z) - K_{zz}(x,z) - uK(x,z)\right]e^{ikz}\,dz = 0. \tag{6.26}$$

In order to satisfy Equation (6.26), it is sufficient that

$$u(x) = -2dK(x,x)/dx$$
$$K_{xx}(x,z) - K_{zz}(x,z) - u(x)K(x,z) = 0. \tag{6.27}$$

The first of these equations gives $u(x)$ from a knowledge of $K(x,z)$, and the second indicates how $K(x,z)$ is related to $u(x)$. In practice, it is of interest to construct $K(x,z)$ from properties of the scattering and bound state solutions, as we now proceed to do.

Recall the relationship between ϕ_1 and ϕ_2 in Equation (6.16). Multiplying by a and using the definition of the Jost function in Equations (6.12) gives

$$a(k)\phi_2 = \phi_1^* + b(k)\phi_1 = e^{-ikx} + \int_x^\infty K(x,z)e^{-ikz}\,dz$$
$$+ b(k)e^{ikx} + b(k)\int_x^\infty K(x,z)e^{ikz}\,dz.$$

Defining $K(x,z)=0$ for $z < x$ extends the integral above from $-\infty$ to $+\infty$, allowing introduction of the Fourier transform $\hat{K}(x,k)$ of $K(x,z)$ as

$$\hat{K}(x,k) = \int_{-\infty}^\infty K(x,z)e^{-ikz}\,dz = a\phi_2 - e^{-ikx} - be^{ikx} - b\int_x^\infty K(x,z)e^{ikz}\,dz.$$

$K(x, z)$ is then found from the inverse Fourier transform as

$$K(x,z) = \frac{1}{2\pi}\int_{-\infty}^{\infty} a\phi_2 e^{ikz} dk - \frac{1}{2\pi}\int_{-\infty}^{\infty} e^{ik(z-x)} dk$$
$$- \frac{1}{2\pi}\int_{-\infty}^{\infty} b e^{ik(z+x)} dk - \frac{1}{2\pi}\int_{x}^{\infty}\int_{-\infty}^{\infty} bK(x,y) e^{ik(z+y)} dk\, dy.$$

The above expression can be written in a more compact form by introducing the function

$$B(x+z) \equiv \frac{1}{2\pi}\int_{-\infty}^{\infty} b(k) e^{ik(x+z)} dk, \qquad (6.28)$$

whereupon

$$K(x,z) = \frac{1}{2\pi}\int_{-\infty}^{\infty} a\phi_2 e^{ikz} dk - \delta(z-x)$$
$$- B(x+z) - \int_{x}^{\infty} K(x,y) B(y+z)\, dy,$$

where $\delta(z-x)$ denotes Dirac's delta function. Restricting the above equation to $z > x$ gives

$$K(x,z) + B(x+z) + \int_{x}^{\infty} K(x,y) B(y+z)\, dy = \frac{1}{2\pi}\int_{-\infty}^{\infty} a\phi_2 e^{ikz} dk. \qquad (6.29)$$

The integral on the right-hand side of Equation (6.29) can be computed using Cauchy's residue theorem by extending the integration into the upper half k-plane along a large semicircular contour C; thus

$$I = \frac{1}{2\pi}\oint_{C} a\phi_2 e^{ikz} dk. \qquad (6.30)$$

For large absolute values of k, the potential in the Schrödinger equation can be neglected and the solution ϕ_2 becomes

$$\phi_2 \sim d_1 e^{ikx} + d_2 e^{-ikx}, \quad \text{for Im}(k) \to +\infty.$$

Observing that $z > x$, it is evident that $a(k)\phi_2(x,k)\exp(ikz)$ vanishes along the semicircular contour as Im(k) is extended to plus infinity, and therefore the integral in Equation (6.30) is identical to the right-hand side of Equation (6.29).

At this point, we consider three possibilities in order of increasing difficulty.

No upper half plane (UHP) poles. If $a(k)$ and $b(k)$ have no poles in the UHP of k, Cauchy's theorem implies that the integral in Equation (6.30) vanishes and Equation (6.29) becomes

$$K(x,z) + B(x+z) + \int_{x}^{\infty} K(x,y) B(y+z)\, dy = 0, \qquad (6.31)$$

with $z > x$. As was noted in Chapter 2, this is called the Gel'fand–Levitan equation, which can be employed to compute $u(x)$ from $b(k)$ as follows. (i) Find $B(x+z)$ from the Fourier transform of Equation (6.28). Solve Equation (6.31) for $K(x,z)$. (iii) Calculate $u(x) = -2dK(x,x)/dx$.

One UHP pole. In this case, it is necessary to evaluate the integral I, defined in Equation (6.30). To this end, note first from Equation (6.17) that at $k = i\kappa_1$

$$a(i\kappa_1)\phi_2(x, i\kappa_1) = b(i\kappa_1)\phi_1(x, i\kappa_1),$$

where from the first of Equations (6.12)

$$\phi_1(x, i\kappa_1) = e^{-\kappa_1 x} + \int_x^\infty K(x,z)e^{-\kappa_1 z}\,dz.$$

Thus the integral I has two components: $I_1 + I_2$. The first of these is

$$I_1 = ir_1 e^{-\kappa_1(x+z)},$$

where

$$r_1 = b_{-1,1} = ic_1^2$$

is the residue of $b(k)$ at $k = \kappa_1$. The second contribution to I is

$$I_2 = 12\pi \oint_C \left(\frac{r_1}{k - i\kappa_1}\right) \left[\int_x^\infty K(x,z)e^{-\kappa_1(x+z)}\,dz\right]dk$$

$$= \int_x^\infty ir_1 K(x,z)e^{-\kappa_1(x+z)}\,dz.$$

Can these extra terms be included in the Gel'fand–Levitan equation

$$K(x,z) + B(x+z) + \int_x^\infty K(x,y)B(y+z)\,dy = 0?$$

Yes, if $B(x+z)$ is redefined as

$$B(x+z) = c_1^2 e^{-\kappa_1(x+z)} + \frac{1}{2\pi}\int_{-\infty}^\infty b(k)e^{ik(x+z)}\,dk.$$

N-UHP poles. For N UHP poles of the reflection and transmission coefficients, one can perform the above redefinition N times, arriving at the final result

$$B(x+z) = \sum_{n=1}^N c_n^2 e^{-\kappa_n(x+z)} + \frac{1}{2\pi}\int_{-\infty}^\infty b(k)e^{ik(x+z)}\,dk. \qquad (6.32)$$

A feature of this formulation of the scattering problem is that the UHP poles and their residues can be selected independently of $b(k)$ for real k. Thus we may have:

CASE 1: No UHP poles with $b(k) \neq 0$.
CASE 2: UHP poles with $b(k) = 0$ for all real k.
CASE 3: UHP poles with $b(k) \neq 0$.

6.1.4 Any questions?

At this point in the discussion, an attentive student might be leaning back in her chair, frowning, and thinking: "Wait a minute. I don't get it."

"Yes," says the teacher, "question?"

"Well, uh, I don't get it."

"What's the problem?"

"Look," she says, "I can understand your case 1. There is a potential $u(x)$ that is too weak to trap a bound state, but it is able to scatter incoming waves, incident waves, so there is some reflection, some $b(k)$. And I can understand case 3, because you continue $b(k)$ into the upper half of the k-plane to find the UHP poles and their residues, but..."

"But ...," says the teacher with a sly smile.

"It's case 2 that bothers me. If $b(k)$ is equal to zero for real k, how can it be continued into the upper half plane to find the corresponding poles?"

"Schrödinger's equation is linear, and the bound states are solutions of this equation. Do you doubt that?"

"No..."

"Surely there's no problem in having a bound state present in the solution of Schrödinger's equation without the scattering solution."

"No. No. But why is $b(k)$ equal to zero? Doesn't that mean that $u(x)$ is transparent to radiation? Can a scattering potential be reflectionless? It sounds oxymoronic."

"Reflectionless potentials may seem strange, but they have been of technical interest for many years. Optical scientists have long worked to reduce internal reflections in complex lens systems, and military engineers attempt to make aircraft invisible to radar. Not to mention acoustics, where sound specialists are constructing anechoic chambers."

"Have they been successful?"

"Well, yes, although the aircraft are rather expensive and delicate. They design something like the scattering potential for a bound state because it is, as you say, reflectionless. What's your problem?"

"How are we able to continue a reflection coefficient that is zero on the real k-axis into the upper half plane to find poles corresponding to the bound states? It sounds like a shell game."

"Ah, yes. Let's go back to Equations (6.16) and (6.17). The first of these is a scattering solution. It has an incident wave that is proportional to $1/a(k)$

and a reflected wave that is proportional to $b(k)/a(k)$. The second one—Equation (6.17)—holds just at $k = i\kappa_n$, where $1/a(k)$ is equal to zero but $b(k)/a(k)$ has a finite value. In this limit, at this point on the k-plane, the incident wave can be turned off without the bound state noticing."

"So you continue a nonzero reflection coefficient into the upper half plane in order to learn about the properties of a bound state, corresponding to a reflectionless potential."

"Right."

"And having done so, you assume the reflection coefficient to be zero."

"Right. We just throw it away... like an empty beer can."

[*Nervous laughter*]

"But that means" continues the student, "that if you calculate the residue of an upper half plane pole and then let the scattering potential become reflectionless, your result will be different from first making the potential reflectionless and then calculating the residue. Doesn't it?"]

"You got it. And an example of such a limit is given in Problems 7 and 8, which are part of the homework assigned for this week."

[*Groans*]

"Perhaps I should say a word about the normalization coefficients for the bound states, the c_n^2, which play a key role in Equation (6.32). If the bound states have the asymptotic behavior $\exp(-\kappa_n x)$ as $x \to +\infty$, then the coefficients are defined as in Equation (6.13). If you use normalized bound state eigenfunctions, then their behavior as $x \to +\infty$ will be as $c_n \exp(-\kappa_n x)$. In either case, of course, the residue of the corresponding pole of the reflection coefficient is equal to ic_n^2."

"Why is that important?" another student asks.

"When the potential is reflectionless, it's convenient to determine the residues from the properties of the bound state."

[*Pause*]

"So where are we?"

"Although the path has been different, we've arrived at the same conclusions as at the end of Chapter 2, with the advantage that the relationship between normalization constants of the bound states and the residues of the corresponding poles of the reflection coefficients is understood. Let's use these ideas to solve a nonlinear wave equation."

6.2 Inverse scattering method for KdV

6.2.1 *General description*

In the introduction to this chapter, the reader was invited to check that the KdV equation

$$u_t - 6uu_x + u_{xxx} = 0$$

is implied by the operator equation

$$L_t = ML - LM \equiv [M, L],$$

where
$$L = -\frac{\partial^2}{\partial x^2} + u(x,t)$$
and
$$M = -4\frac{\partial^3}{\partial x^3} + 3u\frac{\partial}{\partial x} + 3u_x.$$

In the previous section, we have studied solutions of the eigenvalue problem
$$L\psi = \lambda\psi, \tag{6.33}$$

learning how its spectrum is related to the scattering potential, $u(x,t)$. Differentiating Equation (6.33) with respect to time, one finds
$$L_t\psi + L\psi_t = \lambda_t\psi + \lambda\psi_t,$$

and choosing the time evolution of ψ to be governed by
$$\psi_t = M\psi \tag{6.34}$$

allows the previous equation to be written as
$$\{L_t + [L,M]\}\psi = \lambda_t\psi.$$

Thus if the time evolution of ψ is as in Equation (6.34) and $u(x,t)$ is a solution of KdV, the spectrum of Equation (6.33) is independent of time. This is a most useful property, providing a basis for the ISM of solving the KdV equation.

It is interesting to consider how the time evolution prescribed by Equation (6.34) influences the scattering solutions that were considered in the previous section and are displayed in Figure 6.1. The reflection coefficient $b(k)$ is defined as the ratio of a reflected wave
$$\psi_{\text{refl.}} \sim b(k)e^{ikx}$$
to an incident wave
$$\psi_{\text{inc.}} \sim e^{-ikx}.$$

Because the incident and reflected waves are defined in the limit $x \to +\infty$ where both u and u_x are zero, Equation (6.34) takes the simple form
$$\psi_t = -4\partial^3\psi/\partial x^3,$$

implying that the incident and reflected waves depend on time as
$$\psi_{\text{refl.}} \sim b(k)e^{i(kx+4k^3t)}$$
and
$$\psi_{\text{inc.}} \sim e^{-i(kx+4k^3t)}.$$

Thus the reflection coefficient evolves according to the simple equation

$$b(k, t) = b(k, 0)e^{+8ik^3 t}. \tag{6.35}$$

As bound state solutions are obtained by continuing scattering solutions into the upper half of the k-plane, their normalizing constants (or residues) evolve as

$$c_n^2(t) = c_n^2(0)e^{+8\kappa_n^3 t}. \tag{6.36}$$

With these results in hand, the ISM for the KdV equation can be described as the following sequence of activities.

1. Given an initial condition

$$u(x, 0) = f(x),$$

   compute the bound state eigenvalues (κ_n), their residues $c_n^2(0)$ for $n = 1, 2, \ldots, N$, and $b(k, 0)$ for real values of k. This is called the "scattering data" at $t = 0$.
2. Using Equations (6.35) and (6.36), compute $c_n^2(t)$ and $b(k, t)$, the scattering data at time $t > 0$.
3. From Equation (6.32) of the previous section, use the scattering data at time $t > 0$ to construct

$$B(x + z; t) = \sum_{n=1}^{N} c_n^2(0) e^{-\kappa_n (x+z) + 8\kappa_n^3 t}$$

$$+ \frac{1}{2\pi} \int_{-\infty}^{\infty} b(k, 0) e^{i[k(x+z) + 8k^3 t]} \, dk. \tag{6.37}$$

4. Solve the Gel'fand–Levitan equation

$$K(x, z) + B(x + z) + \int_x^{\infty} K(x, y) B(y + z) \, dy = 0$$

   for $K(x, z; t)$.
5. Compute

$$u(x, t) = -2 \frac{\partial}{\partial x} K(x, x; t).$$

6.2.2 Some examples

A single soliton

As a first example of the ISM, consider the problem treated at the close of Chapter 2, assuming that $u(x, t)$ is to be a solution of KdV, and

$$u(x, 0) = -2\kappa^2 \operatorname{sech}^2 \kappa x.$$

The corresponding Schrödinger equation is
$$\psi_{xx} + (k^2 + 2\kappa^2 \operatorname{sech}^2 \kappa x)\psi = 0,$$
for which $b(k,0)$ is zero so $b(k,t)$ is also zero. There is one bound state eigenvalue at $k = i\kappa$, and the corresponding normalized eigenfunction is
$$\sqrt{\frac{\kappa}{2}} \operatorname{sech} \kappa x \to \sqrt{2\kappa} \exp(-\kappa x) \quad \text{as } x \to +\infty,$$
so $c^2(0) = 2\kappa$. Thus at time $t = 0$
$$B(x+z;0) = 2\kappa e^{-\kappa(x+z)},$$
and
$$B(x+z;t) = 2\kappa e^{-\kappa(x+z)+8\kappa^3 t}.$$
With this reflection, it was observed in Chapter 2 that $K(x,z;t)$ is proportional to $\exp(-\kappa z)$, so the Gel'fand–Levitan equation has the solution
$$K(x,z;t) = -\frac{2\kappa e^{-\kappa(x+z)+8\kappa^3 t}}{1 + e^{-2\kappa x + 8\kappa^3 t}}.$$
Finally
$$u(x,t) = -2\frac{\partial}{\partial x}K(x,x;t) = -2\kappa^2 \operatorname{sech}^2[\kappa(x - 4\kappa^2 t)],$$
which is a single KdV soliton of speed $v = 4\kappa^2$.

A two-soliton collision

The reader may feel that the ISM is an unnecessarily complicated way to determine the formula for a single soliton, and rightly so, but the method provides an efficient means for generating formulas of more than one interacting soliton. To obtain a 2-soliton collision, let
$$u(x,0) = -6\operatorname{sech}^2 x,$$
whereupon the corresponding Schrödinger equation is
$$\psi_{xx} + (k^2 + 6\operatorname{sech}^2 x)\psi = 0.$$
This equation has $b(k) = 0$ and two bound state eigenfunctions, going as $e^{-\kappa x}$ in the limit $x \to \infty$. These are:
$$\psi_1(x) = \tfrac{1}{2} \tanh x \operatorname{sech} x$$
$$\kappa_1 = 1$$
$$c_1^2(0) = 6$$

and
$$\psi_2(x) = \tfrac{1}{4} \operatorname{sech}^2 x$$
$$\kappa_2 = 2$$
$$c_2^2(0) = 12.$$

Thus
$$B(x+z;t) = 6e^{8t-(x+z)} + 12e^{64t-2(x+z)},$$
implying that
$$K(x,z;t) = K_1(x,z;t)e^{-\kappa_1 z} + K_2(x,z;t)e^{-\kappa_2 z},$$
and allowing the Gel'fand–Levitan equation to be solved for
$$K(x,z;t) = \frac{6e^{-z}(e^{72t-5x} - e^{8t-x}) - 12e^{-2z}(e^{64t-2x} + e^{72t-4x})}{1 + 3e^{8t-2x} + 3e^{64t-4x} + 2e^{72t-6x}}.$$

Finally
$$u(x,t) = -2\frac{\partial K(x,x;t)}{\partial x} = -12\frac{3 + 4\cosh(2x-8t) + \cosh(4x-64t)}{[3\cosh(x-28t) + \cosh(3x-36t)]^2}.$$

This is Zabusky's original 2-soliton collision for KdV, which was obtained from a Bäcklund transform in Chapter 3 and plotted in Figure 3.3.

An N-soliton solution

Drazin and Johnson show that if
$$u(x,0) = -N(N+1)\operatorname{sech}^2 x,$$
then $b(k,0) = 0$, and there are N discrete eigenvalues at [10]
$$k = i\kappa_n = in \quad (n = 1, 2, \ldots, N).$$
The corresponding eigenfunctions of the Schrödinger equation are given by expressions of the form
$$\psi_n(x) \propto P_N^n(\tanh x),$$
where $P_N^n(\cdot)$ is an associated Legendre function[3] about which much is known [14]. Thus
$$B(x+z;t) = \sum_{n=1}^{N} c_n^2(0) e^{8n^3 t - n(x+z)},$$

[3] After French mathematician Adrien-Marie Legendre (1752–1833).

and the solution of the Gel'fand–Levitan equation can be written as

$$K(x, z; t) = \sum_{n=1}^{N} v_n(x, t) e^{-nz},$$

where the v_n are elements of a column vector $\mathbf{V}$ that satisfies the matrix equation

$$A\mathbf{V} + \mathbf{D} = 0.$$

In this equation, $\mathbf{D}$ is a column vector with elements $c_n^2(0) \exp(8n^3 t - nx)$, and $A = [a_{mn}]$ is an $N \times N$ matrix with elements given by

$$a_{mn} = \delta_{mn} + \left(\frac{c_m^2(0)}{m+n}\right) e^{8m^3 t - (m+n)x}.$$

Finally,

$$u(x, t) = -2 \frac{\partial^2}{\partial x^2} \log |A|, \tag{6.38}$$

which is a special case of the N-soliton formula given in Appendix B.

Equation (6.38) is a nonlinear superposition of N solitons of the form

$$-2n^2 \operatorname{sech}^2[n(x - 4n^2 t) + \theta_n],$$

adding up to $-N(N+1)\operatorname{sech} x$ at $t = 0$. As $t \to +\infty$, however, they spread out into a procession of sech^2 waves, with the wave of greatest speed and highest amplitude in the lead.

A delta-function initial condition

The above examples all involve reflectionless initial potentials, with $b(k, 0) = b(k, t) = 0$.

An example that is not reflectionless was introduced in Section 2.6.1, by assuming an initial condition

$$u(x, 0) = \alpha \delta(x),$$

for which it was noted that

$$b(k, 0) = \frac{\alpha}{2ik - \alpha}.$$

For $\alpha < 0$, this initial potential supports a normalized bound state

$$\psi(x) = \begin{cases} \sqrt{-\alpha/2}\, e^{+\alpha x/2} & \text{for } x > 0 \\ \sqrt{-\alpha/2}\, e^{-\alpha x/2} & \text{for } x < 0, \end{cases}$$

while for $\alpha > 0$ there is no bound state.

Thus for $\alpha < 0$

$$B(x+z;t) = -\frac{\alpha}{2}e^{\alpha(x+z)/2-\alpha^3 t} + \frac{1}{2\pi}\int_{-\infty}^{\infty}\frac{\alpha e^{i[k(x+z)+8k^3 t]}}{2ik-\alpha}dk.$$

Upon solution of the Gel'fand–Levitan equation, one finds

$$u(x,t) \to -\frac{\alpha^2}{2}\text{sech}^2\left[-\frac{\alpha}{2}(x-\alpha^2 t - x_0)\right] + \text{radiation} \qquad (6.39)$$

as $t \to +\infty$, where the soliton component is generated by the first term of $B(x+z;t)$ and a radiation component by the second term. Because the radiation is not constrained to remain localized, it disperses over the x-axis, eventually becoming small compared with the soliton.

Square-well initial conditions

As we have seen in Figure 1.1, one of the earliest soliton experiments began with a rectangularly shaped heap of water at $t = 0$ that evolved into a soliton in a wave tank according to (we now know) the KdV equation. In the context of the present discussion, such an initial condition takes the form

$$u(x,0) = \begin{cases} -u_0 & \text{for } |x| < x_0/2 \\ 0 & \text{for } |x| > x_0/2. \end{cases} \qquad (6.40)$$

The Schrödinger scattering problem for such a square-well potential is treated in introductory quantum mechanics [18, 27], with the result that there are a finite number of bound states, depending upon the width (x_0) and depth (u_0) of the well. In particular if

$$(N+1)^2 \pi^2 > u_0 x_0^2 > N^2 \pi^2, \qquad (6.41)$$

there are $N+1$ bound states.

Each bound state corresponds to a soliton component in the solution of the KdV equation, a phenomenon readily demonstrated for hydrodynamic waves in a wave tank, where N hydrodynamic solitons are found to emerge from initial square wells of water satisfying the above inequalities [5, 24]. This was first observed by Russell, who published the wave profiles shown in Figure 6.2 and noted [26]:

the genesis... of a compound or double wave of the first order, which immediately breaks down by spontaneous analysis into two, the greater going faster and altogether leaving the smaller.

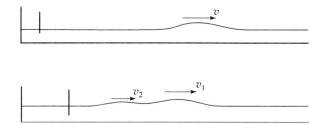

FIG. 6.2. *In the tank experiments of Figure 1.1, John Scott Russell observed that two or more solitary waves could be generated by increasing the quantity of water released. Note that $v_1 > v_2$. (Redrawn from reference [26].)*

6.2.3 Reduction to Fourier analysis in the small amplitude limit

As a computational strategy, the ISM is similar to the Fourier transform method (FM) for linear systems, which was discussed in Chapter 2. In both cases, one goes through the following three steps:

1. The initial condition $u(x,0)$ is expressed as a transformed function.
2. Because the transformed function evolves simply with time, it is readily computed for values of time that are greater than zero.
3. Knowledge of the transformed function at some value of $t > 0$ allows computation of $u(x,t)$.

Interestingly, as the wave amplitude approaches zero and the KdV equation approaches the linear system

$$u_t + u_{xxx} \doteq 0,$$

the ISM reduces to the FM.

To see this recall that under the ISM, the solution is expressed as

$$u(x,t) = -2\frac{\partial}{\partial x}K(x,x;t),$$

where $K(x,z;t)$ is obtained by solving the Gel'fand–Levitan equation. In the small amplitude limit, however, the integral term of this equation involves a product of amplitude variables, becoming small compared with the linear terms. Thus the Gel'fand–Levitan equation reduces to

$$K(x,z;t) \doteq -B(x+z;t),$$

and the solution becomes

$$u(x,t) \doteq 2\frac{\partial}{\partial x}B(2x;t)$$

$$\doteq \frac{1}{\pi}\int_{-\infty}^{\infty}[2\mathrm{i}kb(k,0)]e^{2\mathrm{i}kx+8\mathrm{i}k^3 t}\,dk,$$

where it is assumed that there are no UHP poles of $b(k,0)$. Changing the variable of integration from k to $2k$ and denoting

$$ikb(k/2) \equiv U(k),$$

we have

$$u(x,t) \doteq \frac{1}{2\pi} \int_{-\infty}^{\infty} U(k) e^{i(kx+k^3 t)}\, dk, \tag{6.42}$$

where

$$U(k) = \int_{-\infty}^{\infty} u(x,0) e^{-ikx} dx$$

is a Fourier transform of the initial condition.

The assumption that the reflection coefficient has no UHP poles is not satisfied merely by making the magnitude of $u(x,0)$ sufficiently small. The square-well potential in Equation (6.40), for example, has a bound state for any $u_0 > 0$. In such cases, however, it remains appropriate to use Equation (6.42) because UHP poles of $U(k)$ contribute to $u(x,0)$ where $x > 0$, just as lower half plane poles contribute at $x < 0$. Thus the inverse scattering method for the KdV equation can be viewed as a nonlinear generalization of the FM.

As with the FM, numerical tools are becoming available to compute ISM parameters of the nonlinear spectra [6, 25].

6.3 Two-component scattering theory

In the previous sections, we have reviewed the classical scattering theory for the Schrödinger equation and its application to the ISM for solving the KdV equation. So far the ISM appears of limited importance as we have only used it to solve the KdV equation, but in 1971 Zakharov and Shabat extended the method to the NLS equation [31], employing a matrix generalization of the linear Schrödinger operator. In 1974 Takhtajan and Faddeev presented a matrix ISM for the SG equation [29], and at about the same time Ablowitz, Kaup, Newell, and Segur (AKNS) developed a related scheme that can be used for both the SG and NLS equations, among several others [2, 3]. In this section, the ISM is extended to two-component scattering as developed by AKNS. In order to make the presentation as lucid as possible, we use an outline that resembles the presentation in the previous sections.

6.3.1 Linear theory

In two-component scattering, the operator L and the wave $\boldsymbol{\psi}$ in the eigenvalue equation

$$L\boldsymbol{\psi} = \lambda \boldsymbol{\psi} \tag{6.43}$$

become a 2×2 matrix operator and a vector function; thus

$$L = i \begin{bmatrix} \partial_x & -q(x) \\ -\sigma q^*(x) & -\partial_x \end{bmatrix} \quad \text{and} \quad \boldsymbol{\psi}(x) = \begin{bmatrix} \psi_1(x) \\ \psi_2(x) \end{bmatrix}. \tag{6.44}$$

The parameter σ is a real number (usually ± 1), and $q(x)$ is a complex function of x that plays the role of a potential on which incoming waves can scatter. Also it is assumed that in the limits $x \to \pm\infty$ the potential $q(x)$ vanishes. Note that if $\psi(x; \lambda)$ is a solution of Equation (6.43) for real λ, then $\hat{\psi}(x; \lambda)$ given by

$$\hat{\psi}(x, \lambda) = \hat{I}\psi^*(x; \lambda^*), \quad \text{where} \quad \hat{I} = \begin{bmatrix} 0 & 1 \\ -\sigma & 0 \end{bmatrix}, \tag{6.45}$$

is a linearly independent solution with the same eigenvalue λ.

In the limit $x \to \pm\infty$ the eigenvalue problem (6.43) possesses two linearly independent solutions of the form

$$\mathbf{v}^+ = \begin{bmatrix} 0 \\ 1 \end{bmatrix} e^{+i\lambda x} \quad \text{and} \quad \mathbf{v}^- = \begin{bmatrix} 1 \\ 0 \end{bmatrix} e^{-i\lambda x}.$$

The vector $\mathbf{v}^+(x)$ is a right running wave and $\mathbf{v}^-(x)$ is a left running wave, and we note that $\mathbf{v}^-(x)$ equals $\hat{\mathbf{v}}^+(x)$. In analogy with the Schrödinger equation, eigenvalues $\lambda = \lambda_n$ with positive imaginary parts correspond to bound states, vanishing at $x \to \pm\infty$.

For continuous eigenvalues λ we can write a solution ψ^+ of (6.43) in the form

$$\psi^+(x, \lambda) = \begin{cases} \mathbf{v}^- + b(\lambda)\mathbf{v}^+ & \text{for } x \to +\infty \\ a(\lambda)\mathbf{v}^- & \text{for } x \to -\infty. \end{cases} \tag{6.46}$$

The above expression can be interpreted as a left running wave $\mathbf{v}^-$ hitting the potential q from the right and is partly transmitted into a wave $a\mathbf{v}^-$ with transmission coefficient $a(\lambda)$ and partly reflected into a wave $b\mathbf{v}^+$ with reflection coefficient $b(\lambda)$. On the other hand we can consider a wave coming from the left and hitting the potential. In this case we introduce the function ψ^- according to

$$\psi^-(x, \lambda) = \begin{cases} \mathbf{v}^+ + d(\lambda)\mathbf{v}^- & \text{for } x \to -\infty \\ c(\lambda)\mathbf{v}^+ & \text{for } x \to +\infty, \end{cases} \tag{6.47}$$

where $c(\lambda)$ is a transmission coefficient and $d(\lambda)$ is a reflection coefficient. Because ψ^+ and ψ^- are linearly independent, a general solution ψ of the two-component scattering problem in Equation (6.43) can be written as

$$\psi(x, \lambda) = A(\lambda)\psi^+ + B(\lambda)\psi^- \quad \text{for } x \to \pm\infty.$$

From the definitions of ψ^+ and ψ^- in (6.46) and (6.47) we obtain

$$\psi(x, \lambda) = \begin{cases} A\mathbf{v}^- + (Ab + Bc)\mathbf{v}^+ & \text{for } x \to +\infty \\ (Aa + Bd)\mathbf{v}^- + B\mathbf{v}^+ & \text{for } x \to -\infty. \end{cases}$$

This equation is analogous to Equation (6.8) in Section 6.1.1, and the argument concerning the limiting behavior of A, B, a, b, c, and d as λ approaches a UHP pole of the reflection and transmission coefficients follows directly from the discussion in that section. In order that ψ stays finite as λ approaches a pole, it is necessary that A and B go to zero and the factors $(Ab+Bc)$ and $(Aa+Bd)$ remain finite. This limit of ψ is a bound state solution, for which the reflection and transmission coefficients have poles at λ_n, where $\text{Im}[\lambda_n] > 0$.

With respect to the Schrödinger scattering problem, however, there is one important difference: the poles are not required to be simple. In the present section, it is assumed that all UHP poles are simple, but this assumption is not overly restrictive because a multiple pole can be viewed as a limit of the merging of simple poles.

The limiting solutions $\mathbf{v}^+$ and $\mathbf{v}^-$ are valid in the regions $x \to \pm\infty$ where the potential $q(x)$ vanishes. In the region with nonzero potential, it is appropriate to introduce the following assumptions for solutions of Equation (6.43):

$$\phi^+(x, \lambda) = \begin{bmatrix} \int_x^\infty K_1(x,z)e^{i\lambda z}\, dz \\ e^{i\lambda x} + \int_x^\infty K_2(x,z)e^{i\lambda z}\, dz \end{bmatrix}, \tag{6.48}$$

and

$$\phi^-(x, \lambda) = \begin{bmatrix} e^{-i\lambda x} + \int_{-\infty}^x L_1(x,z)e^{-i\lambda z}\, dz \\ \int_{-\infty}^x L_2(x,z)e^{-i\lambda z}\, dz \end{bmatrix}. \tag{6.49}$$

These are similar to the Jost solutions in Equation (6.12), and it is again assumed that the kernels and their partial derivatives with respect to z vanish as $x \to \pm\infty$ in the above integrals. The function $\phi^-(x,\lambda)$ can be written as a linear combination

$$\phi^-(x, \lambda) = C\hat{\phi}^+(x, \lambda) + D\phi^+(x, \lambda), \tag{6.50}$$

where $\hat{\phi}^+$ is defined as is $\hat{\psi}$ in Equation (6.45).

Dividing by C and considering the limits $x \to \pm\infty$ gives

$$\frac{1}{C}\phi^- = \begin{cases} \mathbf{v}^-(x,\lambda) + D/C\,\mathbf{v}^+(x,\lambda) & \text{for } x \to +\infty \\ 1/C\,\mathbf{v}^-(x,\lambda) & \text{for } x \to -\infty. \end{cases}$$

Comparing with ψ^+ in Equation (6.46) implies that $\phi^- = C\psi^+$, with

$$\frac{1}{C} = a \quad \text{and} \quad \frac{D}{C} = b,$$

and substituting into Equation (6.50) leads to

$$\phi^-(x,\lambda) = \frac{1}{a}\hat{\phi}^+(x,\lambda) + \frac{b}{a}\phi^+(x,\lambda). \tag{6.51}$$

At UHP pole λ_n, corresponding to bound state number n, the two solutions ϕ^- and ϕ^+ are related through

$$\phi^-(x, \lambda_n) = \frac{b_{-1,n}}{a_{-1,n}} \phi^+(x, \lambda_n). \tag{6.52}$$

because $1/a(\lambda_n)$ equals zero.

The relation between the kernels K_1 and K_2 and q is found by considering

$$L\phi^+ = \lambda \phi^+. \tag{6.53}$$

In the right-hand side, perform partial integration of the integrals in ϕ^+, leading to

$$\phi^+ = \begin{bmatrix} 0 \\ 1 \end{bmatrix} e^{i\lambda x} + \frac{i}{\lambda} \begin{bmatrix} K_1(x,x) \\ K_2(x,x) \end{bmatrix} e^{i\lambda x}$$
$$+ \frac{i}{\lambda} \int_x^\infty \begin{bmatrix} K_{1,z}(x,z) \\ K_{2,z}(x,z) \end{bmatrix} e^{i\lambda z} \, dz. \tag{6.54}$$

The left-hand side is calculated as

$$L\phi^+ = \lambda \begin{bmatrix} 0 \\ 1 \end{bmatrix} e^{i\lambda x} + \begin{bmatrix} -iK_1(x,x) \\ iK_2(x,x) \end{bmatrix} e^{i\lambda x}$$
$$+ \int_x^\infty \begin{bmatrix} iK_{1,x}(x,z) \\ -iK_{2,x}(x,z) \end{bmatrix} e^{i\lambda z} \, dz$$
$$+ \begin{bmatrix} -iqe^{i\lambda x} - iq \int_x^\infty K_2(x,z) e^{i\lambda z} \, dz \\ -i\sigma q^* \int_x^\infty K_1(x,z) e^{i\lambda z} \, dz \end{bmatrix}. \tag{6.55}$$

Inserting Equation (6.55) and Equation (6.54) into (6.53) and rearranging yields

$$C_{11}(x) e^{i\lambda x} + \int_x^\infty C_{12}(x,z) e^{i\lambda z} \, dz = 0$$
$$C_{21}(x,x) e^{i\lambda x} + \int_x^\infty C_{22}(x,z) e^{i\lambda z} \, dz = 0,$$

where

$$C_{11} \equiv 2K_1(x,x) + q(x)$$
$$C_{12} \equiv K_{1,z}(x,z) - K_{1,x}(x,z) + qK_2(x,z)$$
$$C_{21} \equiv K_2(x,x) - K_2(x,x) = 0$$
$$C_{22} \equiv K_{2,x}(x,z) + K_{2,z}(x,z) + \sigma q^* K_1(x,z).$$

All coefficients C_{ij} must vanish, leading to the following system of PDEs for the kernels

$$K_{1,x}(x,z) - K_{1,z}(x,z) = qK_2(x,z)$$
$$K_{2,x}(x,z) + K_{2,z}(x,z) = -\sigma q^* K_1(x,z), \tag{6.56}$$

with the additional relation

$$q(x) = -2K_1(x,x). \tag{6.57}$$

Equations (6.56) are valid for all z including $z = x$; thus

$$|q(x)|^2 = \frac{2}{\sigma}\frac{d}{dx}K_2(x,x). \tag{6.58}$$

The system in Equations (6.56) relates the kernels and the potential q. (Note that Equations (6.57) and (6.58) constitute boundary conditions for K_1 and K_2 along the line $z = x$ leading to a unique solution of Equations (6.56).)

In the scattering problem of the Schrödinger equation, there is an integral equation, the Gel'fand–Levitan equation, for the kernel in the solution of the scattered wave. A corresponding integral equation can be deduced for the two-component scattering problem, following the ideas in the previous sections. To see this, start with Equation (6.51), where

$$\phi^+(x,\lambda) \equiv \begin{bmatrix} \phi_1^+(x,\lambda) \\ \phi_2^+(x,\lambda) \end{bmatrix}$$

and

$$\phi^-(x,\lambda) \equiv \begin{bmatrix} \phi_1^-(x,\lambda) \\ \phi_2^-(x,\lambda) \end{bmatrix},$$

and multiply both sides by a to obtain

$$a\phi_1^-(x,\lambda) = \phi_2^{*+}(x,\lambda^*) + b\phi_1^+(x,\lambda) \tag{6.59}$$

and

$$a\phi_2^-(x,\lambda) = -\sigma\phi_1^{*+}(x,\lambda^*) + b\phi_2^+(x,\lambda). \tag{6.60}$$

Divide the left-hand side of Equation (6.59) by 2π, multiply by $\exp(i\lambda z)$, and then integrate with respect to λ from $-\infty$ to $+\infty$. Using the Cauchy theorem along a closed contour extending into the UHP, one finds

$$\text{LHS}(6.59) = \frac{1}{2\pi} \int_{-\infty}^{\infty} a\phi_1^-(x,\lambda)e^{i\lambda z} d\lambda$$

$$= i \sum_{n=1}^{N} \text{Res}[a(\lambda_n)\phi_1^-(x,\lambda_n)e^{+i\lambda_n z}]$$

$$= i \sum_{n=1}^{N} b_{-1,n} \int_{x}^{\infty} K_1(x,y)e^{+i\lambda_n(y+z)} dy. \quad (6.61)$$

In obtaining the last equality, we have used Equation (6.52) and the definition of ϕ^+ in (6.48). The $b_{-1,n}$ are the residues of poles of the reflection coefficient.

Doing the same on the right-hand side leads to

$$\text{RHS}(6.59) = \delta(z-x) + \int_{x}^{\infty} K_2^*(x,y)\delta(z-y) dy$$

$$+ \int_{x}^{\infty} \frac{1}{2\pi} \int_{-\infty}^{\infty} b(\lambda) e^{i\lambda(y+z)} d\lambda K_1(x,y) dy. \quad (6.62)$$

Putting the two sides together gives the integral equation

$$K_2^*(x,z) + \int_{x}^{\infty} B(y+z) K_1(x,y) dy = 0,$$

where $z > x$ and

$$B(x+z) \equiv -i \sum_{n=1}^{N} b_{-1,n} e^{i\lambda_n(x+z)} + \frac{1}{2\pi} \int_{-\infty}^{\infty} b(\lambda) e^{i\lambda(x+z)} d\lambda. \quad (6.63)$$

The same procedure can be applied to Equation (6.60) leading to a second integral equation for the kernels K_1 and K_2

$$-\sigma K_1^*(x,z) + B(x+z) + \int_{x}^{\infty} B(y+z) K_2(x,y) dy = 0. \quad (6.64)$$

Combination of these two integral equations leads to the final result

$$\sigma K_1(x,z) = B^*(x+z) - \int_{x}^{\infty}\int_{x}^{\infty} K_1(x,y) B(y+y') B^*(y'+z) dy\, dy', \quad (6.65)$$

where $B(x+z)$ is determined by the scattering data: $b(\lambda)$ plus $\{\lambda_n\}$ and $\{b_{-1,n}\}$ for $n = 1, 2, \ldots, N$. Solving this integral equation for $K_1(x,z)$ then gives

$$q(x) = -2K_1(x,x).$$

In the next section, we shall see how this theory can be applied in the ISM for several nonlinear PDEs.

6.3.2 ISMs for two-component scattering

Consider a nonlinear equation

$$q_t = N(q) \tag{6.66}$$

that is to be solved by the ISM, using the two-component matrix operator L defined in Equation (6.44). One approach, introduced by Zakharov and Shabat [31], is to find a matrix operator M for which the operator equation $L_t = [M, L]$ implies Equation (6.66), but this is not an easy task.

A more systematic method, introduced by AKNS, begins by noting that Equation (6.43) can be written as [3]

$$\begin{aligned} \psi_{1,x} &= -i\lambda\psi_1 + q(x,t)\psi_2 \\ \psi_{2,x} &= i\lambda\psi_2 - \sigma q^*(x,t)\psi_1. \end{aligned} \tag{6.67}$$

A general time dependence of ψ_1 and ψ_2 can be similarly expressed as

$$\begin{aligned} \psi_{1,t} &= A\psi_1 + B\psi_2 \\ \psi_{2,t} &= C\psi_1 + D\psi_2, \end{aligned} \tag{6.68}$$

where A, B, C, and D are operators yet to be specified. One might expect them to involve derivatives with respect to x, but this is not necessarily so. Why not? Because any x-derivatives that appear on the right-hand side of Equations (6.68) can be eliminated through substitutions from Equations (6.67). Thus a rather general form for the matrix

$$V = \begin{bmatrix} A & B \\ C & D \end{bmatrix}$$

involves elements that are functions only of q, q^*, their x-derivatives, and λ.

There is one additional restriction on the matrix V: the x-derivative of Equation (6.68) must equal the t-derivative of Equation (6.67) in order that ψ be defined as a single valued function of x and t. Assuming $\lambda_t = 0$, this cross-derivative condition allows [3]

$$D = -A$$

and requires

$$\begin{aligned} A_x &= qC + \sigma q^* B \\ B_x + 2i\lambda B &= q_t - 2Aq \\ C_x - 2i\lambda C &= -\sigma q_t^* - 2A\sigma q^*. \end{aligned} \tag{6.69}$$

The problem boils down to this: find solutions of Equations (6.69) that imply Equation (6.66). How do we proceed?

One approach is to assume that A, B, and C are truncated power series in λ. Thus, for example, we could try [4]

$$A = A_0 + A_1\lambda + A_2\lambda^2$$
$$B = B_0 + B_1\lambda + B_2\lambda^2$$
$$C = C_0 + C_1\lambda + C_2\lambda^2,$$

substituting into Equations (6.69), and attempting to balance the resulting equations at each power of λ. This goes as follows:

- Equating the λ^3 terms implies $B_2 = C_2 = 0$.
- Equating the λ^2 terms implies A_2 is a constant.
- Equating the λ terms implies A_1 is a constant that is set to zero.
- Equating the λ^0 terms implies

$$q_t + \frac{A_2}{2}q_{xx} + \sigma A_2 |q|^2 q + \text{constant} = 0.$$

- Finally, choosing $A_2 = -2i$ and $\sigma = +1$ and setting the constant to zero implies the NLS equation

$$iq_t + q_{xx} + 2|q|^2 q = 0.$$

In summary, if the matrix V has the components

$$A = -2i\lambda^2 + iqq^*$$
$$B = 2q\lambda + iq_x$$
$$C = -2q^*\lambda + iq_x^*$$
$$D = +2i\lambda^2 - iqq^*$$

and the scattering operator L is defined as in Equation (6.44) of the previous section (with $\sigma = +1$), then the cross-derivative condition implies the NLS equation. These are the tools needed to solve the NLS equation using the ISM.

Evidently the preceding discussion has only begun to milk the possibilities of relating various forms of the matrix M to nonlinear wave equations. The series expressions for A, B, C, and D can be carried to higher or lower powers, and the constants encountered along the way can be kept (assumed nonzero), introducing additional terms into the resulting nonlinear PDEs. The book by Ablowitz and Segur discusses several such formulations of interest to applied scientists, including the NLS equation and a form of the SG equation [4].

As the SG and NLS equations are of primary interest for applications, they are discussed in the following sections.

6.4 The sine–Gordon equation

As we have learned in Chapter 3, the SG equation

$$\frac{\partial^2 u}{\partial x^2} - \frac{\partial^2 u}{\partial t^2} = \sin u \qquad (6.70)$$

describes the dynamics of a Josephson transmission line and also a one-dimensional system of coupled pendulums. In the first application, (shown in Figure 3.4), u represents the phase difference of the superconducting wave functions across the insulating layer. In the second (see Figure 3.5), u measures the angle of rotation of a pendulum.

To analyze the SG equation using an ISM, it is convenient to transform the variables as[4]

$$x \to \xi = (x+t)/2$$
$$t \to \tau = (x-t)/2$$
$$u(x,t) \to \phi(\xi, \tau) = u(x,t),$$

whereupon Equation (6.70) becomes

$$\frac{\partial^2 \phi}{\partial \xi \partial \tau} = \sin \phi. \qquad (6.71)$$

If the operator L in Equation (6.43) is chosen to be

$$L = i \begin{bmatrix} \partial_\xi & \phi_\xi/2 \\ \phi_\xi/2 & -\partial_\xi \end{bmatrix},$$

the two-component scattering equations can be written as

$$\frac{\partial}{\partial \xi} \begin{bmatrix} \psi_1 \\ \psi_2 \end{bmatrix} = \begin{bmatrix} -i\lambda & -\phi_\xi/2 \\ \phi_\xi/2 & i\lambda \end{bmatrix} \begin{bmatrix} \psi_1 \\ \psi_2 \end{bmatrix}.$$

With the τ-dependence as

$$\frac{\partial}{\partial \tau} \begin{bmatrix} \psi_1 \\ \psi_2 \end{bmatrix} = \frac{i}{4\lambda} \begin{bmatrix} \cos \phi & \sin \phi \\ \sin \phi & -\cos \phi \end{bmatrix} \begin{bmatrix} \psi_1 \\ \psi_2 \end{bmatrix}, \qquad (6.72)$$

it is readily checked that the cross-derivative condition

$$\frac{\partial}{\partial \tau} \frac{\partial}{\partial \xi} \begin{bmatrix} \psi_1 \\ \psi_2 \end{bmatrix} = \frac{\partial}{\partial \xi} \frac{\partial}{\partial \tau} \begin{bmatrix} \psi_1 \\ \psi_2 \end{bmatrix}$$

implies Equation (6.71), providing the basis for an ISM analysis.

[4] These new variables are sometimes called "light cone" coordinates because they point in the direction of characteristic lines on the (x,t)-plane.

Assuming that $\phi \to 0 \pmod{2\pi}$ as $\xi \to \pm\infty$, the time evolution matrix defined in Equation (6.72) takes the simple form

$$V \to \frac{i}{4\lambda} \begin{bmatrix} 1 & 0 \\ 0 & -1 \end{bmatrix};$$

thus the reflection coefficient and the residues of UHP poles evolve with time as

$$b(\lambda, \tau) = b(\lambda, 0) e^{-i\tau/2\lambda}$$

$$r_n(\tau) = r_n(0) e^{-i\tau/2\lambda_n}.$$

From the results of Section 6.3.1, then,

$$\phi_\xi(\xi, \tau) = 4K(\xi, \xi; \tau), \tag{6.73}$$

where $K(\xi, z; \tau)$ is the solution of the integral equation

$$K(\xi, z; \tau) = B^*(\xi + z; \tau) - \int_\xi^\infty \int_\xi^\infty K(\xi, y; \tau) B(y + y'; \tau) B^*(z + y'; \tau) \, dy \, dy' \tag{6.74}$$

with $z > \xi$ and

$$B(\xi + z; \tau) \equiv \frac{1}{2\pi} \int_{-\infty}^\infty b(\lambda, 0) e^{i\lambda(\xi+z) - i\tau/2\lambda} d\lambda$$

$$- i \sum_{n=1}^N r_n(0) e^{i\lambda_n(\xi+z) - i\tau/2\lambda_n} \tag{6.75}$$

is determined from a scattering analysis of the initial conditions.

Let us consider some applications of this formulation.

A single kink or antikink

Perhaps the simplest example is obtained by assuming that the initial potential is reflectionless and has but one bound state, corresponding to a single pole at $(\lambda_1 = i\kappa)$ on the imaginary axis of the upper half λ-plane with residue $r_1 = \pm 2i\kappa$. Thus

$$B(\xi + z; \tau) = \pm 2\kappa e^{-\kappa(\xi+z) - \tau/2\kappa},$$

so

$$K(\xi, z; \tau) = \pm 2\kappa e^{-\kappa(\xi+z) - \tau/2\kappa}$$
$$- 4\kappa^2 e^{-\tau/\kappa} \int_\xi^\infty \int_\xi^\infty K(\xi, y; \tau) e^{-\kappa(y+y')} e^{-\kappa(z+y')} \, dy \, dy'.$$

As $K(\xi, z; \tau) \propto \exp(-\kappa z)$, this integral equation is readily solved for

$$K(\xi, z; \tau) = \pm \frac{2\kappa \exp[-\kappa(\xi + z) - \tau/2\kappa]}{1 + \exp(-4\kappa\xi - \tau/\kappa)}.$$

Thus

$$\frac{\partial \phi}{\partial \xi} = \pm 4K(\xi, \xi; \tau) = \pm 4\kappa \operatorname{sech}(2\kappa\xi + \tau/2\kappa),$$

which integrates to

$$\phi(\xi, \tau) = 4 \arctan\{\exp[\pm(2\kappa\xi + \tau/2\kappa)]\}.$$

Finally one can transform back to the laboratory (x, t)-coordinates, whereupon the corresponding solution of Equation (6.70) is the familiar kink or antikink

$$u(x, t) = 4 \arctan\left[\exp\left(\pm \frac{x - vt}{\sqrt{1 - v^2}}\right)\right],$$

with the laboratory velocity identified as

$$v = \frac{1 - 4\kappa^2}{1 + 4\kappa^2} = \frac{1 + 4\lambda_1^2}{1 - 4\lambda_1^2}. \tag{6.76}$$

Again one might grumble that there are much easier ways to find the expression for an SG kink, but we have also obtained Equation (6.76), relating the velocity of the kink soliton to the locus of a UHP pole of the reflection coefficient at $\lambda_1 = i\kappa$.

Interestingly, at $\lambda_1 = i/2$, the kink is stationary in the laboratory coordinates. More generally, v goes from $+1 \to 0$ as κ goes from $0 \to 1/2$, and v goes from $0 \to -1$ as κ goes from $1/2 \to +\infty$

Square-well initial conditions

As $B(\xi + z; 0)$ was chosen as the starting point, the previous example did not take full advantage of the ISM. Thus it is interesting to study the dynamics of ϕ_ξ evolving from the square-well initial condition [28]

$$\phi_\xi = \begin{cases} A/p & \text{for } 0 \leq \xi \leq p, \\ 0 & \text{for } \xi < 0 \quad \text{and} \quad \xi > p. \end{cases} \tag{6.77}$$

In the context of the superconducting Josephson transmission line shown in Figure 3.4, this corresponds to $A/2\pi$ fluxons uniformly distributed over $0 \leq x \leq 2p$. For the mechanical system of Figure 3.5, it represents $A/2\pi$ helical twists, similarly distributed.

With scattering functions defined at $\tau = 0$ as

$$\text{incident wave} = \begin{bmatrix} 1 \\ 0 \end{bmatrix} e^{-i\lambda\xi},$$

$$\text{reflected wave} = \begin{bmatrix} 0 \\ 1 \end{bmatrix} b(\lambda, 0) e^{+i\lambda\xi},$$

as $\xi \to +\infty$, and

$$\text{transmitted wave} = \begin{bmatrix} 1 \\ 0 \end{bmatrix} a(\lambda) e^{-i\lambda\xi},$$

as $\xi \to -\infty$, it is straightforward to match boundary conditions at $\xi = 0$ and p. The results are three-fold:

1. The reflection coefficient

$$b(\lambda, 0) = \frac{A}{2mp} e^{-2i\lambda p} \left[\frac{\sin(mp)}{\cos(mp) - (i\lambda/m)\sin(mp)} \right],$$

where

$$m^2 \equiv (\lambda^2 + A^2/4p^2).$$

2. For $A > (2N-1)\pi$, there are N bound states at

$$\lambda_n = i\sqrt{(A/2p)^2 - m_n^2},$$

where the m_n are roots of

$$\cot(mp) + \sqrt{\left(\frac{A}{2mp}\right)^2 - 1} = 0.$$

3. The residue of $b(\lambda, 0)$ at λ_n is

$$r_n(0) = i \frac{A^2 + 4\lambda_n^2 p^2}{2Ap(1 - i\lambda_n p)} e^{-2i\lambda_n p}.$$

As we have learned, each bound state corresponds to a soliton of asymptotic speed $1/4\gamma_n^2$ in the (ξ, τ)-coordinates and $(1 + 4\lambda_n^2)/(1 - 4\lambda_n^2)$ back in the laboratory (x, t)-system.

Substituting these scattering data into Equation (6.75) for $B(\xi + z; \tau)$ allows solution of Equation (6.74), leading to $\phi_\xi(\xi, \tau)$ from Equation (6.73).

Although this seems to be a well defined computational procedure, there are certain practical difficulties in carrying it through, including the following. (i) It can be difficult to evaluate the radiative part of the solution, which is generated by the first term of Equation (6.75). Thus numerical tools may be required [25]. (ii) Even if radiation is neglected by arbitrarily setting $b(\lambda, 0) = 0$, solution of Equation (6.74) requires the inversion of an $N \times N$ matrix. For fluxons initiated on real Josephson junctions, N is easily equal to 100 or more, suggesting that one should consider other approaches to obtain numerical solutions, including Whitham's method for slowly modulated periodic waves (which was sketched in Section 3.2.3) and direct numerical integration of Equation (6.70) [28].

Kink-antikink collisions and breathers

In Chapter 3, a Bäcklund transform was used to show that the SG equation has exact solutions representing kink-kink and kink-antikink collisions. Thus starting with Equation (3.84), setting $u_0 = 0$, $(a_1 - a_2)/(a_1 + a_2) = v$, and

$$u_1 = 4\arctan\left[\exp\left(+\frac{x - vt}{\sqrt{1 - v^2}}\right)\right]$$

$$u_2 = 4\arctan\left[\exp\left(+\frac{x + vt}{\sqrt{1 - v^2}}\right)\right],$$

such a solution is

$$u_{\text{kak}}(x, t) = 4\arctan\left[\frac{1}{v}\tan\left(\frac{u_2 - u_1}{4}\right)\right],$$

which takes the simpler form

$$u_{\text{kak}}(x, t) = 4\arctan\left[\frac{\sinh(vt/\sqrt{1 - v^2})}{v\cosh(x/\sqrt{1 - v^2})}\right]. \tag{6.78}$$

With an appropriate choice of the constants c_1 and c_2, this kink-antikink solution can be generated from

$$B(\xi + z; 0) = c_1 e^{-\kappa_1(\xi + z)} + c_2 e^{-\kappa_2(\xi + z)} \tag{6.79}$$

with

$$\kappa_1 = \frac{1}{2}\sqrt{\frac{1 + v}{1 - v}}$$

$$\kappa_2 = \frac{1}{2}\sqrt{\frac{1 - v}{1 + v}}.$$

As $v \to 0$, the UHP poles at $\lambda_1 = i\kappa_1$ and $\lambda_2 = i\kappa_2$ merge to form a double pole at $\lambda = i/2$, at which Equation (6.78) implies that

$$u(x, t) = 4\arctan(t\,\text{sech}\,x). \tag{6.80}$$

It was noted in Chapter 3 that if the velocity parameter in the kink-antikink formula is given the imaginary value

$$v = \frac{i\omega}{\sqrt{1 - \omega^2}}, \quad \omega < 1,$$

then Equation (6.78) becomes the breather

$$u_b(x,t) = 4\arctan\left[\frac{\sqrt{1-\omega^2}}{\omega}\frac{\sin\omega t}{\cosh\sqrt{1-\omega^2}x}\right]. \tag{6.81}$$

In this case, there are two complex UHP poles with equal imaginary parts and real parts of opposite sign, a symmetry that is necessary for $u(x,t)$ to be real.

In the following chapter, a multisoliton perturbation theory is used to study a kink-antikink collision under the influence of dissipation and a source of energy. From this analysis, a critical value of the source is found, below which the kink-antikink collision of Equation (6.78) decays into the breather of Equation (6.81).

6.5 The nonlinear Schrödinger equation

In this section, we consider application of the ISM to the NLS equation

$$i\frac{\partial u}{\partial t} + \frac{\partial^2 u}{\partial x^2} + 2|u|^2 u = 0, \tag{6.82}$$

which was introduced in Chapter 3 as a generic description of wave packets in nonlinear dispersive media. From Section 6.3, this equation is implied by the cross-derivative condition, where

$$L = i\begin{bmatrix}\partial_x & -u(x,t) \\ -u^*(x,t) & -\partial_x\end{bmatrix},$$

and

$$V = \begin{bmatrix}A & B \\ C & -A\end{bmatrix},$$

with

$$A = -2i\lambda^2 + i|u|^2$$
$$B = 2u\lambda + iu_x$$
$$C = -2u^*\lambda + iu_x^*.$$

Assuming that $u(x,t) \to 0$ as $x \to \pm\infty$ implies

$$V \to 2i\lambda^2\begin{bmatrix}-1 & 0 \\ 0 & +1\end{bmatrix},$$

so scattering data for the problem

$$L\begin{bmatrix}\psi_1(x,t) \\ \psi_2(x,t)\end{bmatrix} = \lambda\begin{bmatrix}\psi_1(x,t) \\ \psi_2(x,t)\end{bmatrix}$$

evolve with time as

$$b(\lambda, t) = b(\lambda, 0)e^{+4i\lambda^2 t} \quad \text{and} \quad r_n(t) = r_n(0)e^{+4i\lambda_n^2 t}.$$

Thus solutions of Equation (6.82) are given by

$$u(x,t) = -2K(x,x;t), \tag{6.83}$$

where $K(x,z;t)$ is a solution of the integral equation

$$K(x,z;t) = B^*(x+z;t) - \int_x^\infty \int_x^\infty K(x,y;t)B(y+y';t)B^*(y'+z;t)\,dy\,dy', \tag{6.84}$$

with

$$B(x+z;t) \equiv -i\sum_{n=1}^N r_n(0)e^{i[\lambda_n(x+z)+4\lambda_n^2 t]} + \frac{1}{2\pi}\int_{-\infty}^\infty b(\lambda,0)e^{i[\lambda(x+z)+4\lambda^2 t]}d\lambda. \tag{6.85}$$

In thinking about ISM solutions for the NLS system, recall what is known about corresponding solutions for the KdV and SG equations.

1. All solitons of KdV and kink or antikink SG solitons are real functions that are described by two parameters, fixing the speed and location. In these cases, it is natural to find that corresponding poles of the reflection coefficient lie on the positive imaginary axis with purely imaginary residues. The locus of a pole in the UHP determines the speed of the soliton and its residue provides the position parameter.
2. Breather solitons of the SG equation are described by four parameters: (i) speed, (ii) oscillation frequency, (iii) location, and (iv) phase of the oscillation. To insure that the breather is a real function of x and t, these two poles and their residues must have the symmetry $\lambda_1 = -\lambda_2^*$ and $r_1 = -r_2^*$, thus providing a total of four parameters in the scattering problem.

An NLS soliton is like an SG breather in the sense that it also contains four parameters (fixing speed, frequency, location, and phase), but it is not constrained to be a real function of x and t. Thus the corresponding pole can be anywhere in the upper half of the λ-plane and its residue can be an arbitrary complex number. To explore this situation in detail, let us assume that $b(\lambda, 0) = 0$ and take a single simple pole at

$$\lambda_1 = \xi + i\eta, \quad \eta > 0,$$

with

$$r_1 = \alpha + i\beta,$$

whereupon

$$B(x+z;t) = (\beta - i\alpha)\exp[(-\eta + i\xi)(x+z) - 8\xi\eta t + 4i(\xi^2 - \eta^2)t].$$

Observing that

$$K(x,z;t) \propto \exp[-(\eta + i\xi)z]$$

allows Equation (6.84) to be solved for $K(x,z;t)$, and Equation (6.83) then gives the expression

$$u(x,t) = 2\eta \exp[-2i\xi x - 4i(\xi^2 - \eta^2)t + i\theta_2]\operatorname{sech}[2\eta(x+4\xi t) + \theta_1]$$

for an NLS soliton. The parameters θ_1 and θ_2 are defined as

$$e^{\theta_1} = \frac{2\eta}{\sqrt{\beta^2 + \alpha^2}}$$

$$e^{i\theta_2} = \frac{-\beta + i\alpha}{\sqrt{\beta^2 + \alpha^2}},$$

showing how the position and phase of the soliton are related to the real and imaginary parts of the residue. Note that this equation agrees with Equation (3.106) upon identifying the amplitude as $a = 2\eta$ and the envelope velocity as -4ξ.

Finally, we mention again that poles of the transmission and reflection coefficients need not be simple for the two-component scattering operator. In the first publication on the ISM for the NLS equation, Zakharov and Shabat showed that a double pole leads to two solitons separated by a distance that increases with time as $\log(4\eta^2 t)$ [31]. A similar behavior is observed in Equation (6.80), which represents a kink and an antikink separated by a distance of $2\log 2t$.

6.6 Conservation laws

Because the spatial behavior of a transmitted wave is identical to that of an incident wave for the scattering problems considered in this chapter, it follows that $a(k)$ is a constant of the motion under the dynamics of an ISM analysis. This fact can be used to generate a countably infinite number of independent conservation laws of the form

$$\frac{\partial D}{\partial t} + \frac{\partial F}{\partial x} = 0,$$

where D is the density of a conserved quantity, and F is its flow.

In this section we consider the generation of infinite sets of such laws from ISMs based on both the linear Schrödinger scattering and two-component matrix scattering.

6.6.1 Conservation laws for the KdV equation

For scattering that is governed by the linear Schrödinger equation, Equations (6.5) and (6.12) imply that

$$\psi^+(x,k) = a(k)\phi_2(x,k)$$
$$= a(k)\left[e^{-ikx} + \int_{-\infty}^{x} L(x,z)e^{-ikz}\,dz\right].$$

Without loss of generality, this function can be written as [21]

$$\psi^+(x,k) = a(k)e^{-ikx+\alpha(x,k)},$$

and substitution into Schrödinger's equation implies

$$\alpha_{xx} - 2ik\alpha_x + \alpha_x^2 - u(x,t) = 0.$$

As the magnitude of k becomes large, $\alpha_x \to 0$ so it can be expressed as the series

$$\alpha_x(x,k) \sim \sum_{n=1}^{\infty} \frac{\beta_n(x)}{(2ik)^n},$$

which is readily solved for the β_n by equating coefficients at each power of k. Thus

$$\beta_1 = -u$$
$$\beta_2 = -u_x$$
$$\beta_3 = u^2 - u_{xx},$$

and in general

$$\beta_n = \beta_{n-1,x} + \sum_{m=1}^{n-2} \beta_m \beta_{n-m-1}.$$

Now consider the situation as $x \to +\infty$. From Equation (6.5),

$$\psi^+(x,k) \to e^{-ikx} + b(k)e^{ikx}.$$

Letting $|k| \to \infty$ with $\text{Im}(k) > 0$ simplifies this behavior to

$$\psi^+(x,k) \to e^{-ikx},$$

implying in turn that

$$a(k)e^{\alpha(x,k)} \to 1.$$

Thus in this limit ($x \to +\infty$)

$$-\log a(k) = \int_{-\infty}^{\infty} \alpha_x(x,k)\, dx$$

$$= \sum_{n=1}^{\infty} (2\mathrm{i}k)^{-n} \int_{-\infty}^{\infty} \beta_n(x)\, dx.$$

Because this condition holds at each power of k, the β_n are revealed as a countably infinite number of conserved densities for any nonlinear wave system that is integrable using the linear Schrödinger operator.

These conserved densities can be used to construct a set of conservation laws for the KdV equation that are both nontrivial and independent. Starting with $\beta_1 = u$, the corresponding conservation law is

$$\frac{\partial u}{\partial t} + \frac{\partial}{\partial x}(u_{xx} - 3u^2) = 0,$$

which is recognized as a restatement of the original KdV equation. Thus

$$D_1 = u \quad \text{and} \quad F_1 = u_{xx} - 3u^2.$$

With $\beta_2 = u_x$, the corresponding conserved quantity is zero for any $u(x,t)$ that goes to zero as $x \to \pm\infty$. As this tells us nothing of interest about solutions of the KdV equation, we toss it onto a pile marked "trivial."

With $\beta_3 = u^2 - u_{xx}$, it is evident that the u_{xx} term is a trivial addition; thus the next nontrivial conservation law is

$$\frac{\partial u^2}{\partial t} + \frac{\partial}{\partial x}(2uu_{xx} - u_x^2 - 4u^3) = 0,$$

which can be identified as the KdV equation multiplied by $2u$. Therefore

$$D_2 = u^2 \quad \text{and} \quad F_2 = 2uu_{xx} - u_x^2 - 4u^3,$$

which are independent of D_1 and F_1.

With

$$\beta_4 = 3uu_x - u^2 u_x + u_x u_{xx} - u_{xxx},$$

we recognize again a density that is trivial because it can be written as a perfect derivative with respect to x. It turns out that all of the β_n with even subscripts are trivial in this sense.

Proceeding in this manner, Miura et al. [21] have constructed the first 10 independent conservations laws for KdV—with ever increasing algebraic difficulty—and shown that there are a countably infinite number of them.

6.6.2 Conserved densities for matrix scattering

For the two-component scattering problem defined in Equation (6.43), consider the solution (from Equation (6.49)) [31]

$$\phi_1^-(x,\lambda) = e^{-i\lambda x} + \int_{-\infty}^{x} L_1(x,z) e^{-i\lambda z}\, dz,$$

which can be written as

$$\phi_1^-(x,\lambda) = e^{-i\lambda x + \alpha(x,\lambda)}.$$

The scattering equations require that

$$\alpha_{xx} - (2i\lambda + q_x \cdot q)\alpha_x + \alpha_x^2 + \sigma|q|^2 = 0, \qquad (6.86)$$

and Equations (6.48) and (6.59) imply that in the limit $x \to +\infty$

$$a(\lambda)\phi_1^-(x,\lambda) \to e^{-i\lambda x},$$

or

$$a(\lambda) e^{\alpha(x,\lambda)} \to 1.$$

For $|\lambda| \to \infty$, $\alpha_x \to 0$ so

$$\alpha_x(x,\lambda) \sim \sum_{n=1}^{\infty} \frac{\beta_n(x)}{(2i\lambda)^n}.$$

Assuming $\sigma = +1$, Equation (6.86) requires

$$\beta_1 = |q|^2$$
$$\beta_2 = q q_x^*$$
$$\beta_3 = q q_{xx}^* + |q|^4$$
$$\vdots$$
$$\beta_{n+1} = q\left(\frac{\beta_n}{q}\right)_x + \sum_{i+j=n} \beta_i \beta_j.$$

Taking advantage of the fact that $a(\lambda)$ is a constant of the motion implies that as $x \to +\infty$,

$$-\log a(\lambda) \to \sum_{n=1}^{\infty} (2i\lambda)^{-n} \int_{-\infty}^{\infty} \beta_n(x)\, dx,$$

indicating that the β_n are conserved densities. The first three of these give the nontrivial densities

$$D_1 = |q|^2$$
$$D_2 = i(qq_x^* - q^*q_x) \qquad (6.87)$$
$$D_3 = |q_x|^2 - |q|^4,$$

corresponding respectively to the mass, momentum, and energy of a solution to the NLS equation.

6.7 Summary

Over the past two decades, many books on the ISM and its relation to the theory of solitons have appeared [1, 4, 7–12, 17, 20, 22, 23]. The intent of this chapter is to supplement these works by introducing the subject to beginning students of nonlinear phenomena and by smoothing the paths for those who are interested in mastering this material through further study and research. To this end, an effort has been made to clearly present the theoretical structures for ISMs based on the linear Schrödinger operator and on two-component (matrix) scattering operators. Examples and problems have been chosen to display various facets of inverse scattering theory and also to indicate how the theory can be applied to hydrodynamic solitons, magnetic fluxons on long (superconducting) Josephson transmission lines, and dispersive nonlinear wave packets.

6.8 Problems

1. The Schrödinger eigenvalue problem is

$$L\phi(x) = \lambda\phi(x), \quad \text{where } L = -\frac{d^2}{dx^2} + u(x).$$

(a) Show that the operator L is self-adjoint, that is

$$\int_{-\infty}^{\infty} \psi(x)^* L\phi(x)\, dx = \int_{-\infty}^{\infty} [L\psi(x)]^* \phi(x)\, dx,$$

where ϕ and ψ are arbitrary functions that vanish in $x = \pm\infty$. [Hint: integrate by parts.]

(b) Let λ_n be the nth eigenvalue of L with the associated eigenfunction $\phi_n(x)$. Substitute ψ and ϕ by ϕ_n in the above integral and prove that the eigenvalue λ_n is real.

(c) Substitute ψ by ϕ_m and ϕ by ϕ_n in the above integral and use the result from part (b) to prove the orthogonality property of eigenfunctions

$$\int_{-\infty}^{\infty} \phi_m^*(x)\phi_n(x)\, dx = 0 \quad \text{for } \lambda_m \neq \lambda_n.$$

2. The Wronskian W of two arbitrary functions ψ and ϕ is defined by
$$W(\psi;\phi) \equiv \psi(x)\phi'(x) - \psi'(x)\phi(x).$$
Show that $W(\psi;\phi)$ is independent of x if ψ and ϕ are solutions of the Schrödinger equation for a common value of k^2. [Hint: compute dW/dx.]

3. Prove the energy conservation equations (6.7) by evaluating the Wronskians $W(\psi^+;\psi^{+*})$ and $W(\psi^-;\psi^{-*})$ at $x \to \pm\infty$.

4. (a) In the scattering equations (6.5) and (6.6), show that $a = c$ by calculating $W(\psi^+;\psi^-)$ in the limits $x \to \pm\infty$. (Note: this is an example of the reciprocity theorem.)

   (b) By calculating $W(\psi^+;\psi^{-*})$ in the limits $x \to \pm\infty$, show that $|b| = |d|$.

5. Consider the Schrödinger equation (6.3) with a square-well potential u given by
$$u(x) = \begin{cases} -a & \text{for } -\ell \leq x \leq +\ell, \\ 0 & \text{otherwise,} \end{cases}$$
where a is a positive constant and ℓ measures the width of the well. For negative eigenvalues λ in the interval $-a \leq \lambda < 0$, the associated bound state $\phi(x,\lambda)$ can be written in the form
$$\phi(x,\lambda) = \begin{cases} A_1 e^{\kappa_1(x+l)} & \text{for } x \leq -\ell, \\ A_2 e^{i\kappa_2(x+l)} + B_2 e^{-i\kappa_2(x+l)} & \text{for } -\ell \leq x \leq +\ell, \\ B_3 e^{-\kappa_1(x-l)} & \text{for } \ell \leq x. \end{cases}$$

   (a) Determine κ_1 and κ_2 as function of λ.

   (b) Show that bound state eigenvalues are determined from the transcendental equation
$$\tan(2\ell\sqrt{a+\lambda}) = \frac{2\sqrt{-\lambda}\sqrt{a+\lambda}}{2\lambda+a}.$$
   [Hint: Require that $\phi(x)$ and $\phi'(x)$ be continuous at $x=\pm\ell$.]

   (c) Sketch the bound states $\phi(x,\lambda)$.

   (d) For $a = 9$ and $\ell = 1$ use your favorite graphics software to plot the difference
$$\tan(2\ell\sqrt{a+\lambda}) - \frac{2\sqrt{-\lambda}\sqrt{a+\lambda}}{2\lambda+a}$$
   as a function of λ. How many bound states do you find?

6. For $\lambda > 0$, the Schrödinger equation with the square-well potential given in the previous problem possesses scattering solutions of the form
$$\phi(x,\lambda) = \begin{cases} A_1 e^{i\kappa_1(x+\ell)} + B_1 e^{-i\kappa_1(x+\ell)} & \text{for } x \leq -\ell, \\ A_2 e^{i\kappa_2(x+\ell)} + B_2 e^{-i\kappa_2(x+\ell)} & \text{for } -\ell \leq x \leq +\ell, \\ A_3 e^{i\kappa_1(x-\ell)} + B_3 e^{-i\kappa_1(x-\ell)} & \text{for } \ell \leq x. \end{cases}$$

(a) Determine κ_1 and κ_2 as functions of λ.

(b) Choose $A_1 = 0$ and $B_3 = 1$ and find A_3.

(c) How is ϕ related to ψ^+ in the formulas below Equation (6.4)?

(d) Plot the reflection coefficient $|A_3|^2$ as a function of λ using your favorite graphics software.

7. Let $\phi(x)$ be a solution of the Schrödinger eigenvalue problem in Equation (6.3) with the potential [17]

$$u(x) = -a\operatorname{sech}^2(x).$$

(a) Insert the transformation $\phi(x) = A\operatorname{sech}^b(x)\,y(x)$ and show that y satisfies

$$y''(x) - 2b\tanh(x)y'(x) + (a - b - b^2)\operatorname{sech}^2(x)y(x) = 0,$$

after setting $b^2 = -k^2$.

(b) Introduce the transformation $z = \frac{1}{2}[1 - \tanh(x)]$ and $v(z) = y(x)$. Show that $v(z)$ satisfies the hypergeometric differential equation

$$z(1-z)v''(z) + [\gamma - (\alpha + \beta + 1)z]v'(z) - \alpha\beta v(z) = 0,$$

and express α, β, and γ in terms of a and b.

(c) The bounded solution of the hypergeometric differential equation is written as $F(\alpha, \beta; \gamma; z)$ and can be found from the hypergeometric series

$$F(\alpha, \beta; \gamma; z) = 1 + \frac{\alpha\beta}{\gamma}z + \frac{\alpha(\alpha+1)\beta(\beta+1)}{\gamma(\gamma+1)\cdot 1 \cdot 2}z^2 + \cdots.$$

Let $b = -ik$ and use the hypergeometric series to show in the limit $x \to +\infty$ that $\phi(x, k)$ has the form ce^{ikx}. What is c?

(d) The following relation holds:

$$F(\alpha, \beta; \gamma; z) = A_1 F(\alpha, \beta; \alpha + \beta - \gamma + 1; 1 - z)$$
$$+ (1-z)^{\gamma - \alpha - \beta} B_1 F(\gamma - \alpha, \gamma - \beta; \gamma - \alpha - \beta + 1; 1 - z),$$

where

$$A_1 = \frac{\Gamma(\gamma)\Gamma(\gamma - \alpha - \beta)}{\Gamma(\gamma - \alpha)\Gamma(\gamma - \beta)}, \quad \text{and} \quad B_1 = \frac{\Gamma(\gamma)\Gamma(\alpha + \beta - \gamma)}{\Gamma(\alpha)\Gamma(\beta)}.$$

$\Gamma(x)$ is the gamma function. Use this identity to find the limiting behavior of $\phi(x)$ as $x \to -\infty$.

8. In the previous problem, the reflection coefficient for a wave that is incident from the left, as in Equation (6.6), is
$$d(k,a) = A_1/B_1.$$

   (a) Show that this reflection coefficient vanishes for all values of k when
   $$\sqrt{a + \tfrac{1}{4}} = \tfrac{1}{2} + n, \quad n = 0, 1, 2, \ldots,$$
   by using the identity
   $$\Gamma\left(\tfrac{1}{2} - x\right) \Gamma\left(\tfrac{1}{2} + x\right) = \frac{\pi}{\cos(\pi x)}.$$

   (b) Check that $d(k, a)$ has a pole at $k = i$.

   (c) Show that
   $$\lim_{a \to 2} \lim_{k \to i} [(k - i)d(k,a)] \neq \lim_{k \to i} \lim_{a \to 2} [(k - i)d(k,a)].$$

   (d) Demonstrate the result of (c) with appropriate graphical constructions.

9. Assume that the transmission and reflection coefficients have poles of order $m \geq 1$ at $k = i\kappa_n$. The Laurent series for the transmission coefficient $a(k)$ reads
$$a(k) = \frac{a_{-m,n}}{(k - i\kappa_n)^m} + \frac{a_{-(m-1),n}}{(k - i\kappa_n)^{m-1}} + \cdots$$
and for the reflection coefficient, it is
$$b(k) = \frac{b_{-m,n}}{(k - i\kappa_n)^m} + \frac{b_{-(m-1,n)}}{(k - i\kappa_n)^{m-1}} + \cdots.$$

   (a) Prove that the limiting value of $a(k)b(k)/a'(k)$ diverges for $k \to i\kappa_n$ in the case $m \geq 2$. (See Equation (6.14).)

   (b) Show that the limiting value of $a(k)b(k)/a'(k)$ equals $-b_{-1,n}$ for $k \to i\kappa_n$ in the case $m = 1$.

10. Use a phase plane analysis of Equation (6.3) to show that the UHP poles of $a(k)$ are simple.

11. From a sketch on the (x, z)-plane, indicate how one would use the second of Equations (6.27) and the boundary conditions below Equation (6.12) to determine $K(x, z)$ from $u(x)$.

12. Show in detail that the definite integral of Equation (6.29) is equal to the Cauchy integral of Equation (6.30).

13. Discuss the relationship between the derivation of the Gel'fand–Levitan equation in Section 6.1.3 and that presented in Section 2.6.2. What are the relative advantages and disadvantages of the two approaches?

14. How would you answer the questions from the puzzled student at the end of Section 6.1.3?

15. Assuming that
$$B(x+z;0) = 2\kappa_1 e^{-\kappa_1(x+z)} + 2\kappa_2 e^{-\kappa_2(x+z)},$$
solve the Gel'fand–Levitan equation for the corresponding two-soliton solution of KdV.

16. From the second term of $B(x+z;t)$ in Equation (6.37), discuss the behavior of the radiation in the limit of small amplitude.

17. (a) Describe reflectionless scattering for the Schrödinger equation in the (ψ_x, ψ) phase plane.

    (b) Show that $u(x,0) = -2\kappa^2 \operatorname{sech}^2 \kappa x$ is a reflectionless potential.

    (c) Repeat (b) for $u(x,0) = -N(N+1)\operatorname{sech}^2 x$, where N is a positive integer.

    (d) Discuss the "radiation" term in Equation (6.39).

18. (a) Solve the Schrödinger scattering problem for the square-well defined in Equation (6.40).

    (b) Verify the condition for generation of N KdV solitons given in Equation (6.41).

19. From a sketch on the (x,z)-plane, show how one would use Equations (6.56) to determine $K_1(x,z)$ and $K_2(x,z)$ from $q(x)$.

20. Show that the relation between $K_1(x,z)$ and $K_2(x,z)$ given in Equation (6.64) follows from Equation (6.60).

21. Use the concept of "causality in pseudotime" (which was introduced in Chapter 2) to rederive Equation (6.65).

22. Consider a two-component scattering problem of the form
$$\psi_{1,x} = -i\lambda\psi_1 + u(x,t)\psi_2$$
$$\psi_{2,x} = -\psi_1 + i\lambda\psi_2$$
and
$$\psi_{1,x} = A\psi_1 + B\psi_2$$
$$\psi_{2,x} = C\psi_1 - A\psi_2,$$

with

$$A = -4i\lambda^3 + 2iu\lambda - u_x$$
$$B = 4u\lambda^2 + 2iu_x\lambda - 2u^2 - u_{xx}$$
$$C = -4\lambda^2 + 2u.$$

(a) Show that the scattering problem is the same as for the Schrödinger operator.

(b) Show that the cross-derivative condition implies the KdV equation.

23. Show that the two-component ISM presented in Section 6.3.2 reduces to Fourier analysis as the magnitude of $q(x,t)$ becomes small.

24. (a) If ψ is an N-component vector and U and V are $N \times N$ matrices for which

$$\psi_x = U\psi$$
$$\psi_t = V\psi,$$

show that the cross-derivative condition implies

$$U_t - V_x = [V, U].$$

(b) If L is the two-component matrix operator defined in Equation (6.44), how is this cross-derivative condition related to Equation (6.1)? (See reference [12] for a detailed discussion of this formulation.)

25. (a) Solve the scattering problem for the square-well potential defined in Equation (6.77).

(b) Derive the corresponding scattering data at $t = 0$.

26. Assuming that
$$B(\xi + z; 0) = e^{-\kappa(\xi+z)}$$
in the integral equation (6.74), find $u(x,t)$.

27. Repeat the previous problem for
$$B(\xi + z; 0) = 2ae^{-a(\xi+z)} + (2/a)e^{-(\xi+z)/a}.$$

28. For the SG equation (6.70), find an initial profile that will evolve into a breather with frequency $\omega = \sqrt{3}/2$.

29. Discuss the numerical problem of finding the inverse of an $N \times N$ matrix, where $N = 10$, $N = 100$, and $N = 1000$.

30. Use Equation (6.76) to find the locus of the UHP poles corresponding to the SG breather solution of Equation (6.81).

31. Find values of the constants c_1 and c_2 in Equation (6.79) that generate the kink-antikink collision of Equation (6.78).

32. (a) Plot Equation (6.80) as a function of x and t for $-\infty < x < +\infty$ and $0 \le t < +\infty$.

    (b) Show that this function is a limiting case of both the kink-antikink collision of Equation (6.78) and the breather of Equation (6.81).

33. Without going through the details, indicate how to use the integral equation defined in Equation (6.84) to obtain a general N-soliton formula for the NLS equation.

34. (a) Construct a general solution for the NLS equation that corresponds to simple poles of the transmission and reflection coefficients at $\lambda_1 = \xi_1 + i\eta_1$ and $\lambda_2 = \xi_2 + i\eta_2$.

    (b) Discuss the behavior of this solution as $\lambda_1 \to \lambda_2$, so the transmission and reflection coefficients have a double pole.

35. (a) Given the operators [19, 31]

$$L = i \begin{bmatrix} 1-p & 0 & 0 \\ 0 & 1+p & 0 \\ 0 & 0 & 1+p \end{bmatrix} \frac{\partial}{\partial x} + \begin{bmatrix} 0 & u_1 & u_2 \\ u_1^* & 0 & 0 \\ u_2^* & 0 & 0 \end{bmatrix}$$

and

$$M = -p\frac{\partial^2}{\partial x^2} + \begin{bmatrix} -(|u_1|^2 + |u_2|^2)/(1-p) & -iu_{1,x} & -iu_{2,x} \\ iu_{1,x}^* & |u_1|^2/(1+p) & u_2 u_1^*/(1+p) \\ iu_{2,x}^* & u_2^* u_1/(1+p) & |u_2|^2/(1+p) \end{bmatrix},$$

show that the operator equation

$$L_t = i[L, M]$$

implies the coupled NLS equations

$$iu_{1,t} + u_{1,xx} + \left(\frac{2}{1-p^2}\right) u_1(|u_1|^2 + |u_2|^2) = 0$$

$$iu_{2,t} + u_{2,xx} + \left(\frac{2}{1-p^2}\right) u_2(|u_1|^2 + |u_2|^2) = 0.$$

(b) Can this system be extended to higher order?

(c) Suggest an application for such a system.

36. (a) For the coupled system defined in the previous problem, discuss a scattering solution of the form

$$\text{incident wave} = \begin{bmatrix} 1 \\ 0 \\ 0 \end{bmatrix} e^{-i\lambda x},$$

$$\text{reflected wave no. 1} = \begin{bmatrix} 0 \\ 1 \\ 0 \end{bmatrix} b_1(\lambda, 0) e^{+i\lambda x},$$

$$\text{reflected wave no. 2} = \begin{bmatrix} 0 \\ 0 \\ 1 \end{bmatrix} b_2(\lambda, 0) e^{+i\lambda x},$$

as $x \to +\infty$, and

$$\text{transmitted wave} = \begin{bmatrix} 1 \\ 0 \\ 0 \end{bmatrix} a(\lambda) e^{-i\lambda x},$$

as $x \to -\infty$, and similarly for the other channel.

(b) Can you suggest initial conditions that support a bound state? What do the solitons look like?

37. Find the conserved density (D_3) and flow (F_3) for the KdV equation.

38. Repeat the previous problem for D_4 and F_4.

39. Construct the conservation laws corresponding to the densities listed in Equations (6.87) for the NLS equation.

40. (a) For the SG equation in "light-cone" coordinates, $\phi_{\xi\tau} = \phi$, show that the following conservation laws hold [16]:

$$(\phi_\xi^2)_\tau + (2\cos\phi)_\xi = 0$$
$$(\phi_\xi^4 - 4\phi_{\xi\xi}^2)_\tau + (4\phi_\xi^2 \cos\phi)_\xi = 0.$$

(b) Write the corresponding conservation laws in the laboratory (x, t) system.

(c) Derive these laws using the method outlined in Section 6.6.2.

(d) Can you find two more nontrivial conservation laws?

REFERENCES

1. M J Ablowitz and P A Clarkson. *Solitons, Nonlinear Evolution Equations and Inverse Scattering.* Cambridge University Press, Cambridge, 1991.
2. M J Ablowitz, D J Kaup, A C Newell, and H Segur. The initial value solution for the sine–Gordon equation. *Phys. Rev. Lett.* 30 (1973) 1262–1264.
3. M J Ablowitz, D J Kaup, A C Newell, and H Segur. The inverse scattering transform Fourier analysis for nonlinear problems. *Stud. Appl. Math.* 53 (1974) 294–315.
4. M J Ablowitz and H Segur. *Solitons and the Inverse Scattering Transform.* SIAM, Philadelphia, 1981.
5. A Bettini, T A Minelli, and D Pascoli. Solitons in undergraduate laboratory. *Am. J. Phys.* 51 (1983) 977–984.
6. G Boffetta and A R Osborne. Computing of the direct scattering transform for the nonlinear Schrödinger equation. *J. Comput. Phys.* 102 (1992) 252–264.
7. F Calogero and A Degasperis. *Spectral Transform and Solitons.* North–Holland, Amsterdam, 1982.
8. R K Dodd, J C Eilbeck, J D Gibbon, and H C Morris. *Solitons and Nonlinear Wave Equations.* Academic Press, London, 1982.
9. P G Drazin. *Solitons.* Cambridge University Press, Cambridge, 1983.
10. P G Drazin and R S Johnson. *Solitons: An Introduction.* Cambridge University Press, Cambridge, 1989.
11. G Eilenberger. *Solitons: Mathematical Method for Physicists.* Springer-Verlag, Berlin, 1981.
12. L D Faddeev and L A Takhtajan. *Hamiltonian Methods in the Theory of Solitons.* Springer-Verlag, Berlin, 1987.
13. C S Gardner, J M Greene, M D Kruskal, and R M Miura. Method for solving the Korteweg–de Vries equation. *Phys. Rev. Lett.* 19 (1967) 1095–1097.
14. I S Gradshteyn and I M Ryzhik. *Table of Integrals, Series, and Products.* Academic Press, New York, 1980.
15. R Jost. Über die falschen Nullstellen der Eigenwerte der S–Matrix. *Helv. Phys. Acta* 20 (1947) 256–266.
16. G L Lamb, Jr. Analytical descriptions of ultrashort optical pulse propagation in a resonant medium. *Rev. Mod. Phys.* 43 (1971) 99–124.
17. G L Lamb, Jr. *Elements of Soliton Theory.* John Wiley, New York, 1980.
18. L D Landau and E M Lifshitz. *Quantum Mechanics.* Pergamon, Oxford, 1965.
19. S V Manakov. On the theory of two-dimensional stationary self-focusing of electromagnetic waves. *Sov. Phys. JETP* 38 (1974) 248–253.
20. R Meinel, G Neugebauer, and H Steudel. *Solitonen.* Akademie-Verlag, Berlin, 1991.
21. R M Miura, C S Gardner, and M D Kruskal. Korteweg–de Vries equation and generalizations. II. Existence of conservation laws and constants of motion. *J. Math. Phys.* 9 (1968) 1204–1209.
22. A C Newell. *Solitons in Mathematics and Physics.* SIAM, Philadelphia, 1985.
23. S Novikov, S V Manakov, L P Pitaevskii, and V E Zakharov. *Theory of Solitons: The Inverse Scattering Method.* Consultants Bureau, New York, 1984.
24. M Olsen, H Smith, and A C Scott. Solitons in a wave tank. *Am. J. Phys.* 52 (1984) 826–830.

25. E A Overman II, D W McLaughlin, and A R Bishop. Coherence and chaos in the driven damped sine–Gordon equation: Measurement of the soliton spectrum. *Physica D* 19 (1986) 1–41.
26. J Scott Russell. Report on waves. *Br. Ass. Adv. Sci.* Rep. 14 (1844) 311–392.
27. L I Schiff. *Quantum Mechanics*. 3rd edition. McGraw-Hill, New York, 1968.
28. A C Scott, F Y F Chu, and S A Reible. Magnetic flux propagation on a Josephson transmission line. *J. Appl. Phys.* 47 (1976) 3272–3286.
29. L A Takhtajan and L D Faddeev. Essentially nonlinear one-dimensional model of classical field theory. *Theor. Math. Phys.* 21 (1974) 1046–1057.
30. D D Vvedensky. *Partial Differential Equations*. Addison-Wesley, Wokingham, 1993.
31. V E Zakharov and A B Shabat. Exact theory of two-dimensional self-focusing and one-dimensional self-modulation of waves in nonlinear media. *Sov. Phys. JETP* 34 (1972) 62–69.

7
PERTURBATION THEORY

In the previous chapters of this book, we have considered a number of exact solutions for nonlinear dynamic equations. These include the famous solitons, of course, and N-soliton formulas, but other sorts of traveling-waves have been examined, including those on lattices and nonlinear diffusion systems. That such solutions exist is an important fact, involving a shift in paradigm that has overtaken the fields of applied science since the 1970s.

While some of these exact results are directly useful in particular applications, one often finds that the dynamic system of experimental interest is not precisely (say) the sine–Gordon equation or the FitzHugh–Nagumo system. It may differ by terms that represent additional mechanisms for the storage or dissipation of energy or other approximately conserved quantities. How do we proceed in such cases? Do we throw up our hands and say that the system of interest can only be studied numerically? Not at all; we turn to the subject of this chapter.

The term itself is somewhat misleading, for perturbation theory is not a single, unified approach to more complicated problems; rather it is a bag of tricks, guided by a strategic plan. In order to master the theory, one must appreciate many if not most of the tricks and viscerally understand the strategy. The aim of this chapter is to expose the general approach to perturbation theory by presenting some of the examples with which I am familiar.

From a general perspective, the strategy of it goes like this. Suppose we have an equation (linear or nonlinear) that we know how to solve exactly, say

$$E(\phi) = 0,$$

where ϕ is a known exact solution. Assuming that some additional effect is taken into consideration that can be modeled by adding a small term to the right-hand side, our system becomes

$$E(\tilde{\phi}) = \varepsilon R(\tilde{\phi}).$$

In this notation, $\varepsilon R(\tilde{\phi})$ is the perturbation, $\tilde{\phi}$ is the perturbed solution, and ε is a small parameter that allows us (theoretically or numerically) to turn the perturbation on or off.

In many cases it is reasonable to assume that

$$\tilde{\phi} \to \phi \quad \text{as} \quad \varepsilon \to 0.$$

The author thanks M.P. Sørensen for writing the first drafts of Sections 7.2.2, 7.3.3, and 7.4.2.

If ε is small enough, we may be able to estimate the difference $(\tilde{\phi} - \phi)$ between the solution to be found $(\tilde{\phi})$ and the known solution (ϕ) from the study of a linear equation that is tangential to the original equation at ϕ. There are several ways to execute this strategy.

7.1 Perturbed matrices

Let us begin with a simple example: the 2×2 matrix

$$M_0 = \begin{bmatrix} 4 & 2 \\ -3 & -1 \end{bmatrix}.$$

The normalized eigenvectors (ϕ_0) and associated eigenvalues (λ_0) of this matrix satisfy the equation

$$M_0 \phi_0 = \lambda_0 \phi_0, \qquad (7.1)$$

and are found to be

$$\lambda_0^{(1)} = 1, \quad \text{and} \quad \phi_0^{(1)} = \frac{1}{\sqrt{13}} \begin{bmatrix} 2 \\ -3 \end{bmatrix},$$

and

$$\lambda_0^{(2)} = 2, \quad \text{and} \quad \phi_0^{(2)} = \frac{1}{\sqrt{2}} \begin{bmatrix} 1 \\ -1 \end{bmatrix}.$$

Consider next a perturbation of the matrix M_0 to

$$M = \begin{bmatrix} 4 & (2+\varepsilon) \\ -3 & -1 \end{bmatrix}, \qquad (7.2)$$

which can also be solved exactly. The eigenvalues of M are

$$\lambda^{(1)}, \lambda^{(2)} = \tfrac{1}{2} \left[3 \pm \sqrt{1 - 12\varepsilon} \right];$$

thus,

$$\lambda^{(1)} = 1 + 3\varepsilon + \mathrm{O}(\varepsilon^2) \quad \text{and} \quad \lambda^{(2)} = 2 - 3\varepsilon + \mathrm{O}(\varepsilon^2). \qquad (7.3)$$

To obtain the approximations to the eigenvalues given in Equations (7.3) without directly solving the matrix M, note first that Equation (7.2) can be written as

$$M = M_0 + \varepsilon R, \qquad (7.4)$$

where

$$R = \begin{bmatrix} 0 & 1 \\ 0 & 0 \end{bmatrix}.$$

Assuming that the eigenvectors and associated eigenvalues satisfying the equation

$$M\phi = \lambda\phi \qquad (7.5)$$

can be represented as power series in ε, write

$$\lambda = \lambda_0 + \varepsilon\lambda_1 + \varepsilon^2\lambda_2 + \cdots, \tag{7.6}$$

and

$$\boldsymbol{\phi} = \boldsymbol{\phi}_0 + \varepsilon\boldsymbol{\phi}_1 + \varepsilon^2\boldsymbol{\phi}_2 + \cdots. \tag{7.7}$$

For small enough values of ε, these series are expected to converge.

Substituting Equations (7.4), (7.6), and (7.7) into Equation (7.5) leads to an equation with terms in powers of the perturbation parameter ε. As this equation must hold for all values of ε, we require the terms at each power of ε to be equal. Thus:

- Equating constant terms implies

$$[M_0 - \lambda_0]\boldsymbol{\phi}_0 = 0. \tag{7.8}$$

- Equating terms first order in ε implies

$$[M_0 - \lambda_0]\boldsymbol{\phi}_1 = \lambda_1\boldsymbol{\phi}_0 - R\boldsymbol{\phi}_0. \tag{7.9}$$

- Equating terms of second order in ε implies

$$[M_0 - \lambda_0]\boldsymbol{\phi}_2 = \lambda_2\boldsymbol{\phi}_0 + \lambda_1\boldsymbol{\phi}_1 - R\boldsymbol{\phi}_1. \tag{7.10}$$

$$\vdots$$

- Equating terms of nth order in ε implies

$$[M_0 - \lambda_0]\boldsymbol{\phi}_n = \lambda_n\boldsymbol{\phi}_0 + \cdots + \lambda_1\boldsymbol{\phi}_{n-1} - R\boldsymbol{\phi}_{n-1}. \tag{7.11}$$

Evidently Equation (7.8) is identical to Equation (7.1), so everything on the right-hand side of Equation (7.9) is known except λ_1. By the Fredholm theorem for matrices, a necessary and sufficient condition for Equation (7.9) to have a solution is that its right-hand side be orthogonal to all the null vectors of the adjoint matrix

$$[M_0 - \lambda_0]^\dagger = [M_0 - \lambda_0]^T.$$

To satisfy this orthogonality condition, λ_1 must take the value

$$\lambda_1 = \frac{(\boldsymbol{\psi}, R\boldsymbol{\phi}_0)}{(\boldsymbol{\psi}, \boldsymbol{\phi}_0)}, \tag{7.12}$$

where

$$[M_0 - \lambda_0]^T \boldsymbol{\psi} = 0,$$

and $(\cdot, \cdot)$ indicates an inner product.

For $\lambda_0 = 1$, there is only one null vector of $[M_0 - \lambda_0]^T$, and it is

$$\psi = \begin{bmatrix} 1 \\ 1 \end{bmatrix},$$

so from Equation (7.12), $\lambda_1 = +3$. Thus

$$\lambda^{(1)} = 1 + 3\varepsilon + O(\varepsilon^2)$$

in agreement with the first of Equations (7.3).

For $\lambda_0 = 2$,

$$\psi = \begin{bmatrix} 3 \\ 2 \end{bmatrix},$$

and $\lambda_1 = -3$. Thus

$$\lambda^{(2)} = 2 - 3\varepsilon + O(\varepsilon^2)$$

in agreement with the second of Equations (7.3).

In this example, the Fredholm theorem was used to compute the change in eigenvalues of a perturbed matrix to first order in the perturbation without computing and factoring the perturbed determinant. Fredholm's theorem is used in a similar way for many perturbation calculations.

7.2 A damped harmonic oscillator

7.2.1 *Energy analysis*

Consider the equation

$$\frac{d^2x}{dt^2} + \varepsilon \frac{dx}{dt} + x = 0, \qquad (7.13)$$

with the initial conditions $x(0) = 1$ and $\dot{x}(0) = 0$. Using methods sketched in Chapter 2, the exact solution is readily found to be

$$x(t) = e^{-\varepsilon t/2} \left[\cos\left(t\sqrt{1 - \varepsilon^2/4}\right) + \frac{\varepsilon/2}{\sqrt{1 - \varepsilon^2/4}} \sin\left(t\sqrt{1 - \varepsilon^2/4}\right) \right], \qquad (7.14)$$

so—as with simple matrices of the previous section—perturbation theory is not needed to solve this problem, but we use it for practice.

The amplitude of the exact sinusoidal solution in Equation (7.14) is evidently

$$A(t) = \frac{e^{-\varepsilon t/2}}{\sqrt{1 - \varepsilon^2/4}} = e^{-\varepsilon t/2}[1 + O(\varepsilon^2)]. \qquad (7.15)$$

Let's see how this result can be obtained without making reference to the exact solution.

With $\varepsilon = 0$, Equation (7.14) reduces to $x(t) = \cos t$, for which the energy

$$E = (\dot{x}^2 + x^2)/2 = A(0)^2/2 = \tfrac{1}{2} \qquad (7.16)$$

is conserved.

With $0 < \varepsilon \ll 1$, Equation (7.13) implies that the instantaneous value of the time derivative of E is

$$\frac{dE}{dt} \doteq -\varepsilon \dot{x}^2 = A\frac{dA}{dt},$$

where order ε changes to the oscillation waveform have been ignored. Because $\dot{x} = A(t)\sin t$, averaging $-\varepsilon \dot{x}^2$ over a cycle of oscillation implies that the oscillation amplitude $A(t)$ is governed by the first-order ODE

$$\frac{dA}{dt} \doteq -\frac{\varepsilon}{2}A, \tag{7.17}$$

with the initial condition $A(0) = 1$. Integration of Equation (7.17) with this initial condition leads to

$$A(t) = e^{-\varepsilon t/2}[1 + \mathrm{O}(\varepsilon^2)] \tag{7.18}$$

in agreement with Equation (7.15).

Thus a simple energy argument can be used to determine the time dependence of the amplitude of a slightly damped harmonic oscillator without solving the original equation. In this development, no advantage was taken of the fact that Equation (7.13) is linear, so the method is readily extended to the study of anharmonic oscillators and solitons. This idea is the key to the method of averaging, introduced by van der Pol in 1934 [26] and used to analyze quasiharmonic lattices in Section 5.4.1.

7.2.2 *Multiple time scales*

Before undertaking perturbation analyses of soliton equations, let us consider a more systematic perturbation procedure for finding approximate solutions of the damped harmonic oscillator [13, 19–21]. Besides depending on time, a solution $x(t;\varepsilon)$ of Equation (7.13) also depends on the parameter ε, and a series expansion with respect to ε can be written as

$$\begin{aligned} x(t;\varepsilon) &= x(t;0) + \frac{\partial x(t;0)}{\partial \varepsilon}\varepsilon + \frac{1}{2}\frac{\partial^2 x(t;0)}{\partial \varepsilon^2}\varepsilon^2 + \mathrm{O}(\varepsilon^3) \\ &= x_0(t) + \varepsilon x_1(t) + \varepsilon^2 x_2(t) + \mathrm{O}(\varepsilon^3), \end{aligned} \tag{7.19}$$

with the notation

$$x_n(t) = \frac{1}{n!}\frac{\partial^n x(t;0)}{\partial \varepsilon^n}, \quad n = 1, 2, \ldots.$$

Insertion of the expansion (7.19) into (7.13) leads to a polynomial in ε that equals zero; thus each coefficient of ε^n must be zero.

For $\varepsilon \to 0$, the term $\varepsilon x_1(t)$ is a small correction to $x_0(t)$ so at $\mathrm{O}(\varepsilon)$ we get the following system of equations:

$$\frac{d^2 x_0}{dt^2} + x_0 = 0,$$

$$\frac{d^2 x_1}{dt^2} + x_1 = -\frac{dx_0}{dt}.$$

Solving first for x_0 and then with respect to x_1 using the initial conditions $x(0) = 1$ and $\dot{x}(0) = 0$ gives

$$x_0(t) = \cos t,$$
$$x_1(t) = \tfrac{1}{2}\sin t - \tfrac{1}{2} t \cos t.$$

Note that in $x_1(t)$ appears the term $t \cos t$, which grows without bound as t approaches infinity. Such a term is called "secular" because it eventually violates the assumption that x_1 is a small correction to x_0, leaving the approximation valid only for times less than order $1/\varepsilon$.[1]

There are two reasons that the correction term becomes large as time progresses: the amplitude of the perturbed solution decreases slowly due to the factor $e^{-\varepsilon t/2}$, and the perturbed solution oscillates at a frequency slightly different from that of x_0. Because the zero-order solution $x_0(t)$ does not change its amplitude and frequency, the correction term in x_1 tries to keep up with the increasing discrepancy between $x_0(t)$ and the exact solution. For times larger than $\mathrm{O}(1/\varepsilon)$ the approximate solution fails.

We can fix this problem by noting that the amplitude decreases slowly with time and that the frequency $\omega = \sqrt{1 - \varepsilon^2/4}$ is also slightly decreased. Thus slow changes in the amplitude and frequency are allowed by writing an approximate solution in the form

$$x(t) = A(\varepsilon t)\cos[\omega(\varepsilon)t] + \varepsilon B(\varepsilon t)\sin[\omega(\varepsilon)t]. \tag{7.20}$$

Inserting into Equation (7.13), expanding as $\omega(\varepsilon) = 1 + \varepsilon k_1 + \varepsilon^2 k_2$, and ordering powers of ε gives

$$x(t) = e^{-\varepsilon t/2}\cos\left[\left(1 - \tfrac{1}{8}\varepsilon^2\right)t\right] + \tfrac{1}{2}\varepsilon e^{-\varepsilon t/2}\sin\left[\left(1 - \tfrac{1}{8}\varepsilon^2\right)t\right] + \mathrm{O}(\varepsilon^3), \tag{7.21}$$

---

[1] In ordinary English, the adjective secular implies a temporal—rather than spiritual—quality. In the field of celestial mechanics, the term was coined to indicate unstable planetary orbits. (Were such orbits deemed out of tune with the cosmos?) More prosaically the term has been used by applied mathematicians to indicate terms that grow with time.

corresponding to

$$k_1 = 0$$
$$k_2 = -\tfrac{1}{8}$$
$$A(\varepsilon t) = e^{-\varepsilon t/2}$$
$$B(\varepsilon t) = \tfrac{1}{2}e^{-\varepsilon t/2}.$$

This result is better than the first trial, and a surprisingly good approximation to the exact solution even for ε of order 1.

This guess for the form of the approximate solution was guided by the exact solution. Normally we will not know an exact solution and therefore a procedure is needed that does not use a priori knowledge. To this end, note that the amplitude depends on the slow time scale εt. Similarly there is a dependence on $\varepsilon^2 t$ in the argument of the cosine and the sine functions, suggesting scaled time variables of the form

$$T_0 = t, \quad T_1 = \varepsilon t, \quad T_2 = \varepsilon^2 t, \quad \text{etc.} \tag{7.22}$$

In terms of these "stretched variables," a solution of the damped harmonic oscillator can be written as

$$x(t) = x(T_0, T_1, T_2, \ldots).$$

With this formulation, expressions for time derivatives in terms of T_0, T_1, T_2, etc. can be computed, and a hierarchy of ODE systems can be written for powers of ε. Solving each system in succession, leads to increasingly close approximations to the exact solution. The details of this "multiple-scale analysis" are carried out in Appendix G, leading in a methodical way to Equation (7.21).

As the analysis was to order $O(\varepsilon^2)$, stretched time variables up to the same order were employed. In other words, only the variables T_0, T_1, and T_2 were used in order to have a consistent perturbation analysis. If the expansion is to order $O(\varepsilon^n)$, the scaled variables T_i for $i = 0, 1, \ldots, n$ are required.

7.3 Energy analysis of soliton dynamics

In Chapter 3, several classical soliton equations were treated, and a common feature of these systems is that they conserve energy. When we turn our interest toward real world problems, this property is often lost because of dissipative effects and forcing terms; thus the pure soliton equations must be modified by adding terms that describe such perturbations. In such cases, a judicious combination of perturbation theory and numerical analysis proves useful for developing both a qualitative and a quantitative understanding of the solutions.

7.3.1 Korteweg–de Vries solitons

Consider using the energy analysis sketched in the previous section to study the perturbed Korteweg–de Vries (KdV) equation. With a perturbation added to the right-hand side, this equation reads[2]

$$u_t + uu_x + u_{xxx} = \varepsilon R, \tag{7.23}$$

where ε is a small parameter, and εR represents terms that can depend explicitly on x, t, u, and partial derivatives of u.

With $\varepsilon = 0$, a single soliton solution of the pure KdV equation is

$$u(x,t) = 3v\,\text{sech}^2\left(\frac{\sqrt{v}}{2}(x-vt)\right). \tag{7.24}$$

By introducing the function w defined by $w_x \equiv u$, the KdV equation can be derived from the Lagrangian density (see Appendix A)

$$\mathcal{L} = \tfrac{1}{2}w_x w_t + \tfrac{1}{6}w_x^3 - \tfrac{1}{2}w_{xx}^2, \tag{7.25}$$

and the associated Hamiltonian density is

$$\mathcal{H} = w_t\frac{\partial \mathcal{L}}{\partial w_t} - \mathcal{L} = -\tfrac{1}{6}w_x^3 + \tfrac{1}{2}w_{xx}^2, \tag{7.26}$$

representing the energy density of the system.

The underlying idea in the energy approach to soliton dynamics is to consider variations of the total energy due to perturbations that describe loss and power input, assuming that the term in εR is sufficiently small that order ε changes of the single soliton in Equation (7.24) can be neglected. Then the effect of the perturbation is to slowly vary the velocity v with time. For a given perturbation, this assumption should always be checked by numerical computations. (Note that this analysis does not allow for creation and destruction of solitons, and we neglect possible small amplitude oscillatory waves that may emerge from the perturbation.)

For the KdV equation the total energy is

$$H = \int_{-\infty}^{\infty} \mathcal{H}\,dx. \tag{7.27}$$

The time derivative of H is

$$\frac{dH}{dt} = \int_{-\infty}^{\infty}(w_x w_{xx} + w_{xxxx})w_t\,dx = \int_{-\infty}^{\infty}(-w_{xt} + \varepsilon R)w_t\,dx, \tag{7.28}$$

---

[2]Because the factor of "-6" on the uu_x term is of no particular value in the perturbation analysis, a different normalization of the KdV equation is used here.

after integration by parts with the assumption that derivatives of w go to zero as $x \to \pm\infty$. Thus

$$\frac{dH}{dt} = \varepsilon \int_{-\infty}^{\infty} Rw_t \, dx. \tag{7.29}$$

Inserting (7.24) into (7.27) and evaluating the integrals by neglecting the slow time variation of v leads to

$$H = -\frac{36}{5}v^{5/2}. \tag{7.30}$$

Differentiating Equation (7.30) with respect to time, noting $w_t = -vw_x = -vu$, and equating to Equation (7.29) yields the result

$$\frac{dv}{dt} = +\frac{\varepsilon}{18\sqrt{v}} \int_{-\infty}^{\infty} Ru \, dx. \tag{7.31}$$

As an example of the usefulness of this expression, consider a perturbation of the form

$$\varepsilon R = -\varepsilon u, \tag{7.32}$$

representing a gradual increase of depth in the application of KdV to shallow water waves.

From Equation (7.31), the evolution of the soliton velocity is determined by the first-order ODE

$$\frac{dv}{dt} \doteq -\frac{4}{3}\varepsilon v, \tag{7.33}$$

where "$\doteq$" reminds us that order ε changes in the soliton shape have been neglected in the estimate.

Noting from Equation (7.24) that v is proportional to the soliton amplitude, Equation (7.33) implies that this amplitude evolves with time as

$$A(t) \doteq A(0)e^{-4\varepsilon t/3}, \tag{7.34}$$

a result that was obtained in 1970 by Ott and Sudan [23] and confirmed by numerical studies of Knickerbocker and Newell a decade later [15].

While this example seems to validate the use of energy analysis, some questions linger. The only property of the energy that we have used is the fact that it is constant for the unperturbed system, but the KdV equation—being integrable—has an infinite number of such constants of the motion. Would we get the same result by using any of them?

Let's try another, for example

$$Q_2 = \int_{-\infty}^{\infty} u^2 \, dx,$$

which is readily seen to be a constant for u, a solution of the unperturbed KdV equation. Using the same arguments as above, we arrive again at Equation (7.33), implying that the soliton amplitude decays as in Equation (7.34).

Considering
$$Q_1 = \int_{-\infty}^{\infty} u\, dx,$$
however, we find that
$$\frac{dv}{dt} \doteq -2\varepsilon v,$$
implying that
$$A(t) \doteq A(0)e^{-2\varepsilon t} \tag{7.35}$$
in disagreement with both Equation (7.34) and numerical computations [15]. What is going on?

This paradox of the "missing mass" has been considered by several researchers, and its resolution is related to the emergence of a "shelf" of finite length and constant amplitude in the wake of the soliton [2, 8, 10, 15, 16, 22]. Because the height of this shelf is of order ε and its length is proportional to time, it can absorb a difference of order εt between the estimates of Equations (7.34) and (7.35). Calculations based on both H and Q_2 manage to give the correct result because they are not influenced to order ε by the shelf. From a more general perspective, formation of the shelf can be viewed as radiation from the soliton that is induced by the perturbation. In the wake of this example, the reader will recognize that it is prudent to supplement an energy analysis with numerical solutions to see what really transpires.

Finally this perturbation calculation shows that the organization of energy into a soliton does not increase its lifetime under the dissipative perturbation of Equation (7.32). On the contrary, the lifetime of a small-amplitude wave is $1/\varepsilon$, whereas from Equation (7.34) the lifetime of a soliton has decreased by 25% to $3/4\varepsilon$.

7.3.2 Sine–Gordon solitons

For the sine–Gordon (SG) equation consider
$$u_{tt} - u_{xx} + \sin u = \varepsilon R, \tag{7.36}$$
where εR represents a perturbation. At $\varepsilon = 0$, the single kink solution of the pure SG equation reads
$$u(x,t) = 4\arctan\left[\exp\left(\frac{x - x_0 - vt}{\sqrt{1 - v^2}}\right)\right]. \tag{7.37}$$

Although this unperturbed waveform contains two parameters (v and x_0), the corresponding energy does not depend on x_0, which merely locates the kink at $t = 0$. If the perturbation (R) does not depend on x, therefore, the energy analysis sketched in the previous section can be used.

The unperturbed SG equation can be derived from the Lagrangian density
$$\mathcal{L} = \tfrac{1}{2}u_t^2 - \tfrac{1}{2}u_x^2 - (1 - \cos u),$$

with the associated Hamiltonian density (see Appendix A)

$$\mathcal{H} = u_t \frac{\partial \mathcal{L}}{\partial u_t} - \mathcal{L} = \tfrac{1}{2} u_t^2 + \tfrac{1}{2} u_x^2 + (1 - \cos u). \tag{7.38}$$

As for the KdV equation, it is interesting to consider the total energy H given by Equation (7.27). Assuming slow time variation of the velocity v and using Equations (7.27) and (7.36), the time derivative of H becomes

$$\frac{dH}{dt} = \int_{-\infty}^{\infty} (u_{tt} - u_{xx} + \sin u) u_t dx = \varepsilon \int_{-\infty}^{\infty} R u_t dx. \tag{7.39}$$

(Note that for the pure SG equation ($\varepsilon = 0$), the energy is conserved.) Insertion of the single kink solution into Equation (7.27) and neglecting the slow variation of v leads to

$$H = \frac{8}{\sqrt{1-v^2}}. \tag{7.40}$$

Experimentalists are interested in perturbation terms representing loss and energy input of the form [30]

$$\varepsilon R = -\alpha u_t + \beta u_{xxt} - \gamma, \tag{7.41}$$

because such a perturbation accurately describes the dissipation of magnetic flux on the long superconducting Josephson tunnel junction, shown in Figure 7.1.

The term in α represents loss due to tunneling of normal (i.e. not superconducting) electrons through the junction barrier. The βu_{xxt} term is also a loss term resulting from the finite penetration depth of the electromagnetic field into the superconducting films [27]. A spatially uniform bias current is represented by γ.

Using the perturbation in Equation (7.41), the integral in Equation (7.39) for the single kink solution is evaluated, keeping v fixed, thus neglecting the slow time variation of v. The time derivative of H on the left-hand side is then determined from Equation (7.40), allowing slow time variation of the velocity v. Thus a first-order ODE for the soliton speed is [18]

$$\frac{dv}{dt} = \frac{\pi \gamma}{4}(1 - v^2)^{3/2} - \alpha v (1 - v^2) - \frac{\beta}{3} v, \tag{7.42}$$

which has been confirmed by many numerical and experimental studies.

From this equation, we can interpret the effects of the perturbative terms on the kink dynamics. The two loss terms in α and β will slow down a right propagating kink as well as a left propagating kink. A positive external bias term γ drives the kink to the right, and a negative bias drives the kink to the left. If the above calculation is carried through for the antikink, a minus sign must be added to γ in Equation (7.42). Thus a positive bias will drive the antikink to the left, and a negative bias will drive the antikink to the right.

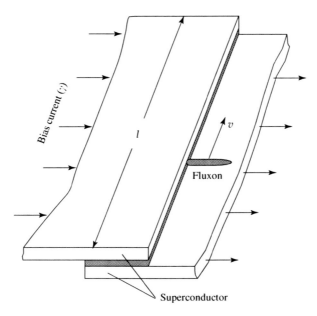

FIG. 7.1. *A (superconducting) Josephson junction of length l with a fluxon (kink) traveling in the x-direction at velocity v.*

On a sufficiently long junction, Equation (7.42) possesses an equilibrium solution v_∞ for $t \to \infty$, determined from the condition $\dot{v} = 0$. With $\beta = 0$, this equation can be solved explicitly as

$$v_\infty = \left[1 + (4\alpha/\pi\gamma)^2\right]^{-1/2}. \tag{7.43}$$

This stationary velocity expresses a balance of power between the dissipative effects and the power input from the bias (γ) term. It is customary to call v_∞ the "power balance velocity."

In applications, the space dimension is often finite and boundary conditions must be imposed. An important case is the overlap geometry shown in Figure 7.1, with open circuit boundary conditions:

$$u_x(0, t) = u_x(l, t) = 0.$$

When a kink is reflected from an open-circuit boundary, it becomes an antikink; thus the system shown in Figure 7.1 can be employed as a fluxon oscillator. In such an oscillator, the sequence of dynamics is as follows: (i) a kink travels to the right, gathering energy from γ, (ii) it is reflected at an open-circuit end as an antikink, (iii) the antikink travels to the left, again gathering energy from γ, (iv) it is reflected as a kink at the other open-circuit end, (v) and so on.

To zero order in the perturbations (i.e. assuming α, β, and γ are zero), Equation (3.80) describes such an oscillation in terms of elliptic functions, providing the basis for an energy analysis [5]. Then the output power (P_out) equals the bias input power minus dissipation. Using Equations (7.39) and (7.41), this power-balance condition is

$$P_\text{out} = \left| \gamma \int_0^l u_t\, dx \right| - \int_0^l (\alpha u_t^2 + \beta u_{xt}^2)\, dx. \qquad (7.44)$$

Under this formulation, output power is obtained by allowing energy to radiate from one end of the device, and the amplitude (A) in Equation (3.80) adjusts itself so that Equation (7.44) is satisfied. As γ can be a direct current, we have a device for transforming energy from a battery to oscillations at frequencies approaching the terahertz range.

7.3.3 Nonlinear Schrödinger solitons

Perturbations play an important role in applications of the nonlinear Schrödinger (NLS) equation and must be included for a complete description of a physical system; thus a perturbed NLS equation is written as

$$iu_t + u_{xx} + 2|u|^2 u = \varepsilon R. \qquad (7.45)$$

With $\varepsilon = 0$, recall from Equation (3.106) that an exact solution is the single soliton

$$u(x,t) = a\,\text{sech}[a(x - v_e t - x_0)] \exp\left[i\tfrac{1}{2}v_e x + \left(a^2 - \tfrac{1}{4}v_e^2\right) t - i\sigma_0\right],$$

where a is the amplitude and v_e is the envelope velocity. Also, x_0 and σ_0 are constants that center the envelope and carrier respectively. For subsequent analysis, it is convenient to define $\xi \equiv v_e/2$ and write the soliton as

$$u(x,t) = a\,\text{sech}(a\theta) \exp(i\xi\theta + i\sigma), \qquad (7.46)$$

with

$$\theta = x - 2\xi t - x_0 \quad \text{and} \quad \sigma = (a^2 + \xi^2)t - \sigma_0 + \xi x_0.$$

Note that this expression possesses four independent parameters, namely a, ξ, x_0, and σ_0, but x_0 and σ_0 merely locate the soliton and set the phase of its carrier, so they do not enter in calculations of energy.

A Lagrangian density for the unperturbed NLS equation is

$$\mathcal{L} = \frac{i}{2}(u^* u_t - u_t^* u) - |u_x|^2 + |u|^4, \qquad (7.47)$$

where u^* is the complex conjugate of u. The variables u and u^* are independent because u comprises two independent real functions, namely its real part and its imaginary part. Variation of $\mathcal{L}$ with respect to u^* gives the pure NLS equation and variation with respect to u gives the complex conjugate equation.

The momentum densities are

$$\pi = \frac{\partial \mathcal{L}}{\partial u_t} = +\frac{i}{2}u^*$$

$$\pi^* = \frac{\partial \mathcal{L}}{\partial u_t^*} = -\frac{i}{2}u,$$

so the Hamiltonian density is

$$\mathcal{H} = u_t \pi + u_t^* \pi^* - \mathcal{L} = |u_x|^2 - |u|^4$$

and the total energy is

$$H = \int_{-\infty}^{\infty} \mathcal{H}\, dx. \tag{7.48}$$

Using the perturbed NLS equation in Equation (7.45) we find

$$\frac{dH}{dt} = -\int_{-\infty}^{\infty} (u_{xx} + 2|u|^2 u) u_t^* \, dx + \text{cc}$$

$$= -\varepsilon \int_{-\infty}^{\infty} R u_t^* \, dx + \text{cc}, \tag{7.49}$$

where "cc" stands for the complex conjugate of the preceding terms.

For the single soliton solution in Equation (7.46), the Hamiltonian in Equation (7.48) becomes

$$H = 2a\xi^2 - \tfrac{2}{3}a^3, \tag{7.50}$$

so the time derivative of the energy is also

$$\frac{dH}{dt} = 2(\xi^2 - a^2)\frac{da}{dt} + 4a\xi\frac{d\xi}{dt}. \tag{7.51}$$

At this point, a difficulty with the energy method arises from the fact that Equation (7.51) contains two time derivatives (a_t and ξ_t); thus we are unable to determine both of them from the single condition of Equation (7.49). Setting the envelope velocity of the soliton to zero by assuming $\xi = 0$ and choosing perturbations that do not change the envelope velocity, however, gives

$$\frac{dH}{dt} = -2a^2 \frac{da}{dt} = -\varepsilon \int_{-\infty}^{\infty} R u_t^* \, dx + \text{cc}. \tag{7.52}$$

As an example, consider the perturbation

$$R = -iu,$$

which seems unlikely to move the soliton (why?) and implies that a small amplitude sinusoidal wave solution of the NLS equation will decay as $\exp(-\varepsilon t)$. From Equation (7.52), the amplitude of a stationary soliton is governed by

$$\frac{da}{dt} = -2\varepsilon a,$$

requiring that

$$a(t) = a(0)e^{-2\varepsilon t}.$$

Thus a simple dissipative perturbation damps the amplitude of a stationary NLS soliton at twice the rate of a small amplitude sinusoid.

7.4 More general soliton analyses

In the present section, some of the disadvantages of the energy approach are resolved by considering two examples: a multiple scale perturbation analysis of SG solitons and an averaged Lagrangian analysis of NLS solitons. These methods are not limited to the study of soliton systems but can be applied to other nonlinear PDEs possessing solitary wave solutions.

7.4.1 Multiple scale analysis of an SG kink

As an example of a multiple scale analysis, consider the perturbed SG equation

$$u_{tt} - u_{xx} + \sin u = \varepsilon R.$$

The idea of multiple scale analysis is to temporarily increase the number of independent variables, allowing separate consideration of fast variables

$$T_0 = t \quad \text{and} \quad X_0 = x$$

(corresponding to those of the unperturbed solution) and slow variables

$$T_1 = \varepsilon t \quad \text{and} \quad X_1 = \varepsilon x.$$

To this end a solution to the perturbed problem is written as

$$u(x,t) = u(X_0, T_0, X_1, T_1),$$

where X_0 and T_0 directly replace x and t in the unperturbed solution. The variables X_1 and T_1 introduce slow (order ε) variations in the parameters of the unperturbed system. With this formulation, the partial derivatives transform as

$$\frac{\partial}{\partial t} \to \frac{\partial}{\partial T_0} + \varepsilon \frac{\partial}{\partial T_1}$$
$$\frac{\partial}{\partial x} \to \frac{\partial}{\partial X_0} + \varepsilon \frac{\partial}{\partial X_1}.$$
(7.53)

The added flexibility in the variables allows adjustment of the slow variations in the parameters to keep the perturbed solution bounded.

For a localized traveling-wave, slow changes with respect to space and time are interchangeable; thus it is convenient to assume that

$$u(x,t) = u(X_0, T_0, T_1).$$

Furthermore, the standard form of the SG equation includes a term of the form $\partial^2 u/\partial t^2$, which would lead to a contribution from the second derivative with respect to the slow time variable T_1. To avoid expanding the order of analysis, it is convenient to write the perturbed SG equation as the first-order system

$$\frac{\partial}{\partial t}\begin{bmatrix} u \\ u_t \end{bmatrix} + \begin{bmatrix} 0 & -1 \\ -\partial_{xx} + \sin(\cdot) & 0 \end{bmatrix}\begin{bmatrix} u \\ u_t \end{bmatrix} = \varepsilon \begin{bmatrix} 0 \\ R \end{bmatrix}. \tag{7.54}$$

Assuming

$$u = u_0 + \varepsilon u_1$$

(the sum of a soliton wave with slowly varying parameters and a dressing term that modifies the wave shape), introducing the fast and slow variables, and separating powers of ε, yields two equations:

$$u_{0,T_0 T_0} - u_{0,X_0 X_0} + \sin u_0 = 0 \tag{7.55}$$

and

$$\frac{\partial}{\partial T_0}\begin{bmatrix} u_1 \\ u_{1,T_0} \end{bmatrix} + \begin{bmatrix} 0 & -1 \\ (-\partial_{X_0 X_0} + \cos u_0) & 0 \end{bmatrix}\begin{bmatrix} u_1 \\ u_{1,T_0} \end{bmatrix} = \begin{bmatrix} -u_{0,T_1} \\ R - u_{0,T_0 T_1} \end{bmatrix}. \tag{7.56}$$

The analysis has been performed to order $O(\varepsilon)$. If more terms are to be included, additional scaled variables must be introduced. Thus to make the expansion to order $O(\varepsilon^2)$, the scaled variable $T_2 = \varepsilon^2 t$ must be employed in order to make the calculation consistent.

From Equation (7.55), we see that u_0 solves the pure SG equation as expected, but the parameters entering the solution u_0 depend on T_1. It is this slow (T_1) dependence of the parameters of u_0 that introduces the extra terms ($-u_{0,T_1}$ and $-u_{0,T_0 T_1}$) on the right-hand side of Equation (7.56).

As a simple example, consider a single kink (or antikink)

$$u_0 = \pm 4 \arctan\left[\exp\left(\frac{X_0 - X(T_0, T_1)}{\sqrt{1 - v^2(T_1)}}\right)\right], \tag{7.57}$$

where

$$\frac{\partial X}{\partial T_0} = v(T_1),$$

and $\partial X/\partial T_1$ is to be determined from the analysis.

In Equation (7.56), the dressing term u_1 solves a linear equation with both a parametric excitation $\cos(u_0)u_1$ and a driving term on the right-hand side. The parametric excitation results from the dynamics of the first-order solution u_0 and the driving term results from the perturbative terms in R together with the dynamics of u_0. Let us write the equation for u_1 in the form

$$L \begin{bmatrix} u_1 \\ u_{1,T_0} \end{bmatrix} = \mathcal{F},$$

where

$$L \equiv \begin{bmatrix} \partial_{T_0} & -1 \\ (-\partial_{X_0 X_0} + \cos u_0) & \partial_{T_0} \end{bmatrix} \quad (7.58)$$

and

$$\mathcal{F} = \begin{bmatrix} -u_{0,v}\dot{v} - u_{0,X}\dot{X} \\ R - u_{0,T_0 v}\dot{v} - u_{0,T_0 X}\dot{X} \end{bmatrix}. \quad (7.59)$$

In the expression for $\mathcal{F}$,

$$\dot{v} \equiv \frac{\partial v}{\partial T_1} \quad \text{and} \quad \dot{X} \equiv \frac{\partial X}{\partial T_1},$$

and

$$u_{0,T_1} = u_{0,v}\dot{v} + u_{0,X}\dot{X}$$
$$u_{0,T_0 T_1} = u_{0,T_0 v}\dot{v} + u_{0,T_0 X}\dot{X}.$$

From Fredholm's theorem, the equation for u_1 has a unique solution only if solutions ψ to the equation $L^\dagger \psi = 0$ are orthogonal to $\mathcal{F}$, where $L^\dagger$ is the adjoint operator of L.

To determine the adjoint operator, consider vector functions

$$\mathbf{w} = \begin{bmatrix} w_1 \\ w_2 \end{bmatrix} \quad \text{and} \quad \mathbf{u} = \begin{bmatrix} u_1 \\ u_2 \end{bmatrix}$$

that are: (i) defined in the region $-\infty < X_0 < +\infty$ and $-\infty < T_0 < +\infty$, (ii) twice differentiable, and (iii) vanish as X_0 and $T_0 \to \pm\infty$. Under these conditions, the adjoint operator $L^\dagger$ is defined by

$$\int_{-\infty}^{\infty}\int_{-\infty}^{\infty} \mathbf{w}^T L \mathbf{u}\, dX_0 dT_0 = \int_{-\infty}^{\infty}\int_{-\infty}^{\infty} (L^\dagger \mathbf{w})^T \mathbf{u}\, dX_0 dT_0. \quad (7.60)$$

Integration by parts in this definition leads to

$$L^\dagger = \begin{bmatrix} -\partial_{T_0} & (-\partial_{X_0 X_0} + \cos u_0) \\ -1 & -\partial_{T_0} \end{bmatrix}, \quad (7.61)$$

from which it is seen that the equation

$$L^\dagger \psi = 0$$

has two solutions:

$$\psi_1 = \begin{bmatrix} u_{0,T_0 T_0} \\ -u_{0,T_0} \end{bmatrix} \quad \text{and} \quad \psi_2 = \begin{bmatrix} u_{0,T_0 v} \\ -u_{0,v} \end{bmatrix}.$$

Requiring that ψ_1 be orthogonal to $\mathcal{F}$ implies

$$\left[\int_{-\infty}^{\infty} (u_{0,T_0 v} u_{0,T_0} - u_{0,v} u_{0,T_0 T_0}) dX_0 \right] \frac{dv}{dT_1}$$
$$+ \left[\int_{-\infty}^{\infty} (u_{0,T_0 X} u_{0,T_0} - u_{0,X} u_{0,T_0 T_0}) dX_0 \right] \frac{\partial X}{\partial T_1}$$
$$= \int_{-\infty}^{\infty} R u_{0,T_0} dX_0. \tag{7.62}$$

If u_0 is a single kink, u_{0,T_0} and $u_{0,X}$ are even functions of $X_0 - X$, whereas $u_{0,T_0 X}$ and $u_{0,T_0 T_0}$ are odd functions. Thus the coefficient of $\partial X/\partial T_1$ vanishes, and (on the substitution $u_{0,T_0 T_0} = u_{0,X_0 X_0} - \sin u_0$ and an integration by parts) the first orthogonality condition reduces to

$$\left[\frac{d}{dv} \int_{-\infty}^{\infty} \left(\tfrac{1}{2} u_{0,T_0}^2 + \tfrac{1}{2} u_{0,X_0}^2 + 1 - \cos u_0 \right) dX_0 \right] \frac{dv}{dT_1} = \int_{-\infty}^{\infty} R u_{0,T_0} dX_0. \tag{7.63}$$

This is just the energy equation

$$\frac{dH}{dt} = \varepsilon \int_{-\infty}^{\infty} R u_t dx,$$

obtained less formally in Section 7.3.2.

On requiring that ψ_2 be orthogonal to $\mathcal{F}$, one notices that the coefficient of dv/dT_1 vanishes identically, leaving the condition

$$\left[\int_{-\infty}^{\infty} (u_{0,T_0 X} u_{0,v} - u_{0,X} u_{0,T_0 v}) dX_0 \right] \frac{\partial X}{\partial T_1} = \int_{-\infty}^{\infty} R u_{0,v} dX_0,$$

which (on noting that $u_{0,X} = -u_{0,X_0}$ and integrating by parts) can be written as

$$\left[\frac{d}{dv} \int_{-\infty}^{\infty} u_{0,X_0} u_{0,T_0} dX_0 \right] \frac{\partial X}{\partial T_1} = \int_{-\infty}^{\infty} R u_{0,v} dX_0. \tag{7.64}$$

From Equation (7.57), $u_{0,v}$ is an odd function of X_0-X. If the perturbation—R—is an even function of $X_0 - X$, then the right-hand side of Equation (7.64) reduces to zero, implying that $\partial X/\partial T_1 = 0$. In this case, our result is equivalent to the energy analysis of Section 7.3.2.

For the kink or antikink waveforms indicated in Equation (7.57), Equations (7.63) and (7.64) take the form

$$\frac{dv}{dT_1} = \pm \tfrac{1}{4}(1-v^2)^{3/2} I_1 \qquad (7.65)$$

and

$$\frac{\partial X}{\partial T_1} = -\tfrac{1}{4}v(1-v^2) I_2, \qquad (7.66)$$

where

$$I_1 \equiv \int_{-\infty}^{\infty} R[u_0(\Theta)]\operatorname{sech}\Theta\, d\Theta,$$

$$I_2 \equiv \int_{-\infty}^{\infty} R[u_0(\Theta)]\Theta\operatorname{sech}\Theta\, d\Theta,$$

$$u_0(\Theta) = \pm 4\arctan\left(e^{\Theta}\right),$$

and

$$\Theta \equiv \frac{X_0 - X(T_0, T_1)}{\sqrt{1 - v^2(T_1)}}.$$

Equations (7.65) and (7.66) give the slow time variations of $v(T_1)$ and $X(T_0, T_1)$. Returning to the original independent variables (x and t), they can be written

$$\begin{aligned}\frac{dv}{dt} &= \pm \tfrac{1}{4}\varepsilon(1-v^2)^{3/2} I_1 \\ \frac{dX}{dt} &= v(t) - \tfrac{1}{4}\varepsilon v(1-v^2) I_2.\end{aligned} \qquad (7.67)$$

The first of these equations can be integrated to obtain $v(t)$, and the second then implies

$$X(t) = \int_0^t v(t')\,dt' + x_0(t),$$

where

$$\frac{dx_0}{dt} = -\tfrac{1}{4}\varepsilon v(1-v^2) I_2. \qquad (7.68)$$

In the unperturbed problem, x_0 is the location of the kink at $t = 0$. With $\varepsilon \neq 0$, this becomes a virtual location, moving at order ε and contributing to the motion of the kink in the final result

$$u_0(x,t) = 4\arctan\left[\exp\left(\frac{x - \int_0^t v(t')\,dt' - x_0(t)}{\sqrt{1-v^2(t)}}\right)\right], \qquad (7.69)$$

where $v(t)$ is obtained from integration of the first of Equations (7.67) and $x_0(t)$ from Equation (7.68).

In general, the integrals I_1 and I_2 can be functions of both time and the location of the kink, but for the perturbation of Equation (7.41), $R[u_0(\Theta)]$ is an even function of Θ. Thus $I_2 = 0$ and x_0 is a constant, in agreement with the energy analysis of Section 7.3.2.

7.4.2 Variational analysis of an NLS soliton

As a second example of a more general solitary wave analysis, a variational technique is used to determine the slow evolution with x and t of the parameters describing an NLS soliton solution under the influence of perturbative terms. These parameters are called collective coordinates, and the idea is to insert a soliton solution into the Lagrangian density, evaluate an averaged Lagrangian function by keeping the collective coordinates fixed during integration, and finally vary the resulting averaged Lagrangian with respect to the collective coordinates [3]. This strategy is related to that of Whitham's method, considered in Chapter 3.

Although perturbations may destroy the Hamiltonian property of the pure soliton equations, dissipative effects and external nonconservative forces can be accounted for in the variation of a Lagrangian function by introducing generalized forces associated with the generalized coordinates. In the following, this procedure is employed by introducing a generalized force for each collective coordinate as in classical mechanics.

To illustrate the method, consider the perturbed NLS equation,

$$iu_t + u_{xx} + 2|u|^2 u = \varepsilon R,$$

in the context of the single soliton solution

$$u(x,t) = a \operatorname{sech}(a\theta) \exp(\mathrm{i}\xi\theta + \mathrm{i}\sigma), \qquad (7.70)$$

where

$$\theta \equiv x - \alpha.$$

In all, there are four independent parameters (or collective coordinates) in this expression, of which a and ξ are constant when $u(x,t)$ is a solution of the unperturbed equation.

The unperturbed NLS equation (and its complex conjugate) can be derived from the Lagrangian density

$$\mathcal{L} = \frac{\mathrm{i}}{2}(u^* u_t - u_t^* u) - |u_x|^2 + |u|^4,$$

from which a total Lagrangian function

$$L = \int_{-\infty}^{\infty} \mathcal{L} \, dx \qquad (7.71)$$

can be defined. With this formulation, it is straightforward to show that
$$L = \tfrac{2}{3}a^3 - 2a\xi^2 + 2a\xi\, d\alpha/dt - 2a\, d\sigma/dt, \tag{7.72}$$
and by varying this expression for L with respect to a, ξ, α, and σ, one finds that
$$da/dt = 0, \quad d\xi/dt = 0,$$
$$d\alpha/dt = 2\xi, \quad d\sigma/dt = a^2 + \xi^2.$$

With a perturbation turned on, these collective coordinates become more interesting functions of time. It is typographically convenient to represent them as
$$[a, \xi, \alpha, \sigma] = [y_1(t), y_2(t), y_3(t), y_4(t)], \tag{7.73}$$
so that
$$L = \int_{-\infty}^{\infty} \mathcal{L}(u, u^*, u_t, u_t^*, u_x, u_x^*)\, dx$$
$$= L(y_i, y_{i,t}), \quad i = 1,2,3,4.$$

On substituting this latter expression into Equation (A.2), it is necessary to evaluate
$$\frac{\partial L}{\partial y_i} - \frac{d}{dt}\left(\frac{\partial L}{\partial y_{i,t}}\right).$$

Let us consider these two terms separately. Interchanging the order of integration and differentiation, the first becomes
$$\frac{\partial L}{\partial y_i} = \frac{\partial}{\partial y_i} \int_{-\infty}^{\infty} \mathcal{L}\, dx$$
$$= \int_{-\infty}^{\infty} \left[\frac{\partial \mathcal{L}}{\partial u}\frac{\partial u}{\partial y_i} + \frac{\partial \mathcal{L}}{\partial u_t}\frac{\partial u_t}{\partial y_i} + \frac{\partial \mathcal{L}}{\partial u_x}\frac{\partial u_x}{\partial y_i}\right] dx + \text{cc}$$
$$= \int_{-\infty}^{\infty} \left[\frac{\partial \mathcal{L}}{\partial u}\frac{\partial u}{\partial y_i} + \frac{\partial \mathcal{L}}{\partial u_t}\frac{\partial u_t}{\partial y_i} - \frac{\partial}{\partial x}\left(\frac{\partial \mathcal{L}}{\partial u_x}\right)\frac{\partial u}{\partial y_i}\right] dx + \text{cc},$$
where an integration by parts on x has taken advantage of the fact that u and its derivatives go to zero as $x \to \pm\infty$.

In evaluating the second term, note that L depends on $y_{i,t}$ only through the u_t and u_t^* terms in $\mathcal{L}$; thus
$$\frac{d}{dt}\left(\frac{\partial L}{\partial y_{i,t}}\right) = \int_{-\infty}^{\infty} \frac{\partial}{\partial t}\left[\frac{\partial \mathcal{L}}{\partial u_t}\frac{\partial u_t}{\partial y_{i,t}}\right] dx + \text{cc}$$
$$= \int_{-\infty}^{\infty} \frac{\partial}{\partial t}\left[\frac{\partial \mathcal{L}}{\partial u_t}\frac{\partial u}{\partial y_i}\right] dx + \text{cc}$$
$$= \int_{-\infty}^{\infty} \left[\frac{\partial}{\partial t}\left(\frac{\partial \mathcal{L}}{\partial u_t}\right)\frac{\partial u}{\partial y_i} + \frac{\partial \mathcal{L}}{\partial u_t}\frac{\partial u_t}{\partial y_i}\right] dx + \text{cc}.$$

and
$$\frac{\partial L}{\partial y_i} - \frac{d}{dt}\left(\frac{\partial L}{\partial y_{i,t}}\right) = \int_{-\infty}^{\infty}\left[\frac{\partial \mathcal{L}}{\partial u} - \frac{\partial}{\partial t}\left(\frac{\partial \mathcal{L}}{\partial u_t}\right) - \frac{\partial}{\partial x}\left(\frac{\partial \mathcal{L}}{\partial u_x}\right)\right]\frac{\partial u}{\partial y_i}dx + \text{cc}.$$

As the term in the square brackets is recognized as
$$[\cdot] = -iu_t^* + u_{xx}^* + 2|u|^2 u^* = \varepsilon R^*,$$

we obtain a perturbed Lagrange–Euler equation
$$\frac{\partial L}{\partial y_i} - \frac{d}{dt}\left(\frac{\partial L}{\partial y_{i,t}}\right) = \varepsilon \int_{-\infty}^{\infty} R\frac{\partial u^*}{\partial y_i}dx + \text{cc}. \tag{7.74}$$

Equation (7.74) is a useful result. It provides ODEs for the time dependence of the collective coordinates $(a, \xi, \alpha, \text{and } \sigma)$ under the influence of the generalized forces that are introduced by the perturbation. Because the analysis neglects changes in the shape of the soliton induced by the perturbation, these ODEs are correct to $O(\varepsilon)$.

As a first example, consider the perturbation to be
$$R = -iu(x,t).$$

Then a straightforward evaluation of Equations (7.74) yields the ODEs
$$da/dt \doteq -2\varepsilon a, \quad d\xi/dt \doteq 0,$$
$$d\alpha/dt = 2\xi, \quad d\sigma/dt = a^2 + \xi^2.$$

where "$\doteq$" implies that corrections of order ε^2 have been neglected.

Consider next the perturbation
$$\varepsilon R = -i\Gamma u + \frac{ig_0 u}{1 + p/p_s}, \tag{7.75}$$

describing light propagation through optical fiber amplifiers that are used for communication systems. In this application, u is the electric field, $-i\Gamma u$ is a linear loss term, and the term in g_0 represents gain, including intensity saturation at large values of a power p, defined by
$$p = \int_{-\infty}^{\infty} |u|^2 dx.$$

The parameter p_s is a saturation power level that is characteristic for a given fiber amplifier.

For this perturbation, $\varepsilon \sim \Gamma$ and the dynamical equations become
$$\frac{da}{dt} \doteq -2\Gamma a + \frac{2g_0 a}{1 + 2a^2/p_s}, \quad \frac{d\xi}{dt} \doteq 0$$
$$\frac{d\alpha}{dt} \doteq 2\xi, \quad \frac{d\beta}{dt} \doteq a^2 + \xi^2. \tag{7.76}$$

For $g_0 = 0$, the above differential equations are solved as

$$a = a_0 e^{-2\Gamma t}$$

$$\xi = \xi_0 \text{ (a constant)},$$

where a_0 is the initial amplitude of the soliton. From these equations, two facts emerge:

- The amplitude of a moving NLS soliton is damped at twice the rate of a low amplitude sinusoid.
- The perturbation defined in Equation (7.75) does not influence the envelope velocity of the soliton to first-order in the damping rate.

For $g_0 > 0$, the term $-2\Gamma a$ damps the light pulse as it travels along the fiber and the term $+2g_0 a/(1+2a^2/p_s)$ amplifies it. Notice that as the amplitude increases the gain decreases, leading to gain saturation. Thus, one can define a power balance condition by requiring $a_t = 0$, implying an equilibrium amplitude a_∞ given as

$$a_\infty = \sqrt{\frac{(g_0 - \Gamma)p_s}{2\Gamma}}. \tag{7.77}$$

From this equation it is seen that g_0 must be larger than Γ for a finite light pulse to be sustained in the fiber amplifier.

7.5 Multisoliton perturbation theory

One of the more remarkable results of soliton theory has been the discovery of multisoliton formulas, which are exact solutions for a variety of nonlinear wave equations with physical applications. Given such an exact solution, it is interesting to consider how it might be altered by the addition of a small perturbing term. At least five such effects are to be expected: (i) changes in the speeds of the solitons, (ii) changes in their relative phases, (iii) changes in the shapes of the solitons, (iv) generation of radiation, and (v) creation or destruction of solitons. The aim of this section is to sketch how to calculate these phenomena.

One approach to multisoliton perturbation theory is to allow for slow variations of the scattering parameters of the inverse scattering theory, which was discussed in the previous chapter. While feasible [8, 10], this approach requires that the analyst be familiar with the details of the inverse scattering method, and many are not. Thus Keener and McLaughlin have shown how soliton perturbation analysis can be couched directly in terms of the soliton speeds and phases, and the amplitudes of the associated radiation field [11, 12]. From this perspective, the situation is as follows:

1. If one is interested in calculating order ε modulations of the soliton speeds and their relative phases, only a multisoliton formula is needed to carry through the analysis. This theory is discussed in the present section.

2. From the perspective of perturbation theory, the creation or destruction of solitons is a difficult subject that has not yet been satisfactorily resolved. The decay of an SG kink and antikink into a breather is one case that can be treated, and we relate this example to multisoliton perturbation theory.
3. To calculate radiation from one or more solitons or order ε changes in their shapes (which can be viewed as a "near field" that moves with the speed of the soliton), it is necessary to make use of additional information that is not contained in a pure multisoliton formula. Although one may obtain this additional information from inverse transform theory, it is more readily derived from a Bäcklund transform. In Appendix H, such a development is sketched.

7.5.1 *General theory*

To focus the discussion, consider the perturbed SG equation given in Equation (7.36) [18] and written here as the first-order system

$$\partial_t \begin{bmatrix} u \\ u_t \end{bmatrix} + \begin{bmatrix} 0 & -1 \\ -\partial_{xx} + \sin(\cdot) & 0 \end{bmatrix} \begin{bmatrix} u \\ u_t \end{bmatrix} = \varepsilon \begin{bmatrix} 0 \\ R \end{bmatrix}, \qquad (7.78)$$

where $0 < \varepsilon \ll 1$, and recall that the perturbation—R—might depend on $u(x,t)$, its time or space derivatives, and explicitly on x and t.

Introducing the vector notation

$$\mathbf{W} \equiv \begin{bmatrix} u \\ u_t \end{bmatrix},$$

Equation (7.78) takes the compact form

$$\frac{\partial}{\partial t}\mathbf{W} + N(\mathbf{W}) = \varepsilon \mathbf{f}, \qquad (7.79)$$

where

$$\mathbf{f} \equiv \begin{bmatrix} 0 \\ R \end{bmatrix}$$

and the nonlinear operator $N(\cdot)$ is defined as

$$N(\mathbf{W}) \equiv \begin{bmatrix} 0 & -1 \\ -\partial_{xx} + \sin(\cdot) & 0 \end{bmatrix}.$$

When $\varepsilon = 0$, a pure multisoliton wave is of the general form

$$\mathbf{W}_{(N+M+2L)} = \mathbf{W}(x, v_1 t + x_1, \ldots, v_{(N+M+2L)} t + x_{(N+M+2L)}; \mathbf{v}), \qquad (7.80)$$

where $\mathbf{v}$ is an $(N+M+2L)$-dimensional constant vector, whose components fix the velocities of the N kinks and M antikinks of the wave and the $2L$ frequencies and velocities of the breathers. The parameters $\mathbf{x} = (x_1, \ldots, x_{(N+M+2L)})$ fix the positions of the kinks, antikinks, breathers, and also the phases of the breathers. The multisoliton waveform given in Equation (7.80) has $2(N + M + 2L)$ parameters that we will refer to collectively as a vector $\mathbf{p} = (\mathbf{v}, \mathbf{x})$. (Note that time t enters the multisoliton waveform $\mathbf{W}$ only through the combinations $v_i t$ ($i = 1, 2, \ldots, N + M + 2L$), although the velocity parameters v_i may enter in other places as well.)

The first step in our perturbation analysis is to assume that the solution $\mathbf{W}$ of Equation (7.78) is

$$\mathbf{W} = \mathbf{W}_0 + \varepsilon \mathbf{W}_1 + O(\varepsilon^2), \tag{7.81}$$

where $\mathbf{W}_0$ is the pure multisoliton state represented by Equation (7.80) except that all the parameters in $\mathbf{p}$ are allowed to vary slowly with time in the following manner:

$$\mathbf{W}_0 = \mathbf{W}_{(N+M+2L)}[x, X_1(t) + x_1(t), \ldots,$$
$$X_{(N+M+2L)}(t) + x_{(N+M+2L)}(t); \mathbf{v}(t)], \tag{7.82}$$

where $dX_i/dt = v_i$.

A few comments may be appropriate to clarify this formulation of the problem.

1. First of all, the temporal modulation in the parameters $\mathbf{p}(t)$ is induced by the perturbation $\varepsilon \mathbf{f}$. Because the perturbation is $O(\varepsilon)$, $d\mathbf{p}/dt = O(\varepsilon)$. At this stage, the modulation is not known: it must be computed.
2. Next, some care is needed to correctly identify the appropriate "velocity" parameters $(v_1, \ldots, v_{(N+M+2L)})$ before replacing $v_i t$ with $X_i(t)$. One approach to this identification begins with the asymptotic form of $\mathbf{W}_0$ for large t. In this limit, $\mathbf{W}_0$ will be decomposed into a collection of kinks, antikinks, and breathers that can be treated individually.
3. The appropriate identification for the frequency of a breather is obtained if one recognizes that the breather waveform is an analytic continuation in parameter space of the doublet waveform for a kink-antikink pair.

As a concrete example of a zero-order solution, consider a single kink ($N = 1$, $M = L = 0$) that was studied as an example of multiple scale analysis in Section 7.4.1. Thus from Equation (7.69)

$$\mathbf{W}_0 = \begin{bmatrix} 4 \arctan\left[\exp\left(\frac{x - X_0(t) - x_0(t)}{\sqrt{1 - v^2(t)}}\right)\right] \\ \frac{-2v(t)}{\sqrt{1 - v^2(t)}} \operatorname{sech}\left(\frac{x - X_0(t) - x_0(t)}{\sqrt{1 - v^2(t)}}\right) \end{bmatrix}, \tag{7.83}$$

with

$$X_0(t) \equiv \int_0^t v(t') dt'.$$

The second step in the perturbation scheme is to find the ODEs that govern the modulations of the parameters $\mathbf{p}(t)$. This calculation begins with the insertion of the perturbation expansion of Equation (7.81) into the nonlinear system of Equation (7.78). Linearization about $\mathbf{W}_0$ then yields

$$L\mathbf{W}_1 = \mathcal{F}(\mathbf{W}_0), \quad \mathbf{W}_1(t=0) = \mathbf{0}, \tag{7.84}$$

where the linear operator L is

$$L \equiv \begin{bmatrix} \partial_t & -1 \\ (-\partial_{xx} + \cos u_0) & \partial_t \end{bmatrix}. \tag{7.85}$$

Thus the first-order correction $\mathbf{W}_1$ satisfies a linear, variable coefficient, non-homogeneous initial value problem. The choice of vanishing initial data is equivalent to the assumption that at $t = 0$ the wave is the pure soliton state $\mathbf{W}_0$.

In Equation (7.84), the effective source $\mathcal{F}$ is given by

$$\mathcal{F} = \mathbf{f}(\mathbf{W}_0) - \frac{1}{\varepsilon} \sum_{j=1}^{2(N+M+2L)} \frac{\partial \mathbf{W}_0}{\partial p_j} \frac{dp_j}{dt}, \tag{7.86}$$

where, in the calculations of the partial derivatives, $X_1(t), \ldots, X_{(N+M+2L)}(t)$ are held fixed. Note that the effective source includes corrections to the perturbation $\mathbf{f}$ because—as the parameters of $\mathbf{W}_0$ vary with time—$\mathbf{W}_0$ is not an exact solution of the unperturbed SG system. In the calculation of this effective source, $dp_j/dt = \mathrm{O}(\varepsilon)$, so the correction is $\mathrm{O}(1)$ and not $\mathrm{O}(1/\varepsilon)$ as the notation makes it appear.

Equation (7.84) for $\mathbf{W}_1$ is a linear initial value problem that can be explicitly solved using a Green function. In particular, this Green function is useful for the calculation of the radiation from an accelerating kink, but if only the modulations of the solitons are of interest, the calculation is much more straightforward.

Generally the solution $\mathbf{W}_1$ of Equation (7.84) grows with time. When this occurs, the first-order correction $\varepsilon \mathbf{W}_1$ becomes large and the approximation of Equation (7.81) becomes invalid. The origin of such growth is a resonance between the source $\mathcal{F}$ and the Green function for the operator defined in Equation (7.85).

To avoid the growth, the modulation of the parameters $\{p_j\}$ must be chosen to eliminate this resonance. The worst secularity (linear growth in t) arises from a resonance between solitons in the effective source $\mathcal{F}(\mathbf{W}_0)$ and the same solitons in the Green function. This leading secularity will be eliminated if the effective source is orthogonal to a certain finite dimensional subspace of the null space of the operator

$$L^\dagger = -\begin{bmatrix} \partial_t & (\partial_{xx} - \cos u_0) \\ 1 & \partial_t \end{bmatrix}, \tag{7.87}$$

the adjoint of L.

It is convenient to denote the null spaces of L and $L^\dagger$ by $\mathcal{N}(L)$ and $\mathcal{N}(L^\dagger)$, and the discrete subspaces by $\mathcal{N}_{\mathrm{d}}(L)$ and $\mathcal{N}_{\mathrm{d}}(L^\dagger)$. These subspaces have dimension $2(N+M+2L)$.

The idea is to select the modulation of the $2(N+M+2L)$ parameters $\{p_j\}$ in a manner that forces the effective source $\mathcal{F}(\mathbf{W}_0)$ to be orthogonal to $\mathcal{N}_{\mathrm{d}}(L^\dagger)$. Thus

$$\mathcal{F}(\mathbf{W}_0) \perp \mathcal{N}_{\mathrm{d}}(L^\dagger),$$

which is recognized as a statement of the solvability condition as required by Fredholm's theorem.

This orthogonality condition is actually a set of $2(N+M+2L)$ first-order ODEs that determine the responses of the solitons to the perturbation εR. To obtain these equations, let $\{\mathbf{b}_j(x), j = 1, 2, \ldots, 2(N+M+2L)\}$ denote a basis for the discrete subspace $\mathcal{N}_{\mathrm{d}}(L^\dagger)$. Then—using the definition of the effective source given in Equation (7.86)—the orthogonality condition takes the form

$$\sum_{k=1}^{2(N+M+2L)} \left(\mathbf{b}_j, \frac{\partial \mathbf{W}_0}{\partial p_k}\right) \frac{dp_k}{dt} = \varepsilon(\mathbf{b}_j, \mathbf{f}), \tag{7.88}$$

where $j = 1, 2, \ldots, 2(N+M+2L)$, and

$$(\mathbf{F}, \mathbf{G}) \equiv \int_{-\infty}^{\infty} \mathbf{F}^T(x)\mathbf{G}(x)dx$$

is an inner product notation for integrals over x.

Equations (7.88) are a system of $2(N+M+2L)$ first-order ODEs for slow temporal variations of the parameters of the kinks, antikinks, and breathers. At this point, we lack only a basis for $\mathcal{N}_{\mathrm{d}}(L^\dagger)$, but this can be calculated directly from the multisoliton waveform $\mathbf{W}_0$, so the modulations of the p_j depend only on the multisoliton waveform.

To find the basis $\mathbf{b}_j$ of the subspace $\mathcal{N}_{\mathrm{d}}(L^\dagger)$, note first that it is needed only through O(1) in ε. This allows us to freeze the time dependence of the $p_j(t)$ when computing $\{\mathbf{b}_j(x)\}$. Thus, in the computation of the basis, we can use the unperturbed solution of Equation (7.80) in place of Equation (7.82).

Note next that members of $\mathcal{N}_{\mathrm{d}}(L)$ can be generated by differentiating $\mathbf{W}_0$ with respect to the $2(N+M+2L)$ parameters $\mathbf{p}$. To see this, observe that Equation (7.78) has no explicit dependence on these parameters; therefore differentiation of Equation (7.78) with respect to p_j becomes simply

$$L\frac{d\mathbf{W}_0}{dp_j} = 0.$$

In this manner, $2(N+M+2L)$ independent elements

$$\frac{d\mathbf{W}_0}{dp_j} \in \mathcal{N}_{\mathrm{d}}(L)$$

can be generated. To find elements of $\mathcal{N}_d(L^\dagger)$ note that if $\mathbf{V} \in \mathcal{N}_d(L)$, then $J\mathbf{V} \in \mathcal{N}_d(L^\dagger)$, where

$$J \equiv \begin{bmatrix} 0 & -1 \\ 1 & 0 \end{bmatrix}. \tag{7.89}$$

Therefore, to leading order in ε,

$$\mathcal{N}_d(L^\dagger) = \mathrm{span}\left\{J\frac{d\mathbf{W}_0}{dp_j};\ j = 1, 2, \ldots, 2(N+M+2L)\right\}.$$

Finally, because

$$\mathbf{W}_0 \simeq \mathbf{W}(x, v_1 t + x_1, \ldots, v_{(N+M+2L)} t + x_{(N+M+2L)}; \mathbf{v}),$$

it is evident that

$$\frac{d\mathbf{W}_0}{dv_j} \simeq t\frac{\partial \mathbf{W}_0}{\partial x_j} + \frac{\partial \mathbf{W}_0}{\partial v_j},$$

where the $\{X_j\}$ are held fixed in computing the partial derivatives. Thus, for fixed t as functions of x,

$$\mathrm{span}\left\{J\frac{d\mathbf{W}_0}{dp_j}\right\} = \mathrm{span}\left\{J\frac{\partial \mathbf{W}_0}{\partial p_j}\right\}$$

and Equations (7.88) take the form

$$\sum_{k=1}^{2(N+M+2L)} \left(J\frac{\partial \mathbf{W}_0}{\partial p_j}, \frac{\partial \mathbf{W}_0}{\partial p_k}\right) \frac{dp_k}{dt} = \varepsilon \left(J\frac{\partial \mathbf{W}_0}{\partial p_j}, \mathbf{f}\right), \tag{7.90}$$

where $j = 1, 2, \ldots, 2(N+M+2L)$. Equations (7.90) are the main result of this section.

With this set of ODEs, the response of the velocity and location parameters of the solitons to a perturbation can be computed directly from the pure multisoliton waveform: no additional information is needed. From a broader perspective, our perturbation theory has reduced an infinite dimensional system (the original PDE) to a finite dimensional system. The reduced system is a set of ODEs for the salient features of the solution: the soliton positions and velocities.

7.5.2 *Kink-antikink collisions*

As a specific example of multisoliton perturbation theory, consider a kink-antikink collision of the perturbed SG equation

$$u_{tt} - u_{xx} + \sin u = -\alpha u_t - \gamma, \tag{7.91}$$

with $-\infty < x < +\infty$. In this example, α is a small parameter, representing dissipative effects in a long Josephson junction transmission line, and γ is a bias parameter providing energy input. It is assumed that the kink and antikink have velocities v and $-v$ respectively.

If γ is sufficiently large, the velocity (v) will decrease to some minimum value during the collision and then rise again to the power balance velocity (v_∞) defined in Equation (7.43). In this case, the kink-antikink pair survives the collision. If γ is reduced below a critical value (γ_c) the velocity falls to zero and the kink-antikink pair becomes an oscillating breather which can no longer absorb bias energy and eventually dies out.

The critical value of bias can be estimated through the following energy argument. The total kinetic energy of the incoming kink and anitkink traveling at v_∞ is

$$E_{\text{kin}} = 2 \times \tfrac{1}{2} \times 8 \times v_\infty^2 = \frac{8}{1 + (4\alpha/\pi\gamma)^2},$$

where v_∞ is given in Equation (7.43). The energy loss during the collision (E_{loss}) has been computed by Pedersen et al. as

$$E_{\text{loss}} = 4\pi^2 \alpha f(v),$$

where $f(v)$ is a function that decreases smoothly from 2 at $v = 0$ to 1 at $v = 1$ [24]. (Kivshar and Malomed have estimated the energy loss as $8\pi^2 \alpha$ [14].) Defining the critical bias (γ_c) as that for which $E_{\text{kin}} = E_{\text{loss}}$ gives

$$\gamma_c^2 = \frac{16\alpha^3 f}{2 - \pi^2 \alpha f}.$$

For $\alpha \ll 1/\pi^2$, this implies

$$2\sqrt{2}\alpha^{3/2} < \gamma_c < 4\alpha^{3/2}. \tag{7.92}$$

Let us see how this energy estimate agrees with a more exact perturbation calculation.

For the perturbation expansion $\mathbf{W} = \mathbf{W}_0 + \alpha \mathbf{W}_1 + \cdots$, where

$$\mathbf{W}_0 = \begin{bmatrix} u_0 \\ u_{0,t} \end{bmatrix},$$

an appropriate form for the slow modulation of the kink-antikink solution is

$$u_0 = 4 \arctan \left[\frac{\sinh\left(\int_0^t v(t')dt' + x_0(t)/\sqrt{1-v^2}\right)}{v \cosh\left(x - x_1(t)/\sqrt{1-v^2}\right)} \right],$$

together with a corresponding formula for the second component of $\mathbf{W}_0$. This expression has only three independent parameters (v, x_0, and x_1) rather than the four that might be expected for a two-kink solution because it contains the assumption that the magnitudes of the kink and antikink velocities are of equal magnitude.

The first-order perturbation

$$\mathbf{W}_1 = \begin{bmatrix} u_1 \\ u_{1,t} \end{bmatrix}$$

is generated by the effective source

$$\mathcal{F}(\mathbf{W}_0) = \frac{1}{\alpha}\begin{bmatrix} -u_{0,v}\dot{v} + u_{0,x}\dot{x}_1 - u_{0,x_0}\dot{x}_0 \\ -\alpha u_{0,t} - \gamma - u_{0,vt}\dot{v} + u_{0,xt}\dot{x}_1 - u_{0,x_0t}\dot{x}_0 \end{bmatrix},$$

which must be orthogonal to the discrete null space of $L^\dagger$. Three independent elements of $\mathcal{N}_d(L^\dagger)$ are found by differentiating u_0 with respect to x, t, and v. These are

$$\mathbf{b}_1 = \begin{bmatrix} u_{0,xt} \\ -u_{0,x} \end{bmatrix}, \quad \mathbf{b}_2 = \begin{bmatrix} u_{0,tt} \\ -u_{0,t} \end{bmatrix}, \quad \text{and} \quad \mathbf{b}_3 = \begin{bmatrix} u_{0,vt} \\ -u_{0,v} \end{bmatrix}.$$

If $x_1 = 0$, both $u_{0,xt}$ and $u_{0,x}$ are odd functions of x, so the orthogonality condition $(\mathcal{F},\mathbf{b}_1) = 0$ is satisfied automatically, without placing constraints on $\dot{v}$, $\dot{x}_1$, or $\dot{x}_0$. Also the coefficients of $\dot{x}_1$ from both $(\mathcal{F},\mathbf{b}_2)$ and $(\mathcal{F},\mathbf{b}_3)$ are zero; thus the parameter x_1 can be set to zero, centering the kink-antikink collision on the origin.

After eliminating terms that are zero, the conditions $(\mathcal{F},\mathbf{b}_2) = 0$ and $(\mathcal{F},\mathbf{b}_3) = 0$ become

$$\left(\int_{-\infty}^{\infty}(u_{0,tt}u_{0,v} - u_{0,vt}u_{0,t})dx\right)\dot{v} = \int_{-\infty}^{\infty}(\alpha u_{0,t}^2 + \gamma u_{0,t})dx \qquad (7.93)$$

and

$$\left(\int_{-\infty}^{\infty}(u_{0,vt}u_{x_0} - u_{0,x_0t}u_{0,v})dx\right)\dot{x}_0 = \int_{-\infty}^{\infty}(\alpha u_{0,v}u_{0,t} + \gamma u_{0,v})dx. \qquad (7.94)$$

Writing $u_{0,tt} = u_{0,xx} - \sin u_0$ and integrating by parts, the first of these equations is recognized as the energy equation

$$\frac{dH(u_0)}{dv}\dot{v} = -\int_{-\infty}^{\infty}(\alpha u_{0,t}^2 + \gamma u_{0,t})dx,$$

where

$$H(u_0) = \frac{16}{\sqrt{1-v^2}}.$$

From this condition, v can be determined from the ODE

$$\frac{dv}{d\Theta} = \frac{\sqrt{1-v^2}}{v}F(v,\Theta;\alpha,\gamma), \quad \text{where } \Theta \equiv \frac{\int_0^t v(t')dt' + x_0(t)}{\sqrt{1-v^2}} \qquad (7.95)$$

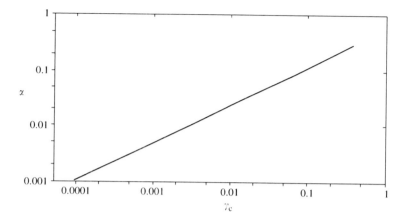

FIG. 7.2. *The relationship between γ_c and α for kink-antikink annihilation, calculated from Equation (7.95).*

and

$$F = \gamma\pi(1-v^2)^{3/2}\frac{\cosh\Theta}{4\sqrt{\sinh^2\Theta+v^2}}$$
$$-\alpha v^3(1-v^2)\frac{\cosh^2\Theta}{\sinh^2\Theta+v^2}\left[\frac{1}{v^2}-\frac{\log|(\sqrt{\sinh^2\Theta+v^2}-\sinh\Theta)/v|}{\sinh\Theta\sqrt{\sinh^2\Theta+v^2}}\right]. \quad (7.96)$$

From numerical integrations of this equation for the initial condition that $v = v_\infty$ at $t = -\infty$, the critical value of bias (γ_c) has been determined at which v falls to zero and kink-antikink annihilation occurs. The results, plotted in Figure 7.2, are in close agreement with numerical integration of Equation (7.91) up to $\alpha \approx 0.3$ [18], and they fall within the range of Equation (7.92).

7.5.3 *Radiation from a fluxon*

It is shown in Appendix H that radiation from solitons can be computed to first order as

$$\mathbf{W}_1(x,t) = \int_0^t\int_{-\infty}^\infty \mathcal{G}(x,t|x',t')\boldsymbol{\mathcal{F}}(x',t')dx'dt', \quad (7.97)$$

where $\mathcal{G}(x,t|x',t')$ is a radiative Green function and $\boldsymbol{\mathcal{F}}(x',t')$ is a source. Here we use this formulation to study the radiation from a Josephson junction fluxon that is propagating through a region of periodic spatial modulation (periodic changes of the width of the junction, perhaps) [18].

Under the influence of forcing and dissipative effects, the kink will have an average speed (v_m) about which it wobbles with frequency

$$\omega_0 = 2\pi v_m/a,$$

where a is the period of spatial modulation. Thus in Equation (7.97), the source will be of the form of a Fourier series in ω_0:

$$\mathcal{F}(x',t') = \mathbf{F}_0(x' - v_m t')$$
$$+ \mathbf{F}_1(x' - v_m t')e^{i\omega_0 t'} + \mathbf{F}_1^*(x' - v_m t')e^{-i\omega_0 t'}$$
$$+ \text{higher harmonics.}$$

Consider first the average term

$$\mathbf{W}_{10}(x,t)$$
$$= \frac{1}{4\pi i} \int_{-\infty}^{\infty} d\lambda \int_0^t dt' \int_{-\infty}^{\infty} dx' \frac{\exp\{-i[k(\lambda)(x - x') + \omega(\lambda)(t - t')]\}}{\lambda(\zeta^2 - \lambda^2)^2}$$
$$\times [g_{ij}]\mathbf{F}_0(x' - v_m t'), \qquad (7.98)$$

where the Green function components are defined below Equation (H.8) and λ is a spectral parameter that allows integration over frequency. With the t' integration from 0 to t, this expression gives a transient response induced by turning $\mathbf{F}_0$ on at $t = 0$. To find the steady-state response, the t' integration is evaluated as

$$\lim_{t \to \infty} \int_{-t}^{t} dt' = \int_{-\infty}^{\infty} dt'.$$

Integration over x' involves the factors

$$\int_{-\infty}^{\infty} dx' F(x' - v_m t') \exp(ikx') \propto \exp(ikv_m t'),$$

so the integration over t' then has the factor

$$\int_{-\infty}^{\infty} dt' \exp[i(kv_m + \omega)t'] = 2\pi\delta(kv_m + \omega).$$

Thus $\omega = -kv_m$ in Equation (7.98), implying that $\mathbf{W}_{10}(x,t)$ is a function only of $x - v_m t$. In other words, the steady-state perturbation rides with the kink making an order ε correction to its shape which can be neglected in first-order calculations of radiation.

Consider next the fundamental harmonic in $\mathcal{F}$

$$\mathbf{W}_{11}(x,t)$$
$$= \frac{1}{4\pi\mathrm{i}} \int_{-\infty}^{\infty} d\lambda \int_{-\infty}^{\infty} dt' \int_{-\infty}^{\infty} dx' \frac{\exp\{-\mathrm{i}[k(\lambda)(x-x')+\omega(\lambda)(t-t')]\}}{\lambda(\zeta^2-\lambda^2)^2}$$
$$\times [g_{ij}] \left[\mathbf{F}_1(x'-v_m t')e^{\mathrm{i}\omega_0 t'} + \mathbf{F}_1^*(x'-v_m t')e^{-\mathrm{i}\omega_0 t'}\right]. \tag{7.99}$$

Again the integration over x' introduces a factor $\exp(\mathrm{i}kv_m t')$, so the t' integration becomes

$$\int_{-\infty}^{\infty} \exp\{\mathrm{i}[kv_m+\omega\pm\omega_0]t'\}dt' = 2\pi\delta(kv_m+\omega\pm\omega_0),$$

driving radiation at the Doppler shifted frequencies[3]

$$\omega = -kv_m \pm \omega_0. \tag{7.100}$$

The emitted radiation is constrained by a dispersion relation that can be written in physical units as

$$\omega^2 = k^2 c^2 + \omega_J^2, \tag{7.101}$$

where (in Josephson jargon) c is the "Swihart velocity" and ω_J is the "plasma frequency" of a junction. Solving Equations (7.100) and (7.101) for ω leads to the radiation frequency

$$\omega = \omega_0 \frac{1+(v_m/c)\sqrt{1-(\omega_J^2/\omega_0^2)(1-v_m^2/c^2)}}{1-v_m^2/c^2},$$

which approaches $2\omega_0/(1-v_m^2/c^2)$ as the average fluxon speed approaches the Swihart velocity.

7.6 Neural perturbations

One of the aims of this book is to emphasize that solitons—while important in the theory of emergent structures—comprise only part of the picture. Thus Chapter 4 was devoted to the investigation of reaction-diffusion systems where coherent structures emerge from nonlinear interactions between sources of energy and dissipative effects. In this section, we consider how perturbation theory can be applied to neural systems which are described by reaction-diffusion equations.

---

[3] For Christian Johann Doppler (1803–1853), an Austrian physicist and mathematician.

7.6.1 The FitzHugh–Nagumo system

In Section 4.3.2, we considered the FitzHugh–Nagumo (F–N) model of nerve impulse propagation, for which the traveling-wave speed (v) is shown as a function of the temperature parameter (ε) in Figure 4.12. Here is presented a perturbation analysis developed by Casten et al. that allows computation of this curve in the range $0 < \varepsilon \ll 1$ [4].

First assume traveling-wave solutions of Equations (4.25) and (4.26) with $b = 0$ and $c = 0$, which become

$$\frac{\partial^2 V}{\partial \xi^2} + v\frac{\partial V}{\partial \xi} = F(V) + R,$$

$$\frac{\partial R}{\partial \xi} = -\frac{\varepsilon V}{v}.$$

(The assumption that b and c are zero is made to simplify the algebra. All of these computations can be carried through for the general case.) Next, express v, V, and R as power series in ε; thus

$$v = v_0 + \varepsilon v_1 + \varepsilon^2 v_2 + \cdots$$
$$V = V_0 + \varepsilon V_1 + \varepsilon^2 V_2 + \cdots$$
$$R = R_0 + \varepsilon R_1 + \varepsilon^2 R_2 + \cdots . \tag{7.102}$$

Equating terms at equal powers of ε leads to

$$\frac{d^2 V_0}{d\xi^2} + v_0 \frac{dV_0}{d\xi} - [F(V_0) + R_0] = 0$$

$$\frac{dR_0}{d\xi} = 0 \tag{7.103}$$

and

$$\frac{d^2 V_1}{d\xi^2} + v_0 \frac{dV_1}{d\xi} - V_1 F'(V_0) = R_1 - v_1 \frac{dV_0}{d\xi} \tag{7.104}$$

$$\frac{dR_1}{d\xi} = -\frac{V_0}{v_0}. \tag{7.105}$$

Assuming that the pulse is propagating into a region of zero recovery variable (as indicated in Figure 4.10), then $R_0 = 0$, and from our previous studies of the Z–F equation in Section 4.1.1, Equation (7.103) has the solution

$$V_0(\xi) = \left[1 + \exp\left(\xi/\sqrt{2}\right)\right]^{-1}. \tag{7.106}$$

From integration of Equation (7.105),

$$R_1(\xi) = \frac{1}{v_0} \int_\xi^\infty V_0(\xi') d\xi'.$$

Thus, everything on the right-hand side of Equation (7.104) is known except the value of v_1, which can be selected to satisfy the Fredholm condition for this equation to have a solution.

To this end, Equation (7.104) can be written in the form

$$LV_1 = R_1 - v_1 \frac{dV_0}{d\xi},$$

where the operator

$$L \equiv \frac{d^2}{d\xi^2} + v_0 \frac{d}{d\xi} - F'(V_0).$$

From differentiation of Equation (7.103) with respect to ξ, $dV_0/d\xi$ is recognized as a null function of L. As the adjoint of L is

$$L^\dagger = \frac{d^2}{d\xi^2} - v_0 \frac{d}{d\xi} - F'(V_0),$$

it has a null function ψ, where

$$L^\dagger \psi = 0,$$

and

$$\psi = e^{v_0 \xi} \frac{dV_0}{d\xi}.$$

From the Fredholm theorem, a necessary condition for Equation (7.104) to have a solution is that its right-hand side must be orthogonal to ψ. Thus

$$v_1 = \frac{\int_{-\infty}^{\infty} \left[\int_{\xi}^{\infty} V_0(\xi') d\xi' \right] (dV_0/d\xi) e^{v_0 \xi} d\xi}{v_0 \int_{-\infty}^{\infty} (dV_0/d\xi)^2 e^{v_0 \xi} d\xi},$$

and the traveling-wave velocity is

$$v = v_0 + \left(\frac{\varepsilon}{v_0} \right) \frac{\int_{-\infty}^{\infty} \left[\int_{\xi}^{\infty} V_0(\xi') d\xi' \right] (dV_0/d\xi) e^{v_0 \xi} d\xi}{\int_{-\infty}^{\infty} (dV_0/d\xi)^2 e^{v_0 \xi} d\xi} + O(\varepsilon^2). \qquad (7.107)$$

The first term on the right-hand side of this expression gives the traveling-wave velocity when ε and the recovery variable (R) are both equal to zero. The second term, which is negative, gives an $O(\varepsilon)$ correction to the traveling-wave velocity, assuming that the pulse retains its zero-order shape. The final (unevaluated) term accounts for variations in the traveling-wave velocity that arise from changes in the pulse shape.

This procedure is not appropriate for the unstable impulse of low velocity with $v_0 = 0$, but with a perturbation expansion of the form [4]

$$V = V_0 + \sqrt{\varepsilon} V_1 + \varepsilon V_2 + \cdots$$

$$R = \sqrt{\varepsilon} R_1 + \varepsilon R_2 + \cdots$$

$$v = \sqrt{\varepsilon} v_1 + \varepsilon v_2 + \cdots,$$

it follows that
$$v = \sqrt{\varepsilon} v_1 + O(\varepsilon), \tag{7.108}$$

where
$$v_1 = + \left[\frac{\int_{-\infty}^{\infty} V_0^2 d\xi}{\int_{-\infty}^{\infty} (dV_0/d\xi)^2 d\xi} \right]^{1/2}.$$

With the locus of traveling-wave solutions for a F–N system plotted on the (v, ε) parameter plane (as in Figure 4.12), Equations (7.107) and (7.108) indicate the asymptotic behavior near $\varepsilon = 0$.

7.6.2 Electrodynamic (ephaptic) coupling of nerves

The previous section presented a perturbation analysis of the F–N description of nerve impulse propagation, using as a small parameter the reciprocal of a time constant governing the recovery variable. We now consider two nerve fibers that are parallel and relatively close together so the electrodynamics on one fiber can influence that on the other and vice versa. Such parallel fiber interactions have been studied by electrophysiologists since the 1940s [9, 28]. Termed "ephaptic" (as opposed to "synaptic"), these interactions provide a means of communication between appropriately oriented axons and dendrites [7, 29].

The system of interest is represented by the coupled F–N equations [31]

$$\begin{aligned}
\frac{\partial V_1}{\partial t} &= (1-\alpha)\frac{\partial^2 V_1}{\partial x^2} - \alpha \frac{\partial^2 V_2}{\partial x^2} - F(V_1) - R_1 \\
\frac{\partial R_1}{\partial t} &= \varepsilon V_1 \\
\frac{\partial V_2}{\partial t} &= (1-\alpha)\frac{\partial^2 V_2}{\partial x^2} - \alpha \frac{\partial^2 V_1}{\partial x^2} - F(V_2) - R_2 \\
\frac{\partial R_2}{\partial t} &= \varepsilon V_2,
\end{aligned} \tag{7.109}$$

where the subscripts refer to variables on fibers #1 and #2.

The present analysis does not assume that ε is small; rather it takes a value of about 0.1, corresponding to a normal nerve impulse. The small parameter is α, which measures the coupling between nerve impulses on the two fibers. This coupling arises because the external currents associated with an impulse on one fiber influence the membrane potential of the other fiber, and vice-versa. As $\alpha \to 0$, the fibers become uncoupled and nerve impulses on one fiber propagate independently of those on the other.

To begin the perturbation analysis, assume traveling-wave solutions of the form
$$V_k(x,t) = V_k(\xi) = V_k(x - vt), \quad k = 1, 2.$$

where v is the propagation velocity of two impulses traveling in synchronism. Then Equations (7.109) become a set of ODEs that can be written

$$-v\frac{dV_1}{d\xi} = (1-\alpha)\frac{d^2V_1}{d\xi^2} - \alpha\frac{d^2V_2}{d\xi^2} - F(V_1) - R_1$$

$$-v\frac{dR_1}{d\xi} = \varepsilon V_1$$

$$-v\frac{dV_2}{d\xi} = (1-\alpha)\frac{d^2V_2}{d\xi^2} - \alpha\frac{d^2V_1}{d\xi^2} - F(V_2) - R_2$$

$$-v\frac{dR_2}{d\xi} = \varepsilon V_2.$$

A solution $[V_1(\xi), V_2(\xi)]$ will consist of two impulses, one on each fiber, moving with the same velocity. Because $\alpha \ll 1$, this solution is written as the power series

$$V_k = V_{k0} + \alpha V_{k1} + \alpha^2 V_{k2} + \cdots, \quad k = 1, 2.$$

The velocity is also written as

$$v(k) = v_0 + \alpha v_1^{(k)} + \alpha^2 v_2^{(k)} + \cdots,$$

where it is provisionally assumed that solutions may have different velocities on the two fibers. Eliminating R_1 and R_2 yields

$$\frac{d^3V_{k0}}{d\xi^3} + v_0\frac{d^2V_{k0}}{d\xi^2} - F'(V_{k0})\frac{dV_{k0}}{d\xi} + \frac{\varepsilon}{v_0}V_{k0} = 0 \qquad (7.110)$$

and

$$\frac{d^3V_{k1}}{d\xi^3} + v_0\frac{d^2V_{k1}}{d\xi^2} - F'(V_{k0})\frac{dV_{k1}}{d\xi} - \left(F''(V_{k0})\frac{dV_{k0}}{d\xi} - \frac{\varepsilon}{v_0}\right)V_{k1}$$
$$= v_1^{(k)}\left(\frac{\varepsilon}{v_0^2}V_{k0} - \frac{d^2V_{k0}}{d\xi^2}\right) + \frac{d^3V_{10}}{d\xi^3} + \frac{d^3V_{20}}{d\xi^3}, \qquad (7.111)$$

where $k = 1, 2$. Here

$$F'(V_{k0}) \equiv \left[\frac{dF(V_k)}{dV_k}\right]_{V_k = V_{k0}},$$

and similarly for $F''(V_{k0})$.

The original coupled nonlinear equations have been reduced to the uncoupled nonlinear system of Equations (7.110) and the linear system of Equations (7.111) for first-order corrections. Notice that Equations (7.111) are uncoupled in the V_{k1}, and the right-hand side terms involve only the zero order solutions and the two first-order velocity perturbations: $v_1^{(1)}$ and $v_1^{(2)}$. Thus each of Equations (7.111) can be written as an nonhomogeneous, linear operator equation

$$L_k V_{k1} = f_k.$$

From Fredholm's theorem, a necessary condition for a solution to exist is that the inner product

$$(y_k, f_k) \equiv \int_{-\infty}^{\infty} y_k(\xi) f_k(\xi) d\xi = 0,$$

where y_k is a solution of

$$L_k^\dagger y_k = 0, \qquad (7.112)$$

and $L_k^\dagger$ is the adjoint of L_k with respect to the inner product.

Because the adjoint of

$$-F'(V_{k0}) \frac{d}{d\xi} - F''(V_{k0}) \frac{dV_{k0}}{d\xi}$$

is

$$+F'(V_{k0}) \frac{d}{d\xi},$$

Equations (7.112) become

$$\frac{d^3 y_k}{d\xi^3} - v_0 \frac{d^2 y_k}{d\xi^2} - F'(V_{k0}) \frac{dy_k}{d\xi} - \frac{\varepsilon}{v_0} y_k = 0, \quad k = 1, 2. \qquad (7.113)$$

From solutions of Equations (7.113), one can compute inner products with the right-hand sides of Equations (7.111) and obtain useful expressions for the first-order perturbations of the traveling-waves, the $v_1^{(k)}$. To effect this calculation, proceed as follows:

1. Solve Equation (7.110) for $V_{k0}(\xi)$. This function will have the shape of a normal nerve impulse.
2. Solve Equation (7.113) for $y_k(\xi)$, the solution of the adjoint problem. This function has the qualitative shape of $V_{k0}(-\xi)$.
3. Assume that V_{20} differs from V_{10} by a translation of δ in the traveling-wave variable ξ. Thus

$$V_{20}(\xi) = V_{10}(\xi - \delta)$$
$$y_2(\xi) = y_1(\xi - \delta),$$

implying that the impulse on fiber #2 is leading the impulse on fiber #1 by a distance δ.

The Fredholm solvability condition for Equations (7.111) then requires that the inner products of the right-hand sides with the y_k be zero. This in turn implies

$$Nv_1^{(k)} = \int_{-\infty}^{\infty} y_k \left(\frac{d^3 V_{10}}{d\xi^3} + \frac{d^3 V_{20}}{d\xi^3} \right) d\xi,$$

where

$$N \equiv \int_{-\infty}^{\infty} y_1 \left(\frac{d^2 V_{10}}{d\xi^2} - \frac{\varepsilon}{v_0^2} V_{10} \right) d\xi = \int_{-\infty}^{\infty} y_2 \left(\frac{d^2 V_{20}}{d\xi^2} - \frac{\varepsilon}{v_0^2} V_{20} \right) d\xi.$$

To first-order in α, the condition for a traveling-wave solution is

$$v(1) = v_0 + \alpha v_1^{(1)} = v(2) = v_0 + \alpha v_1^{(2)}$$

or

$$v_1^{(1)} = v_1^{(2)}. \tag{7.114}$$

In Figure 7.3, first-order velocity corrections are sketched as functions of δ, (the displacement of impulse #2 with respect to #1), computed in the above manner for a typical F–N impulse [17].

Five solutions of Equation (7.114) are displayed, but only three of these (denoted by black circles) are stable in the following sense. An increase in δ, implying that impulse #2 advances with respect to impulse #1, causes the speed of impulse #1 to become greater than that of impulse #2. This causes the change in δ to decrease, implying stability. By a corresponding argument, the intersections denoted by the open circles are unstable.

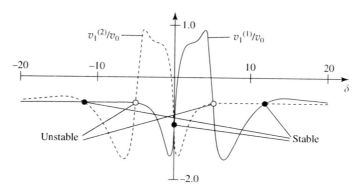

FIG. 7.3. *First-order corrections to impulse speeds under the influence of mutual coupling. (Data courtesy of Steve Luzader [17].)*

Although the foregoing analysis was developed under the assumption that both nerve impulses travel at the same speed, the results obtained suggest the ODE

$$\frac{d\delta}{dt} = v_1^{(2)}(\delta) - v_1^{(1)}(\delta).$$

Interestingly, this dependence has been confirmed by numerical studies of the full PDEs for the coupled system [6].

Such locking of the speeds and positions of nerve impulses on parallel fibers corresponds to temporal locking of the phases and frequencies of weakly coupled nonlinear oscillators. Observed and described in 1940 [9], impulse interactions on closely situated axons and dendrites may play functional roles in the information processing systems of living organisms [7, 28, 31]. For additional details, see references [1, 29], where applications of corresponding analyses to coupled Z–F equations, to M–C systems, and to coupled myelinated fibers are described. This approach can also be used for coupled soliton systems, with interacting fluxons on stacked Josephson junction oscillators being of particular interest [25, 32].

7.7 Summary

Although the subject of perturbation theory has not been covered in the course of this chapter, it may have been uncovered for the interested reader, making clear the general strategy and sketching some ways it can be used.

Our discussion began with introductory applications of the theory to perturbations of small matrices and damped linear oscillators, both of which can be solved exactly. Energy methods were introduced for the oscillator, followed by a discussion of multiple time scale analysis. Although included to present the underlying approach with a minimum of conceptual difficulty, these examples are not trivial. Finding the determinants of large matrices is a challenging numerical problem, and the method of averaging is useful in the design and analysis of approximately sinusoidal, nonlinear oscillators.

The first applications to nonlinear systems were in Section 7.3, where the slow variation of wave energy was used to obtain a first-order ODE for changes in the speed or amplitude of solitary waves under the influence of a perturbation. Although solitons of the KdV, SG, and NLS equations were presented as examples, such an energy analysis can be applied to any solitary wave that is governed by a single parameter and conserves energy in the $\varepsilon = 0$ limit, thereby allowing the wave velocity to vary over a continuous range of values.

The upside of energy analysis is that it is physically motivated. The downside is two-fold: it is inconvenient for solutions with more than one slowly varying parameter, and there is no means for estimating the effects of the perturbation on the shape of the solitary wave. To deal with the former difficulty, it is possible to use either the method of multiple time scales or an averaged Lagrangian approach, as described in Section 7.4.

Although soliton equations and waveforms have been used as examples, nothing in the foregoing discussions has required the insights or analytical power of the inverse scattering method (ISM), described in the previous chapter. In Section 7.5, some results of soliton theory are used to develop analytical tools that are powerful enough to handle perturbations of N-soliton waveforms. This approach has two appealing features: familiarity with the ISM theory is not required to implement the analysis, and explicit Green functions can be obtained for calculating first-order distortion of, and radiation from, an N-soliton waveform. Although the formulas appearing in these studies are somewhat complicated, this reflects the intricacy of the calculations involved.

Soliton perturbation theory has attracted much research effort over the past two decades, most of which is beyond the scope of this introductory discussion. Thus the reader interested in further soliton studies is directed to the extensive review prepared by Kivshar and Malomed [14]. A major theme of this book, however, has been to show that the purview of nonlinear wave theory extends beyond those systems that can be derived from Lagrangian functionals. To underscore this point, Section 7.6 describes applications of perturbation theory to an important model for biological nerve conduction: the F–N system. The chapter closes with a problem at the frontier of modern research in nonlinear wave dynamics: transverse coupling between solitary waves on parallel systems.

7.8 Problems

1. Repeat the calculations of Section 7.1 for the following perturbation matrices:

$$R = \begin{bmatrix} 0 & 0 \\ 2 & 0 \end{bmatrix}, \quad \begin{bmatrix} 0 & 1 \\ 0 & 1 \end{bmatrix}, \begin{bmatrix} 1 & 1 \\ 1 & 0 \end{bmatrix}, \quad \text{and} \quad \begin{bmatrix} 0 & 0 \\ 1 & 1 \end{bmatrix}.$$

2. Derive the hierarchy of matrix perturbation Equations (7.8) through (7.11).

3. Show that the adjoint of a square matrix with real elements is equal to its transpose.

4. (a) For

$$R = \begin{bmatrix} 0 & 1 \\ 0 & 0 \end{bmatrix}$$

and M as in Section 7.1, calculate $\phi_1^{(1)}$ and $\phi_1^{(2)}$ from Equation (7.9).

(b) Use them to obtain $\lambda_2^{(1)}$ and $\lambda_2^{(2)}$ from Equation (7.10).

(c) Compare your result with the exact solution of Equation (7.5).

5. (a) Find all the eigenvalues and eigenvectors of the 100×100 matrix

$$M_0 = \text{diag}[1, 2, 3, \ldots, 98, 99, 100].$$

(b) Use the ideas of Section 7.1 to derive a general expression to order ε for all the eigenvalues of the 100×100 matrix

$$M = M_0 + \varepsilon R,$$

where R is an arbitrary matrix of dimension 100×100.

(c) Compare the numerical efficiency of the perturbative approach to a direct solution of the equation $\det(M) = 0$.

6. Repeat the previous problem for the order ε^2 dependence of the eigenvalues.

7. Use the Laplace transform method to derive the exact solution of the damped harmonic oscillator in Equation (7.14). [Hint: see Chapter 2.]

8. Show that $dE/dt = 0$, where E is the harmonic oscillator energy defined in Equation (7.16).

9. Why is the bracket in Equation (7.18) of the form $[1 + O(\varepsilon^2)]$ instead of $[1 + O(\varepsilon)]$?

10. (a) Use the perturbation calculation of Section 7.2.1 to find $A(t)$ when Equation (7.13) is changed to

$$\frac{d^2 x}{dt^2} + \varepsilon(x^2 - 1)\frac{dx}{dt} + x = 0,$$

with $x(0) = b$, $\dot{x}(0) = 0$, and $0 < \varepsilon \ll 1$.

(b) Support your result with a phase plane analysis of the system.

11. (a) Check the algebra leading from Equation (7.20) to (7.21) for a damped harmonic oscillator.

(b) Compare Equation (7.21) with the exact solution of Equation (7.13).

12. Derive Equations (G.4) for perturbations of a damped harmonic oscillator.

13. Use the Lagrange–Euler variational equations of Appendix A to obtain the unperturbed KdV equation from the Lagrangian density given in Equation (7.25).

14. Show that the Hamiltonian (H) that is defined in Equations (7.27) and (7.26) is independent of time for the unperturbed KdV equation.

15. (a) Show that the quantity

$$Q_2 = \int_{-\infty}^{\infty} u^2 \, dx$$

is a constant when u is a solution of the unperturbed KdV equation.

(b) Use this fact to derive Equation (7.34).

PROBLEMS

16. (a) Show that the quantity
$$Q_1 = \int_{-\infty}^{\infty} u\,dx$$
is a constant when u is a solution of the unperturbed KdV equation.

    (b) Use this fact to derive Equation (7.35).

17. Consider a KdV equation for which the perturbation
$$\varepsilon R = -\varepsilon u$$
is switched on at time $t = 0$, just as the soliton passes the point $x = 0$. Compute the dimensions of a shelf, extending from $x = 0$ to the rear of the soliton that accounts for the difference in mass decrement calculated in the two previous problems.

18. Use energy analysis to study the evolution of the perturbed KdV equation
$$u_t + uu_x + u_{xxx} = \varepsilon u_{xx},$$
where $-\infty < x < +\infty$, $0 \le t < +\infty$, and
$$u(x,0) = 3\,\text{sech}^2(x/2).$$

19. (a) Study the system of the previous problem using the techniques of traveling-wave analysis that were presented in Chapter 3.

    (b) Compare results of the two approaches.

20. (a) From the SG energy density $\mathcal{H}$ given in Equation (7.38) and the conservation law
$$\frac{\partial \mathcal{H}}{\partial t} + \frac{\partial \mathcal{P}}{\partial x} = 0,$$
find an expression for the power flow ($\mathcal{P}$) within a general solution ϕ. [Hint: see Appendix A.]

    (b) For the kink solution $\phi = 4\arctan(\exp\Theta)$, where $\Theta \equiv (x - vt)/\sqrt{1-v^2}$, compute the total energy passing the point $x = 0$ as
$$\int_{-\infty}^{\infty} \mathcal{P}\,dt.$$

    (c) Compute the kink energy as
$$\int_{-\infty}^{\infty} \mathcal{H}\,dx.$$
and compare your result with that obtained in part (b).

21. Derive Equation (7.42) for the perturbed speed of a Josephson junction fluxon.

22. (a) Show that Equation (7.43) for the power balance of a fluxon follows from Equation (7.42) with $\beta = 0$.

   (b) Discuss solutions of Equation (7.42) for various initial conditions on v and the following assumptions about the parameters: (i) $\gamma = 0$, (ii) $\beta = 0$, (iii) $\alpha = 0$, and (iv) $\alpha = 0$ and $\beta = 0$.

23. Derive the NLS equation from the Lagrangian density given in Equation (7.47).

24. Verify the expression for NLS energy given in Equation (7.50).

25. Consider the problem
$$iu_t + u_{xx} + 2|u|^2 u = i\varepsilon u,$$
where $-\infty < x < +\infty$, $t \geq 0$, and
$$u(x, 0) = a \operatorname{sech} ax.$$

   (a) Present a physical argument that the center of mass of the solution should remain fixed under the perturbation.

   (b) For $0 < \varepsilon \ll 1$, find an $O(\varepsilon)$ estimate for $u(x,t)$.

26. Show that the adjoint of L in Equation (7.58) is as in Equation (7.61).

27. Verify the calculations leading to Equations (7.63) and (7.64) for the perturbed speed and location of an SG kink.

28. On the (v, X) phase plane, discuss solutions of Equations (7.67) for the perturbation
$$R = \delta(x) \sin \phi.$$

29. Consider the unperturbed NLS equation
$$iu_t + u_{xx} + 2|u|^2 u = 0.$$

   (a) Show that a Lagrangian density is
$$\mathcal{L} = \frac{i}{2}(u^* u_t - u u_t^*) - |u_x|^2 + |u|^4.$$

   (b) Assume a solitary wave solution of the form
$$u(x, t) = a \operatorname{sech}(a\theta) \exp(i\xi\theta + i\sigma),$$
where $\theta = x + \alpha$, and a, ξ, α, and σ are allowed to be functions of time. Show that
$$L \equiv \int_{-\infty}^{\infty} \mathcal{L} dx$$
$$= \tfrac{2}{3} a^3 - 2a\xi^2 - 2a\xi\alpha_t - 2a\sigma_t.$$

(c) Vary L with respect to a, ξ, α, and σ, showing that

$$\frac{da}{dt} = 0, \quad \frac{d\xi}{dt} = 0, \quad \frac{d\alpha}{dt} = -2\xi, \quad \text{and} \quad \frac{d\sigma}{dt} = a^2 + \xi^2.$$

30. Consider the perturbed NLS equation

$$iu_t + u_{xx} + 2|u|^2 u = \varepsilon R,$$

with the solitary wave solution

$$u(x,t) = a \operatorname{sech}(a\theta) \exp(i\xi\theta + i\sigma),$$

where the parameters are defined as in the previous problem. With $R = -iu$, vary the expression for L obtained in the previous problem, demonstrating that to order ε

$$\frac{da}{dt} = -2\varepsilon a, \quad \frac{d\xi}{dt} = 0, \quad \frac{d\alpha}{dt} = -2\xi, \quad \text{and} \quad \frac{d\sigma}{dt} = a^2 + \xi^2.$$

31. Repeat the previous problem for $R = iu_{xx}$.

32. Repeat the previous problem for $R = -i|u|^2 u$.

33. Sketch a family of solution trajectories for Equation (7.76) with: $\Gamma > g_0$, and $\Gamma < g_0$.

34. Use perturbation theory to study the system

$$iu_t + u_{xx} + 2|u|^2 u = -i\varepsilon u + \alpha e^{i\omega t},$$

where $0 < \varepsilon \ll 1$ and $|\alpha| \ll 1$. [Hint: recall that the first order response to both ε and α is linear.]

35. Describe the behaviors of the system

$$iu_t + u_{xx} + 2|u|^2 u = i(\alpha_1 u + \alpha_2 u_x + \alpha_3 u_{xx})$$

for small values of the α_j in the $(\alpha_1, \alpha_2, \alpha_3)$ parameter space.

36. Substitute Equation (7.81) into Equation (7.78) and derive the linear operator for SG perturbations given in Equation (7.85).

37. Show that $L^\dagger$ given in Equation (7.87) is the adjoint of the operator L defined in Equation (7.85).

38. If $\mathbf{V}$ is a null function of the operator L defined in Equation (7.85), show that $J\mathbf{V}$ is a null function of its adjoint $L^\dagger$ given in Equation (7.87), where the 2×2 matrix J is defined in Equation (7.89).

39. Assume that the perturbed SG system of Equation (7.78) has the single-kink solution of Equation (7.83). Use Equations (7.90) to show that the parameters $v(t)$ and $x_0(t)$ are governed by the ODEs

$$\frac{dv}{dt} = -\frac{\varepsilon}{4}(1-v^2)\int_{-\infty}^{\infty} R[\phi_0(\Theta,x,t)]\mathrm{sech}\Theta\, dx$$

and

$$\frac{dx_0}{dt} = -\frac{\varepsilon}{4}v\sqrt{1-v^2}\int_{-\infty}^{\infty} R[\phi_0(\Theta,x,t)]\Theta\mathrm{sech}\Theta\, dx,$$

where

$$\Theta = \Theta(x,t) \equiv +\frac{x - \int_0^t v(t')dt' - x_0(t)}{\sqrt{1-v^2(t)}}.$$

40. In the previous problem, assume that the perturbation depends only on x; thus $R = R(x)$ and is independent of $\phi_0(x,t)$ and its derivatives. Show that the kink velocity obeys Newton's second law in the form

$$\frac{dv}{dt} = -\frac{\partial U}{\partial X},$$

where $U \equiv \dfrac{\varepsilon(1-v^2)^{3/2}}{8}\int_{-\infty}^{\infty} R(\Theta\sqrt{1-v^2} - X)\phi_0(\Theta)d\Theta$,

and $X = X(t) \equiv \int_0^t v(t')dt' + x_0$

define the location of the kink as a function of time.

41. Show that Equation (7.93) can be interpreted as a condition of power balance.

42. Sketch solution trajectories of Equation (7.95) for: (i) $\gamma > \gamma_c$, and (ii) $\gamma < \gamma_c$.

43. Show by direct substitution of Equation (H.1) into Equation (7.84) that the SG Green function—$\mathcal{G}(x,t|x',t')$—must satisfy the conditions in Equations (H.2) and (H.3).

44. Consider the forced Klein–Gordon equation

$$\frac{\partial^2 \phi}{\partial t^2} - \frac{\partial^2 \phi}{\partial x^2} + \phi = \varepsilon R,$$

with the initial conditions $\phi(x,0) = 0$ and $\phi_t(x,0) = 0$.

(a) Write it as a first-order matrix system for the variable

$$\mathbf{W} \equiv \begin{bmatrix} \phi \\ \phi_t \end{bmatrix}.$$

(b) Assuming that $\mathbf{W} = \mathbf{W}_0 + \varepsilon \mathbf{W}_1 + O(\varepsilon^2)$, find $\mathbf{W}_0$ and construct an integral expression for $\mathbf{W}_1$ using the zero-soliton matrix Green function defined in Equation (H.5).

(c) Check that your solution is correct.

45. Verify that the single-kink radiative Green function defined in Equation (H.8) satisfies the conditions of Equations (H.2) and (H.3).

46. Discuss (without going through the details) a strategy for constructing the radiative Green function for a two-kink interaction on the SG equation.

47. Show from Equation (7.98) that in the steady state limit $(t \to \infty)$
$$\mathbf{W}_{10} = \mathbf{W}_{10}(x - v_m t).$$

48. For a niobium–tin Josephson transmission line described by the equation
$$\frac{\partial^2 \phi}{\partial x^2} - \frac{1}{c^2}\frac{\partial^2 \phi}{\partial t^2} - \frac{\alpha}{c\lambda_J}\frac{\partial \phi}{\partial t} = \frac{1}{\lambda_J^2}\sin\phi$$
(where ϕ is the normalized magnetic flux density), some physical parameters have been measured as in the following table [30].

Device	#1	#2	units
c	1.76×10^9	2.3×10^9	cm/s
λ_J	0.0263	0.127	cm
α	5.55×10^{-3}	5.52×10^{-3}	—

Use these parameters and the concepts presented in Section 7.5.3 to discuss the design of oscillators for which the radiation frequency is: $10^{10}, 10^{11}$, and 10^{12} Hz.

49. (a) Show that the shape of the trailing edge of the pulse does not influence the order ε correction term to the traveling-wave speed in Equation (7.107).

(b) Does the first-order correction from the recovery variable (potassium conductance) increase or decrease the traveling-wave velocity? Explain your answer.

(c) Evaluate Equation (7.107) for $F(V) = V(V-a)(V-1)$.

(d) Discuss the behavior of the traveling-wave velocity (v) as the parameter $a \to \frac{1}{2}$ from below.

50. Consider the F–N equation with
$$F(V) = V(V - 0.25)(V - 1).$$

(a) Use Equations (7.107) and (7.108) to plot an $O(\varepsilon)$ approximation to the locus of traveling-wave solutions on a (v, ε) parameter plane.

(b) How does the result of (a) compare with the true locus?

51. Repeat the previous problem for

$$F(V) = \begin{cases} 0 & \text{for } V < a \\ V - 1 & \text{for } V > a. \end{cases}$$

52. Consider a F–N impulse that is propagating into a region where the recovery variable $R = R_0 > 0$. How does the traveling-wave velocity depend on R_0?

53. (a) With $\varepsilon = 0$ in the F–N system, show that there is a stationary (not moving) pulse shaped solution with $V_0(\xi) = V_0(x)$, where

$$d^2 V_0/d\xi^2 = F(V_0).$$

(b) Assuming a perturbation expansion of the form

$$V = V_0 + \sqrt{\varepsilon} V_1 + \cdots$$
$$R = \sqrt{\varepsilon} R_1 + \cdots$$
$$v = \sqrt{\varepsilon} v_1 + \cdots,$$

show that

$$v_1 = \left[\frac{\int_{-\infty}^{\infty} V_0^2 d\xi}{\int_{-\infty}^{\infty} (dV_0/d\xi)^2 d\xi} \right]^{1/2}.$$

(c) Discuss the behavior of your result for $F(V) = V(V-a)(V-1)$ as $a \to \frac{1}{2}$ from below.

54. (a) Starting with Equations (7.109) in the Z–F limit ($\varepsilon = 0$), carry through the computations of Section 7.6.2.

(b) How does the expression for $v^{(1)}(\delta)$ change? [Hint: see reference [1].]

(c) Sketch a corresponding version of Figure 7.3.

55. Use the M–C nerve model of Section 4.3.1 to sketch the conditions for ephaptic coupling in Figure 7.3. [Hint: see reference [29].]

56. (a) Write out Equations (7.110) and (7.111) for coupled nerve impulses with

$$F(V) = V(V-a)(V-1).$$

(b) Check that the linear operators on the left hand side of Equations (7.113) are adjoints of those on the left-hand side of Equations (7.111).

57. Formulate and solve a perturbation problem that is of interest to you.

REFERENCES

1. S Binczak, J C Eilbeck, and A C Scott. Ephaptic coupling of myelinated nerve fibers, *Physica D* 148 (2001) 159–174.
2. S Chandler and R S Johnson. On the asymptotic solutions of the perturbed KdV equation using the inverse scattering method. *Phys. Lett. A* 86 (1981) 337–340.
3. J G Caputo, N Flytzanis, and M P Sørensen. The ring laser configuration studied by collective coordinates. *J. Opt. Soc. Am. B* 12 (1995) 139–145.
4. R G Casten, H Cohen, and P A Lagerstrom. Perturbation analysis of an approximation to Hodgkin-Huxley theory. *Q. Appl. Math.* 32 (1975) 356–402.
5. G Costabile, R D Parmentier, B Savo, D W McLaughlin, and A C Scott. Exact solutions of the SG equation describing oscillations in a long (but finite) Josephson junction. *Appl. Phys. Lett.* 32 (1978) 587–589.
6. J C Eilbeck, S D Luzader, and A C Scott. Pulse evolution on coupled nerve fibers. *Bull. Math. Biol.* 43 (1981) 389–400.
7. J G R Jefferys. Nonsynaptic modulation of neuronal activity in the brain: Electric currents and extracellular ions. *Physiol. Rev.* 75 (1995) 689–723.
8. V I Karpman and E M Maslov. Structure of tails produced under the action of perturbations on solitons. *Sov. Phys. JETP* 48 (1978) 252–259.
9. B Katz and O H Schmitt. Electric interaction between two adjacent nerve fibers. *J. Physiol.* 97 (1940) 471–488.
10. D J Kaup and A C Newell. Solitons as particles and oscillators in slowly varying media. *Proc. R. Soc. (London) A* 361 (1978) 413–466.
11. J P Keener and D W McLaughlin. A Green function for a linear equation associated with solitons. *J. Math. Phys.* 18 (1977) 2008–2013.
12. J P Keener and D W McLaughlin. Solitons under perturbations. *Phys. Rev. A* 16 (1977) 777–790.
13. J Kevorkian and J D Cole. *Multiple Scale and Singular Perturbation Methods.* Springer-Verlag, New York, 1996.
14. Y S Kivshar and B A Malomed. Dynamics of solitons in nearly integrable systems. *Rev. Mod. Phys.* 61 (1989) 763–915.
15. C J Knickerbocker and A C Newell. Shelves and the Korteweg–de Vries equation. *J. Fluid Mech.* 98 (1980) 803–818.
16. G L Lamb, Jr. *Elements of Soliton Theory.* John Wiley, New York, 1980.
17. S D Luzader. *Neurophysics of Parallel Nerve Fibers.* PhD thesis, University of Wisconsin, Madison, 1979.
18. D W McLaughlin and A C Scott. Perturbation analysis of fluxon dynamics. *Phys. Rev. A* 18 (1978) 1652–1680.
19. A H Nayfeh. *Perturbation Methods.* John Wiley, New York, 1973.
20. A H Nayfeh and D T Mook. *Nonlinear Oscillations.* John Wiley & Sons, New York, 1979.
21. A H Nayfeh. *Introduction to Perturbation Techniques.* John Wiley & Sons, New York, 1981.
22. A C Newell. *Solitons in Mathematics and Physics.* SIAM, Philadelphia, 1985.
23. E Ott and R N Sudan. Damping of solitary waves. *Phys. Fluids* 13 (1970) 1432–1434.
24. N F Pedersen, M R Samuelsen, and D Welner. Soliton annihilation in the perturbed SG system. *Phys. Rev. B* 30 (1984) 4057–4059.

25. A Petraglia, A V Ustinov, N F Pedersen, and S Sakai. Numerical study of fluxon dynamics in a system of two-stacked Josephson junctions. *J. Appl. Phys.* 77 (1995) 1171–1177.
26. B van der Pol. The nonlinear theory of electric oscillations. *Proc. IRE* 22 (1934) 1051–1086.
27. A C Scott. Distributed device applications of the superconducting tunnel junction. *Solid State Electron.* 7 (1964) 137–146.
28. A C Scott. The electrophysics of a nerve fiber. *Rev. Mod. Phys.* 47 (1975) 487–533.
29. A C Scott. *Neuroscience: A Mathematical Primer*. Springer-Verlag, New York, 2002.
30. A C Scott, F Y F Chu, and S A Reible. Magnetic-flux propagation on a Josephson transmission line. *J. Appl. Phys.* 47 (1976) 3272–3286.
31. A C Scott and S D Luzader. Coupled solitary waves in neurophysics. *Phys. Scr.* 20 (1979) 395–401.
32. A C Scott and A Petraglia. Flux interactions on stacked Josephson junctions. *Phys. Lett. A* 211 (1996) 161–167.

8
QUANTUM LATTICE SOLITONS

In general, the soliton is a classical phenomenon. The calculation of quantum corrections to John Scott Russell's "Wave of Translation" is unnecessary; and for a tsunami or Jupiter's Great Red Spot, such an effort would be absurd. Yet there are solitons that emerge at the level of molecules and molecular crystals, and for these the constraints of quantum theory are important. The aim of this chapter is to introduce the theory of quantum lattice solitons in a manner that is helpful for understanding such experimental studies.

Because the anharmonicity in a molecule arises from the properties of its chemical bonds, we begin with a review of a mass-spring oscillator with a spring that is weakly nonlinear. From this perspective, the introduction of quantum theory presents little analytical difficulty, and yields results that can be generalized to applications that are motivated by measurements of vibrational spectra in molecules, molecular crystals, and biomolecules.

The chapter closes with some comments on the general theory of quantum lattices that may suggest directions for future research.

8.1 Quantum oscillators

8.1.1 *A classical nonlinear oscillator*

Consider a mass M that is attached to a rigid support by a nonlinear spring as shown in Figure 8.1. The nonlinearity of the spring can be expressed by the

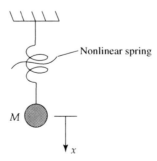

FIG. 8.1. *A simple nonlinear oscillator.*

potential function
$$V(x) = \frac{Kx^2}{2} - \frac{\alpha x^4}{4}, \qquad (8.1)$$
where x measures the extension of the spring from its equilibrium position and K is the linear spring constant. For $\alpha > 0$, the final term in Equation (8.1) introduces a "soft" nonlinearity, meaning that the restoring force of the spring is less than proportional to its extension, as is the case for interatomic bonds.

From Newton's second law, the extension of the spring obeys the second-order ODE
$$M\frac{d^2x}{dt^2} + Kx - \alpha x^3 = 0. \qquad (8.2)$$
Solution trajectories of this equation can be understood by defining the momentum
$$p \equiv M\frac{dx}{dt}$$
and noting that the Hamiltonian (H)
$$H(p,x) = \frac{p^2}{2M} + V(x) \qquad (8.3)$$
is independent of time along a solution trajectory and equal to the total energy (E) of the system. Thus
$$H(p,x) = E, \qquad (8.4)$$
and Equation (8.2) is equivalent to the two first-order ODEs
$$\begin{aligned} \frac{dx}{dt} &= \frac{\partial H}{\partial p} = \frac{p}{M} \\ \frac{dp}{dt} &= -\frac{\partial H}{\partial x} = -Kx + \alpha x^3. \end{aligned} \qquad (8.5)$$

Assuming $\alpha = 0$, the system has the solution
$$x = A\cos\omega t,$$
with energy
$$E = \tfrac{1}{2}KA^2,$$
where
$$\omega \equiv \sqrt{K/M} \text{ rad/s}.$$
The energy can take on any value in the range
$$E \geq 0,$$
but at the level of atomic physics this conclusion is inconsistent with the facts of nature.

8.1.2 The birth of quantum theory

At the beginning of the twentieth century, it became evident that one may not assume an oscillator to have any value for its energy. This counter-intuitive observation was forced upon physics through attempts by the German theoretical physicist Max Planck to calculate the radiation spectrum from an idealized entity called a "black body," having a uniform temperature (T) and reflecting nothing [52, 67].

A black body can be modeled by a large cavity with a small hole in the side. Light that enters the hole is almost entirely absorbed within the cavity, leaving none to be reflected. Planck's studies of the radiation intensity of a black body as a function of its temperature convinced him that electromagnetic oscillators within the cavity could not assume any level of energy.

To follow his reasoning, note that the number of electromagnetic modes of oscillation per unit volume that lie between frequency ω and $\omega + d\omega$ is

$$\mathcal{N}(\omega)d\omega = \frac{\omega^2}{\pi^2 c^3}d\omega,$$

where c is the velocity of light. If one assumes equipartition of energy among these linear modes, each receiving its allotted kT of energy (where k is Boltzmann's constant[1]), the radiation intensity as a function of frequency and temperature becomes

$$\rho(\omega, T) \propto \omega^2 T. \tag{8.6}$$

This classical result is in contradiction with experimental observations. To see this, imagine that you are peeking into an oven through a small hole. When the oven is cold, you see nothing; the inside looks totally black. As it becomes heated, infrared radiation appears (you can first sense it with your cheek), becoming a deep red glow as the proper temperature is attained. In scientific terms, this means that the radiation intensity attains a maximum value at some frequency, and this frequency of maximum intensity increases with temperature.

Such an observation is clearly at variance with Equation (8.6). It was this stark disagreement between a simple classical calculation and experimental reality that led Planck to introduce a revolutionary assumption in 1901. The energy of an oscillator can only take the special values

$$E = 0, \hbar\omega, 2\hbar\omega, 3\hbar\omega, \ldots, n\hbar\omega \ldots,$$

where $\hbar$ (called "h-bar") is a physical quantity that is now referred to as "Planck's constant" [52].

---

[1] After Austrian physicist Ludwig Boltzmann (1844–1906). His suicide has been attributed to a state of depression resulting from the intense scientific war between the atomists and the energists at the turn of the century.

With this assumption, the probability of the oscillator being at energy $n\hbar\omega$ is proportional to $\exp(-n\hbar\omega/kT)$, so the average energy of the oscillator is

$$\langle E \rangle = \frac{\sum_{n=1}^{\infty} n\hbar\omega e^{-n\hbar\omega/kT}}{\sum_{n=0}^{\infty} e^{-n\hbar\omega/kT}} = \frac{\hbar\omega}{e^{\hbar\omega/kT}-1}.$$

Thus the intensity of radiation in a black body becomes

$$\rho(\omega,T) = \frac{2\hbar\omega^3/\pi c^3}{e^{\hbar\omega/kT}-1}, \tag{8.7}$$

in units of joules per hertz volume. This expression has a maximum value as a function of ω, and this maximum increases with temperature in accord with our experience with an oven. Moreover, there is close agreement with experimental observations if

$$\hbar = 1.054\,572\,7 \times 10^{-34} \text{ J/s}$$
$$k = 1.380\,658 \times 10^{-23} \text{ J/K}$$
$$c = 299\,792\,458 \text{ m/s},$$

the best current values [11].

Even after being awarded a Nobel Prize for it, however, Planck was unhappy with this result. Why should the energy of an oscillator take only certain values? It seemed crazy.

In 1913, a young Danish physicist named Niels Bohr, working at the Cavendish Laboratory in Cambridge, suggested that a similarly *ad hoc* quantization of energy could be applied to the electron orbits in the nuclear atom, giving agreement with observations of the spectrum of atomic hydrogen [8], and in 1925 Werner Heisenberg proposed a theory with physical variables represented by matrices instead of numbers. The big break, however, came in December of 1925 through the revolutionary work of Erwin Schrödinger. When the first of his several manuscripts arrived at *Annalen der Physik* on January 27, 1926, Planck wrote that he had read it "like an eager child hearing the solution of a riddle that had plagued me for a long time." Why was he so impressed?

Schrödinger proposed that Equation (8.4) for conservation of energy can be viewed as the operator equation [60]

$$\hat{H}(\hat{p},\hat{x})\Psi = \hat{E}\Psi, \tag{8.8}$$

with

$$\hat{p} \equiv -i\hbar\frac{\partial}{\partial x},$$
$$\hat{x} \equiv x,$$
$$\hat{E} \equiv +i\hbar\frac{\partial}{\partial t},$$

$\Psi(x,t)$ being a "wave function" upon which the operators act.

QUANTUM OSCILLATORS

For a stationary solution of Equation (8.8), varying with time only as

$$e^{-iE_n t/\hbar},$$

Equation (8.8) reduces to the eigenvalue equation

$$\hat{H}(\hat{p},\hat{x})\psi_n = E_n \psi_n, \tag{8.9}$$

called "Schrödinger's equation", where ψ_n is an eigenfunction, and E_n is its corresponding energy eigenvalue [60]. Thus, allowed values of oscillator energy are eigenvalues of the energy operator $\hat{H}(\hat{p},\hat{x})$.

Because Equation (8.8) is linear, general solutions can be constructed from the eigenfunctions of Equation (8.9) as "wave packets" of the form

$$\Psi(x,t) = \sum_n c_n \psi_n(x) e^{-iE_n t/\hbar}, \tag{8.10}$$

where the summation is interpreted as an integral over continuous ranges of the eigenvalue E, and the c_n can be complex. How is this wave packet to be related to experimental observations? Nowadays physicists agree with Max Born that the probability of finding the oscillator position between x_1 and x_2 depends on time as

$$\int_{x_1}^{x_2} |\Psi(x,t)|^2 \, dx,$$

implying a probability density

$$P(x,t) = |\Psi(x,t)|^2. \tag{8.11}$$

Thus $|\Psi(x,t)|^2 \, dx$ is the probability that the system lies between x and $x + dx$. In connection with this formulation, the following points can be made:

1. Equation (8.10) can be viewed as a generalized Fourier transform of the solution. This perspective leads directly to the Heisenberg relationship, stating that uncertainty in momentum (Δp) times uncertainty in position (Δx) must be at least of the order of Planck's constant

$$\Delta p \times \Delta x \geq \hbar. \tag{8.12}$$

2. The operators $\hat{p}$ and $\hat{x}$ do not commute. In fact

$$\hat{x}\hat{p} - \hat{p}\hat{x} = i\hbar. \tag{8.13}$$

3. It is also evident from the time dependence, of the components in the wave packet of Equation (8.10) that uncertainty in time (Δt) and the range of the energy eigenvalues (ΔE) are related by

$$\Delta E \times \Delta t \geq \hbar. \tag{8.14}$$

4. The c_n in Equation (8.10) are arbitrary complex constants chosen so $P(x, 0)$ corresponds to one's knowledge of the initial data. In examples where the wave packet disperses, it is possible to reduce the uncertainties in position and momentum to the minimum indicated in Equation (8.12) by making appropriate measurements at some later time. Such a "wave packet collapse" is not described by Schrödinger's equation (or any other theory) and remains one of the disturbing features of quantum mechanics.
5. The importance of Equation (8.9) for modern science cannot be overemphasized. Applied to the dynamics of interacting atoms as studied by Born and Oppenheimer [9], Schrödinger's equation becomes the fundamental equation of chemistry, describing how the atomic elements organize themselves into chemical molecules.

8.1.3 A quantum linear oscillator

Return now to the subject of Section 8.1.1—the oscillator of Figure 8.1—and consider its description from the perspective of quantum theory. Upon understanding the quantum behavior of a simple nonlinear oscillator, we shall be on the way to appreciating the quantum structure of a soliton.

Let us begin with the nonlinear parameter α equal to zero so Equation (8.2) reduces to a linear oscillator. Then the time independent Schrödinger equation takes the form

$$\frac{\hbar^2}{2M}\frac{d^2\psi}{dx^2} + \left(E - \tfrac{1}{2}Kx^2\right)\psi = 0, \tag{8.15}$$

which has been studied in great detail over the past seven decades. Its energy eigenvalues are given by

$$E_n = (n + \tfrac{1}{2})\hbar\omega, \tag{8.16}$$

where $\omega = \sqrt{K/M}$ is the frequency of the corresponding classical oscillator.

For our purposes, it is convenient to notice that the classical energy may also be written in the form

$$H = \omega|A|^2, \tag{8.17}$$

where

$$A \equiv \frac{(p - iM\omega x)}{\sqrt{2M\omega}}$$

$$A^* \equiv \frac{(p + iM\omega x)}{\sqrt{2M\omega}},$$

obey the dynamic equations

$$i\frac{dA}{dt} = \omega A \quad \text{and} \quad -i\frac{dA^*}{dt} = \omega A^*. \tag{8.18}$$

Equations (8.18) are evidently satisfied by

$$A(t) = A(0)\, e^{-i\omega t}, \tag{8.19}$$

called a "rotating-wave" form of the solution.

Under Schrödinger's quantization, A and A^* become operators that are conveniently normalized as [39]

$$\begin{aligned} A \to b &\equiv \frac{(\hat{p} - iM\omega\hat{x})}{\sqrt{2M\hbar\omega}} \\ A^* \to b^\dagger &\equiv \frac{(\hat{p} + iM\omega\hat{x})}{\sqrt{2M\hbar\omega}}. \end{aligned} \tag{8.20}$$

With these definitions, it follows from Equation (8.13) that

$$bb^\dagger - b^\dagger b = 1. \tag{8.21}$$

Also the classical energy becomes the energy operator

$$\hat{H} = \frac{\hbar\omega}{2}(bb^\dagger + b^\dagger b) = \hbar\omega(b^\dagger b + \tfrac{1}{2}), \tag{8.22}$$

where the average of the operators $bb^\dagger$ and $b^\dagger b$ is taken because the classical form of the Hamiltonian—given by Equation (8.17)—does not specify the ordering of the factors A and A^*.

Comparing Equations (8.16) and (8.22), it follows that

$$b^\dagger b \psi_n = n \psi_n \,;$$

in other words, $b^\dagger b$ is an operator that counts the number of the harmonic oscillator eigenfunctions. For this reason it is convenient to name it the "number operator"

$$\hat{N} \equiv b^\dagger b, \tag{8.23}$$

and the quanta counted by $\hat{N}$ are often called "bosons."[2]

The actions of b and $b^\dagger$ on eigenfunctions of $\hat{N}$ can be obtained through the following argument [14]. Assume that the number operator has an eigenfunction ψ_n with eigenvalue n; that is to say,

$$\hat{N}\psi_n = n\psi_n.$$

From the commutation relation of Equation (8.21),

$$\hat{N}b = b(\hat{N} - 1)\,;$$

---

[2]After the Indian physicist Satyendra Nath Bose (1894–1974), who formulated the statistics of photons.

thus
$$\hat{N}(b\psi_n) = b(\hat{N} - 1)\psi_n = (n-1)(b\psi_n),$$

implying that if ψ_n is an eigenfunction of $\hat{N}$ with eigenvalue n, then $b\psi_n$ is also an eigenfunction of $\hat{N}$ with eigenvalue $n-1$. In other words, b acts as a "lowering operator" for eigenfunctions of $\hat{N}$. The eigenvalue of $\hat{N}$ is lowered by one unit under the action of b. Using a similar argument, $b^\dagger$ acts as a "raising operator" on eigenfunctions of $\hat{N}$, increasing the eigenvalue of $\tilde{N}$ by one unit. From Equation (8.22), b and $b^\dagger$ are also respectively lowering and raising operators for the eigenfunctions of $\hat{H}$, and the eigenfunctions of $\hat{N}$ are identical to those of $\hat{H}$.

What do these eigenfunctions look like? It is easily checked that

$$\psi_0 \propto \exp\left(-M\omega x^2/2\hbar\right)$$

is an energy eigenfunction of Equation (8.15) with eigenvalue $E_0 = \hbar\omega/2$, and that $b\psi_0 = 0$. Thus ψ_0 is the lowest eigenfunction of $\hat{N}$, with eigenvalue $n = 0$. Defining

$$y \equiv \sqrt{(M\omega/\hbar)}\, x,$$

other eigenfunctions can be written as

$$\psi_n(y) = \frac{e^{-y^2/2} H_n(y)}{\sqrt{(2^n n! \sqrt{\pi})}}, \tag{8.24}$$

where the $H_n(y)$ are Hermite polynomials,[3] and the functions are normalized so

$$\int_{-\infty}^{\infty} \psi_n^2\, dx = 1.$$

The first five of these eigenfunctions (with $H_0 = 1$, $H_1 = 2y$, $H_2 = 4y^2 - 2$, $H_3 = 8y^3 - 12y$, and $H_4 = 16y^4 - 48y^2 + 12$) are plotted as in Figure 8.2. In laboratory units, the eigenfunctions are

$$\psi_n(x) = \frac{(M\omega/\pi\hbar)^{1/4}}{\sqrt{2^n n!}} \exp\left(-\frac{M\omega x^2}{2\hbar}\right) H_n\left(\sqrt{\frac{M\omega}{\hbar}}\, x\right).$$

Finally, the actions of b and $b^\dagger$ on these eigenfunctions are summarized as follows:

$$b\psi_n = \sqrt{n}\, \psi_{n-1}$$
$$b\psi_0 = 0 \tag{8.25}$$
$$b^\dagger \psi_n = \sqrt{n+1}\, \psi_{n+1}.$$

---

[3]For French mathematician Charles Hermite (1822–1901).

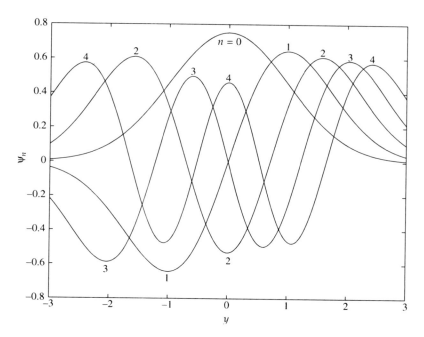

FIG. 8.2. *The first five eigenfunctions of a simple harmonic oscillator, plotted from Equation (8.24).*

8.1.4 *The rotating-wave approximation*

Consider next the nonlinear oscillator of Section 8.1.1 from the perspective of quantum theory. The direct approach to this problem is to augment Equation (8.15) with the $\alpha x^4/4$ term to

$$\frac{\hbar^2}{2M}\frac{d^2\psi_n}{dx^2} + \left(E_n - \tfrac{1}{2}Kx^2 + \tfrac{1}{4}\alpha x^4\right)\psi_n = 0 \qquad (8.26)$$

and determine the eigenfunctions and corresponding eigenvalues. While one may solve this equation numerically, it is convenient to turn to an approximate analysis in order to obtain analytical results.

Defining the operators b and $b^\dagger$ as in Equations (8.20),

$$\tfrac{1}{4}\alpha x^4 = \frac{\alpha\hbar^2}{16M^2\omega^2}(b^\dagger - b)^4,$$

but the operator $(b^\dagger - b)^4$ requires some care in its interpretation. Constants aside, it is a quantum representation of the classical expression

$$(A^* - A)^4 = (A^*)^4 - 4(A^*)^3 A + 6|A|^4 - 4A^* A^3 + A^4,$$

where from Equation (8.19)

$$A \sim e^{-i\omega t} + O(\alpha).$$

From the classical perspective, all the terms in $(A^* - A)^4$ are oscillating at either frequency 2ω or 4ω except for the $6|A|^4$ term, which varies slowly in time with the amplitude of A. This is the rotating-wave approximation (RWA), which we met in Chapter 5. Under the RWA, only the $6|A|^4$ term is retained; the others are neglected because they are small and have zero average values. Thus the classical Hamiltonian is

$$H \doteq \omega|A|^2 - \frac{3}{8}\frac{\alpha}{M^2\omega^2}|A|^4.$$

In quantizing $|A|^4$, we take the average of all six orderings of the operators b and $b^\dagger$, called "symmetric ordering" [33]. Thus

$$|A|^4 \rightarrow \tfrac{1}{6}(b^\dagger b^\dagger bb + b^\dagger bb^\dagger b + b^\dagger bbb^\dagger + bbb^\dagger b^\dagger + bb^\dagger bb^\dagger + bb^\dagger b^\dagger b),$$

which, through applications of Equation (8.21), can be put into the form

$$b^\dagger bb^\dagger b + b^\dagger b + \frac{1}{2} = \hat{N}^2 + \hat{N} + \frac{1}{2},$$

where $\hat{N} \equiv b^\dagger b$. Thus the quantum energy operator becomes

$$\hat{H} \doteq \left(\hbar\omega - \frac{\gamma}{2}\right)\left(\hat{N} + \frac{1}{2}\right) - \frac{\gamma}{2}\hat{N}^2, \tag{8.27}$$

with

$$\gamma \equiv \frac{3\alpha\hbar^2}{4M^2\omega^2}.$$

As Equation (8.27) is the main result of this subsection, it seems appropriate to pause and emphasize that it is an approximate Hamiltonian, derived under the rotating-wave assumption. As such, it has certain advantages and disadvantages that should be noted.

1. Perhaps the most important analytical convenience of Equation (8.27) is that it commutes with the number operator $\hat{N} \equiv b^\dagger b$, where the properties of b and $b^\dagger$ are given in Equations (8.25). Thus the x-dependence of the energy eigenfunctions is assumed to be the same as for a linear oscillator. If the energy eigenfunctions are numbered from zero with the integer n, the corresponding energy eigenvalues are given by

$$E_n = \left(\hbar\omega - \frac{\gamma}{2}\right)\left(n + \frac{1}{2}\right) - \frac{\gamma}{2}n^2. \tag{8.28}$$

2. The primary disadvantage of the approximation in Equation (8.27) is evident from inspection of Equation (8.26). The x-dependences of the eigenfunctions are not those for a linear oscillator; they differ by corrections of order α. These corrections allow transitions from the ground state ($n = 0$) to $n = 2, 3, \ldots$, an "overtone spectrum." In other words, Equation (8.27) correctly gives the frequencies (in hertz) of the overtone spectrum as

$$\nu_n = \frac{E_n - E_0}{2\pi\hbar} = \frac{1}{2\pi}\left[\left(\omega - \frac{\gamma}{2\hbar}\right)n - \frac{\gamma}{2\hbar}n^2\right], \qquad (8.29)$$

but it fails to predict the intensities of these transitions.

3. The RWA necessarily ignores the effects of odd powers of x in the potential. Thus, for example,

$$x^3 \sim (A^* - A)^3 = (A^*)^3 - 3(A^*)^2 A + 3A^* A^2 - A^3,$$

with no slowly varying terms to be retained.

4. The interplay between $\hbar\omega$ and γ in the first term of Equation (8.28) is illusory. From Equation (8.29), molecular spectroscopists will measure a function of the form

$$\nu_n = An - Bn^2,$$

called the "Birge–Sponer relation" [7, 19]. They would then view the measured value of A as the fundamental frequency of the oscillator.

5. Finally we note that the rotating-wave approximation repairs a defect in our original formulation of Schrödinger's equation in (8.26). The potential

$$V(x) = \frac{K}{2}x^2 - \frac{\alpha}{4}x^4$$

has a minimum value at $x = 0$ and maxima at $x = \pm\sqrt{K/\alpha}$. Thus—in a strict sense—the original model does not have localized quantum eigenfunctions; the particle will eventually leak out through the potential barrier. The RWA analysis avoids this difficulty because Equation (8.27) does have localized eigenfunctions.

As the above nonlinear potential implies the experimentally observed Birge–Sponer relation, it is an interesting characterization of an interatomic potential, indicating that the bond will break when x reaches the value $\sqrt{K/\alpha}$. But how does one calculate the values of K and α for the forces between atoms of a real molecule?

8.1.5 *The Born–Oppenheimer approximation*

In 1927, Robert Oppenheimer,[4] a fledgling US physicist working with Born, had an idea. To calculate the interatomic forces of a molecule, he suggested, why

---

[4]Two decades later, he became widely known as the scientific director of the atom bomb project at Los Alamos.

not first use Schrödinger's equation to find electronic wave functions assuming the atomic nuclei to be stationary? Then the energy of the system would be known as a function of the positions of the nuclei, allowing computation of the interatomic forces. Born immediately recognized this to be a good idea and with characteristic thoroughness worked out the details in an early and impressive application of perturbation theory to quantum problems [9, 59].

To see how this goes, consider a molecule that is composed of N atoms of mass M_k ($k = 1, 2, \ldots, N$). Assuming these nuclei to be stationary allows one to write Schrödinger's wave function as the product

$$\psi(\mathbf{r}_j, \mathbf{R}_k) = u(\mathbf{r}_j; \mathbf{R}_k) w(\mathbf{R}_k), \tag{8.30}$$

where $\mathbf{r}_j$ is a $3n$-dimensional vector that measures the positions of the n electrons, and $\mathbf{R}_k$ is a $3N$-dimensional vector for the positions of the atomic nuclei. (In H_2O, for example, $n = 10$ and $N = 3$.)

When the atomic nuclei are allowed to move, Born used a perturbation theory in the small parameter

$$\varepsilon = \sqrt{m_e/M_k},$$

where m_e is the mass of an electron; thus errors in a first-order theory are of the order of $\sqrt{m_e/M_k}$. From this theory, $u(\mathbf{r}_j; \mathbf{R}_k)$ is determined by the approximate Schrödinger equation

$$\hat{H}_{\text{electrons}} u(\mathbf{r}_j; \mathbf{R}_k) \doteq U(\mathbf{R_k}) u(\mathbf{r}_j; \mathbf{R}_k), \tag{8.31}$$

and the factor $w(\mathbf{R}_k)$ is determined by

$$\hat{H}_{\text{nuclei}} w(\mathbf{R}_k) \doteq E w(\mathbf{R}_k). \tag{8.32}$$

Equations (8.31) and (8.32) have the general form of Schrödinger's equation, but the operator bracket

$$\hat{H}_{\text{electrons}} \equiv \left[-\frac{\hbar^2}{2m_e} \sum_{j=1}^{n} \nabla_j^2 + V(\mathbf{r}_j, \mathbf{R}_k) \right]$$

in Equation (8.31) is an energy operator acting on the n electrons only, with the nuclear coordinates ($\mathbf{R}_k$) appearing as parameters. Notice, however, that the potential energy—$V(\mathbf{r}_j, \mathbf{R}_k)$—depends upon all the coordinates in the problem.

The operator bracket

$$\hat{H}_{\text{nuclei}} \equiv \left[-\frac{\hbar^2}{2M_k} \sum_{k=1}^{N} \nabla_k^2 + U(\mathbf{R}_k) \right]$$

in Equation (8.32) is an energy operator for the N nuclei. The potential energy $U(\mathbf{R}_k)$ for the nuclei is the eigenstate energy of the electronic system, depending upon the nuclear coordinates. For a stable molecule, $U_0(\mathbf{R}_k)$ will have a minimum value when the electrons of the molecule are arranged in the n lowest eigenstates of Equation (8.31).

This minimum potential function

$$U_0(\mathbf{R}_k)$$

is called the "Born–Oppenheimer potential energy." Derivatives of this function with respect to the components of $\mathbf{R}_k$ give the Born–Oppenheimer approximation to the forces acting on the atoms of a molecule. Thus if the $3N$ components of $\mathbf{R}_k$ are written as

$$x_1, y_1, z_1, x_2, y_2, z_2, \ldots, x_N, y_N, z_N,$$

then

$$-\partial U_0/\partial x_1$$

is the force acting on the first atom in the x-direction.

For molecules with more than a few atoms, such an *ab initio* calculation of $U_0(\mathbf{R}_k)$ is a difficult numerical problem; thus an approximating function is often assumed to have the form [44]

$$\tilde{U}_0 \approx \frac{1}{2} \sum_{ij} K_b (b - b_0)^2 + \frac{1}{2} \sum_{ij} K_\theta (\theta - \theta_0)^2$$

$$+ \frac{1}{2} \sum_{ij} K_\phi [1 + \cos(n\phi - \delta)] + \sum_{lm} \left[\frac{A}{r^{12}} - \frac{C}{r^6} + \frac{q_l q_m}{Dr} \right], \qquad (8.33)$$

where the parameters are selected to fit a variety of experimental and numerical data. Typical sets of parameters in currently available computer codes have the following features:

1. The only interactions considered are between pairs of atoms.
2. The first three summations are over pairs of atoms that share a covalent bond, indicated by the indices i and j. The first two are quadratic potentials leading to linear forces.
3. The lengths of covalent bonds are indicated by $b = b_{ij}$.
4. Bending angles of covalent bonds are indicated by $\theta = \theta_{ij}$.
5. Twisting angles of covalent bonds are indicated by $\phi = \phi_{ij}$.
6. The fourth (last) summation is over pairs of atoms that do not share a covalent bond, indicated by the indices l and m. The scalar distance between atoms l and m is indicated by $r = r_{lm}$.
7. The first two terms in the last summation describe a "six–twelve potential," which becomes large (i.e. repulsive) when the separation between the two molecules (r) is small and goes to zero for large r. With

appropriate choices for the parameters A and C, these terms can represent either a van der Waals interaction[5] or a hydrogen bond.

8. The last term in the last summation accounts for longer range electrostatic interactions between atoms. In this case the partial charges (q_l and q_m) must be estimated as well as the dielectric screening, D.

Given an approximate Born–Oppenheimer potential with the form indicated in Equation (8.33), one might either integrate Newton's classical equations or solve the quantum problem of Equation (8.32). Because the quantum approach is far more difficult than the corresponding classical calculations, it should only be undertaken when quantum corrections to the classical analysis are large enough to be experimentally observed. (This caution is ignored by those who would use quantum theory to explain the intricate nonlinear dynamics of the human brain [63].)

8.1.6 Dirac's notation

A convenient notation for quantum mechanical wave functions that was introduced by Dirac [14] is widely used and merits our attention. It is best presented in the context of eigenfunctions of the quantum harmonic oscillator. Briefly, the energy eigenfunctions of Equation (8.15) are written in the form

$$\psi_0(x) \equiv |0\rangle$$
$$\psi_1(x) \equiv |1\rangle$$
$$\psi_2(x) \equiv |2\rangle$$
$$\vdots$$
$$\psi_n(x) \equiv |n\rangle,$$

each of which is called a "ket." Thus the fact that ψ_n is an energy eigenfunction is indicated by the symbol "$|\cdot\rangle$" with the corresponding eigenvalue of the number operator carried inside.

For each ket, there is a corresponding "bra" (indicated as "$\langle \cdot |$") so the inner product

$$(\psi_n, \psi_{n'}) \equiv \int_{-\infty}^{\infty} \psi_n^*(x)\psi_{n'}(x)\,dx$$

is written compactly as the "braket"

$$\langle n|n'\rangle.$$

With the normalization $\langle n|n\rangle = 1$, it is evident that

$$\langle n|n'\rangle = \delta_{n,n'},$$

[5] Formulated by Dutch physical chemist Johannes Diderik van der Waals (1837–1923).

and Equations (8.25) take the form

$$b|n\rangle = \sqrt{n}\,|n-1\rangle$$
$$b|0\rangle = 0$$
$$b^\dagger|n\rangle = \sqrt{n+1}\,|n+1\rangle.$$

Although Dirac's notation is neither necessary nor sufficient to comprehend quantum theory, it is widely used and typographically convenient. In the following sections, we shall consistently (if somewhat redundantly) write wave functions as kets, using the notation

$$\psi \equiv |\psi\rangle.$$

8.1.7 Pump-probe measurements

As an application of the above ideas, it is instructive to analyze the pump-probe measurement shown in Figure 8.3. In this experiment, a probe beam is used to observe the spectrum of a sample both with and without the prior action of a pump pulse. If the sample were to respond in an entirely linear manner, the classical superposition theorem implies that the response to the probe would be the same whether the pump is present or not. In other words, a pump-probe measurement should respond only to nonlinear behavior of the sample. Let us see how this conclusion follows from quantum theory.

Classically, the ω component of the probe beam can be represented as

$$\cos\omega t \propto \left(e^{i\omega t} + e^{-i\omega t}\right).$$

From Equation (8.20) this corresponds to the quantum operators

$$(b^\dagger + b),$$

which act on all modes of the sample at frequency ω.

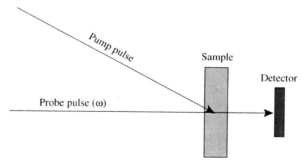

FIG. 8.3. *Sketch of a pump-probe experiment.*

Assuming that the sample is entirely linear and the pump has not been turned on, sample modes at ω will be excited by the raising operator of the probe ($b^\dagger$) from the ground state ($n = 0$) to the first ($n = 1$) level. The absorption from the probe can be computed from "Fermi's Golden Rule" as

$$KP(0)|\langle 1|b^\dagger|0\rangle|^2 Q(1) = K,$$

where $P(0)$ is the probability of the ground state being occupied (this is unity) and $Q(1)$ is the probability of the first level being empty (this is also unity). In this formulation, $|\langle 1|b^\dagger|0\rangle|^2$ is the probability of the system making a transition from $|0\rangle$ to $|1\rangle$ under the action of the raising operator ($b^\dagger$) introduced by the probe. Finally, K is a proportionality constant that accounts for the number of sample modes seen by the probe.

If the linear mode is irradiated by the pump prior to its interrogation by the probe, some of the higher levels will be occupied, as indicated in the following table.

Level (n)	$P(n)$	$Q(n)$
N	p_N	$(1 - p_N)$
...	...	...
2	p_2	$(1 - p_2)$
1	p_1	$(1 - p_1)$
0	$(1 - p_1 - p_2 - \cdots - p_N)$	$(p_1 + p_2 + \cdots + p_N)$

Here p_n is the probability of state $n \geq 1$ being occupied, and $(1 - p_n)$ is the probability of it being empty. To compute the total absorption from the probe, it is necessary to consider all possible upward and downward transitions among these states, using Fermi's Golden Rule and the facts that

$$|\langle n+1|b^\dagger|n\rangle|^2 = n+1 \quad \text{and} \quad |\langle n-1|b|n\rangle|^2 = n.$$

Thus the net absorption from the probe is

$$K[(N+1)p_N + Np_{N-1}(1 - p_N) + \cdots$$
$$+ 3p_2(1 - p_3) + 2p_1(1 - p_2) + (1 - p_1 - p_2 - \cdots - p_N)(1 - p_1)]$$
$$- K[Np_N(1 - p_{N-1}) + \cdots + 2p_2(1 - p_1) + p_1(p_1 + p_2 + \cdots + p_N)]$$
$$= K,$$

where the first bracket accounts for all the upward transitions induced by $b^\dagger$ and the second bracket accounts for all the downward transitions induced by b. Although this quantum expression for the net absorption appears complicated, it reduces to K, which is equal to the probe absorption when the pump is not applied.

Presently, pump-probe spectroscopy is an important tool for studying the anharmonic properties of molecular crystals for two reasons: any changes in probe absorption (with and without the pump) necessarily stem from nonlinear features of the sample, and such experiments can be performed in any spectral range from infrared to ultraviolet frequencies with a time resolution that is a small fraction of a picosecond [20, 48].

8.2 Self-trapping in the dihalomethanes

Methane is a biologically important molecule with the formula CH_4, indicating a structure of four hydrogens attached at equal angles to a tetravalent carbon atom. In the dihalomethanes (CF_2H_2, CCl_2H_2, CBr_2H_2, and CI_2H_2), two of these hydrogens are replaced by halides: fluorine, chlorine, bromine, or iodine. For physical chemists, these molecules form little laboratories where interactions between the remaining two CH stretching oscillators can be experimentally investigated [47].

For dichloromethane, the classical picture is like this:

$$\begin{array}{c} \text{Cl} \\ \text{H} \longleftrightarrow \text{C} \leftrightarrow \text{H} \\ \text{Cl} \end{array}$$

where the two CH stretching oscillations are each anharmonic as was described in the previous section. In addition, these two oscillators can interact with each other through both mechanical and electromagnetic fields. For a sufficiently large value of the anharmonicity parameter, it is possible for the energy to be concentrated on one or the other of the CH bonds as is indicated in the above diagram.

Before beginning the analysis, let us consider how the information that physical chemists gather is presented. An important tool is the spectrograph, used to measure infrared transmission and absorption, and also the Raman scattering[6] of laser beams as they pass through a sample of the material under study. In such measurements, the data are ultimately related to the spacings of lines that have been etched onto a diffraction grating in the spectrograph. If the experimenter were to present the data as frequencies, for example, the numbers would change with improved estimates for the speed of light. A more convenient choice is to use a measure for frequency that depends upon the wavelength, which is directly related to the spacing of the diffraction grating. Such a unit is called the "wave number" [68].

Basically, the wave number of an electromagnetic wave is the reciprocal of its free space wavelength (λ) measured in centimeters. Thus the frequency in cm^{-1} (physical chemists call it ν) is

$$\nu = \frac{1}{\lambda} \text{ cm}^{-1}.$$

---

[6]The first Asian to be awarded a Nobel prize in science, Indian physicist Chandrasekhra Venkata Raman (1888–1970) discovered this effect by asking: Why is the sky blue?

(In other words, the wave number is equal to the free space propagation number divided by 2π.)

To convert to ordinary frequency, one multiplies by the speed of light; thus

$$\frac{\omega}{2\pi} = \frac{c}{\lambda} \text{ Hz}$$

and the corresponding quantum energy is

$$\hbar\omega = \frac{2\pi\hbar c}{\lambda} \text{ J}.$$

To express the energy in electron-volts, divide by the electronic charge [11]

$$e = 1.6021773 \times 10^{-19} \text{ C},$$

so

$$\frac{\hbar\omega}{e} = \left(\frac{2\pi\hbar c}{e}\right)\frac{1}{\lambda}.$$

Thus a handy conversion factor for thinking about experimental data is:

$$\text{frequency in cm}^{-1} = 8065.541 \times \text{energy in electron volts}.$$

As one becomes familiar with these units, it is convenient to use the same symbols for frequency and energy, viewing them as expressed in cm^{-1}, electron-volts, joules, hertz, or radians per second, as appropriate.

8.2.1 Classical analysis

In a RWA, the classical Hamiltonian that governs our two interacting CH oscillators is

$$H = \omega_0(|A_1|^2 + |A_2|^2) - \frac{\gamma}{2}(|A_1|^4 + |A_2|^4) - \varepsilon(A_1 A_2^* + A_2 A_1^*), \tag{8.34}$$

implying the dynamic equations

$$\begin{aligned} i\frac{dA_1}{dt} &= \omega_0 A_1 - \gamma|A_1|^2 A_1 - \varepsilon A_2 \\ i\frac{dA_2}{dt} &= \omega_0 A_2 - \gamma|A_2|^2 A_2 - \varepsilon A_1. \end{aligned} \tag{8.35}$$

For notational convenience, it is here assumed that A_1 and A_2 are unitless, while d/dt, ω_0, γ, and ε are all expressed in units of cm^{-1} with conversions to energy or frequency units whenever that is appropriate.

With $\varepsilon = 0$, A_1 and A_2 are the complex mode amplitudes of two uncoupled anharmonic oscillators of the sort that was considered in the previous section. The aim now is to understand the relationship between the classical and quantum behaviors of the system when $\varepsilon \neq 0$.

Under the classical dynamics of Equation (8.35), the Hamiltonian of Equation (8.34) is conserved. It is easily verified that another conserved quantity is the number

$$N = |A_1|^2 + |A_2|^2. \tag{8.36}$$

Assume that the system of Equations (8.35) has a stationary solution of the form

$$\begin{aligned} A_1 &= \phi_1 e^{-i\omega t} \\ A_2 &= \phi_2 e^{-i\omega t}, \end{aligned} \tag{8.37}$$

where ϕ_1 and ϕ_2 are independent of time. Then for a fixed value of N, three classes of solutions are readily demonstrated [18].

Symmetric mode

In this case, the oscillation is of equal amplitude and phase on both oscillators. Thus $\phi_1 = \phi_2 = \sqrt{N/2}$, and

$$\begin{aligned} \omega &= \omega_0 - \varepsilon - \gamma N/2 \\ H &= (\omega_0 - \varepsilon)N - \gamma N^2/4. \end{aligned}$$

In the notation that was introduced for stationary solutions in Chapter 5, this mode is designated as $(\uparrow\uparrow)$. It is unstable for $\gamma > 2\varepsilon/N$.

Antisymmetric mode

In this case, the oscillation is of equal amplitude and out of phase on both oscillators. Thus again $\phi_1 = -\phi_2 = \sqrt{N/2}$, and

$$\begin{aligned} \omega &= \omega_0 + \varepsilon - \gamma N/2 \\ H &= (\omega_0 + \varepsilon)N - \gamma N^2/4, \end{aligned}$$

indicated as $(\uparrow\downarrow)$. This mode is stable for positive values of γ.

Local mode

In this case, the amplitude on one mode is larger than the other; thus

$$\begin{aligned} \phi_1 &= \left\{ \frac{N}{2} \left[1 \pm \left(1 - \frac{4\varepsilon^2}{N^2\gamma^2}\right)^{1/2} \right] \right\}^{1/2} \\ \phi_2 &= \left\{ \frac{N}{2} \left[1 \mp \left(1 - \frac{4\varepsilon^2}{N^2\gamma^2}\right)^{1/2} \right] \right\}^{1/2}, \end{aligned} \tag{8.38}$$

with the designations $(\uparrow \; \cdot)$ or $(\cdot \; \uparrow)$ in the notation of Chapter 5, indicating that one of the oscillator amplitudes goes to zero as γ becomes much greater than ε. The local mode requires that

$$\gamma \geq 2\varepsilon/N \tag{8.39}$$

and has

$$\omega = \omega_0 - \gamma N$$
$$H = \omega_0 N - \gamma N^2/2 - \varepsilon^2/\gamma.$$

As γ is increased from small values, the stable symmetric mode $(\uparrow\uparrow)$ bifurcates into an unstable symmetric mode and two stable local modes: $(\uparrow \; \cdot)$ and $(\cdot \; \uparrow)$. This is a standard "pitchfork bifurcation" [49] located at $\gamma = 2\varepsilon/N$. The concentration of vibrational energy on one of the CH bonds at a sufficiently large value of the anharmonicity parameter γ is an elementary example of a lattice soliton. As the local mode is clearly present in the nonlinear classical analysis, it is interesting to ask how it is represented by the linear structure of quantum theory.

8.2.2 Quantum analysis

Being a chemical molecule, dichloromethane requires quantum theory for its precise description. To provide corresponding data, physical chemists measure lines of absorption at particular energies [47], interpreted as differences between eigenvalues of an appropriate energy operator. Following Schrödinger [60], we let the Hamiltonian and number, defined in Equations (8.34) and (8.36), become operators

$$N \to \hat{N} \quad \text{and} \quad H \to \hat{H},$$

where

$$\hat{N} = b_1^\dagger b_1 + b_2^\dagger b_2 \tag{8.40}$$

and

$$\hat{H} = (\omega_0 - \gamma/2)\hat{N} - \frac{\gamma}{2}[(b_1^\dagger b_1)^2 + (b_2^\dagger b_2)^2] - \varepsilon(b_1^\dagger b_2 + b_2^\dagger b_1). \tag{8.41}$$

In deriving these operators, we have set

$$A_1 \to b_1$$
$$A_2 \to b_2$$
$$A_1^* \to b_1^\dagger$$
$$A_2^* \to b_2^\dagger$$

so b_1 and b_2 are lowering operators for the first and second oscillators, respectively. Similarly, $b_1^\dagger$ and $b_2^\dagger$ are raising operators for the two oscillators. In these expressions for $\hat{N}$ and $\hat{H}$, the constant terms have been dropped; thus they measure number and energy with respect to the lowest (ground state) eigenvalues.

To this point the quantum problem for a dihalomethane model has been formulated. To solve the problem, we must find the eigenvalues and eigenfunctions of $\hat{H}$. How do we proceed?

First of all, there is a matter of notation. If the two oscillators are uncoupled ($\varepsilon = 0$), Dirac's expression for an eigenfunction is $|n_1\rangle|n_2\rangle$, where n_1 and n_2 are the number levels of the two oscillators. For typographical convenience, this is written as

$$|n_1\rangle|n_2\rangle \equiv |n_1, n_2\rangle,$$

whereupon

$$b_1 |n_1, n_2\rangle = \sqrt{n_1}\, |n_1 - 1, n_2\rangle$$
$$b_2 |n_1, n_2\rangle = \sqrt{n_2}\, |n_1, n_2 - 1\rangle$$
$$b_1^\dagger |n_1, n_2\rangle = \sqrt{n_1 + 1}\, |n_1 + 1, n_2\rangle$$
$$b_2^\dagger |n_1, n_2\rangle = \sqrt{n_2 + 1}\, |n_1, n_2 + 1\rangle.$$

If $\varepsilon \neq 0$, $|n_1, n_2\rangle$ is no longer an eigenfunction of the energy operator because the two CH stretching modes interact through mechanical and electromagnetic forces. Thus it seems reasonable to try a combination of such expressions, but what combination?

To answer this question, note that the operators $\hat{N}$ and $\hat{H}$ commute, implying that a nondegenerate eigenfunction of $\hat{H}$ will also be an eigenfunction of $\hat{N}$. (Can you show this?) For a fixed value of

$$n = n_1 + n_2,$$

the most general eigenfunction of $\hat{N}$ is

$$|\psi_n\rangle = c_1|n, 0\rangle + c_2|n - 1, 1\rangle + \cdots + c_n|1, n - 1\rangle + c_{n+1}|0, n\rangle, \qquad (8.42)$$

where it is evident that

$$\hat{N}|\psi_n\rangle = n|\psi_n\rangle.$$

In this expression for $|\psi_n\rangle$, the c_j are $n + 1$ arbitrary complex constants that can be determined by demanding that $|\psi_n\rangle$ be also an eigenfunction of $\hat{H}$. That is to say, the requirement

$$\hat{H}|\psi_n\rangle = E|\psi_n\rangle \qquad (8.43)$$

fixes the values of the c_j.

Thus Equation (8.43) can be viewed as $n + 1$ equations for the coefficients of ket products: $|n_1, n_2\rangle$. All such equations can be written in matrix form as

$$\mathcal{H}\mathbf{c} = E\mathbf{c},$$

where **c** is the column vector $\mathrm{col}(c_1, c_2, \ldots, c_{n+1})$ and

$$\mathcal{H} \equiv H_0 - \varepsilon V. \tag{8.44}$$

The eigenvalues of the matrix $\mathcal{H}$ are the energy eigenvalues, and the corresponding eigenvectors together with Equation (8.42) determine the energy eigenfunctions. This approach to the solution of problems in quantum soliton theory is called the "number state method" (NSM) [23].

The matrix $\mathcal{H}$ can be described as follows.

1. With $\varepsilon = 0$, $\mathcal{H} = H_0$, a diagonal matrix of the form

$$H_0 = \mathrm{diag}[\alpha_0, \alpha_1, \alpha_2, \ldots, \alpha_2, \alpha_1, \alpha_0], \tag{8.45}$$

where

$$\alpha_j \equiv \omega_0 n - \frac{\gamma}{2}(n + n^2 - 2nj + 2j^2).$$

2. With $\varepsilon \neq 0$, off-diagonal elements are introduced into $\mathcal{H}$ through the additional matrix $-\varepsilon V$, where

$$V = \begin{bmatrix} 0 & \sqrt{n} & 0 & 0 & \cdot & 0 \\ \sqrt{n} & 0 & \sqrt{2(n-1)} & 0 & \cdot & 0 \\ 0 & \sqrt{2(n-1)} & 0 & \sqrt{3(n-2)} & \cdot & 0 \\ \cdot & \cdot & \cdot & \cdot & \cdot & \cdot \\ 0 & \cdot & \sqrt{3(n-2)} & 0 & \sqrt{2(n-1)} & 0 \\ 0 & \cdot & 0 & \sqrt{2(n-1)} & 0 & \sqrt{n} \\ 0 & \cdot & 0 & 0 & \sqrt{n} & 0 \end{bmatrix}$$

Thus $\mathcal{H} = H_0 - \varepsilon V$ is an $(n+1) \times (n+1)$, real, nonsingular matrix with symmetry about both diagonals. From a perturbation theory in small ε/γ, its lowest two eigenvalues can be shown to differ by [5]

$$\Delta E = \frac{2n\varepsilon}{(n-1)!}\left(\frac{\varepsilon}{\gamma}\right)^{n-1} + O(\varepsilon^{n+1}/\gamma^n), \tag{8.46}$$

corresponding to eigenfunctions of $\hat{H}$ of the form

$$|\psi_n\rangle^{\pm} = \tfrac{1}{\sqrt{2}}(|n,0\rangle \pm |0,n\rangle) + O(\varepsilon/\gamma). \tag{8.47}$$

We are now in a position to compare our quantum results with those from the classical analysis. At first glance, the two pictures seem to be quite different. From Equations (8.38), the classical analysis predicts energy localization in one or the other oscillator for a sufficiently large value of the anharmonicity parameter γ, while the quantum eigenfunctions—given in Equation (8.47)—have an equal probability of finding energy on either oscillator. It was this seeming contradiction that led to rejections of the local mode theory of physical chemistry, even though experimental evidence spoke strongly in its favor [35].

If we consider our quantum mechanical solution to be a wave packet with the form

$$|\Psi_n(t)\rangle = \tfrac{1}{\sqrt{2}}|\psi_n\rangle^+ e^{-iE^+ t} + \tfrac{1}{\sqrt{2}}|\psi_n\rangle^- e^{-iE^- t} + \mathrm{O}(\varepsilon/\gamma), \qquad (8.48)$$

then

$$|\Psi_n(0)\rangle = |n, 0\rangle + \mathrm{O}(\varepsilon/\gamma),$$

indicating that most of the energy is localized in oscillator #1 at time $t = 0$. How long will it remain there? Until the two main components in the wave packet $|\Psi_n(t)\rangle$ change their phase. This occurs in a "tunneling time" of order

$$\tau \sim \frac{\pi\hbar}{\Delta E},$$

where $\Delta E \equiv |E^+ - E^-|$ is measured in joules.

From numerical computations of the eigenvalues of $\mathcal{H}$, it is observed that the $\mathrm{O}(\varepsilon^{n+1}/\gamma^n)$ correction in Equation (8.46) is always negative [5] so

$$\Delta E < \frac{2n\varepsilon^n}{(n-1)!\gamma^{n-1}},$$

and the tunneling time τ for the initially localized energy to move from one oscillator to the other is

$$\tau > \frac{\pi\hbar(n-1)!\gamma^{n-1}}{2n\varepsilon^n}. \qquad (8.49)$$

For $n \geq 2$ and

$$\gamma > 2\varepsilon/n,$$

this lower bound on the tunneling time grows rapidly with n, in agreement with Equation (8.39). Thus there is no contradiction between the predictions of quantum and classical analysis.

In addition to its practical value for the interpretation of vibrational spectra of the dihalomethanes (and other molecules with two identical degrees of freedom), this example has several implications for an understanding of the relationship between quantum mechanics and classical nonlinear dynamics.

1. There is a broad parallel between the strategies of the classical and quantum analyses. To obtain Equations (8.38), a fixed value was assumed for the classical number N, allowing us to find stationary solutions as defined in Equation (8.37). Similarly in the quantum analysis, we assumed a fixed value for the eigenvalue of the corresponding number operator $\hat{N}$, allowing construction of the wave packet in Equation (8.48).
2. From the factor $(n-1)!$ in the numerator of Equation (8.49), the tunneling time grows rapidly with the principal quantum number n, easily

becoming longer than the duration of any conceivable measurement time. (It is sometimes asserted that without a mechanism for "collapse of the wave packet" the classical world would become fuzzy, with cannon balls degenerating into clouds of probability and cats being simultaneously dead and alive. Such claims neglect the arithmetic of Equation (8.49), which implies tunneling times for local modes that are longer than the age of the universe.)

3. Although the Hamiltonian operator defined in Equation (8.41) is symmetric with respect to the interchange of indices "1" and "2," a classical local mode—as in Equations (8.38)—is not. The very strong increase of tunneling time with n implied by Equation (8.49) provides a clear example of "symmetry breaking" as one proceeds from the quantum to the classical realm.

8.2.3 Comparison with experiments

In the previous subsection, we have derived several formulas; now the moment of truth is at hand. Does our quantum analysis bear any relation to experimental reality? Fortunately, measurements on the dihalomethanes by Ole Sonnich Mortensen and his colleagues provide a wealth of relevant data [47]. For each molecule, they have measured 15 transition energies between CH stretching vibrations, and our formulation contains only three parameters: ω_0, γ, and ε. A least squares fit of the eigenvalues of $\mathcal{H}$ to these spectral lines yields the optimum values for the parameters (in cm^{-1}) of the three molecules given in the following table.

Molecule:	CCl_2H_2	CBr_2H_2	CI_2H_2
ε	29.54	32.80	33.69
γ	127.44	125.45	124.25
$\omega_0 - \gamma/2$	3083.79	3089.53	3068.74
RMS errors	9.0	6.9	8.0

How good is the agreement? The last row of entries records the root mean square (RMS) errors in matching the spectral transitions for each molecule. By setting the three parameters ω_0, γ, and ε to the values indicated, our theory matches the fifteen observed lines to better than 0.3%.

Although this compares well with the more standard analysis of Mortensen et al., it is based upon approximations that ignore the following effects.

1. *Transition intensities.* In the quantized rotating-wave approximation, the eigenfunctions of a single bond are assumed to be the $|n\rangle$, identical to those of an harmonic oscillator. Because a simple harmonic oscillator allows transitions only between adjacent levels,

$$|n\rangle \to |n \pm 1\rangle,$$

the same restriction is imposed upon our model. In reality, as is noted from Equation (8.26), the wave functions of an anharmonic oscillator have different x-dependences than those of an harmonic oscillator. This allows transitions from the ground state to all other levels; thus

$$|0\rangle \to |1\rangle$$
$$|0\rangle \to |2\rangle$$
$$|0\rangle \to |3\rangle$$
$$\vdots$$
$$|0\rangle \to |n\rangle$$

with diminishing intensity as $|n\rangle$ increases. To its discredit, our model predicts zero for the intensities of all of these "overtone" transitions beyond the first.

2. *Line widths.* The overtone bands in our model have line widths given by the expression for ΔE in Equation (8.46), but observed line widths are much larger. Real CH stretching oscillations have been studied in great detail, both theoretically and experimentally, by Quack and his colleagues, on methane-like molecules of the class CHX_3 [31,43,53]. These works show how the CH stretching oscillation interacts with transverse bending modes of about half the stretching frequency in a process called Fermi resonance. (To appreciate this phenomenon, think of a child on a swing, pumping his or her legs at twice its natural frequency to gain amplitude.[7]) This interaction results in line widths of the CH stretch overtone bands of about 200 cm^{-1}.

8.3 Boson lattices

In the previous two sections, we have studied the quantum theory of nonlinear oscillators with one and two degrees of freedom. During the course of this effort, the basic tools needed to investigate any number of freedoms have been introduced, but numerical difficulties grow combinatorially with the number of quanta and the number of freedoms. It is important to understand such limits in order to recognize those problems that can actually be solved and those that cannot.

8.3.1 The discrete self-trapping equation

To appreciate the numerical difficulties that arise in the quantum analysis of nonlinear lattices, let us throw caution to the wind and jump from two degrees of freedom to an arbitrarily large number, with unrestricted energetic couplings.

---
[7]Electrical engineers use varactor diodes to produce the same effect.

An appropriate generalization of the system in the previous section is then the discrete self-trapping (DST) equation [18]

$$\left(i\frac{d}{dt} - \omega_0\right)\mathbf{A} + \gamma \text{diag}(|A_1|^2, |A_2|^2, \ldots, |A_f|^2)\mathbf{A} + \varepsilon M\mathbf{A} = 0, \quad (8.50)$$

where the vector

$$\mathbf{A} \equiv \text{col}(A_1, A_2, \ldots, A_f)$$

represents the complex mode amplitudes of f anharmonic, interacting degrees of freedom. Dispersive interactions between the degrees of freedom are introduced through the real, symmetric, $f \times f$ matrix

$$\varepsilon M = [\varepsilon m_{jk}]$$

and anharmonicity through the diagonal matrix of the second term. If the real parameter $\gamma = 0$ and the real parameter $\varepsilon \neq 0$, Equation (8.50) represents a system of coupled linear oscillators. If, on the other hand, $\varepsilon = 0$ and $\gamma \neq 0$, Equation (8.50) represents a system of noninteracting, anharmonic oscillators. We are primarily interested in the general case where both $\varepsilon \neq 0$ and $\gamma \neq 0$. (As in the previous section, A is taken to be unitless, while d/dt, ω_0, γ, and ε are measured in units of cm^{-1}.)

Solutions of the classical DST have two constants of the motion: the number

$$N = \sum_{j=1}^{f} |A_j|^2 \quad (8.51)$$

and the energy

$$H = \omega_0 N - \frac{\gamma}{2}\sum_{j=1}^{f} |A_j|^4 - \varepsilon \sum_{j \neq k} m_{jk} A_j^* A_k. \quad (8.52)$$

In the previous section, a special example of this system was considered with $f = 2$ and

$$M = \begin{bmatrix} 0 & 1 \\ 1 & 0 \end{bmatrix}.$$

Thus under quantization

$$N \to \hat{N} \quad \text{and} \quad H \to \hat{H},$$

where

$$\hat{N} = \sum_{j=1}^{f} b_j^\dagger b_j \quad (8.53)$$

and
$$\hat{H} = (\omega_0 - \gamma/2)\hat{N} - \frac{\gamma}{2}\sum_{j=1}^{f}(b_j^\dagger b_j)^2 - \varepsilon \sum_{j \neq k} m_{jk} b_j^\dagger b_k. \qquad (8.54)$$

As in the previous section, $b_j^\dagger$ and b_j are boson raising and lowering operators for the jth freedom, and constant terms have been dropped in the definitions of $\hat{N}$ and $\hat{H}$.

Again we apply the NSM to find eigenvalues and eigenfunctions of the energy operator. Because $\hat{N}$ and $\hat{H}$ commute, nondegenerate eigenfunctions of $\hat{H}$ are eigenfunctions of $\hat{N}$. The most general eigenfunction of $\hat{N}$ with eigenvalue n is a linear combination of terms of the form

$$|n_1\rangle|n_2\rangle \ldots |n_f\rangle \equiv |n_1, n_2, \ldots, n_f\rangle,$$

with
$$n = n_1 + n_2 + \cdots + n_f.$$

The number of such terms is equal to the number of ways that n quanta can be put into f modes of oscillation (or the number of ways that n beans can be placed in f jars). This is

$$p = \frac{(n+f-1)!}{(f-1)!\, n!}, \qquad (8.55)$$

a function that grows rapidly with n and f.

The number p is important because it places an upper bound on the size of the matrices. For example, if $f = 3$ and $n = 4$, $p = 15$ and the most general eigenfunction of $\hat{N}$ is

$$\begin{aligned}|\psi_4\rangle = {}& c_1|400\rangle + c_2|040\rangle + c_3|004\rangle + c_4|310\rangle + c_5|301\rangle \\ & + c_6|130\rangle + c_7|031\rangle + c_8|103\rangle + c_9|013\rangle + c_{10}|220\rangle \\ & + c_{11}|202\rangle + c_{12}|022\rangle + c_{13}|211\rangle + c_{14}|121\rangle + c_{15}|112\rangle.\end{aligned}$$

Because a general wave function $|\psi_n\rangle$ of this sort is an eigenfunction of $\hat{N}$, we must choose the constants $(c_1 \ldots c_p)$ so that

$$\hat{H}|\psi_n\rangle = E|\psi_n\rangle.$$

Equating coefficients of the kets—$|n_1, n_2, \ldots, n_f\rangle$—leads to the matrix equation

$$\mathcal{H}_n \mathbf{c} = E\mathbf{c},$$

where $\mathbf{c} \equiv \mathrm{col}\,(c_1, c_2, \ldots, c_p)$ and $\mathcal{H}_n$ is a symmetric, $p \times p$ matrix with real components.

From a cursory inspection of Equation (8.55), it is clear that the dimension of $\mathcal{H}$ may easily become inconveniently large. Thus it is of interest to consider means for block-diagonalization of $\mathcal{H}$ to reduce the numerical task. This can be accomplished by taking advantage of additional symmetries of the problem, which will also be those of the energy operator $\hat{H}$. If a molecule has a certain symmetry, for example, $\hat{H}$ will commute with the operators of the corresponding symmetry group, and nondegenerate eigenfunctions of $\hat{H}$ must also be eigenfunctions of those symmetry operators.

A convenient example is the molecule benzene (C_6H_6) with the six carbon atoms arranged in a hexagonal ring as in Figure 1.3. If the DST equation is used to represent the six C–H stretching oscillators of the molecule, the matrix M takes the form

$$M = \begin{bmatrix} 0 & 1 & a & b & a & 1 \\ 1 & 0 & 1 & a & b & a \\ a & 1 & 0 & 1 & a & b \\ b & a & 1 & 0 & 1 & a \\ a & b & a & 1 & 0 & 1 \\ 1 & a & b & a & 1 & 0 \end{bmatrix}.$$

As $f = 6$,

$$p = \frac{(n+5)!}{5!\,n!}.$$

One need not diagonalize a matrix of this size, however, because a general eigenfunction of both $\hat{N}$ (with eigenvalue n) and the translation operator $\hat{T}$ (with eigenvalue τ) can be constructed as

$$|\psi_{n,\tau}\rangle = c_1 \sum_{j=1}^{6} (\hat{T}/\tau)^{j-1} |n, 0, 0, 0, 0, 0\rangle$$

$$+ c_2 \sum_{j=1}^{6} (\hat{T}/\tau)^{j-1} |n-1, 1, 0, 0, 0, 0\rangle$$

$$+ \cdots + c_{p/6} \sum_{j=1}^{6} (\hat{T}/\tau)^{j-1} |n-5, 1, 1, 1, 1, 1\rangle,$$

where

$$\hat{T}|n_1, n_2, n_3, n_4, n_5, n_6\rangle = |n_2, n_3, n_4, n_5, n_6, n_1\rangle \tag{8.56}$$

and

$$\hat{T}|\psi_{n,\tau}\rangle = \tau|\psi_{n,\tau}\rangle. \tag{8.57}$$

Evidently this formulation reduces the size of the matrix to be diagonalized from p to $p/6$. (If $p/6$ is not an integer, then the next higher integer value is used in constructing the reduced matrix.)

At the tenth overtone of the fundamental CH stretching oscillation (the highest that has been observed experimentally) $n = 11$ so $p = 4368$ and $p/6 = 728$. This reduction of the dimension of the problem is important not merely because it reduces the size of the matrix to be solved; it also reduces the size of the matrix that must be constructed.

As in the discussions related to Equations (8.46) and (8.48) of the previous section, it turns out that for $\gamma > \varepsilon$ the lowest f eigenvalues of $\mathcal{H}_n$ lie within an energy range

$$\Delta E = \frac{(\Delta\lambda)n\varepsilon^n}{(n-1)!\gamma^{n-1}} + \mathrm{O}\left(\frac{\varepsilon^{n+1}}{\gamma^n}\right), \tag{8.58}$$

where $\Delta\lambda$ is the maximum splitting between eigenvalues of the matrix $M = [m_{jk}]$, appearing in Equation (8.50) [5].

We have seen in Section 8.2.3 that $\gamma > \varepsilon$ for a CH stretching oscillation in the dihalomethanes. This implies that ΔE, as defined in Equation (8.58), is small enough for the six lowest energy eigenvalues to be perceived experimentally as a single overtone band for each value of $n \geq 2$.

Choosing the DST parameters to be

$$\omega_0 - \gamma/2 = 3098.01\,\mathrm{cm}^{-1}$$
$$\gamma = 117.45\,\mathrm{cm}^{-1}$$
$$\varepsilon = -4.0\,\mathrm{cm}^{-1}$$
$$a = 0.25$$
$$b = -0.25$$

gives good agreement with the four components of the fundamental spectrum (two are degenerate) and 10 overtones of liquid benzene: a total of 14 spectral lines [64].

These overtone bands can be viewed as experimental evidence for quantum lattice solitons (or local modes) at the molecular level. Interestingly, they were reported in the 1920s (see Figure 1.3) [19].

8.3.2 A lattice nonlinear Schrödinger equation

Consider a lattice nonlinear Schrödinger equation of the form

$$\left(\mathrm{i}\frac{d}{dt} - \omega_0\right)A_j + \varepsilon(A_{j+1} + A_{j-1}) + \gamma|A_j|^2 A_j = 0,$$

with $j = 1, 2, 3, \ldots, f$ and periodic boundary conditions so $A_{j+f} = A_j$. Introducing the transformation

$$A_j = \phi_j e^{-\mathrm{i}(\omega_0 - \gamma)t},$$

and quantizing by letting $\phi_j \to b_j$ and $\phi_j^* \to b_j^\dagger$, leads to the quantum discrete nonlinear Schrödinger (QDNLS) equation[8]

$$i\frac{db_j}{dt} + \varepsilon(b_{j+1} + b_{j-1}) + \gamma b_j^\dagger b_j b_j = 0. \tag{8.59}$$

This equation is related to the number operator

$$\hat{N} = \sum_{j=1}^{f} b_j^\dagger b_j \tag{8.60}$$

and the energy operator

$$\hat{H} = -\sum_{j=1}^{f}[\varepsilon(b_j^\dagger b_{j+1} + b_j^\dagger b_{j-1}) + \frac{\gamma}{2}b_j^\dagger b_j^\dagger b_j b_j]. \tag{8.61}$$

Because $\hat{H}$ commutes with both $\hat{N}$ and the translation operator $\hat{T}$ that was defined in Equation (8.56), a quantum analysis is begun by constructing a general eigenfunction of both $\hat{N}$ and $\hat{T}$. For $n=1$, this takes the form

$$|\psi_{1,\tau}\rangle = c_1 \sum_{j=1}^{f} (\hat{T}/\tau)^{j-1}|100\cdots 00\rangle$$

so

$$E_1 = -\varepsilon(\tau + \tau^{-1}) = -2\varepsilon\cos k.$$

Here

$$\tau = e^{ik}$$

is the eigenvalue of $\hat{T}$ that corresponds to the propagation number

$$k = 2\pi\nu/f,$$

where for f odd

$$\nu = 0, \pm 1, \pm 2, \ldots, \pm(f-1)/2.$$

This expression for the energy eigenvalues can be written as

$$E_1(k) = E_1(0) + \frac{k^2}{2m^*} + \mathrm{O}(k^4),$$

where

$$m^* \equiv 1/2\varepsilon$$

[8] This formulation is sometimes called the "Heisenberg picture," in which the wave function remains constant while the operators are viewed as functions of time. In this chapter, discussions are entirely in the "Schrödinger picture," where the operators are constant and the wave functions depend on time.

is called the "effective mass" because the group velocity (V_g) of a wave packet is

$$V_g = \frac{dE_1(k)}{dk} \doteq \frac{k}{m^*},$$

making the kinetic energy term

$$\frac{k^2}{2m^*} \doteq \frac{1}{2} m^* V_g^2$$

at low values of the group velocity.

With f odd and $n = 2$, a general eigenfunction of both $\hat{N}$ and $\hat{T}$ is

$$|\psi_{2,\tau}\rangle = c_1 \sum_{j=1}^{f} (\hat{T}/\tau)^{j-1} |20 \cdots 0\rangle + c_2 \sum_{j=1}^{f} (\hat{T}/\tau)^{j-1} |110 \cdots 0\rangle$$

$$+ \cdots + c_{(f+1)/2} \sum_{j=1}^{f} (\hat{T}/\tau)^{j-1} |10 \cdots 010 \cdots 00\rangle. \quad (8.62)$$

Requiring that

$$\hat{H}|\psi_{2,\tau}\rangle = E|\psi_{2,\tau}\rangle$$

and equating the coefficients of kets leads to the $p \times p$ block-diagonalized matrix

$$\mathcal{H}_2 = \mathrm{diag}[\mathcal{Q}(2, \tau_1), \mathcal{Q}(2, \tau_2), \ldots, \mathcal{Q}(2, \tau_f)].$$

For a particular value of τ, energy eigenvalues are obtained from the matrix equation

$$\mathcal{Q}(2, \tau)\mathbf{c} = E\mathbf{c},$$

where $\mathbf{c} \equiv \mathrm{col}(c_1, c_2, \ldots, c_{(f+1)/2})$, $\mathcal{Q}(2, \tau)$ is the $[(f+1)/2] \times [(f+1)/2]$ tridiagonal matrix

$$\mathcal{Q}(2, \tau) = -\begin{bmatrix} \gamma & q^* \sqrt{2} & 0 & \cdot & \cdot & \cdot \\ q\sqrt{2} & 0 & q^* & 0 & \cdot & 0 \\ \cdot & q & 0 & q^* & \cdot & \cdot \\ \cdot & \cdot & \cdot & \cdot & \cdot & \cdot \\ \cdot & \cdot & \cdot & q & 0 & q^* \\ \cdot & \cdot & \cdot & \cdot & q & p \end{bmatrix}, \quad (8.63)$$

and

$$q \equiv 1 + \tau \quad \text{and} \quad p \equiv \tau^{(f+1)/2} + \tau^{(f-1)/2}.$$

Much can be learned about the QDNLS equation from an examination of the eigenvalues of this matrix. Using $\tau = \exp(ik)$, these eigenvalues are plotted in Figure 8.4 as a function of k in the limit of large f. This plot shows two main types of eigenvalues. First there is a continuum band which is bounded by $\pm 4\varepsilon \cos(k/2)$

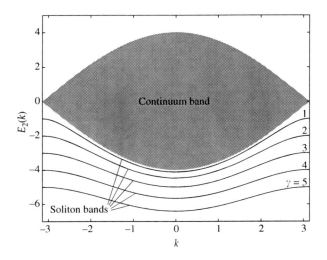

FIG. 8.4. *Energy eigenvalues of the QDNLS equation in the* $f \to \infty$ *limit, with* $\varepsilon = 1$ *and several values of* γ.

and is insensitive to the value of γ. As $f \to \infty$, the density of eigenvalues within this region increases without bound; thus the band is indicated on the figure by a gray shading. Second, there is a single band comprising f eigenvalues, which lies below the continuum band by an amount that depends upon γ. In Figure 8.4, this lower band is plotted for several values of γ.

Examination of the corresponding eigenfunctions shows that for the lower band the coefficient c_1 in Equation (8.62) is large compared to the other coefficients; thus this band indicates that the system prefers to be in a state where the two quanta sit on the same site. (For an eigenfunction of $\hat{H}$, there is an equal probability of finding the two quanta together on any of the f sites.) This localization of energy corresponds to the soliton solution of the classical DNLS equation; hence the name "soliton" band.

Near $k = 0$, it is clear from Figure 8.4 that the energies of states in the soliton band are given by an expression of the form

$$E_2(k) = E_2(0) + k^2/2m^* + O(k^4). \tag{8.64}$$

In the limit $f \to \infty$, the soliton band has energy [65]

$$E_2(k) = -\sqrt{\gamma^2 + 16\varepsilon^2 \cos^2(k/2)}. \tag{8.65}$$

Defining the "soliton binding energy" (E_b) as the difference between the energy of the soliton band at $k = 0$ and the bottom of the continuum band implies that

$$E_b = \sqrt{\gamma^2 + 16\varepsilon^2} - 4\varepsilon,$$

and
$$m^* = \sqrt{(\gamma^2 + 16\varepsilon^2)}/4\varepsilon^2$$

Proceeding in this manner, it is possible—in principle—to construct block-diagonalized Hamiltonian matrices for any value of the quantum number n. As was noted in the previous section, however, the sizes of these blocks may grow beyond the limits of computational convenience (or even possibility) so it is of interest to consider approximate calculations that are useful in asymptotic limits. With this motivation, the following results have been obtained.

Small γ, large f, arbitrary n

If the anharmonicity parameter $\gamma \ll \varepsilon$, the size of a classical DNLS soliton is about $24\varepsilon/(n+1)\gamma$ lattice spacings. Thus making

$$f \gg \frac{24\varepsilon}{(n+1)\gamma}$$

insures that a classical soliton will fit onto the lattice. Using the above methods, the binding energy has been calculated to be $\gamma^2/8\varepsilon$ for $n = 2$ and $\gamma^2/2\varepsilon$ for $n = 3$ [46]. Both of these results agree with the expression

$$E_{\rm b} = \frac{\gamma^2}{48\varepsilon} n(n^2 - 1), \tag{8.66}$$

derived for the continuum NLS equation from considerations of quantum field theory [36]. It is reasonable, therefore, to assume that Equation (8.66) gives the correct binding energy for a large QDNLS soliton on an infinite lattice at any value of n.

To calculate the effective mass of a QDNLS soliton, note that for $\gamma \ll \varepsilon$ the classical problem is nearly linear, and the single quantum energy is $E_1(k) \doteq -2\varepsilon \cos k$. Thus for n quanta with propagation numbers $k_1, k_2, \ldots, k_n$, the energy is just the sum $-2\varepsilon \sum_j \cos k_j$. For the n-quantum wave function, the lowest energy is found for $k_1 = k_2 = \cdots = k_n$; so as $\gamma \to 0$

$$E_n(k) \to -2n\varepsilon \cos(k/n), \tag{8.67}$$

and
$$m^* \to n/2\varepsilon. \tag{8.68}$$

In other words, the large DNLS soliton discussed in Section 5.2.2 can be viewed as a condensation of n bosons.

Large γ, arbitrary f, arbitrary n

In the limit of large γ, it is evident for $n = 2$ that the dominant element of the matrix $\mathcal{Q}(2,\tau)$ displayed in Equation (8.63) is the 2×2 submatrix in the upper left-hand corner. For arbitrary values of n, perturbation theory in small

ε/γ shows that to calculate the leading k-dependent terms it is only necessary to consider the sequence of interactions [65]

$$|n\rangle \leftrightarrow |n-1,1\rangle \leftrightarrow |n-2,2\rangle \leftrightarrow \cdots \leftrightarrow |2,n-2\rangle \leftrightarrow |1,n-1\rangle \leftrightarrow |n\rangle,$$

requiring an approximate wave function

$$|\psi_n\rangle \doteq \frac{1}{\sqrt{f}} \left\{ c_1 \sum_{j=1}^{f} (\hat{T}/\tau)^{j-1} |n0\cdots 0\rangle + c_2 \sum_{j=1}^{f} (\hat{T}/\tau)^{j-1} |(n-1)10\cdots 0\rangle \right.$$

$$+ c_3 \sum_{j=1}^{f} (\hat{T}/\tau)^{j-1} |(n-1)00\cdots 1\rangle$$

$$\left. + \cdots + c_n \sum_{j=1}^{f} (\hat{T}/\tau)^{j-1} |(n/2)(n/2)0\cdots 0\rangle \right\}$$

for n even and a similar expression for n odd. Using this approximate wave function, one finds the lowest eigenvalue of the dominant $n \times n$ submatrix of $-\mathcal{Q}(n,\tau)$ to define a soliton band as

$$E_n(k) \doteq -\frac{1}{2}n(n-1)\gamma - \left(\frac{2n\varepsilon^n}{(n-1)!\gamma^{n-1}}\right)\cos k, \qquad (8.69)$$

where the symbol "$\doteq$" indicates that the first term on the right-hand side is correct to $\mathrm{O}(\varepsilon/\gamma)$ while the second (k-dependent) term is correct to $\mathrm{O}(\varepsilon^{n-1}/\gamma^{n-1})$. In other words, terms of $\mathrm{O}(\varepsilon/\gamma)$ that do not depend on k have been dropped.

Thus for the QDNLS at large γ, the soliton binding energy and effective mass are

$$E_\mathrm{b} \doteq \tfrac{1}{2}n(n-1)\gamma \qquad (8.70)$$

and

$$m^* \doteq \frac{(n-1)!\gamma^{n-1}}{2n\varepsilon^n}. \qquad (8.71)$$

8.3.3 Soliton wave packets

We are now in a position to consider the quantum description of a lattice soliton. The picture that emerges from our study of the QDNLS system with f degrees of freedom (lattice sites) and translational symmetry is as follows. For each value of the principal quantum number n and propagation number k, there is a lowest energy eigenvalue. These f lowest eigenvalues lie on a band

$$E = E_n(k)$$

as is shown on Figure 8.4. The propagation number

$$k = 2\pi\nu/f, \qquad (8.72)$$

where $\nu = 0, \pm 1, \ldots, \pm(f/2 - 1), f/2$ for f even and $0, \pm 1, \ldots, \pm(f-1)/2$ for f odd.

Each energy eigenvalue corresponds to a pure eigenstate, written in Dirac's notation as $|\psi_n(k)\rangle$ and normalized as

$$\langle \psi_n(k)|\psi_n(k)\rangle = 1.$$

Wave packet solutions of the time dependent Schrödinger equation

$$i\frac{d}{dt}|\Psi(t)\rangle = \hat{H}|\Psi(t)\rangle \tag{8.73}$$

can be constructed as double sums over the principal quantum number (n) and the propagation number (k); thus

$$|\Psi(t)\rangle = \sum_{n=1}^{\infty} a_n \sum_k G_n(k)|\psi_n(k)\rangle e^{-iE_n(k)t}, \tag{8.74}$$

where k takes the f values between $-\pi$ and π that are indicated in Equation (8.72) and shown in Figure 8.4. Because it is necessary that

$$\langle \Psi(t)|\Psi(t)\rangle = 1,$$

both the a_n and the $G_n(k)$ can be chosen as sets of complex numbers that satisfy the normalization conditions

$$\sum_n |a_n|^2 = 1 \quad \text{and} \quad \sum_k |G_n(k)|^2 = 1.$$

The wave function of Equation (8.74) is a soliton wave packet.

It is important to notice that Equation (8.74) does not represent the most general wave function that satisfies Equation (8.73) because it is constructed only from eigenfunctions with eigenvalues on the soliton bands. It is evident from Figure 8.4 that the system has other eigenstates (the continuum band) that are excluded from $|\Psi(t)\rangle$ as defined in Equation (8.74). Thus the soliton wave packet is characterized by two interdependent properties. First, for given values of k and n, $|\Psi(t)\rangle$ has the lowest energy, and second, under the same conditions, $|\Psi(t)\rangle$ has the highest probability of quanta being located near each other. These two properties are the basis for referring to $|\Psi(t)\rangle$ as a "soliton wave packet."

This section concludes with several comments:

1. If the soliton is known to be located in a particular region of the lattice, $G_n(k)$ is essentially a discrete Fourier transform of the pulse shape, implying that uncertainties in position (Δj) and in propagation number (Δk) are constrained by

$$\Delta k \times \Delta j \sim 1,$$

the uncertainty relationship of Heisenberg mentioned in Section 8.1.2. If the $G_n(k)$ are adjusted to narrow the wave packet and thereby locate the soliton more precisely, a wider band of k-states (momentum) must necessarily be used.

2. For the description of spectral measurements, each wave packet is formed for a particular value of the principal quantum number (say $n = n_0$); thus
$$a_n = \delta_{n,n_0}.$$

3. In measurements involving infrared or Raman spectra on molecular crystals, the wavelength of the interacting radiation is much larger than the dimensions of a unit cell, so it is usually appropriate to assume that $k \approx 0$.

4. For wave packets that are close to classical solitons, $n \gg 1$, and it is necessary to employ a range of values for the principal quantum number. This can be accomplished using an approximation described in the following section.

8.3.4 The Hartree approximation

Let us return to the DST equation that was introduced in Section 8.3.1 and consider the problem of constructing its quantum wave function for relatively large values of the principal quantum number (or number of bosons) n. A direct approach is often not feasible because the dimension of the matrix $\mathcal{H}_n$—given in Equation (8.55)—is too large. In this case, it is appropriate to consider the "Hartree approximation" (HA),[9] with the wave function for each boson calculated in the average field of all the others [74].

We start with the energy operator of Equation (8.54) in the "Schrödinger picture," implying that the state vector $|\psi(t)\rangle$ is time dependent, and the quantum operators are those at time $t = 0$. Schrödinger's equation for the state vector is then
$$i\frac{d}{dt}|\psi(t)\rangle = \hat{H}|\psi(t)\rangle,$$
and the most general n-quantum (boson) state vector can be written as
$$|\psi_n(t)\rangle = \frac{1}{\sqrt{n!}} \sum_{j_1=1}^{f} \sum_{j_2=1}^{f} \cdots \sum_{j_n=1}^{f} \theta_n(j_1, j_2, \ldots, j_n, t) b_{j_1}^\dagger b_{j_2}^\dagger \ldots b_{j_n}^\dagger |0\rangle, \qquad (8.75)$$

where $|0\rangle \equiv |0\rangle_1 |0\rangle_2 \cdots |0\rangle_f$ is the vacuum state. The θ_n are f^n time dependent coefficients of corresponding number states. For example, if $f = 2$ and $n = 3$, $\theta_3(2,1,2,t)$ indicates that the first boson is put onto the second freedom, the second boson is put on the first freedom, and the third boson is put on the second freedom; thus it is a coefficient of the number state $|1\rangle|2\rangle$.

---
[9] After Douglas R. Hartree (1897-1958), a British physicist who pioneered this method for the computation of electronic states in atoms.

In other words, $\theta_n(j_1, j_2, \ldots, j_n, t)$ is an n-boson wave function, normalized as

$$\sum_{j_1=1}^{f} \sum_{j_2=1}^{f} \cdots \sum_{j_n=1}^{f} |\theta_n(j_1, j_2, \ldots, j_n, t)|^2 = 1. \tag{8.76}$$

Substituting the state vector into the Schrödinger equation and using the boson commutation relations $[b_j, b_k^\dagger] = \delta_{jk}$ gives the following equation for the n-boson wave function:

$$\left(i\frac{d}{dt} - \tilde{\omega}_0\right)\theta_n(j_1, j_2, \ldots, j_n, t) + \sum_{k=1}^{f}[m_{j_1,k}\theta_n(k, j_2, j_3, \ldots, j_n, t)$$
$$+ m_{j_2,k}\theta_n(j_1, k, j_3, \ldots, j_n, t) + \cdots + m_{j_n,k}\theta_n(j_1, j_2, \ldots, k, t)]$$
$$+ \gamma \sum_{l=1}^{n} \sum_{m>l}^{n} \delta_{j_l,j_m}\theta_n(j_1, \ldots, j_l, \ldots, j_m, \ldots, j_n, t) = 0, \tag{8.77}$$

where $\tilde{\omega}_0 \equiv (\omega_0 - \gamma)$. Equation (8.77) is the Schrödinger equation for a system of bosons at f discrete sites (freedoms) with linear coupling (m_{jk})—a quantum version of the DST equation, which was introduced in Section 5.2.3. The last term on the left-hand side indicates a discrete delta function interaction between pairs of bosons, stemming from the fact that it is energetically favorable to have several bosons at a single site of the lattice.

As was noted above, the θ_n in Equation (8.77) are f^n time dependent coefficients of corresponding number states, but not all the θ_n are independent because bosons are indistinguishable quanta. For example, if $f = 2$ and $n = 2$, Equation (8.75) becomes

$$|\psi_2(t)\rangle = \theta_2(1, 1, t)|2\rangle|0\rangle + \theta_2(2, 2, t)|0\rangle|2\rangle$$
$$+ \frac{1}{\sqrt{2}}[\theta_2(1, 2, t) + \theta_2(2, 1, t)]|1\rangle|1\rangle. \tag{8.78}$$

In the number state formulation of Section 8.3.1, on the other hand, the most general eigenfunction of the boson number operator ($\hat{N}$) was constructed as

$$|\psi_2\rangle = c_1|2\rangle|0\rangle + c_2|1\rangle|1\rangle + c_3|0\rangle|2\rangle,$$

with c_1, c_2, and c_3 chosen so $|\psi_2\rangle$ is also an eigenstate of $\hat{H}$ with eigenvalue E. Time dependence is then introduced by multiplying each energy eigenfunction by the factor e^{-iEt}.

In Equation (8.78), $\theta_2(1, 2, t)$ is necessarily equal to $\theta_2(2, 1, t)$ because there is no physical difference between putting the first boson on the first freedom and the second on the second and putting the first on the second and the second on the first. Thus the true order of the system is equal to the number of ways that n bosons can be placed on f freedoms.

So far this is merely a reformulation of the problem, no approximations having been made. Recognizing that the order of the quantum system may be inconveniently large, we turn now to the HA, in which it is assumed that the n-boson wave function $\theta_n(j_1, \ldots, j_n, t)$ can be written as a product of the form

$$\theta_n^{(H)}(j_1, \ldots, j_n, t) \doteq \prod_{k=1}^{n} \Phi_{n,j_k}(t), \tag{8.79}$$

satisfying the above mentioned symmetry condition for a many-boson wave function

$$\theta_n(j_1, \ldots, j_l, \ldots, j_m, \ldots, j_n, t) = \theta_n(j_1, \ldots, j_m, \ldots, j_l, \ldots, j_n, t).$$

The physical idea of the HA is that each boson feels the same mean field potential due to all the other bosons; thus the many-body wave function can be approximated as a product of identical single-boson wave functions $\Phi_{n,j_k}(t)$ with $j_k = 1, 2, \ldots, f$ and $k = 1, 2, \ldots, n$ labeling the boson. As the single-boson wave function is assumed to be identical for each boson, we may drop the superfluous k label and simply use the notation $\Phi_{n,j}(t)$, where $j = 1, 2, \ldots, f$.

With the HA wave function in Equation (8.79), the n-boson state vector in Equation (8.75) becomes

$$|\psi_n(t)\rangle^{(H)} = \frac{1}{\sqrt{n!}} \left[\sum_{j=1}^{f} \Phi_{n,j}(t) b_j^\dagger \right]^n |0\rangle, \tag{8.80}$$

and from Equation (8.76) the normalization condition is

$$\sum_{j=1}^{f} |\Phi_{n,j}(t)|^2 = 1. \tag{8.81}$$

To obtain an equation of motion for $\Phi_{n,j}(t)$, note that Equation (8.77) for $\theta_n(j_1, j_2, \ldots, j_n, t)$ can be obtained by finding extrema of the functional

$$S = \int_{-\infty}^{\infty} dt \sum_{j_1=1}^{f} \sum_{j_2=1}^{f} \cdots \sum_{j_n=1}^{f} \theta_n^* \left\{ \left(i\frac{d}{dt} - \tilde{\omega}_0 \right) \theta_n(j_1, j_2, \ldots, j_n, t) \right.$$

$$+ \sum_{k=1}^{f} [m_{j_1,k} \theta_n(k, j_2, j_3, \ldots, j_n, t) + m_{j_2,k} \theta_n(j_1, k, j_3, \ldots, j_n, t) + \cdots$$

$$+ m_{j_n,k} \theta_n(j_1, j_2, \ldots, k, t)]$$

$$\left. + \gamma \sum_{l=1}^{n} \sum_{m>l}^{n} \delta_{j_l,j_m} \theta_n(j_1, \ldots, j_l, \ldots, j_m, \ldots, j_n, t) \right\} \tag{8.82}$$

as $\partial S/\partial \theta^* = 0$.

Substituting the HA wave function from Equation (8.79) into Equation (8.82) and using the normalization condition of Equation (8.81) gives

$$S^{(\mathrm{H})} = n \int_{-\infty}^{\infty} dt \sum_{j=1}^{f} \left\{ \Phi_{n,j}^{*} \left[i\frac{d\Phi_{n,j}}{dt} - \tilde{\omega}_0 \Phi_{n,j} + \sum_{k=1}^{f} m_{jk} \Phi_{n,k} \right] + \frac{\gamma}{2}(n-1)|\Phi_{n,j}|^4 \right\}.$$

Requiring

$$\frac{\partial S^{(\mathrm{H})}}{\partial \Phi_{n,j}^{*}} = 0$$

for the Hartree solution yields the following equation for the effective single-boson wave function:

$$i\frac{d\Phi_{n,j}}{dt} - \tilde{\omega}_0 \Phi_{n,j} + \sum_{k=1}^{f} \varepsilon m_{jk} \Phi_{n,k} + \gamma(n-1)|\Phi_{n,j}|^2 \Phi_{n,j} = 0. \quad (8.83)$$

Equation (8.83) is the main result of this section. Together with Equation (8.79), it enables one to construct $\theta_n^{(\mathrm{H})}(j_1, \ldots, j_n, t)$, the HA to the many-boson wave function $\theta_n(j_1, \ldots, j_n, t)$. It is interesting to notice that Equation (8.83) is closely related to the classical DST in Equation (8.50), which was considered in Chapter 5. The differences are these:

1. the solution of Equation (8.83) is constrained by the normalization condition of Equation (8.81),
2. the nonlinear parameter γ is multiplied by the factor $(n-1)$, and
3. the site frequency in Equation (8.50) has changed from ω_0 to $\tilde{\omega}_0 \equiv \omega_0 - \gamma$ in Equation (8.83).

Note that Equation (8.83) can be written in Hamiltonian form as

$$i\frac{d\Phi_{n,j}}{dt} = \frac{\partial h_n}{\partial \Phi_{n,j}^{*}},$$

where

$$h_n = \sum_{j=1}^{f} \left[\tilde{\omega}_0 |\Phi_{n,j}|^2 - \frac{\gamma}{2}(n-1)|\Phi_{n,j}|^4 - \varepsilon \sum_{k=1}^{f} \Phi_{n,j}^{*} m_{jk} \Phi_{n,k} \right] \quad (8.84)$$

is the effective single-boson Hamiltonian in the presence of the other $(n-1)$ bosons.

Having derived these formulas for the HA, it is appropriate to ask: How good is it? One test is to compare the exact expression for the binding energy of a soliton on an infinite QDNLS lattice with $\gamma \ll \varepsilon$ given in Equation (8.66) with that obtained from the HA. Thus

$$E_b^{\text{exact}} = \frac{\gamma^2}{48\varepsilon} n(n^2 - 1)$$

$$E_b^{\text{H}} = \frac{\gamma^2}{48\varepsilon} n(n-1)^2.$$

As expected, the binding energy in the HA is less than that for the exact solution. For $n > 1$, the fractional error

$$\frac{E_b^{\text{exact}} - E_b^{\text{H}}}{E_b^{\text{exact}}} = \frac{2}{n+1}$$

equals $2/3$ at $n = 2$ and $1/2$ at $n = 3$. For a relatively small number of degrees of freedom, however, this error estimate is unduly pessimistic. With two and three freedoms and any value of the principal quantum number, the lowest Hartree energy gives a better estimate of the lowest energy eigenvalue [74].

Finally, consider the HA as a means for constructing wave packets that approximate the dynamics of classical particles. This question was first addressed by Schrödinger in 1926, who described a nondispersive packet for the simple harmonic oscillator that follows the classical motion [61].

From our present perspective, it is convenient to construct a wave function that is an eigenvector of the lowering operator. Given a complex number A, such an eigenvector is

$$|A\rangle \equiv e^{-|A|^2/2} \sum_{n=0}^{\infty} \frac{A^n}{\sqrt{n!}} |n\rangle, \qquad (8.85)$$

for which

$$b|A\rangle = A|A\rangle \qquad (8.86)$$

and

$$\langle A|A \rangle = 1.$$

In the first of Equations (8.20), we quantized an oscillator as

$$A \to b,$$

replacing the rotating-wave amplitude by a bosonic lowering operator. Starting from the quantum picture, on the other hand, we can now construct a wave function $|A\rangle$ that has the property

$$\langle A|b|A\rangle = A,$$

establishing the correspondence between the quantum lowering operator and the classical rotating-wave amplitude in the direction

$$b \to A.$$

In the language of quantum theory, a wave function that satisfies the condition of Equation (8.86) is called a "coherent state" [28, 29]. Such a state is the closest quantum approximation to the classical solution of the unquantized problem. Generalizations of the coherent state formulation to other quantum systems have been developed and reviewed by Perelomov [50], and are summarized in a book by Makhankov [41].

To construct a coherent state wave packet that corresponds to a soliton, it is necessary to know the boson eigenfunctions (the $|\psi_n(k)\rangle$) that go with each of the coefficients—the a_n in Equation (8.74). Although it is possible in principle to compute these using the number state technique sketched in Section 8.3.2, it is often impossible in practice because the dimensions of the matrices to be constructed and diagonalized become too large.

It is just in this situation that the HA becomes most useful. As $n \to \infty$, Equation (8.79) becomes an ever better approximation to the true wave function. Also calculation of the approximate single boson wave function involves only the solution of a classical nonlinear lattice system, Equation (8.83) becoming no more difficult to solve as n increases.

8.4 More general quanta

The NSM, introduced in Section 8.2 and exercised in Section 8.3, can be used in a variety of quantum systems where the fundamental quanta are not standard bosons. Several examples of such applications are presented in this section, where the analytical methods employed are similar to those described above; thus the discussions are less detailed. It is hoped, however, that these sketches will impress the reader with the flexibility of the NSM and suggest other directions for related research.

8.4.1 The Ablowitz–Ladik equation

In this section, we quantize the classical system

$$\mathrm{i}\frac{dA_j}{dt} + (A_{j+1} - 2A_j + A_{j-1}) + \frac{\gamma}{2}|A_j|^2(A_{j+1} + A_{j-1}) = 0, \tag{8.87}$$

where $j = 1, 2, \ldots, f$ and the boundary conditions are periodic. Furthermore, by assuming the dispersive energy (ε) to be unity, we are in effect measuring all energies in units of ε. Introduced by Ablowitz and Ladik in 1976, this is a true soliton system in the sense that it has an associated linear scattering operator providing the basis for an inverse scattering method (ISM) of solution [1, 2].

Defining a nonstandard Poisson bracket (as in Chapter 5) to be

$$\{\phi, \psi\} = \sum_{j=1}^{f} \left(\frac{\partial \phi}{\partial A_j} \frac{\partial \psi}{\partial A_j^*} - \frac{\partial \psi}{\partial A_j} \frac{\partial \phi}{\partial A_j^*} \right) \left(1 + \frac{\gamma}{2}|A_j|^2 \right)$$

and a Hamiltonian as

$$H = -\sum_{j=1}^{f}\left[A_j^*(A_{j+1}+A_{j-1}) - \frac{4}{\gamma}\log\left(1+\frac{\gamma}{2}|A_j|^2\right)\right],$$

Equation (8.87) can be written

$$i\frac{dA_j}{dt} = \{A_j, H\}.$$

Because a fundamental rule of quantum theory is that Poisson brackets become commutators [14], this formulation suggests the quantization [38]

$$H \to \hat{H} = -\sum_{j=1}^{f}\left[b_j^\dagger(b_{j+1}+b_{j-1}) - 2\frac{\log(1+(\gamma/2)b_j^\dagger b_j)}{\log(1+(\gamma/2))}\right], \tag{8.88}$$

where $b_j^\dagger$ and b_j are raising and lowering operators satisfying the commutation relations

$$[b_j, b_k^\dagger] = \left(1 + \frac{\gamma}{2}b_j^\dagger b_k\right)\delta_{jk}, \quad [b_j^\dagger, b_k^\dagger] = [b_j, b_k] = 0, \tag{8.89}$$

and

$$\log\left(1+\frac{\gamma}{2}b_j^\dagger b_j\right) \equiv \frac{\gamma}{2}b_j^\dagger b_j - \frac{1}{2}\left(\frac{\gamma}{2}b_j^\dagger b_j\right)^2 + \cdots.$$

These commutation relations can be viewed as deformations of the standard ones for bosons, obtained by shrinking $\gamma/2$ to zero. For $\gamma \neq 0$ it is customary to call the basic quantum components "q-deformed bosons" or "q-bosons," where $q = \gamma/2$ is the "deformation parameter."

Using an argument similar to that presented in Section 8.1.3 for standard bosons, one can define number states for which [54]

$$b^\dagger|n\rangle = \left\{\frac{2}{\gamma}\left[\left(1+\frac{\gamma}{2}\right)^{n+1}-1\right]\right\}^{1/2}|n+1\rangle$$

$$b|n\rangle = \left\{\frac{2}{\gamma}\left[\left(1+\frac{\gamma}{2}\right)^{n}-1\right]\right\}^{1/2}|n-1\rangle \tag{8.90}$$

$$b|0\rangle = 0.$$

The energy operator defined in Equation (8.88) commutes with the q-boson number operator

$$\hat{N} = \sum_{j=1}^{f}\frac{\log(1+(\gamma/2)b_j^\dagger b_j)}{\log(1+(\gamma/2))}, \tag{8.91}$$

where

$$\hat{N}|n_1, n_2, \ldots, n_f\rangle = (n_1 + n_2 + \cdots + n_f)|n_1, n_2, \ldots, n_f\rangle.$$

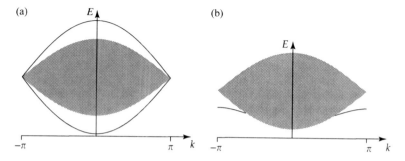

FIG. 8.5. *Two quantum energy eigenvalues against propagation number for (a) the quantum Ablowitz–Ladik equation and (b) the fermionic polaron model with $\gamma < 2$.*

Because this number operator has the same properties as the standard boson number operator, one can solve the quantum Ablowitz–Ladik (QAL) problem using the NSM, just as in Section 8.3.2 for the QDNLS, leading to the two quantum spectrum sketched in Figure 8.5(a) [65]. The upper band in this figure represents classically unstable solutions that change phase by π radians between adjacent lattice sites.

The lower band becomes identical to that for QDNLS as $\gamma \to 0$. For large γ, on the other hand,

$$E_n(k) = -2\cos(k/n)\left(\frac{\gamma}{2}\right)^{(n-1)/2} \tag{8.92}$$

so for $n > 1$ the soliton binding energy

$$E_b = 2\left(\frac{\gamma}{2}\right)^{(n-1)/2}, \tag{8.93}$$

and the effective mass

$$m^* = \frac{n^2}{2}\left(\frac{2}{\gamma}\right)^{(n-1)/2}. \tag{8.94}$$

Solitary wave solutions of the classical DNLS equation become pinned to the lattice at a sufficiently large value of γ, whereas solitons of the AL equation can propagate at any speed. How is this difference in classical behaviors represented in the quantum picture? Compare the maximum value of the group velocity dE/dk in the two cases:

$$V_m^{\text{QDNLS}} = \frac{2n}{(n-1)!}\gamma^{(1-n)}$$

$$V_m^{\text{QAL}} = \frac{2}{n}\sin(\pi/n)\left(\frac{\gamma}{2}\right)^{(n-1)/2}.$$

Evidently in the classical limit where $n \to \infty$

$$V_m^{\text{QDNLS}} \to 0$$
$$V_m^{\text{QAL}} \to \infty,$$

indicating that the DNLS soliton becomes pinned, whereas the AL soliton does not. This is another example of the means that linear quantum theory employs to reproduce phenomena that emerge from nonlinear aspects of the corresponding classical theory.

In comparing the behavior of solitary wave solutions of QDNL and solitons of QAL, one should remain aware that both are constructed as quantum wave packets as indicated in Equation (8.74). Thus both wave packets will spread out in space, as time advances, due to the uncertainty in momentum (and therefore speed) that is given in Equation (8.12). In both cases, quantum theory destroys the "wave of permanent profile" that is so characteristic of classical nonlinear wave dynamics.

8.4.2 Salerno's equation

The NSM has an interesting and important property: it can be extended to any energy operator that commutes with a previously employed number operator. An example of this property—suggested by Salerno [54]—starts with the commutation relations for quanta of the AL system in Equations (8.89). Writing the q-boson number operator in a slightly more general form as[10]

$$\hat{N} = \sum_{j=1}^{f} \frac{\log(1 + \epsilon/\eta b_j^\dagger b_j)}{\log(1 + \epsilon/\eta)}, \tag{8.95}$$

he showed that a commuting energy operator is

$$\hat{H} = -\sum_{j=1}^{f} \left[b_j^\dagger (b_{j+1} + b_{j-1}) - \frac{2 + \omega_0 + \eta}{\log(1 + \epsilon/\eta)} \log\left(1 + \frac{\epsilon}{\eta} b_j^\dagger b_j\right) + \eta b_j^\dagger b_j \right]. \tag{8.96}$$

Requiring the parameters η and ϵ to be related as

$$\eta = 2\epsilon/(\gamma - \epsilon),$$

$\hat{H}$ has two interesting limits:

1. *The AL limit.* With $\omega_0 = 0$ and $\epsilon \to 0$,

$$\eta \to 0 \quad \text{and} \quad \frac{\epsilon}{\eta} \to \frac{\gamma}{2},$$

so $\hat{H}$ reduces to the AL energy operator defined in Equation (8.88).

---

[10]The reader should be careful not to confuse the symbol "ϵ," which is a parameter in Salerno's equation, with "ε," which has been used as a measure of dispersion.

2. *The QDNLS limit.* With $\omega_0 = \gamma/2$ and $\epsilon \to \gamma$

$$\epsilon/\eta \to 0,$$

and $\hat{H}$ becomes the QDNLS Hamiltonian

$$\hat{H} \to -\sum_{j=1}^{f} \left[b_j^\dagger (b_{j+1} - 2b_j + b_{j-1}) + \frac{\gamma}{2} b_j^\dagger b_j^\dagger b_j b_j \right].$$

That the energy operator in Equation (8.96) is more than a curiosity is seen by comparing the corresponding classical equation

$$i\frac{dA_j}{dt} - (\omega_0 + 2 - \epsilon |A_j|^2) A_j + \left(1 + \frac{\epsilon}{\eta} |A_j|^2\right) (A_{j+1} + A_{j-1}) = 0 \qquad (8.97)$$

with the simplest equation describing the propagation of an excitation on a molecular crystal

$$i\frac{dA_j}{dt} - \Omega A_j + J(A_{j+1} + A_{j-1}) = 0. \qquad (8.98)$$

Here A_j is the complex mode amplitude of a molecular vibration of frequency Ω, and J is the nearest neighbor resonance interaction energy. As we shall see in Section 8.5, assuming coupling of the site oscillators to low frequency phonons leads to a dependence upon local energy of

$$\Omega = \Omega_0 + \Omega_1 |A_j|^2,$$

the anharmonicity of a standard polaron. Similarly, coupling of the resonance interaction to low frequency phonons leads to

$$J = J_0 + J_1 |A_j|^2,$$

as represented by the third term of Equation (8.97). (I am indebted to Nitant Kenkre for pointing this out.)

Finally, Salerno has studied a q-boson lattice with equal coupling among the freedoms ($m_{jk} = 1 - \delta_{jk}$) and $q = -2$. This system behaves as a Bose gas with hard-core repulsion and exhibits Bose–Einstein condensation [55].

8.4.3 A fermionic polaron model

So far we have looked at QDST systems where the quanta obey bosonic commutation relations, and at the QAL and quantum Salerno equations for which the basic quanta are q-bosons. In this section, a quantum model based on "fermions" is considered. Such systems have raising and lowering operators ($a_j^\dagger$ and a_j) with the commutation relations [14]

$$\begin{aligned} a_j a_k^\dagger + a_k^\dagger a_j &= \delta_{jk} \\ a_j a_k + a_k a_j &= 0 \\ a_j^\dagger a_k^\dagger + a_k^\dagger a_j^\dagger &= 0. \end{aligned} \qquad (8.99)$$

These are just the standard boson commutation relations with the "−" signs changed to "+" signs. For this reason, the left hand sides of Equations (8.99) are sometimes called "anticommutators."

Consider eigenvalues of the operator

$$\hat{N} \equiv a^\dagger a.$$

From the second and third of equations (8.99), $aa = a^\dagger a^\dagger = 0$. Thus

$$\hat{N}^2 = a^\dagger a a^\dagger a = a^\dagger(1 - a^\dagger a)a = \hat{N}$$

so if $\hat{N}|n\rangle = n|n\rangle$,

$$n^2|n\rangle = \hat{N}^2|n\rangle = \hat{N}|n\rangle = n|n\rangle,$$

implying that $n^2 = n$. Thus $\hat{N}$ has only two eigenvalues: $n = 0$ and $n = 1$. Assuming the corresponding eigenstates to be nondegenerate allows only two eigenstates, $|0\rangle$ and $|1\rangle$, with

$$\hat{N}|0\rangle = 0$$
$$\hat{N}|1\rangle = |1\rangle,$$

indicating that $\hat{N}$ is a number operator. From these properties, it follows that

$$a^\dagger|0\rangle = |1\rangle$$
$$a|1\rangle = |0\rangle$$
$$a|0\rangle = 0$$
$$a^\dagger|1\rangle = 0.$$

We now know enough about fermions to study the energy operator

$$\hat{H} = \sum_{j=1}^{f}[\omega_0 a_j^\dagger a_j - J(a_j^\dagger a_{j+1} + a_j^\dagger a_{j-1}) + V a_j^\dagger a_j a_{j+1}^\dagger a_{j+1}], \tag{8.100}$$

with $j = 1, 2, \ldots, f$ and periodic boundary conditions, proposed to describe the dynamics of an extra electron in a one-dimensional crystal [42]. In this model, ω_0 is a site energy, J is a nearest neighbor interaction energy and V is an anharmonic parameter, arising from electron coupling to optical vibrations of the crystal.

Evidently $\hat{H}$ commutes with the number operator

$$\hat{N} = \sum_{j=1}^{f} a_j^\dagger a_j, \tag{8.101}$$

so the NSM can be used to find eigenfunctions and eigenvalues of $\hat{H}$, but there is an extra wrinkle.

Because of the anticommutation relations, the order in which the fermions are inserted into the chain is important [14]. Thus we define normal ordering to be that where the fermions are inserted from left to right. For example,

$$a_2^\dagger a_1^\dagger |00\rangle = |11\rangle, \quad \text{but} \quad a_1^\dagger a_2^\dagger |00\rangle = -|11\rangle \quad \text{and} \quad \hat{T}|1001\rangle = -|1100\rangle.$$

With these details in mind, let us use the number state method to solve the fermionic polaron system. First of all, the order of the matrix to be solved is limited by the number of ways that n fermions can be placed on f freedoms with no more than one per freedom, so

$$p = \frac{f!}{n!(f-n)!}.$$

Taking advantage of the f-fold translational symmetry reduces the order to

$$\frac{p}{f} = \frac{(f-1)!}{n!(f-n)!}.$$

With f odd and $n = 2$, a general eigenfunction of both $\hat{N}$ and $\hat{T}$ takes the form

$$|\psi_{2,\tau}\rangle = c_1 \sum_{j=1}^{f} (\hat{T}/\tau)^{j-1}|110\cdots 0\rangle + c_2 \sum_{j=1}^{f} (\hat{T}/\tau)^{j-1}|1010\cdots 0\rangle$$

$$+ \cdots + c_{(f-1)/2} \sum_{j=1}^{f} (\hat{T}/\tau)^{j-1}|10\cdots 010\cdots 00\rangle, \qquad (8.102)$$

differing from that given in Equation (8.62) because no more than one fermion can be assigned to any one degree of freedom. For typographical convenience, set $\omega_0 = 0$ and $\gamma \equiv -V/J$ in Equation (8.100). Then requiring

$$H|\psi_{2,\tau}\rangle = E|\psi_{2,\tau}\rangle$$

and equating the coefficients of kets leads to the matrix equation

$$\mathcal{Q}(2,\tau)\mathbf{c} = E\mathbf{c},$$

where $\mathbf{c} \equiv \text{col}(c_1, c_2, \ldots, c_{(f-1)/2})$ and $\mathcal{Q}(2,\tau)$ is the $[(f-1)/2] \times [(f-1)/2]$ tridiagonal matrix

$$\mathcal{Q}(2,\tau) = -\begin{bmatrix} \gamma & q^* & 0 & \cdot & \cdot & \cdot \\ q & 0 & q^* & 0 & \cdot & 0 \\ \cdot & q & 0 & q^* & \cdot & \cdot \\ \cdot & \cdot & \cdot & \cdot & \cdot & \cdot \\ \cdot & \cdot & \cdot & q & 0 & q^* \\ \cdot & \cdot & \cdot & \cdot & q & p \end{bmatrix}, \qquad (8.103)$$

and
$$q \equiv 1+\tau \quad \text{and} \quad p \equiv -\left[\tau^{(f+1)/2} + \tau^{(f-1)/2}\right].$$

In the limit $f \to \infty$ and $\gamma > 2\cos(k/2)$, the soliton band has energy

$$E_2(k) = -\left[\gamma + \frac{4}{\gamma}\cos^2\left(\frac{k}{2}\right)\right]. \tag{8.104}$$

At $k = 0$, the bottom of the continuum band is at -4; thus if $\gamma > 2$, the resulting eigenvalue plot appears as in Figure 8.4, but for $0 < \gamma < 2$, the central part of the soliton band merges with the continuum band as shown in Figure 8.5(b).

In the large γ limit with both f and n arbitrary,

$$E_n(k) \doteq -(n-1)\gamma - \frac{2}{\gamma^{n-1}}\cos k, \tag{8.105}$$

where the symbol '$\doteq$' has the same meaning as below Equation (8.69). This implies that

$$E_{\mathrm{b}} \doteq (n-1)\gamma \tag{8.106}$$

and

$$m^* \doteq \gamma^{n-1}/2. \tag{8.107}$$

Fermionic quantum models differ from those based on bosons in a fundamental way: they have no classical limit. In a bosonic theory like the QDNLS system, the classical system is approached by allowing the number of bosons in each freedom to become large, constructing a soliton wave packet as in Equation (8.74), and imposing the coherent state conditions indicated in Equations (8.85) and (8.86). For a fermionic system, this procedure is not possible because no more than one quantum can be placed on a freedom.

8.4.4 The Hubbard model

Finally consider the reduced one-dimensional Hubbard model, of interest in connection with the theory of high-T_{c} superconductivity [45]. In one-dimension, it can be defined by the energy operator

$$\hat{H} = -\sum_{j=1}^{f}[a_j^\dagger(a_{j+1}+a_{j-1}) + b_j^\dagger(b_{j+1}+b_{j-1}) + \gamma a_j^\dagger a_j b_j^\dagger b_j], \tag{8.108}$$

with periodic boundary conditions. In this Hamiltonian, the $a_j^\dagger(a_j)$ and $b_j^\dagger(b_j)$ are raising (lowering) operators for different electronic spin states, both obeying fermionic anticommutation relations.

In developing the NSM for the Hubbard model, one can proceed as in the previous section but with additional structure to account for the two spin states. Thus elementary product states take the form

$$\begin{bmatrix} |n_{a1} & n_{a2} & \cdots & n_{af}\rangle \\ |n_{b1} & n_{b2} & \cdots & n_{bf}\rangle \end{bmatrix},$$

where each column corresponds to a degree of freedom. The f entries in the upper (lower) row are either "1" or "0" depending upon whether there is or is not a fermion in the "spin up" ("spin down") position at that lattice site. Thus $a_j^\dagger (b_j^\dagger)$ generates a fermion in the upper (lower) row.

The Hamiltonian operator commutes with the number operators

$$\hat{N}_a = \sum_{j=1}^{f} a_j^\dagger a_j \quad \text{and} \quad \hat{N}_b = \sum_{j=1}^{f} b_j^\dagger b_j \qquad (8.109)$$

for the upper and lower rows, respectively, implying that $\hat{H}$ cannot switch quanta between rows. Eigenfunctions of $\hat{N}_a$ and $\hat{N}_b$ have number eigenvalues

$$n_a = n_{a1} + n_{a2} + \cdots + n_{af}$$
$$n_b = n_{b1} + n_{b2} + \cdots + n_{bf}.$$

Again consider the case $n = n_a + n_b = 2$. The cases $n_a = 2$ and $n_b = 2$ are trivial. For any f and $n_a = n_b = 1$, a general eigenfunction of $\hat{N}_a$, $\hat{N}_b$, and $\hat{T}$ is

$$|\psi_{2,\tau}\rangle = c_1 \sum_{j=1}^{f} (\hat{T}/\tau)^{j-1} \begin{bmatrix} |1 & 0 & \cdots & 0\rangle \\ |1 & 0 & \cdots & 0\rangle \end{bmatrix} + c_2 \sum_{j=1}^{f} (\hat{T}/\tau)^{j-1} \begin{bmatrix} |1 & 0 & \cdots & 0\rangle \\ |0 & 1 & \cdots & 0\rangle \end{bmatrix}$$
$$+ \cdots + c_f \sum_{j=1}^{f} (\hat{T}/\tau)^{j-1} \begin{bmatrix} |1 & 0 & \cdots & 0\rangle \\ |0 & 0 & \cdots & 1\rangle \end{bmatrix}.$$

Energy eigenvalues are determined by the $f \times f$ matrix

$$\mathcal{Q}(2,\tau) = - \begin{bmatrix} \gamma & q^* & 0 & \cdot & \cdot & q \\ q & 0 & q^* & 0 & \cdot & 0 \\ 0 & q & 0 & q^* & \cdot & \cdot \\ \cdot & \cdot & \cdot & \cdot & \cdot & \cdot \\ \cdot & \cdot & \cdot & q & 0 & q^* \\ q^* & \cdot & \cdot & \cdot & q & 0 \end{bmatrix}, \qquad (8.110)$$

where

$$q \equiv 1 + \tau.$$

For $f \gg 1$, interestingly, the eigenvalues of this matrix—a fermion model—are identical to those of the matrix in Equation (8.63) for the QDNLS—a boson model [17]. Thus the soliton band appears as in Figure 8.4.

In more standard notation, a simplified version of the Hubbard energy operator is

$$\hat{H} = -t \sum_{\sigma, j \neq k}^{f} c_{j\sigma}^\dagger c_{k\sigma} + U \sum_{j=1}^{f} n_{j\uparrow} n_{j\downarrow},$$

where σ is a spin state variable, taking the two values $\sigma = \uparrow$ or $\downarrow$. In addition, $c_{j\sigma}^\dagger (c_{j\sigma})$ are fermionic raising (lowering) operators, t is a dispersive energy, $n_{j\uparrow} \equiv c_{j\uparrow}^\dagger c_{j\uparrow}$, and U is an anharmonic energy that must be negative for a superconductive state to form. A simplification of this model is that the dispersive energy is assumed to be the same between all lattice sites, implying that the Hamiltonian is invariant to any permutation of the lattice.

Ideally, one wishes to study superconductivity for a three-dimensional lattice of (say) m sites in each direction, where the total number of freedoms $f = m^3$. At each of these sites, there are four possible states: $(0\ 0)$, $(\uparrow 0)$, $(0\ \downarrow)$, and $(\uparrow\downarrow)$, the $(\uparrow\uparrow)$ and $(\downarrow\downarrow)$ states being forbidden by the structure of the Hamiltonian. Thus for a system of $m \times m \times m$ lattice sites, there are 4^{m^3} terms in the most general wave function. This number grows rapidly with m, so it is desirable to take advantage of all possible symmetries. Recently, Salerno has made progress in reducing the order of the problem for finite lattices through the introduction of additional symmetries into the form of the wave function [34, 56–58].

8.5 Energy transport in protein

A polaron-like mechanism for the localization and transport of vibrational energy in protein was proposed in 1973 by Davydov [13]. Referring to the atomic structure of an alpha-helix region of protein shown in Figure 5.4, this mechanism can be described classically as follows. Vibrational energy of the CO stretching (or Amide-I) oscillators that is localized on the helix acts through a phonon coupling effect to distort the structure of the helix. The distortion reacts, again through phonon coupling, to trap the Amide-I oscillation energy, thereby preventing its dispersion.

From the perspective of biochemistry, it is an appealing means for energy storage and transport because the amount of energy involved is about equal to that from the hydrolysis of adenosine triphosphate (ATP) into adenosine diphosphate (ADP) by the chemical reaction

$$\text{ATP}^{4-} + \text{H}_2\text{O} \to \text{ADP}^{3-} + \text{HPO}_4^{2-} + \text{H}^+.$$

Under normal physiological conditions (temperature and pH), approximately 10 kcal/mol or 0.42 eV of free energy are released by this reaction, or about two quanta of Amide-I oscillator energy.

8.5.1 Dynamic equations

As shown in Figure 5.4, the alpha-helix region of protein is a chain of amino acids held in helical shape by longitudinal hydrogen bonds. There are three channels running along the helix with the structure

$$\cdots \text{H} - \text{N} - \text{C}\!\!=\!\!\text{O} \cdots \text{H} - \text{N} - \text{C}\!\!=\!\!\text{O} \cdots \text{H} - \text{N} - \text{C}\!\!=\!\!\text{O} \cdots,$$

Davydov proposed a quantum mechanical description of this system with the energy operator comprising three components:

$$\hat{H} = \hat{H}_{\text{ex}} + \hat{H}_{\text{ph}} + \hat{H}_{\text{int}}. \qquad (8.111)$$

The first of these, which he called the "exciton" operator, is

$$\hat{H}_{\text{ex}} = \sum_{j,\alpha} \left[E_0 b_{j\alpha}^\dagger b_{j\alpha} - J \left(b_{j\alpha}^\dagger b_{j+1,\alpha} + b_{j\alpha}^\dagger b_{j-1,\alpha} \right) \right.$$
$$\left. + L \left(b_{j\alpha}^\dagger b_{j,\alpha+1} + b_{j\alpha}^\dagger b_{j,\alpha-1} \right) \right]. \qquad (8.112)$$

With reference to Figure 5.4, the subscript j $(= 1, 2, \ldots, f)$ counts molecules along a single channel and the subscript α $(= 1, 2, 3)$ specifies a particular channel; thus the index pair (j, α) selects an individual amino acid. The $b_{j\alpha}^\dagger$ $(b_{j\alpha})$ are boson raising (lowering) operators for the Amide-I (CO stretching) oscillators, E_0 is a quantum of the Amide-I site energy, $-J$ is the nearest neighbor dipole-dipole coupling energy along a channel, and $+L$ is the nearest neighbor dipole-dipole coupling energy between channels.[11]

The phonon energy operator is

$$\hat{H}_{\text{ph}} = \frac{1}{2} \sum_{j,\alpha} [\hat{p}_{j\alpha}^2/M + w(\hat{u}_{j+1,\alpha} - \hat{u}_{j\alpha})^2],$$

where $\hat{p}_{j\alpha}$ and $\hat{u}_{j\alpha}$ are respectively the momentum and position operators for longitudinal displacement of an amino acid, M is the mass of an amino acid, and w is the spring constant of a hydrogen bond.

Finally the exciton–phonon interaction operator is

$$\hat{H}_{\text{int}} = \sum_{j,\alpha} \chi(\hat{u}_{j+1,\alpha} - \hat{u}_{j\alpha}) b_{j\alpha}^\dagger b_{j\alpha},$$

where χ is an exciton–phonon coupling parameter that determines the level of anharmonicity in the corresponding classical problem. From comparison with the first term on the right hand side of Equation (8.112), it is evident that

$$\chi = dE_0/dR, \qquad (8.113)$$

R being the length of the hydrogen bond directly adjacent to a particular Amide-I oscillator.

As the quantum polaron system cannot be exactly solved [32], Davydov proposed an analysis that is based upon a product trial wave function

$$|\psi\rangle \doteq |\Psi\rangle|\Phi\rangle \qquad (8.114)$$

---

[11] In this section, the variables and parameters are expressed in standard units.

where $|\Psi\rangle$ describes a single quantum excitation ($n = 1$) of the Amide-I system as

$$|\Psi\rangle = \sum_{j,\alpha} a_{j\alpha}(t) b_{j\alpha}^\dagger |0\rangle_{\text{ex}},$$

and $|0\rangle_{\text{ex}} = |0, 0, \ldots, 0\rangle$ is the vacuum state of the Amide-I oscillators. Also $|\Phi\rangle$ is a coherent phonon state for which

$$\langle \Phi | \hat{u}_{j\alpha} | \Phi \rangle = \beta_{j\alpha}(t)$$
$$\langle \Phi | \hat{p}_{j\alpha} | \Phi \rangle = \pi_{j\alpha}(t) = M \dot{\beta}_{j\alpha}.$$

In these expressions, $a_{j\alpha}(t)$ is a complex number representing the probability amplitude for finding a quantum of Amide-I energy in a particular amino acid, and $\beta_{j\alpha}(t)$ and $\pi_{j\alpha}(t)$ are respectively the quantum averages of the longitudinal displacement and momentum operators of an amino acid molecule.

The average value of $\hat{H}$ can be minimized with respect to the product wave function by calculating $\langle \psi | \hat{H} | \psi \rangle = H(a_{j\alpha}, a_{j\alpha}^*, \beta_{j\alpha}, \pi_{j\alpha})$ and requiring that the corresponding Hamilton equations

$$i\hbar \dot{a}_{j\alpha} = [E_0 + W + \chi(\beta_{j+1,\alpha} - \beta_{j\alpha})] a_{j\alpha}$$
$$- J(a_{j+1,\alpha} + a_{j-1,\alpha}) + L(a_{j,\alpha+1} + a_{j,\alpha-1}) \quad (8.115)$$
$$M\ddot{\beta}_{j\alpha} - w(\beta_{j+1,\alpha} - 2\beta_{j\alpha} + \beta_{j-1,\alpha}) = \chi(|a_{j\alpha}|^2 - |a_{j-1,\alpha}|^2)$$

be satisfied, where

$$W = \frac{1}{2} \sum_{j,\alpha} [M \dot{\beta}_{j\alpha}^2 + w(\beta_{j+1,\alpha} - \beta_{j\alpha})^2]$$

is the total phonon energy. For a single quantum state, the normalization condition $\langle \psi | \psi \rangle = 1$ requires $\sum_{j,\alpha} |a_{j\alpha}|^2 = 1$.

Under certain approximations, Equations (8.115) reduce to a discrete nonlinear Schrödinger equation. According to this formulation, each soliton transports an amount of energy approximately equal to E_0 (a quantum of CO stretching oscillation) along the alpha-helix.

The solution of the second of Equations (8.115) can be divided into two components: a particular solution that represents the "phonon dressing" of the soliton, and a solution of the homogeneous equation that can represent the effects of ambient temperature. At low temperature the second component can be neglected, but at biological temperatures it may lead to attenuation of the soliton [62].

Although the Amide-I oscillations on the alpha-helix described above involve several approximations (mechanical interactions between oscillators are neglected as are dipole-dipole interactions beyond nearest neighbors and more complex phonon spectra), it is often convenient to use an even simpler model for theoretical studies. Such a model is motivated by a symmetric longitudinal mode for which

$$\hat{H} = \hat{H}_{\text{ex}} + \hat{H}_{\text{ph}} + \hat{H}_{\text{int}}, \qquad (8.116)$$

where

$$\hat{H}_{\text{ex}} = \sum_{j=1}^{f} [(E_0 + 2L)b_j^\dagger b_j - J(b_j^\dagger b_{j+1} + b_j^\dagger b_{j-1})]$$

$$\hat{H}_{\text{ph}} = \frac{1}{2} \sum_{j=1}^{f} [\hat{p}_j^2/\tilde{M} + \tilde{w}(\hat{u}_{j+1} - \hat{u}_j)^2]$$

$$\hat{H}_{\text{int}} = \chi \sum_{j=1}^{f} (\hat{u}_{j+1} - \hat{u}_j) b_j^\dagger b_j,$$

and the bending and twisting of the alpha-helix have been eliminated by assumption. Comparison of Equations (8.111) and (8.116) shows that

$$\tilde{w} = 3w \quad \text{and} \quad \tilde{M} = 3M.$$

The product wave function of Equation (8.114) can be extended to include an arbitrary number (n) of Amide-I quanta by introducing the approximate Hartree wave function that was discussed in Section 8.3.4; thus

$$|\Psi\rangle = \frac{1}{\sqrt{n!}} \left[\sum_{j=1}^{f} a_j(t) b_j^\dagger \right]^n |0\rangle_{\text{ex}}.$$

Under the rotating transformation

$$a_j(t) = \phi_j(t) e^{-it(E_0 + 2L + W)/\hbar},$$

W being the total phonon energy, the system becomes

$$i\hbar \frac{d\phi_j}{dt} = \chi(\beta_{j+1} - \beta_j)\phi_j - J(\phi_{j+1} + \phi_{j-1})$$

$$\tilde{M} \frac{d^2 \beta_j}{dt^2} - \tilde{w}(\beta_{j+1} - 2\beta_j + \beta_{j-1}) = n\chi(|\phi_j|^2 - |\phi_{j-1}|^2), \qquad (8.117)$$

with

$$\sum_{j=1}^{f} |\phi_j|^2 = 1.$$

The best currently available parameter values for the model are summarized in Table 8.1 [62].

Table 8.1. *Parameter values for Equations (8.117).*

Parameter	Value	Units
χ	35–62	Piconewtons
$\tilde{w}$	39–59	Newtons/meter
$\tilde{M}$	5.7×10^{-25}	kg
J	7.8	cm^{-1}

As was discussed in Chapter 5, this system reduces to a DNLS equation in the "adiabatic approximation," implying neglect of the phonon kinetic energy, as is appropriate for a solitary wave that moves slowly compared with the sound speed. Thus after dropping the $\tilde{M}\ddot{\beta}_j$ term in the second of Equations (8.117), it becomes

$$\beta_{j+1} - \beta_j \doteq -\frac{n\chi}{\tilde{w}}|\phi_j|^2$$

so the first equation can be written as

$$i\hbar\frac{d\phi_j}{dt} + \frac{n\chi^2}{\tilde{w}}|\phi_j|^2\phi_j + J(\phi_{j+1} + \phi_{j-1}) \doteq 0, \qquad (8.118)$$

where '$\doteq$' indicates the adiabatic approximation. Each soliton of Equation (8.118) transports n quanta (bosons) of the Amide-I (CO stretching) oscillation. (Note that the '$\hbar$' can be removed from this equation if the ϕ_j are assumed to be unitless and units of cm^{-1} are used for d/dt, $\chi^2/\tilde{w}$, and J.)

8.5.2 *Experimental observations*

Detailed spectrographic studies of natural proteins are problematic because each amino acid is situated in a particular molecular environment that alters its site frequency in ways that are difficult to predict, even from a knowledge of the atomic coordinates. Thus the CO stretching resonance (Amide-I) typically appears as a broad band, spread over several tens of cm^{-1} by the pseudo-random nature of the globular protein structure. This spectral spreading makes it difficult for the experimenter to detect and interpret the subtle changes of structure that would be produced by a molecular soliton.

In the early 1970s, Careri began an experimental investigation of protein dynamics, aiming to circumvent some of the problems arising in the study of real protein by looking at hydrogen bonded molecular crystals that can be regarded as "model proteins," such as crystalline acetanilide (ACN): ($CH_3CONHC_6H_5$). The similarity of bond lengths and angles for the peptide group (HNCO) suggested that the dynamic behavior of ACN might provide clues to the corresponding

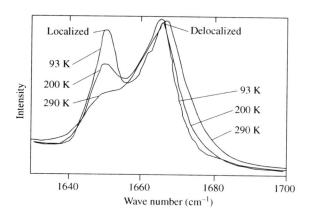

FIG. 8.6. *Infrared absorption spectra of crystalline ACN in the Amide-I (CO stretching) region at three different temperatures. Notice both the anomalous peak at 1650 cm^{-1}, indicating a self-trapped (soliton) state, and the standard (delocalized) peak at 1665 cm^{-1}. (Data courtesy of Peter Hamm.)*

behavior of natural protein. This comparison becomes more striking if one notes that both crystalline ACN and alpha-helix feature hydrogen bonded peptide channels with the atomic structure

$$\cdots H - N - C{=}O \cdots H - N - C{=}O \cdots H - N - C{=}O \cdots,$$

the starting point for the theory of Davydov that was sketched in the previous section.

Careri's intuition was rewarded by the discovery of an anomalous resonance peak at 1650 cm^{-1} in crystalline ACN (shown in Figure 8.6), but a decade of effort established only that this peak was an Amide-I (CO stretching) resonance that could not be conventionally assigned. Interestingly, this discovery was made and the subsequent experimental studies were carried on without knowledge of Davydov's theory of self-trapped molecular vibrations. It was after Careri had exhausted all possibilities for conventional assignment of the band that dynamical self-trapping was considered [10].

The basic idea is that the 1650 cm^{-1} band is evidence of a self-trapped Amide-I state arising through interaction with an intramolecular or optical phonon mode. A theoretical model, therefore, is a one-dimensional lattice with a single phonon degree of freedom (sometimes called "Einstein oscillator") of frequency ω_0, coupled to each Amide-I mode through a phonon coupling parameter χ_0. In the spirit of the previous section, an energy operator is defined as

$\hat{H} = \hat{H}_{\text{ex}} + \hat{H}_{\text{ph}} + \hat{H}_{\text{int}}$, where

$$\hat{H}_{\text{ex}} = \sum_{j=1}^{f} [E_0 B_j^\dagger B_j - J(B_j^\dagger B_{j+1} + B_j^\dagger B_{j-1})]$$

$$\hat{H}_{\text{ph}} = \frac{1}{2} \sum_{j=1}^{f} [\hat{p}_j^2/M + K\hat{q}_j^2] \qquad (8.119)$$

$$\hat{H}_{\text{int}} = \chi_0 \sum_{j=1}^{f} \hat{q}_j B_j^\dagger B_j,$$

and $j\ (= 1, 2, \ldots, f)$ counts ACN molecules along a chain. The radian frequency of the Einstein oscillator is

$$\omega_0 = E_0/\hbar = \sqrt{K/M}.$$

As in Section 8.1.3, it is convenient to transform from position and momentum operators ($\hat{q}_j$ and $\hat{p}_j$) describing the Einstein oscillator to boson creation and annihilation operators ($b_j^\dagger$ and b_j). Thus

$$b_j \equiv \frac{1}{\sqrt{2M\hbar\omega_0}} (M\omega_0 \hat{q}_j - i\hat{p}_j)$$

$$b_j^\dagger \equiv \frac{1}{\sqrt{2M\hbar\omega_0}} (M\omega_0 \hat{q}_j + i\hat{p}_j),$$

where $[b_j, b_k^\dagger] = \delta_{jk}$. Under this formulation the phonon and interaction Hamiltonians become

$$\hat{H}_{\text{ph}} = \hbar\omega_0 \sum_{j=1}^{f} \left(b_j^\dagger b_j + \frac{1}{2} \right)$$

$$\hat{H}_{\text{int}} = \lambda_0 \sum_{j=1}^{f} (b_j + b_j^\dagger) B_j^\dagger B_j$$

and

$$\lambda_0 \equiv \sqrt{\frac{\hbar}{2M\omega_0}} \chi_0.$$

At this point, one should note that $B_j^\dagger$ (B_j) are boson raising (lowering) operators for the Amide-I (CO stretching) oscillators at site j, and (as previously noted) $b_j^\dagger$ (b_j) are boson raising (lowering) operators for the corresponding Einstein oscillators that represent optical mode lattice vibrations.

Let us assume a single Amide-I quantum ($n = 1$) and a temperature of zero Kelvin and use Davydov's product assumption to approximate a solution. Thus

$$|\psi\rangle = |\Psi\rangle|\Phi\rangle,$$

where

$$|\Psi\rangle = \sum_{j=1}^{f} a_j(t) B_j^\dagger |0\rangle_{\text{ex}},$$

and $|\Phi\rangle$ is a sum of coherent states for the Einstein oscillators

$$|\Phi\rangle = \sum_{j=1}^{f} \exp[\beta_j(t) b_j^\dagger - \beta_j^*(t) b_j] \, |0\rangle_{\text{ph}}.$$

Treating $\langle\psi|\hat{H}|\psi\rangle$ as a classical Hamiltonian implies the dynamic equations

$$i\hbar \dot{a}_j = [E_0 + W + \lambda_0(\beta_j + \beta_j^*)] a_j - J(a_{j+1} + a_{j-1}),$$

$$i\hbar \dot{\beta}_j = \hbar\omega_0 \beta_j + \lambda_0 |a_j|^2,$$

where W is the total energy of the Einstein oscillators. Assuming that $\dot{\beta}_j = 0$ (the adiabatic assumption) and introducing the rotating transformation

$$a_j = \phi_j e^{-it(E_0+W)/\hbar}$$

leads to the DNLS equation

$$i\hbar \frac{d\phi_j}{dt} + \frac{2\lambda_0^2}{\hbar\omega_0} |\phi_j|^2 \phi_j + J(\phi_{j+1} + \phi_{j-1}) = 0,$$

where it turns out that

$$\frac{2\lambda_0^2}{\hbar\omega_0} = \frac{\chi_0^2}{K} \gg J.$$

To establish this inequality, note first that, from detailed electromagnetic calculations [62], $J = 4\,\text{cm}^{-1}$. Second, if the inequality is satisfied, then it is shown below that the binding energy of a self-trapped state (soliton) with respect to a ($k = 0$) exciton is

$$E_\text{b} \doteq \frac{\chi_0^2}{2K} - 2J. \tag{8.120}$$

Finally, in Figure 8.6, the $1650\,\text{cm}^{-1}$ peak is assigned to a soliton and the peak at $1665\,\text{cm}^{-1}$ to an exciton. This implies that $E_\text{b} = 15\,\text{cm}^{-1}$ so $(\chi_0^2/2K) = 23\,\text{cm}^{-1}$, confirming the original assumption.

Because

$$\frac{2\lambda_0^2}{\hbar\omega_0} \gg J,$$

the results of Chapter 5 imply that the soliton is not mobile but pinned by Peierls forces near one molecule in the lattice; thus

$$M_{\text{sol}} = \infty. \tag{8.121}$$

As J is small compared with the anharmonic parameter (χ_0^2/K), it is also possible to use a perturbation expansion in small J. This approach has two advantages. First, it provides quantitative estimates of the accuracy of Davydov's analysis and the conditions where it breaks down, and second, solutions are obtained when the number of Amide-I quanta n is greater than unity. To this end, it is convenient to write $\hat{H}$ in the form

$$\hat{H} = \sum_{j=1}^{f} \hat{h}_j - J\hat{V} \tag{8.122}$$

where

$$\hat{h}_j = [E_0 + \lambda_0(b_j + b_j^\dagger)]B_j^\dagger B_j + \hbar\omega_0 b_j^\dagger b_j$$
$$\hat{V} = \sum_{j=1}^{f} (B_j^\dagger B_{j+1} + B_j^\dagger B_{j-1}) \tag{8.123}$$

and the ground state phonon energy is ignored.

The operator $\hat{h}_j$ is the energy operator of a "displaced harmonic oscillator" [32] for which exact eigenfunctions are known. Dropping the molecule index j $(= 1, 2, \ldots, f)$ for typographical convenience, these eigenfunctions can be written as

$$\hat{h}|u\rangle = E|u\rangle$$
$$|u\rangle = |n\rangle|\phi(n,\tilde{n})\rangle$$
$$|\phi(n,\tilde{n})\rangle = \sqrt{\tilde{n}!}\exp\left[-\frac{1}{2}\left(\frac{n\lambda_0}{\hbar\omega_0}\right)^2\right] \tag{8.124}$$
$$\times \sum_{m=0}^{\infty} \frac{(-n\lambda_0/\hbar\omega_0)^{m-\tilde{n}}}{\sqrt{m!}} L_{\tilde{n}}^{m-\tilde{n}}\left[\left(\frac{n\lambda_0}{\hbar\omega_0}\right)^2\right]|m\rangle$$
$$E = nE_0 + \tilde{n}\hbar\omega_0 - \frac{n^2\lambda_0^2}{\hbar\omega_0}.$$

In these equations,

$$B^\dagger B|n\rangle = n|n\rangle,$$
$$b^\dagger b|m\rangle = m|m\rangle,$$

and $L_n^m[\cdot]$ is an associated Laguerre polynomial.[12] As $n\lambda_0 \to 0$, $|\phi(n,\tilde{n})\rangle$ reduces to the phonon number state $|\tilde{n}\rangle$, but in general $|\phi(n,\tilde{n})\rangle$ is a sum over all phonon number states.

Assuming periodic boundary conditions, a zero-order estimate (in powers of J) of the wave function is

$$|\psi\rangle = \frac{1}{\sqrt{f}} \sum_{j=1}^{f} e^{ikj} |u_j\rangle, \qquad (8.125)$$

where $k = 2\pi\nu/f$ and $\nu = 0, \pm 1, \ldots, f/2$ (or $\pm(f-1)/2$) for f even (or odd). This wave function is an eigenfunction of the translational symmetry operator.

Next make two assumptions. The first is that

$$n = 1,$$

implying that only a single quantum of Amide-I oscillation is considered. The second is

$$\tilde{n} = 0,$$

implying that the phonon system is in the ground state, as one expects at low temperatures. With these assumptions, a first-order estimate of the energy is

$$E(k) \doteq E_0 - \frac{\lambda_0^2}{\hbar\omega_0} - 2J \exp\left[-\left(\frac{\lambda_0}{\hbar\omega_0}\right)^2\right] \cos k, \qquad (8.126)$$

where "$\doteq$" means that terms of order

$$\frac{K^2 J^2}{\chi_0^4} \sim 1\%$$

are neglected.

Because the wavelength of infrared light is large compared with a lattice spacing, only $k = 0$ exciton states will be observed experimentally. From the first of Equations (8.119), the energy of such a state is

$$E_{\text{ex}}(k = 0) = E_0 - 2J.$$

If a wave packet of the energy eigenfunctions is constructed to localize a quantum of Amide-I energy near a single molecule, then all values of k must be equally represented, and from Equation (8.126) the average energy of the wave packet will be

$$E_{\text{sol}} = E_0 - \lambda_0^2/\hbar\omega_0.$$

[12] After French mathematician Edmond Nicolas Laguerre (1834–1886).

Assigning $E_{\text{ex}}(k=0)$ to the $1665\,\text{cm}^{-1}$ peak in Figure 8.6 and E_{sol} to the $1650\,\text{cm}^{-1}$ peak implies that

$$E_0 - 2J = 1665$$

$$E_0 - \lambda_0^2/\hbar\omega_0 = 1650.$$

As $J = 4\,\text{cm}^{-1}$, we find that

$$\frac{\lambda_0^2}{\hbar\omega_0} = \frac{\chi_0^2}{2K} = 23\,\text{cm}^{-1},$$

in agreement with Davydov's estimate of the anharmonic parameter.

From Equation (8.126), however, note that the binding energy of a $k=0$ soliton is

$$E_{\text{b}} = E_{\text{ex}}(k=0) - E(0)$$

$$= \frac{\chi_0^2}{2K} - 2J\left\{1 - \exp\left[-\left(\frac{\lambda_0}{\hbar\omega_0}\right)^2\right]\right\}, \tag{8.127}$$

which is larger than that of Equation (8.120).

Also from Equation (8.126) the effective mass of a $k=0$ soliton is

$$M_{\text{sol}} = m^* \exp(\lambda_0/\hbar\omega_0)^2, \tag{8.128}$$

where $m^* = \hbar^2/2a^2 J$ is the effective mass of a bare exciton, and less than Davydov's result that $M_{\text{sol}} = \infty$ in Equation (8.121).

These results provide quantitative estimates for some of the errors arising from Davydov's product wave function. Evidently as $(\lambda_0/\hbar\omega_0)$ becomes greater than unity, implying that the interaction energy is larger than the phonon energy, such errors approach zero.

In addition to the translational invariance of Equation (8.125), the exact wave function has two additional properties that are not shared by Davydov's product wave function.

1. *Higher levels of Amide-I excitation* $(n>1)$. From the last of Equations (8.124), one expects to observe an overtone series at the frequencies $\nu(n)$ where $\hbar\nu(n) = E(n) - E(0)$; thus in cm^{-1} the Birge–Sponer relation is

$$\nu(n) \approx 1673n - 23n^2,$$

where the coefficient of n has been chosen so $\nu(1) = 1650\,\text{cm}^{-1}$ as is shown in Figure 8.6. Measured overtone frequencies are given in Table 8.2. If the coefficients are adjusted to [66]

$$\nu(n) = 1674n - 24.7n^2, \tag{8.129}$$

Table 8.2. *A comparison of measured overtone frequencies in ACN with theoretical values. All values are in cm^{-1}.*

n	$\nu(n)$ measured	$\nu(n)$ theoretical
1	1650.0 ± 0.5	1649.3
2	3250 ± 1	3249.2
3	4803 ± 3	4799.7
4	6304 ± 5	6300.8

the measured values are predicted to an RMS difference of $2.4\,cm^{-1}$ (within the uncertainties of experimental observations); thus a better value for the parameter $\lambda_0^2/\hbar\omega_0 = \chi_0^2/2K$ is $24.7\,cm^{-1}$.

2. *Higher levels of phonon excitation* ($\tilde{n} > 0$). With $n = 1$ and $\tilde{n} = 0$, one sees from the third of Equations (8.124) that $|\phi(n,\tilde{n})\rangle$ reduces to a coherent state. What about $\tilde{n} > 0$? For $n = 0$

$$|\phi(0,\tilde{n})\rangle = |\tilde{n}\rangle,$$

a phonon number state with $b^\dagger b |\tilde{n}\rangle = \tilde{n}|\tilde{n}\rangle$. From the considerations of Section 8.1.2, these ground states will be thermally populated when $\hbar\omega_0 \leq kT$. Taking this effect into consideration leads to a quantitative understanding of the temperature dependence of the $1650\,cm^{-1}$ band that is shown in Figure 8.6.

To calculate the temperature dependent intensity of the $1650\,cm^{-1}$ band, one must find the sum of all transitions from the ground states $|\tilde{n}\rangle$ to first excited states $|\phi(1,\tilde{n})\rangle$ with $m = \tilde{n}$. Because the ground states are thermally populated with probabilities[13]

$$P(\tilde{n}) = \left[1 - \exp\left(-\frac{\hbar\omega_0}{kT}\right)\right] \exp\left(-\tilde{n}\frac{\hbar\omega_0}{kT}\right),$$

the temperature dependence of the $1650\,cm^{-1}$ band is given by Fermi's Golden Rule as

$$W(T) = \sum_{\tilde{n}=0}^{\infty} P(\tilde{n}) |\langle \tilde{n}|\phi(1,m)\rangle|^2. \tag{8.130}$$

Using identity number 9.976 from Gradshteyn and Ryzhik [30], this sum is computed to be

$$W(T) = \exp\left[-\left(\frac{\lambda_0}{\hbar\omega_0}\right)^2 \coth\left(\frac{\hbar\omega_0}{2kT}\right)\right] I_0\left[\left(\frac{\lambda_0}{\hbar\omega_0}\right)^2 \operatorname{csch}\left(\frac{\hbar\omega_0}{2kT}\right)\right] \tag{8.131}$$

---

[13] Note that in these equations k is the Boltzmann constant, not to be confused with the propagation number.

Table 8.3. *Normalized integrated Raman intensity values $W(T)/W(0)$ of the 1650 cm^{-1} band in single-crystal ACN. The σ are RMS errors.*

Temperature (K)	σ	Relative intensity	σ
21	4	1.000	0.016
21	4	0.996	0.016
53	4	0.892	0.019
53	4	0.895	0.019
100	10	0.710	0.022
100	10	0.726	0.022
149	6	0.496	0.025
149	6	0.507	0.025
227	12	0.356	0.028
227	12	0.346	0.028
305	8	0.170	0.032
305	8	0.169	0.032

where $I_0[\,\cdot\,]$ is the modified Bessel function[14] of the first kind, of order zero.

Experimental values for $W(T)$ are presented in Table 8.3. Using the value for $\lambda_0^2/\hbar\omega_0 = 24.7\,\mathrm{cm}^{-1}$, the only free parameter is ω_0. Fitting to the experimental data requires the optical mode frequency to be

$$\hbar\omega_0 \sim 75\,\mathrm{cm}^{-1},$$

a value that is reasonable upon comparison with the far infrared lattice phonon spectrum of ACN [10].

8.5.3 Recent comments

As sketched in Section 8.5.1, Davydov's product wave function leads to the classical picture presented in Equations (8.117). There is no quantum uncertainty in this description; the soliton solutions have a precise location at each instant of time. This feature is analogous to the Born-Oppenheimer and the HA, where product assumptions also lead to nonlinear classical theories. In the formulations of Equations (8.122), (8.123), and (8.124), on the other hand, eigenstates are constructed to share the translational symmetry of the lattice. Thus localized solitons are viewed as wave packets of translational invariant wave functions, as was discussed in Section 8.3.3.

Following the original assignment of the Amide-I band in ACN to autolocalization by Careri and his colleagues in 1984, much supporting evidence has accumulated, including observation of the predicted overtones (Table 8.2), agreement of the theory with measurements of the temperature dependence of

---

[14] After German applied mathematician Friedrich Wilhelm Bessel (1784-1846).

the 1650 band intensity (Table 8.3 and Equation (8.131)), and both x-ray and neutron scattering measurements that preclude assignments based on changes in the physical structure [62].

Recently, Edler and Hamm have used the pump-probe technique sketched in Section 8.1.7 to compare the probe responses of the 1665 and 1650 cm^{-1} shown in Figure 8.6 [15]. This approach is effective because the amplitude of a delocalized quantum state is small (implying a linear response which will not be observed in a pump-probe measurement), whereas the amplitude of a localized state is large, implying nonlinear response. Thus pump-probe measurements can be regarded as direct observations of energy localization.

Edler and Hamm used a pump pulse of 150 fs duration centered on 1640 cm^{-1} with a bandwidth of 120 cm^{-1}, and the temperature range of the experiments was from 93 to 293 K. At the lowest temperature (93 K), no response was seen at 1665 cm^{-1}, whereas a pronounced nonlinear response was observed at 1650 cm^{-1}. This confirms the assignments of the 1650 band to self-trapping and of the 1665 band to a delocalized state.

With increasing temperature, the magnitude of the 1650 response decreased in proportion to the decreasing intensity of the band itself, while the response of the 1665 band was found to increase. These measurements show that the 1665 band becomes disorder (Anderson) localized at high temperature, because of thermally induced randomness. This empirical observation in accord with the theoretical analysis by Cruzeiro-Hansson and Takeno [12], suggesting that the 1650 band is also thermally localized at high temperature.

The 1650 probe response disappears in about 2 ps, whereas the recovery of the ground state is over a much longer time ($\sim$35 ps). Thus the system initially relaxes into a state that is either spectroscopically dark or outside the spectral window of the probe beam.

One candidate for a spectrally dark state is as follows. Absorption rate is related to ease of emission (the reciprocity theorem again), and it is important to consider that the self-trapping mechanism described in Section 8.5.2 is based on an *optical* phonon (Einstein oscillator) with a frequency of about 75 cm^{-1}. Davydov's original theory, sketched in Section 8.5.1, invokes *acoustic* phonons, as is evident from the second of Equations (8.115). As Davydov often emphasized, self-trapping by acoustic phonons is expected to have a lower absorption intensity and a longer lifetime because it is restricted by a smaller Franck–Condon factor.[15] In other words, the probability of an acoustic mode transition is reduced because the lattice does not have time to readjust itself to the new configuration. Thus ACN, and natural protein, may absorb radiation into a state that is optically self-trapped and then relax into another state that is acoustically self-trapped, with a longer lifetime. This speculation is supported by recent theoretical studies, based on statistical thermodynamics, showing that the lifetime for solitons trapped

---

[15]After James Franck (1882–1964) and Edward Uhler Condon (1902–1974), respectively a German and an American physicist.

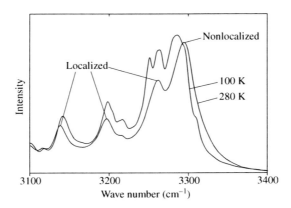

FIG. 8.7. *Infrared absorption spectrum of crystalline ACN in the NH stretching region at two different temperatures. (Data courtesy of Peter Hamm.)*

by low frequency modes increases with amplitude, whereas the corresponding lifetime for high frequency mode trapping decreases [69].

Experimental evidence for solitons on the alpha-helices of natural protein is difficult to obtain because the site frequencies of Amide-I oscillators vary in a pseudorandom manner over several tens of wave numbers [68]. Using a free-electron laser source, however, Austin and his coworkers have measured room-temperature lifetimes of Amide-I oscillations in myoglobin, a globular protein that is mostly alpha-helix. Although this band typically relaxes on a scale of less than a picosecond in protein, these authors found a prolonged lifetime of about 15 ps in the blue wing of the absorption spectrum [75].

Finally, another protein site for self-localized energy storage in protein is the NH stretching band (see Figure 8.7), as was proposed in 1986 by Alexander and Krumhansl [3]. Recent room temperature pump-probe measurements in this region of the ACN spectrum show that the 3295 cm^{-1} mode is delocalized, whereas the modes at 3260, 3195, and 3137 cm^{-1} are localized [16]. Excitation of the localized modes disappears from the spectral window or becomes spectrally dark on a 1 ps timescale, followed by a slow ground state recovery within 18 ps. These local modes are interesting candidates for biological energy storage and transport for two reasons: the band structure depends only weakly on temperature (see Figure 8.7), and the single quantum energy is about equal to the energy released in the hydrolysis of a molecule of ATP. With reference to Figure 5.4, the binding energy is expected to be larger (implying less sensitivity to temperature) for localization of NH stretching because the same proton participates in both the hydrogen bond and the NH oscillation, thereby increasing χ in Equation (8.113). (The neglect of this possibility over the past decade and a half may be partly attributed to the untimely death of Denise Alexander.)

8.6 A quantum lattice sine–Gordon equation

The molecule called "4-methyl-pyridine" (C_6H_7N) forms a molecular crystal that is related to a lattice sine–Gordon equation. For our purposes, the interesting feature of this molecule is the methyl (CH_3) group that can rotate (see Figure 8.8), providing the atomic equivalent of a spinning top.

This rotationally periodic potential and the interactions of adjacent methyl groups have been studied experimentally by Fillaux and Carlile, leading them to propose that the rotational dynamics of the methyl groups are described by the discrete sine–Gordon (DSG) system [26, 27]

$$\frac{d^2 u_j}{dt^2} = u_{j+1} - 2u_j + u_{j-1} - \Gamma^2 \sin u_j, \tag{8.132}$$

where u_j is one-third of the angle of rotation about the C–CH_3 axis, j counts molecules along a one-dimensional chain, and

$$\Gamma = 0.82.$$

As the far infrared spectra of this molecular crystal and its chemical cousins are becoming known, it is interesting to consider a quantum analysis of the DSG system, but this is not a simple task.

Although the continuum version of the SG equation—arising in the limit $\Gamma \to 0$—is integrable and the corresponding quantum theory is well understood, Equation (8.132) does not share these attractive properties [51]. In this situation,

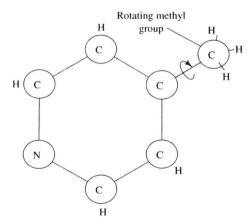

FIG. 8.8. *The atomic structure of 4-methyl-pyridine.*

two approximate quantum analyses have been proposed:

1. Fillaux and Carlile have assumed Γ to be close enough to zero so that results from the continuum quantization of moving SG breathers can be used. Although a reasonable fit to the experimental data is demonstrated, some of the spectral assignments are uncertain, and it cannot be said that this approach is satisfactory.
2. As $\Gamma \to 0$, a stationary breather solution of the DSG system approaches

$$u(x,t) \to 4\arctan\left[\frac{\tan\mu \sin(\Gamma t \cos\mu)}{\cosh(\Gamma x \sin\mu)}\right],$$

encouraging Wattis to seek stationary breather solutions of the form [71]

$$u(x,t) = 4\arctan\left[\frac{\tan\mu \sin\omega t}{\cosh \rho x}\right].$$

In this formulation, ω is the breather frequency, μ sets the amplitude, and ρ establishes a scale for the width of the breather. Using a combination of perturbative and Lagrangian concepts, he has shown how ω and ρ must depend on μ, leading to the dependence of breather energy on the amplitude parameter. From such results, a Bohr–Sommerfeld (BS) quantum condition,[16] requiring

$$\oint \sum_j \dot{u}_j du_j = 2\pi n \hbar, \qquad (8.133)$$

has been used to compute certain energy levels. This analysis led Wattis to calculations of observed energy differences in 4-MP that differ from those of Fillaux and Carlile (FC) in assignments of the principal quantum numbers [70]. However, the agreement of Wattis's calculated spectra is not as good as that obtained by Fillaux and Carlile.

Four possibilities for resolving these discrepancies come to mind. First, of course, it is important to obtain improved analytic descriptions of classical stationary breathers [73]. Second, a generalization by Wattis of his analysis to moving breathers may combine the better aspects of the BS and FC approaches [72]. Third, it may be necessary augment the physical model to include more dispersive interactions between 4-MP molecules than are represented in Equation (8.132). Finally, a quantum system studied by Aubry (called the "boson-Hubbard model"), in which the on-site Hamiltonian is allowed to be an arbitrary polynomial in the boson number operator, may provide a bridge between BS quantization of Equation (8.133) and the NSM [4].

This unfinished problem is noted here as an example of the sort of difficulties and opportunities to be expected in using quantum theory to study the behavior

---

[16] After Danish physicist Niels Bohr (1885–1962) and German physicist Arnold Sommerfeld (1868–1951).

of realistic nonlinear lattices. Although theoretically untidy, such models as the DSG system may be of greater technical interest than pure soliton lattices in the development of molecular devices for the generation and detection in the far infrared (terahertz) range of the electromagnetic spectrum—the "last frontier" of electronics.

8.7 Theoretical perspectives

In this chapter, we have considered several quantum energy operators that correspond to lattice problems of experimental interest. Having obtained such an operator, the next step is to solve it by putting some matrix representation of the Hamiltonian into diagonal form. Beyond perturbation theory, there are two techniques available to perform this diagonalization: using the fact that the Hamiltonian often commutes with a number operator (thereby relating their eigenfunctions) to directly determine the energy eigenfunctions and eigenvalues, and taking advantage of the special analytic properties of systems that can be solved by the classical ISM. The first of these is called the NSM and the second is called the quantum inverse scattering method (QISM). The aim of this section is to clarify the differences between the two approaches, emphasizing their relative strengths and weaknesses for quantum analyses of nonlinear lattices.

The reader will notice that our analyses to this point have been limited to the NSM. The reasons for this are two-fold: the NSM is conceptually more simple, and it has certain technical advantages over the QISM. We shall briefly review the structure of the NSM from a theoretical perspective, introduce and describe the QISM, and then present a critical comparison of the two methods.

8.7.1 Number state method

As we have seen in several examples, the NSM converts the eigenfunction problem

$$\hat{H}|\psi\rangle = E|\psi\rangle$$

to the finite matrix equation

$$\mathcal{H}_n \mathbf{c} = E\mathbf{c}$$

for a particular eigenvalue n of the number operator $\hat{N}$. In other words, the matrix representation of $\hat{H}$ is block-diagonalized into

$$\text{diag}[\mathcal{H}_0, \mathcal{H}_1, \mathcal{H}_2, \ldots, \mathcal{H}_n, \ldots],$$

where each of the blocks has dimension $p \times p$, and

$$p = \frac{(n+f-1)!}{n!\,(f-1)!}$$

for bosons and

$$p = \frac{f!}{n!\,(f-n)!}$$

for fermions.

As p may become inconveniently large, one attempts to use any and all available symmetries to further reduce the size of the blocks $\mathcal{H}_n$. For example, in one-dimensional systems with f freedoms and periodic boundary conditions, both $\hat{H}$ and $\hat{N}$ commute with a translation operator $\hat{T}$, so each $\mathcal{H}_n$ can be further block-diagonalized to

$$\mathcal{H}_n = \text{diag}[\mathcal{Q}(n,\tau_1), \mathcal{Q}(n,\tau_2), \mathcal{Q}(n,\tau_3), \ldots, \mathcal{Q}(n,\tau_f)],$$

where each matrix $\mathcal{Q}$ has dimension[17]

$$\frac{p}{f} \times \frac{p}{f},$$

and the τ_j are eigenfunctions of $\hat{T}$, related to the propagation number k by

$$\tau = e^{ik}.$$

Then one obtains energy eigenvalues $E_n(k)$ as functions of both n and k by numerical diagonalization of $\mathcal{Q}(n,\tau_j)$. From these eigenvalues and their associated eigenfunctions, soliton wave packets can be constructed as in Equation (8.74).

This is the gist of the NSM. It is robust and straightforward: a standard tool available since the early days of quantum theory that takes no notice of the integrability of the underlying classical system.

8.7.2 Quantum inverse scattering method

As described in a recent book by Korepin et al. [37], the QISM is based upon the ISM for classical systems [25]. For one-dimensional, discrete systems (lattices) with f freedoms and periodic boundary conditions, this method can be briefly described as follows.[18]

The space and time evolutions of a scattering function $\phi_j(t)$ ($j = 1, 2, \ldots, f$) are governed by linear operators L_j and M_j, which depend on the dynamical variables and a spectral parameter λ. Thus

$$\phi_{j+1} = L_j \phi_j \quad \text{and} \quad \dot{\phi}_j = M_j \phi_j,$$

and the classical nonlinear equations under consideration are implied by

$$\dot{L}_j = M_{j+1} L_j - L_j M_j.$$

---

[17] If p/f is not an integer, then the next larger integer determines the dimension of $\mathcal{Q}$.

[18] This sketch of the QISM is presented merely as an hors d'oeuvre to whet the reader's appetite. Those with a taste for more should study the comprehensive monograph by Korepin et al.

Defining a "transfer operator" as
$$T \equiv L_f \cdots L_1,$$
it is readily shown that
$$\dot{T} = [M_1, T], \tag{8.134}$$
implying that the time derivative of the trace
$$\frac{d}{dt}\operatorname{tr} T = 0.$$
For the integrable lattices discussed in this chapter,
$$T(\lambda) = \begin{bmatrix} A(\lambda) & B(\lambda) \\ C(\lambda) & D(\lambda) \end{bmatrix},$$
so
$$\operatorname{tr} T(\lambda) = A(\lambda) + D(\lambda)$$
is a constant of the motion, and the coefficients of its power series expansion in λ are classical conserved quantities.

Upon quantization, $\operatorname{tr} T(\lambda)$ becomes an operator, and linear combinations of the coefficients of the expansion of its eigenvalues as a power series in λ give the eigenvalues of energy, propagation number, and so on. The problem is to find the eigenvalues of the quantum operator
$$\operatorname{tr} \hat{T}(\lambda) = \hat{A}(\lambda) + \hat{D}(\lambda).$$

A requirement of the QISM is the existence of a vacuum state $|0\rangle$, having the property
$$\hat{C}(\lambda)|0\rangle = 0.$$
Choosing the n different values for λ
$$\lambda_1, \lambda_2, \ldots \quad \text{and} \quad \lambda_n,$$
one can construct a trial wave function as
$$|\psi(\lambda_1, \ldots, \lambda_n)\rangle = \prod_{j=1}^{n} \hat{B}(\lambda_j)|0\rangle, \tag{8.135}$$
but there is no guarantee that $|\psi(\lambda_1, \ldots, \lambda_n)\rangle$ is an eigenfunction of the trace operator.

However by directly calculating
$$[\hat{A}(\lambda) + \hat{D}(\lambda)]|\psi(\lambda_1, \ldots, \lambda_n)\rangle = \Lambda(\lambda)|\psi(\lambda_1, \ldots, \lambda_n)\rangle + \text{other terms}, \tag{8.136}$$
it is possible to select the λ_j such that the "other terms" are equal to zero. Then the eigenvalues are fixed by the coefficients of a power series expansion of $\Lambda(\lambda)$. After an early calculation by Hans Bethe [6], the conditions that must be satisfied are called the Bethe equations, and the corresponding λ_j are called Bethe parameters.

8.7.3 QISM analysis of the DST dimer

To see how the QISM works in practice, Let us apply it to the DST system with two freedoms that was studied by the NSM in Section 8.2.2 [21]. Starting with the classical system defined in Equation (8.35) and assuming that $\omega_0 = 0$ and $\varepsilon = 1$, one can show that the system is equivalent to

$$\dot{\mathcal{T}} = M\mathcal{T} - \mathcal{T}M, \tag{8.137}$$

where

$$M = \begin{bmatrix} -\lambda & \sqrt{\gamma}A_1^* \\ \sqrt{\gamma}A_2 & 0 \end{bmatrix}.$$

Also

$$\mathcal{T} = L_2 L_1$$

with

$$L_j = \begin{bmatrix} \lambda - i\gamma|A_j|^2 & -\sqrt{\gamma}A_j^* \\ -\sqrt{\gamma}A_j & i \end{bmatrix},$$

and $j = 1, 2$. Then quantizing as

$$\tilde{A}_j \to b_j$$
$$\tilde{A}_j^* \to b_j^\dagger$$

leads to the quantum trace operator

$$\operatorname{tr}\hat{\mathcal{T}}(\lambda) = \hat{A}(\lambda) + \hat{D}(\lambda)$$
$$= \lambda^2 - i\lambda\gamma(b_1^\dagger b_1 + b_2^\dagger b_2) - \gamma^2 b_1^\dagger b_1 b_2^\dagger b_2 + \gamma(b_2^\dagger b_1 + b_1^\dagger b_2) - 1,$$

which is related to the Hamiltonian operator defined in Equation (8.41) by

$$\hat{H} = \frac{1}{\gamma}\left[\frac{1}{2}\left(\frac{d}{d\lambda}\operatorname{tr}\hat{\mathcal{T}}(\lambda)\right)^2 - \operatorname{tr}\hat{\mathcal{T}}(\lambda)\right]_{\lambda=0} - \frac{1}{\gamma}$$
$$= -\frac{\gamma}{2}[(b_1^\dagger b_1)^2 + (b_2^\dagger b_2)^2] - (b_1^\dagger b_2 + b_2^\dagger b_1),$$

where the constant term has been ignored. This implies that the energy eigenvalues are related to eigenvalues of the trace operator by

$$E = \frac{1}{\gamma}\left[\frac{1}{2}\left(\frac{d}{d\lambda}\Lambda(\lambda)\right)^2 - \Lambda(\lambda)\right]_{\lambda=0} - \frac{1}{\gamma}.$$

In this example, the Bethe equations are

$$\lambda_k^2 + \prod_{j \neq k}^n \frac{\lambda_k - \lambda_j - i\gamma}{\lambda_k - \lambda_j + i\gamma}, \quad k = 1, 2, \ldots, n, \tag{8.138}$$

and eigenvalues of the trace operator are given by

$$\Lambda(\lambda) = \lambda^2 \prod_{j=1}^{n} \frac{\lambda - \lambda_j - i\gamma}{\lambda - \lambda_j} - \prod_{j=1}^{n} \frac{\lambda - \lambda_j + i\gamma}{\lambda - \lambda_j}.$$

For $n \geq 2$, Enol'skii and Salerno have shown that the solvability conditions for the Bethe equations are [21–23]

$$\det[P - \beta I] = 0,$$

where P is the $(n+1) \times (n+1)$ matrix

$$P = \begin{bmatrix} q_{0,n} & -i\gamma & 0 & 0 & \cdot & 0 \\ q_{1,n} & q_{1,n-1} & -2i\gamma & 0 & \cdot & 0 \\ q_{2,n} & q_{2,n-1} & q_{2,n-2} & -3i\gamma & \cdot & 0 \\ \cdot & \cdot & \cdot & \cdot & \cdot & \cdot \\ \cdot & \cdot & \cdot & \cdot & \cdot & -ni\gamma \\ q_{n,n} & q_{n,n-1} & q_{n,n-2} & \cdot & \cdot & q_{n,0} \end{bmatrix}.$$

Here

$$q_{k,n-l} \equiv (+i\gamma)^{k-l}\theta(k-l)\binom{n-l}{n-k}(1-\delta_{k,l})$$
$$- \theta(k-l)(-i\gamma)^{k-l-2}\binom{n-l}{n-k-2}[1 - \delta_{k,|l-2|}\theta(l-2)],$$

and $\theta(\cdot)$ is the step function for which $\theta(x) = 1$ if $x \geq 0$ and $\theta(x) = 0$ if $x < 0$. Finally the energy eigenvalues are related to the eigenvalues of P by

$$E_j = \frac{\beta_j}{\gamma} - \frac{\gamma n^2}{2}, \tag{8.139}$$

where $j = 1, 2, \ldots, n+1$. Similar calculations have been carried through for the Ablowitz–Ladik equation of Section 8.4.1 and for the fermionic polaron model of Section 8.4.3 [23].

8.7.4 *Comparison of the NSM and the QISM*

We are now in a position to compare the advantages and disadvantages of the NSM and the QISM. From a general perspective, the QISM appears to provide a basis for the diagonalization of quantum energy operators in integrable lattice systems. Because the classical ISM effectively transforms a lattice problem to f uncoupled oscillators, classical integrability implies a number of conserved quantities equal to the number of lattice points, and this would seem to allow complete diagonalization of the energy operator without resorting to numerical methods.

For lattice systems, the QISM as presently formulated does not achieve this goal. On the contrary, as we have seen in the previous section, a QISM that is based on the Bethe technique can be more demanding technically than the NSM and may yield no more useful results. With respect to the construction of the wave function for a DST dimer, in fact, the QISM results are less useful because it is more difficult to obtain wave functions from Equation (8.135) than from the expansion of Equation (8.42).

Although it seems to have the potential for directly diagonalizing the Hamiltonian (by representing it in ever smaller blocks that correspond to eigenvalues of additional conserved quantities), the QISM is less efficient than the NSM for finding energy eigenvalues of the integrable lattices considered in this chapter. Why is this so? Perhaps because the principle quantum number, introduced in the course of applying the Bethe method, can be interpreted as the eigenvalue of some number operator.

This is not to suggest that a procedure to fully exploit the power of classical integrability in the analysis of quantum problems will not be found. My feeling, however, is that such a technique will be be essentially different from the Bethe method (which merely generates the wave function through n applications of the operator $\hat{B}$ on the vacuum) by introducing additional symmetries of the problem. In a sense, the NSM might be on an incremental path toward this goal, because any additional symmetry that is recognized (translational symmetry, for example) can be built into the original structure of the NSM wave function.

On the other hand, the current version of the QISM offers an advantage over the NSM for lattice calculations in the thermodynamic limit, where both $f \to \infty$ and $n \to \infty$ with

$$\frac{n}{f} \to \text{constant}.$$

Such problems are numerically challenging for the NSM, but with the QISM the Bethe equations can be solved in a continuum approximation [24, 40].

In addition to being easier to use, the NSM has another attractive feature that is not shared by the QISM: it can be generalized to any energy operator that commutes with a solvable number operator. Salerno's equation, discussed in Section 8.4.2, provides an example of this property, where the number operator defined in Equation (8.95) is formally identical to that defined for the integrable Ablowitz–Ladik system in Equation (8.91). Although Salerno's equation is not integrable, it is as easily solved using the NSM as is the energy operator for the integrable Ablowitz–Ladik system.

The DST system is another example. With two degrees of freedom, the classical DST is integrable, and the corresponding QDST can be solved by the QISM, albeit with some difficulty as we have seen. For more than two freedoms, however, the QDST cannot be treated by the QISM, yet the NSM works handily. This property can be stated as follows. Whenever a number operator $\hat{N}$ can be used to block-diagonalize an energy operator $\hat{H}$ by the NSM, the same number operator can be used to block-diagonalize all energy operators that commute with $\hat{N}$.

8.8 Summary

Although this brief survey of quantum lattice solitons has skipped many details, we have obtained a number of experimentally useful results. Because quantum effects appear primarily at the atomic level, our emphasis has been upon understanding the structures of states with few quanta, as is expected in spectral observations on molecules and molecular crystals.

In addition to such problems, we have considered soliton systems with translational symmetry, a property that must be shared by the eigenfunctions of the corresponding quantum system. Wave packets describing quantum corrections to the classical behavior of a soliton are recognized as generalized Fourier transforms composed of extended eigenfunctions, and the quantum uncertainty principle stems directly from the fact that components oscillating at different frequencies will eventually get out of step. In order to represent this dispersion, one needs to know how the energies of the quantum states depend upon their propagation numbers, and to estimate the stability of a soliton, one must consider its binding energy.

Thus the wave packet describing the location of a nonlinear lattice solitary wave or soliton disperses with time just as does that for an electron. From the perspective of quantum theory, one is no more or less a "particle" than the other.

8.9 Problems

1. Sketch and discuss solution trajectories of the classical nonlinear oscillator of Equations (8.5) in the (x, p) phase plane for various values of the parameters M, K, and α.

2. Show that the Hamiltonian defined in Equation (8.3) is independent of time along solution trajectories of Equation (8.2).

3. Check that the number of modes of oscillation per unit volume of an electromagnetic cavity that lie between frequency ω and $\omega + d\omega$ is

$$\mathcal{N}(\omega)d\omega = \frac{\omega^2}{\pi^2 c^3} d\omega,$$

where c is the velocity of light. [Hint: count modes in reciprocal space (k_x, k_y, k_z).]

4. (a) If the energy of an oscillator only takes the values $E_n = n\hbar\omega$ and the probability of the oscillator being at energy $n\hbar\omega$ is $\exp(-n\hbar\omega/kT)$, show that the average energy of the oscillator is

$$\langle E \rangle = \frac{\hbar\omega}{e^{\hbar\omega/kT} - 1}.$$

[Hint: use the fact that in thermal equilibrium the occupation probability of a state of energy E is proportional to $\exp(-E/kT)$.]

(b) Show that $\langle E \rangle \to kT - \hbar\omega/2$ as $\hbar\omega/kT \to 0$.

(c) Assuming that $E_n = (n+1/2)\hbar\omega$, determine $\langle E \rangle$ for $kT \ll \hbar\omega$.

5. (a) From Equation (8.7), plot the black body radiation intensity against frequency for several values of temperature.

   (b) Discuss how Equation (8.7) differs from (8.6).

6. Use the Bohr–Sommerfeld condition

$$\oint p\, dx = 2\pi\hbar n$$

to obtain Planck's quantum levels for a linear harmonic oscillator.

7. Verify the commutation relation of Equation (8.13). [Hint: let the operators act on a wave function.]

8. (a) Develop the quantum description of a mass $M = 1$ that can move only in direction x in a potential field defined by

$$V(x) = \begin{cases} 0 & \text{for } |x| < L \\ \infty & \text{for } |x| > L. \end{cases}$$

[Hint: find the eigenfunctions for which $\psi(x)$ and $\psi(x)_x$ are continuous at $x = \pm L$ and approach zero as $x \to \pm\infty$.]

   (b) If $P(x,0) = \delta(x)$, find $P(x,t)$ as defined in Equation (8.11).

9. (a) Consider the previous problem in the limit $L \to \infty$ and describe the dispersion of $P(x,t)$ as $t \to \infty$.

   (b) What restraints does the requirement that

$$\int_{-\infty}^{\infty} P(x,t)\, dx = 1$$

place on the parameters of $\Psi(x,t)$?

   (c) If the particle is observed to have position $x = 1$ at $t = 100$, discuss the subsequent change of the wave packet.

   (d) Provide a physical interpretation of the "wave packet collapse" that occurred in (c) at $t = 100$?

10. (a) Discuss the previous problem in the context of Fourier transform theory.

    (b) Relate uncertainties in x times uncertainties in $p(\equiv \sqrt{2ME})$ to Planck's constant.

11. Describe solutions of the Schrödinger equation (8.15) in the phase plane $(\psi, d\psi/dx)$.

12. Consider Equation (8.17) to be in the Hamiltonian form

$$H = i\omega X P,$$

where $P \equiv A$ and $X \equiv -iA^*$. Show that Hamilton's equations of motion lead to Equations (8.18).

13. Consider a harmonic oscillator with the unit of length equal to $\sqrt{\hbar/M\omega}$.

    (a) Show that the ground state eigenfunction of Equation (8.15) is

    $$\psi_0 \propto e^{-x^2/2}.$$

    (b) By writing other eigenfunction as $\psi_n \propto u_n \psi_0$, show that u_n is a Hermite polynomial of order n. [Hint: assume polynomials with undetermined coefficients.]

    (c) Find the normalization factors for these eigenfunctions.

14. Show that the commutation relation of Equation (8.21) follows from Equation (8.13).

15. Check that the actions of the raising and lowering operators given in Equations (8.25) hold for eigenfunctions of Equation (8.15).

16. Show that the action of $b^\dagger$ on an eigenfunction of Equation (8.15) with energy eigenvalue $(n + 1/2)\hbar\omega$ is to raise it to an eigenfunction with eigenvalue $(n + 3/2)\hbar\omega$.

17. Formulate the action of a boson raising operator $(b^\dagger)$ as a linear Bäcklund transform.

18. Study eigenfunctions of Equation (8.26) in the phase plane $(\psi, d\psi/dx)$ for various values of the anharmonic parameter. Note the difference between positive values of α, corresponding to a hard spring, and negative values, corresponding to a soft spring.

19. Check the calculations leading to the Hamiltonian of Equation (8.27).

20. Show that symmetric ordering of the operators $b^\dagger$ and b is obtained by evaluating $(b^\dagger - b)^{2m}$ and keeping only terms with the same number of b and $b^\dagger$ operators.

21. In reference [33], it has been shown that symmetric ordering quantization of $|A|^{2m}$ implies

$$|A|^{2m} \to \frac{m!}{2^m} L_m(-2x)$$

for all m, where $L_m(\cdot)$ is a Laguerre polynomial of zero-order and $x^k \equiv (b^\dagger)^k(b)^k$. Use this result to construct the quantum operators in terms of $\hat{N} \equiv b^\dagger b$ for $|A|^6$, $|A|^8$, and $|A|^2 + \alpha|A|^4 + \beta|A|^6 + \gamma|A|^8$.

22. Consider an anharmonic oscillator with the energy

$$E = \tfrac{1}{2}Mx^2 + V(x),$$

where $V(x)$ is the Morse potential

$$V(x) = V_0[e^{2(x-x_0)/a} - 2e^{(x-x_0)/a}].$$

(a) Find the classical Hamiltonian (H) in the RWA.

(b) How does $\hat{H}$ depend on $\hat{N}$ for the quantum representation of this oscillator?

(c) Make a plot of E_n against n for the oscillator eigenfunctions. How well does your plot fit the Birge–Sponer relation?

23. Discuss what an experimentalist would measure on a quantum oscillator that is described by the classical Hamiltonian

$$|A|^2 + \alpha|A|^4 + \beta|A|^6 + \gamma|A|^8.$$

24. Use perturbation theory to find the first-order dependence of the eigenfunctions and eigenvalues in Equation (8.26) on α.

25. Consider two linear operators, $\hat{A}$ and $\hat{B}$, that commute. Show that nondegenerate eigenfunctions of $\hat{A}$ are also eigenfunctions of $\hat{B}$.

26. Find the Boltzmann energy kT in cm^{-1} at room temperature.

27. What is the unit of time in Equations (8.35)? How is this unit related to the time on your wrist watch?

28. Consider the Hamiltonian of Equation (8.34) to be a function of $X_1 \equiv A_1$, $X_2 \equiv A_2$, $P_1 \equiv iA_1^*$, and $P_2 \equiv iA_2^*$. Show that Hamilton's equations lead to Equations (8.35).

29. (a) Show that the DST Hamiltonian defined in Equation (8.34) is conserved under the dynamics of Equations (8.35).

(b) Show that the number defined in Equation (8.36) is conserved under the dynamics of Equations (8.35).

30. (a) Show that a local mode solution of Equations (8.35) has the amplitudes given in Equations (8.38).

(b) Make a plot of ω/ε against $\gamma N/\varepsilon$ for all the stationary solutions of Equations (8.35).

(c) Describe in detail what happens as γ passes through the critical value of $2\varepsilon/n$.

31. Assuming $H = H(N)$ for stationary solutions of the classical Equations (8.35), show that the frequency ω of a stationary solution is given by

$$\omega = dH/dN.$$

32. (a) Derive $\hat{H}$ and $\hat{N}$ in Equations (8.41) and (8.40) from the classical expressions for H and N in Equations (8.34) and (8.36).

    (b) Show that $\hat{H}$ and $\hat{N}$ commute.

33. Check the expressions given for the elements of the matrix $\mathcal{H}$, defined in Equation (8.44).

34. (a) Assume $n = 2$ in Equation (8.44) and find algebraic expressions for the two lowest energy eigenvalues $E^{\pm}$ and the corresponding eigenfunctions $|\psi_n\rangle^{\pm}$. Check that your results agree with Equations (8.46) and (8.47).

    (b) Repeat for $n = 3$.

    (c) Repeat for $n = 4$.

35. Use perturbation theory to obtain Equation (8.46).

36. Consider CH stretch vibrations in the dihalomethanes CCl_2H_2, CBr_2H_2, and CI_2H_2.

    (a) Neglecting the interactions between oscillations, show that the lowest energy eigenvalue against quantum number is

    $$E_0 = \omega_0 n - \frac{\gamma}{2}(n + n^2).$$

    (b) Use this result to estimate $n_{\max}$, the number of quanta needed to break a CH bond.

    (c) Check this estimate of $n_{\max}$ by studying the potential function

    $$V(x) = \frac{k}{2}x^2 - \frac{\alpha}{2}x^4.$$

    (d) Compare $n_{\max}$ for each of the three dihalomethanes. Can you suggest a physical explanation for the differences?

37. (a) From Stirling's formula, show that for large n the tunneling time defined in Equation (8.49) satisfies the inequality

    $$\tau > \frac{2\pi\hbar}{\varepsilon}\left[\frac{(n-1)\gamma}{\varepsilon}\right]^{n-1}.$$

(b) For dichloromethane, make a plot of τ against n for $1 < n < n_{\max}$, as found in the previous problem.

38. Show that the number N and energy H defined in Equations (8.51) and (8.52) remain constant under the dynamics of the DST equation.

39. Show that the number operator $\hat{N}$ and energy operator $\hat{H}$ defined in Equations (8.53) and (8.54) commute.

40. Find the number of ways that n beans can be placed into f jars.

41. Compute p from Equation (8.55) for $f \leq 10$ and $n \leq 10$.

42. Construct the most general eigenfunction of the number operator $\hat{N}$ that is defined in Equation (8.53) for:

   (a) $f = 2$ and $n = 10$.
   (b) $f = 6$ and $n = 2$.
   (c) $f = 6$ and $n = 3$.
   (d) $f = 6$ and $n = 4$.
   (e) $f = 4$ and $n = 4$.

43. Compare the parameters of uncoupled C–H stretching oscillation in benzene with those in the dihalomethanes.

44. Find the matrix corresponding to $\mathcal{Q}(2, \tau)$ in Equation (8.63) for f even.

45. (a) Derive Equation (8.65) for the eigenvalues of the matrix in Equation (8.63) in the limit as $f \to \infty$.

   (b) Show that the maximum group velocity of a wave packet of eigenfunctions with this energy is $4/\sqrt{\gamma^2 + 8}$.

46. Show that the size of a classical soliton of the DNLS system discussed in Section 8.3.2 is about equal to $24/(n+1)\gamma$ lattice spacings for $\gamma \ll 1$.

47. (a) Use Equation (8.67) to show that the maximum group velocity of a large QDNLS wave packet

$$V_m \to 2\varepsilon \sin(\pi/2n) \quad \text{as } \gamma \to 0.$$

   (b) Use Equation (8.69) to find the maximum group velocity of a small QDNLS wave packet.

48. (a) Show that the HA for the solution of the quantum DST equation given in Equation (8.80) implies

$$\langle n_j(t)\rangle^{(H)} \equiv^{(H)} \langle \psi_n(t)|b_j^\dagger b_j|\psi^n(t)\rangle^{(H)} = n|\Phi_{n,j}(t)|^2.$$

(b) Explain this equation in words.

49. (a) For the QNLS studied in Section 8.3.2, show that the HA to the binding energy of a soliton in the limit $\gamma \to 0$ and $f \to \infty$ is

$$E_b^H = \frac{\gamma^2}{48\varepsilon} n(n-1)^2.$$

(b) Why is it to be expected that the binding energy in the HA is less than that for the exact wave function? [Hint: consider the virial theorem.]

50. Consider an approximate wave packet for the QDNLS of the form

$$|\Psi(t)\rangle = e^{-n_0/2} \sum_n \frac{n_0^{n/2}}{\sqrt{n!}} |\psi_n(t)\rangle^{(H)},$$

where $|\psi(t)\rangle^{(H)}$ is a HA to the true wave function.

(a) Use the normalization condition of Equation (8.81) to show that

$$^{(H)}\langle \psi_n(t)|\psi_n(t)\rangle^{(H)} = 1.$$

(b) Show from this that

$$\langle \Psi(t)|\Psi(t)\rangle = 1.$$

(c) As $n \to \infty$, show that

$$b_j|\psi_n(t)\rangle^{(H)} \to \sqrt{n}\,\Phi_{n,j}|\psi_{n-1}(t)\rangle^{(H)},$$

and therefore as $n_0 \to \infty$

$$b_j|\Psi(t)\rangle \to \sqrt{n_0}\,\Phi_{n_0,j}|\Psi(t)\rangle.$$

(d) Noting from Equation (8.83) that the classical solution

$$A_j(t) \doteq \sqrt{n}\Phi_{n,j}(t)$$

for $n \gg 1$, complete the demonstration that as $n_0 \to \infty$,

$$\langle \Psi(t)|b_j|\Psi(t)\rangle \to A_j(t),$$

where $A_j(t)$ is a solution of the classical DNLS equation with the normalization

$$\sum_{j=1}^f |A_j|^2 = n_0.$$

51. Check that the Hamiltonian formulation given for the Ablowitz–Ladik system of Equation (8.87) is correct.

52. Show that the energy and number operators for QAL defined in Equations (8.88) and (8.91) commute.

53. (a) Give a physical interpretation of the upper soliton band for the QAL system in Figure 8.5(a).

    (b) What happens to this band in the continuum limit?

54. (a) Show that the actions of the raising and lowering operators for q-bosons given in Equations (8.90) reduce to those of standard bosons in the limit $\gamma \to 0$.

    (b) Discuss the character of q-bosons as the parameter $q = \gamma/2$ is varied, paying particular attention to the value $q = -2$.

55. Show that the binding energy and effective mass for the QAL system given in Equations (8.93) and (8.94) follow from Equation (8.92).

56. Describe the similarities and differences between bosons, q-bosons, and fermions. Are there other types of quanta?

57. Show that the fermionic energy and number operators defined in Equations (8.100) and (8.101) commute.

58. Show that the number of ways that n beans can be placed in f jars with no more than one bean per jar is $f!/n!(f-n)!$.

59. Demonstrate from Equation (8.104) that for $\gamma > 2$ the binding energy, effective mass, and maximum wave packet velocity for a two-quanta, fermionic polaron are given by

$$E_b = (\gamma - 2)^2/\gamma$$
$$m^* = \gamma/2$$
$$V_m = 2/\gamma.$$

60. Show from Equation (8.105) that as $\gamma \to \infty$ an upper bound on the group velocity of a fermionic wave packet is

$$V_m = 2/\gamma^{n-1}.$$

61. Demonstrate that the energy operators and the number operators for the Hubbard model defined in Equations (8.108) and (8.109) commute.

62. Compute 4^{m^3} vs. m for $m \leq 5$.

63. Is the alpha-helix soliton, as described by Davydov's product assumption of Equation (8.114), constrained by the Heisenberg uncertainty relations? If not, why not?

64. Consider the approximate DNLS equation for solitons on alpha-helix protein that is given in Equation (8.118). Under what conditions is the adiabatic approximation (upon which it is based) valid?

65. Show that
$$m^* = \hbar^2/2a^2 J$$
is the effective mass of a bare exciton on the ACN system defined by Equation (8.119).

66. Assuming the phonon Hamiltonian
$$\hat{H}_{\text{ph}} = \hbar\omega_0 \sum_{j=1}^{f} \left(b_j^\dagger b_j + \tfrac{1}{2}\right)$$
and a coherent state wave function of the form
$$|\Phi\rangle = \sum_{j=1}^{f} \exp\left[\beta_j(t) b_j^\dagger - \beta_j^*(t) b_j\right] |0\rangle_{\text{ph}},$$
show that
$$\langle\Phi|\hat{H}_{\text{ph}}|\Phi\rangle = \hbar\omega_0 \sum_{j=1}^{\infty} \left(|\beta_j|^2 + \tfrac{1}{2}\right).$$

67. (a) Use experimental and numerical results from Section 8.5.2 to compute the difference between Davydov's estimate of the binding energy of a self-trapped state in ACN, given in Equation (8.120), and corresponding results from the perturbation theory in small J, given in Equation (8.127).

   (b) Repeat for estimates of effective mass in Equations (8.121) and (8.128).

   (c) Which pair of results is more accurate? Why?

68. Show that the third of Equations (8.124) for a displaced linear oscillator reduces to a coherent state for $n = 1$ and $\tilde{n} = 0$.

69. Show that Equation (8.131) for the temperature dependence of the absorption intensity at 1650 cm^{-1} in crystalline ACN follows from Equation (8.130).

70. Use Equation (8.131) to calculate the intensity of the 1650 cm^{-1} soliton band in crystalline ACN as a function of temperature. Compare your results with the experimental measurements given in Table 8.3.

71. (a) Verify Equation (8.134) for the time derivative of a lattice transfer operator.

   (b) Show that the trace of the transfer operator is independent of time.

72. For the QISM analysis of the two freedom DST system that is sketched in Section 8.7.3, show that
$$\dot{T} = M_1 T - T M_1,$$
where
$$M_j = \begin{bmatrix} -\lambda & \sqrt{\gamma} A_j^* \\ \sqrt{\gamma} A_{j+1} & 0 \end{bmatrix}, \quad j = 1, 2.$$

418   QUANTUM LATTICE SOLITONS

73. (a) For the two freedom DST system of Section 8.7.3, compute

$$\hat{T} = \begin{bmatrix} \hat{A}(\lambda) & \hat{B}(\lambda) \\ \hat{C}(\lambda) & \hat{D}(\lambda) \end{bmatrix}.$$

(b) Show that $\hat{C}(\lambda)|0\rangle = 0$, where $|0\rangle \equiv |0\rangle_1 |0\rangle_2$.

(c) For $n = 1$, compute $|\psi(\lambda_1)\rangle = \hat{B}(\lambda_1)|0\rangle$.

(d) Show that for

$$[\hat{A}(\lambda) + \hat{D}(\lambda)]|\psi(\lambda_1)\rangle = \Lambda(\lambda)|\psi(\lambda_1)\rangle,$$

λ_1 must satisfy the equation

$$\lambda_1^2 + 1 = 0.$$

(e) Find $\Lambda(\lambda)$ for $n = 1$ and use it to show that the energy eigenvalues are

$$E_1 = \pm 1 - \gamma/2.$$

Discuss the physical meaning of these eigenvalues.

(f) Check that the same energy eigenvalues are given by Equation (8.139).

74. Repeat the previous problem for $n = 2$.

75. Without doing so, describe how to solve the system of Bethe equations given in Equation (8.138) for various values of the integer k.

76. Check that the energy eigenvalues given by Equation (8.139) are the same as those from Equation (8.44) with $\omega_0 - \gamma/2 = 0$.

77. Consider a lattice nonlinear Schrödinger system written in the form

$$i\frac{dA_j}{dt} + \varepsilon(A_{j+1} - 2A_j + A_{j-1}) + \gamma|A_j|^2 A_j = 0,$$

where $j = 1, 2, 3, \ldots, f$, $A_{j+f} = A_j$, and the A_j are unitless so ε and γ have units of seconds^{-1}.

(a) If adjacent points on the lattice are separated by a distance a, show that the dynamics of this system conserves the energy

$$H = \sum_{j=1}^{f} \left[\varepsilon a^2 \left(\frac{A_j^* - A_{j-1}^*}{a} \right) \left(\frac{A_j - A_{j-1}}{a} \right) + 2\varepsilon|A_j|^2 - \frac{\gamma}{2}|A_j|^4 \right]$$

and the number

$$N = \sum_{j=1}^{f} |A_j|^2.$$

(b) For $\gamma \ll \varepsilon$, a classical soliton will be spread out over many lattice points, and the discrete variable ja can be regarded as a continuous variable x. By introducing the rotating transformation

$$A_j(t) \equiv \tilde{A}_j e^{-2i\varepsilon t} \longrightarrow \phi(x,t)\, e^{-2i\varepsilon t},$$

show that $\phi(x,t)$ is governed by the classical nonlinear Schrödinger equation

$$i\frac{d\phi}{dt} + \varepsilon a^2 \frac{\partial^2 \phi}{\partial x^2} + \gamma |\phi|^2 \phi = 0.$$

(c) Letting $f \longrightarrow \infty$ with $-\infty < x < +\infty$, show that solutions of this equation conserve the Hamiltonian

$$H = \int_{-\infty}^{\infty} \left(\varepsilon a \phi_x^* \phi_x - \frac{\gamma}{2a}|\phi|^4\right) dx,$$

and the number

$$N = \int_{-\infty}^{\infty} \frac{1}{a}|\phi|^2\, dx,$$

among others.

78. If the system of the previous problem is quantized by letting

$$\tilde{A}_j\, (\tilde{A}_j^*) \to b_j\, (b_j^\dagger),$$

show that the corresponding energy and number operators become

$$H \to \hat{H} = \sum_{j=1}^{f} \left[\varepsilon a^2 \left(\frac{b_j^\dagger - b_{j-1}^\dagger}{a}\right)\left(\frac{b_j - b_{j-1}}{a}\right) - \frac{\gamma}{2} b_j^\dagger b_j^\dagger b_j b_j\right]$$

and

$$N \to \hat{N} = \sum_{j=1}^{f} b_j^\dagger b_j.$$

79. In the quantum continuum limit of the previous problem, let

$$|\psi_{n,\tau}\rangle \to |\psi(n,k)\rangle,$$

where $\tau = e^{ik}$ is an eigenvalue of the translation operator, $-\pi < k \leq +\pi$, and let

$$b_j \text{ and } b_j^\dagger \to \hat{\phi}(x) \quad \text{and} \quad \hat{\phi}^\dagger(x),$$

called field (lowering and raising) operators.

(a) Show that these field operators obey the commutation relations

$$[\hat{\phi}(x), \hat{\phi}(y)] = [\hat{\phi}^\dagger(x), \hat{\phi}^\dagger(y)] = 0 \quad \text{and} \quad [\hat{\phi}(x), \hat{\phi}^\dagger(y)] = \delta(x-y).$$

(b) In terms of these field operators, show that the quantum lattice operators obtained in the previous problem become

$$\hat{\mathcal{H}} = \int_{-\infty}^{\infty} \left[\varepsilon a \hat{\phi}_x^\dagger \hat{\phi}_x - \frac{\gamma}{2a} \hat{\phi}^\dagger \hat{\phi}^\dagger \hat{\phi} \hat{\phi} \right] dx,$$

and

$$\hat{\mathcal{N}} = \frac{1}{a} \int_{-\infty}^{\infty} \hat{\phi}^\dagger \hat{\phi} \, dx,$$

with eigenvalues

$$\hat{\mathcal{H}} |\psi(n,k)\rangle = E(n,k) |\psi(n,k)\rangle$$

and

$$\hat{\mathcal{N}} |\psi(n,k)\rangle = n |\psi(n,k)\rangle.$$

80. (a) Prove that the energy of a quantum soliton is

$$E(n,k) = n\varepsilon k^2 - \frac{\gamma^2}{48\varepsilon} n(n^2 - 1).$$

(b) Describe $|\psi(n,k)\rangle$ for a quantum soliton.

REFERENCES

1. M J Ablowitz and J F Ladik. Nonlinear differential-difference equations and Fourier analysis. *J. Math. Phys.* 17 (1976) 1011–1018.
2. M J Ablowitz and J F Ladik. A nonlinear difference scheme and inverse scattering. *Stud. Appl. Math.* 55 (1976) 213–229.
3. D M Alexander and J A Krumhansl. Localized excitations in hydrogen-bonded molecular crystals. *Phys. Rev. B* 33 (1986) 7172–7185.
4. S Aubry. Breathers in nonlinear lattices: Existence, linear stability and quantization. *Physica D* 103 (1997) 201–250.
5. L Bernstein, J C Eilbeck, and A C Scott. The quantum theory of local modes in a coupled system of nonlinear oscillators. *Nonlinearity* 3 (1990) 293–323.
6. H Bethe. Zur Theorie der Metalle I. Eigenwerte und Eigenfunktionen der Linearen Atomkette. *Z. Phys.* 71 (1931) 205–226.
7. R T Birge and H Sponer. The heat of dissociation of non-polar molecules. *Phys. Rev.* 28 (1926) 259–283.
8. N Bohr. On the constitution of atoms and molecules. *Philos. Mag.* 26 (1913) 1–25 and 476–502.
9. M Born and J R Oppenheimer. Zur Quantentheorie der Molekeln. *Ann. Phys.* 84 (1927) 457–484.
10. G Careri, U Buontempo, F Galluzzi, A C Scott, E Gratton, and E Shyamsunder. Spectroscopic evidence for Davydov-like solitons in acetanilide. *Phys. Rev. B* 30 (1984) 4689–4702.
11. E R Cohen and B N Taylor. The fundamental physical constants. *Phys. Today* Part 2, August 1994, 9–13.
12. L Cruzeiro-Hansson and S Takeno. Davydov model: The quantum, mixed quantum-classical and full classical systems. *Phys. Rev. E* 56 (1997) 894–906.

13. A S Davydov. *Solitons in Molecular Systems*. 2nd edition. Reidel, Dordrecht, 1991.
14. P A M Dirac. *The Principles of Quantum Mechanics*. Clarendon Press, Oxford, 1958.
15. J Edler and P Hamm. Self-trapping of the amide I band in a peptide model crystal. *J. Chem. Phys.* 117 (2002) 2415–2424.
16. J Edler, P Hamm, and A C Scott. Femtosecond study of self-trapped excitons in crystalline acetanilide. *Phys. Rev. Lett.* 88 (2002) 067403-1–4.
17. J C Eilbeck, H Gilhøj, and A C Scott. Soliton bands in anharmonic lattices. *Phys. Lett. A* 172 (1993) 229–235.
18. J C Eilbeck, P S Lomdahl, and A C Scott. The discrete self-trapping equation. *Physica D* 16 (1985) 318–338.
19. J W Ellis. Molecular absorption spectra of liquids below $3\,\mu$. *Trans. Faraday Soc.* 25 (1929) 888–897.
20. T Elsaesser, S Mukamel, M M Murnanae, and N F Scherer. *Ultrafast Phenomenon XII*. Springer-Verlag, Berlin, 2000.
21. V Z Enol'skii and M Salerno. On the calculation of the energy spectrum of quantum integrable systems. *Phys. Lett. A* 155 (1991) 121–125.
22. V Z Enol'skii, M Salerno, A C Scott, and J C Eilbeck. Alternate quantizations of the discrete self-trapping dimer. *Phys. Scr.* 43 (1991) 229–235.
23. V Z Enol'skii, M Salerno, A C Scott, and J C Eilbeck. There's more than one way to skin Schrödinger's cat. *Physica D* 59 (1992) 1–24.
24. L D Faddeev and L A Takhtajan. Spectrum and scattering of excitations in the one-dimensional isotropic Heisenberg model. *J. Sov. Math.* 24 (1984) 241.
25. L D Faddeev and L A Takhtajan. *Hamiltonian Methods in the Theory of Solitons*. Springer-Verlag, Berlin, 1987.
26. F Fillaux and C J Carlile. Inelastic-neutron-scattering study of methyl tunneling and the quantum sine–Gordon breather in isotopic mixtures of 4-methyl-pyridine at low temperatures. *Phys. Rev. B* 42 (1990) 5990–6006.
27. F Fillaux, C J Carlile, and G J Kearly. Inelastic neutron-scattering study at low temperature of the quantum sine–Gordon breather in 4-methyl-pyridine with partially deuterated methyl groups. *Phys. Rev. B* 44 (1991) 12280–12293.
28. R J Glauber. The quantum theory of optical coherence. *Phys. Rev.* 130 (1963) 2529–2539.
29. R J Glauber. Coherent and incoherent states of the radiation field. *Phys. Rev.* 131 (1963) 2766–2788.
30. I S Gradshteyn and I M Ryzhik. *Table of Integrals, Series, and Products*. Academic Press, New York, 1980.
31. T-K Ha, M Lewerenz, R Marquardt, and M Quack. Overtone intensities and dipole moment surfaces for the isolated CH chromophore in CHD_3 and CHF_3: Experiment and *ab initio* theory. *J. Chem. Phys.* 93 (1990) 7097–7109.
32. H Haken. *Quantum Field Theory of Solids*. North-Holland, Amsterdam, 1976.
33. M H Hays and A C Scott. Quantizing the discrete self-trapping equation. *Phys. Lett.* 188A (1994) 21–26.
34. O J Heilmann and E H Lieb. Violation of the noncrossing rule: The Hubbard Hamiltonian for benzene. *Ann. NY Acad. Sci.* 172 (1971) 584–617.
35. B R Henry and W Siebrand. Anharmonicity in polyatomic molecules: The CH-stretching overtone spectrum of benzene. *J. Chem. Phys.* 49 (1968) 5369–5376.

36. A Klein and F Kreis. Nonlinear Schrödinger equation: A testing ground for the quantization of nonlinear waves. *Phys. Rev. D* 13 (1976) 3282–3294.
37. V E Korepin, N M Bogoliubov, and A G Izergin. *Quantum Inverse Scattering Method and Correlation Functions*. Cambridge University Press, Cambridge, 1993.
38. P P Kulish. Quantum difference nonlinear Schrödinger equation. *Lett. Math. Phys.* 5 (1981) 191–197.
39. W H Louisell. Correspondence between Pierce's coupled mode amplitudes and quantum operators. *J. Appl. Phys.* 33 (1962) 2435–2436.
40. J Lowenstein. Introduction to the Bethe ansatz approach in $(1+1)$ dimensional models. In *Recent Advances in Field Theory and Statistical Mechanics*, J B Zuber and R Stora, eds., North-Holland, Amsterdam, 1984.
41. V G Makhankov. *Soliton Phenomenology*. Kluwer Academic, Dordrecht, 1990.
42. V G Makhankov and V K Fedyanin. Non-linear effects in quasi-one-dimensional models of condensed matter theory. *Phys. Rep.* 104 (1984) 1–86.
43. R Marquardt and M Quack. The wave packet motion and intramolecular vibrational redistribution in CHX_3 molecules under infrared multiphoton excitation. *J. Chem. Phys.* 95 (1991) 4854–4876.
44. J A McCammon and S C Harvey. *Dynamics of Proteins and Nucleic Acids*. Cambridge University Press, Cambridge, 1987.
45. R Micnas, J Ranninger, and S Robaszkiewicz. Superconductivity in narrow-band systems with local attractive interactions. *Rev. Mod. Phys.* 62 (1992) 113–171.
46. P D Miller, A C Scott, J Carr, and J C Eilbeck. Binding energies for discrete nonlinear Schrödinger equations. *Phys. Scr.* 44 (1991) 509–516.
47. O S Mortensen, B R Henry, and M A Mohammadi. The effects of symmetry within the local mode picture: A reanalysis of the overtone spectra of the dihalomethanes. *J. Chem. Phys.* 75 (1981) 4800–4808.
48. S Mukamel. *Principles of Nonlinear Optical Spectroscopy*. Oxford University Press, New York, 1995.
49. A H Nayfeh and B Balachandran. *Applied Nonlinear Dynamics*. John Wiley, New York, 1995.
50. A M Perelomov. Generalized coherent states. *Usp. Phys. Nauk* 123 (1977) 23–55.
51. M Peyrard and M D Kruskal. Kink dynamics in the highly discrete sine–Gordon system. *Physica D* 14 (1984) 88–102.
52. M Planck. Über das Gesetz der Energieverteilung im Normalspectrum. *Ann. Phys.* 4 (1901) 553–563.
53. M Quack. Spectra and dynamics of coupled vibrations in polyatomic molecules. *Annu. Rev. Phys. Chem.* 41 (1990) 839–874.
54. M Salerno. Quantum deformations of the discrete nonlinear Schrödinger equation. *Phys. Rev. A* 46 (1992) 6856–6859.
55. M Salerno. Bose–Einstein condensation in a system of q-bosons. *Phys. Rev. E* 50 (1994) 4528–4530.
56. M Salerno. SO(4)-invariant basis functions for strongly correlated Fermi systems. *Phys. Lett. A* 217 (1996) 269–274.
57. M Salerno. The Hubbard model on a complete graph: Exact analytical results. *Z. Phys. B* 99 (1996) 469–471.

58. M Salerno. Ferromagnetic ground states of the Hubbard model on a complete graph. *Z. Phys. B* 101 (1996) 619–621.
59. L I Schiff. *Quantum Mechanics*. 3rd edition. McGraw-Hill, New York, 1968.
60. E Schrödinger. Quantisierung als Eigenwertproblem. *Ann. Phys.* 79 (1926) 361–376.
61. E Schrödinger. Die stetige Übergang von der Mikro- zur Makromechanik. *Naturwissenschaften* 28 (1926) 664–666.
62. A C Scott. Davydov's soliton. *Phys. Rep.* 217 (1992) 1–67.
63. A C Scott. On quantum theories of the mind. *J. Consciousness Stud.* 3 (1996) 484–491.
64. A C Scott and J C Eilbeck. On the CH stretch overtones of benzene. *Chem. Phys. Lett.* 132 (1986) 23–28.
65. A C Scott, J C Eilbeck, and H Gilhøj. Quantum lattice solitons. *Physica D* 78 (1994) 194–213.
66. A C Scott, E Gratton, E Shyamsunder, and G Careri, The IR overtone spectrum of the vibrational soliton in crystalline acetanilide, *Phys. Rev. B* 32 (1985) 5551–5554.
67. J C Slater. *Quantum theory of Matter*. McGraw-Hill, New York, 1951.
68. A T Tu. *Spectroscopy in Biology*. John Wiley, New York, 1982.
69. A R Vasconcellos, M V Mesquite, and R Luzzi. Statistical thermodynamic approach to vibrational solitary waves in acetanilide. *Phys. Rev. Lett.* 80 (1998) 2008–2011.
70. J A D Wattis. *Analytic Approximations to Solitary Waves on Lattices*. PhD Thesis, Mathematics Department, Heriot–Watt University, 1993.
71. J A D Wattis. Variational approximations to breather modes in the discrete sine–Gordon equation. *Physica D* 82 (1995) 333–339.
72. J A D Wattis. Variational approximations to breathers in the discrete sine–Gordon equation II: Moving breathers and Peierls–Nabarro energies. *Nonlinearity* 8 (1996) 1583–1598.
73. J A D Wattis. Stationary breather modes of generalized nonlinear Klein–Gordon lattices. *J. Phys. A: Math. Gen.* 31 (1998) 3301–3323.
74. E Wright, J C Eilbeck, M H Hays, P D Miller, and A C Scott. The quantum discrete self-trapping equation in the HA. *Physica D* 69 (1993) 18–32.
75. A Xie, L van der Meer, W Hoff, and R H Austin. Long-lived Amide-I vibrational modes in myoglobin. *Phys. Rev. Lett.* 84 (2000) 5435–5438.

9
LOOKING AHEAD

At the end of this introduction to nonlinear science, it seems appropriate to review the ground we have covered and consider where studies of emergent phenomena might be going in the present century. Over the past few decades the landscape of applied science has changed dramatically. Many concepts and ideas that were widely accepted in the 1960s have been unexpectedly superseded by quite different perspectives, altering profoundly our collective understanding of dynamics and opening new vistas into the future. As we have seen, examples of such changes include the following.

Solitons. Before 1965, it was believed that the dynamics of nonlinear partial differential equations were so intricate that analytic solutions would be nearly impossible to find. Now it is known that dozens of such systems, several arising in applications, can be exactly solved using techniques that are based upon recognizing the "soliton" as a new dynamic entity emerging from an underlying PDE system. Important in itself, this development greatly expands the range of perturbation studies by allowing the analyst to choose a fully nonlinear function (say an N-soliton formula) as a zero-order estimate.

Low dimensional chaos. In the 1960s, most applied mathematicians felt that low order systems of ordinary differential equations were so simple to analyze that they did not merit the attentions of serious scientists. Through numerical studies, it unexpectedly became evident that third-order nonlinear systems often exhibit "strange attractors", where the solution trajectory wanders aimlessly within a bounded region, until the end of time. Over the past decade, the related phenomenon of chaos has been widely discussed in the popular press and in several books on nonlinear dynamics.

Nonlinear diffusion. Although discovered and publicly demonstrated in 1906 and theoretically predicted and experimentally verified in the early 1920s, a rediscovery in 1951 of oscillatory dynamics and traveling-waves in solutions of reacting chemicals was refused publication because of the mistaken notion that the phenomenon would violate the second law of thermodynamics. This unreceptive response to an important scientific observation seems doubly strange in retrospect because it occurred at the time that emergent theories based upon nonlinear diffusion were being proposed for nerve impulse propagation and the development of biological form.

Throughout the 1960s, experimental methods and theoretical understanding in this area have gradually improved, and today dynamic pattern formation is accepted as the basis for experimental observations of anomalous heart muscle contractions, prairie fires, fairy rings of mushrooms, spiral lichens, and slime mold dynamics, in addition to the earlier data on chemical waves and nerve membrane dynamics. By shifting the theater of competition from the individual cell to higher levels of organization, related phenomena may introduce the virtue of altruism into biological evolution, mitigating some of its gloomier aspects.

Local modes in molecules. Since the 1920s, there has been experimental evidence suggesting the autolocalization of energy in small molecules, and with the advent of laser based spectroscopy in the late 1960s, this evidence became very strong. In particular, it was observed that the overtone spectra of CH stretching modes in normal benzene (C_6H_6) and of five-deuterated benzene (C_6D_5H) are almost identical. Although more convincing evidence for local modes in small molecules can scarcely be imagined, manuscripts describing the phenomenon were rejected in the 1970s on the grounds that it is theoretically impossible.

Why impossible? The argument goes like this. Molecular vibrations are governed by quantum theory, which is linear in the formulation of Schrödinger. A molecule of benzene has six-fold rotational symmetry about the planar axis; thus the eigenfunctions of Schrödinger's equation must share this symmetry, thereby precluding localization of energy.

Now it is known that the (six for benzene) eigenfunctions of lowest energy—at a given principle quantum number n—are "quasidegenerate," lying within an energy band of range that is inversely proportional to $(n-1)!$, which grows rapidly with n. Because coherence time is inversely proportional to energy band width, a wave packet of eigenfunctions can be constructed to localize vibrational energy on a single bond for a time that is long enough to be experimentally significant. Thus, the constraints of quantum theory are brought into agreement with facts of nature.

From the perspective of nonlinear wave theory, a local mode of the CH stretching oscillation on benzene can be described as the emergence of a pinned soliton on a periodic lattice of six sites.

Vibrational polarons in biomolecules and molecular crystals. In 1973, several scientists independently suggested that a polaron-like mechanism might be responsible for the storage and transport of vibrational energy in large biological molecules, particularly the alpha-helix regions of proteins. In simplest terms, it was proposed that localized intramolecular vibrations would distort the adjacent lattice structure, which would then act as a potential well, trapping the original vibrations. Throughout the 1980s, this idea was extensively studied from a variety of theoretical perspectives, suggesting that polaronic phenomena are to be found in some biomolecules.

Although experimental confirmation in a biological preparation is lacking, a certain "anomalous" infrared band on molecular crystals of acetanilide (a model protein related to Tylenol, a popular analgesic) was assigned to a vibrational polaron in 1984. Initially rejected by many physical chemists on grounds similar to the former criticisms of local modes in molecules, this assignment is now confirmed. The anomalous infrared band in crystalline acetanilide is a lattice soliton, or "discrete breather" in the jargon of modern nonlinear wave theory.

Current research in this area is exciting for two reasons. First, there have been advances in the theory of discrete breathers, leading to many suggestions for possible experimental observations. Second, the advent of infrared pump-probe spectroscopy allows experimentalists to directly determine whether a particular spectral line corresponds to a localized mode.

Dendritic logic in neurons. Throughout most of the present century, it has been assumed by neurophysiologists that the dendritic (ingathering) fibers of a nerve cell are passive, unable to support the active nerve impulses that are known to be characteristic of axonal (outgoing) fibers. As this belief has persisted in the face of evidence that many dendritic membranes are active, one wonders why it is so strongly held.

Lacking a fundamental understanding of the nonlinear diffusion underlying nerve impulse conduction, electrophysiologists have viewed it simply as an "all-or-nothing" process: once launched, an active nerve impulse is assumed to fire all of the active membrane in its path. While this assumption causes no difficulty on an axonal tree (where an impulse on the main trunk propagates outward toward distal branches), the concept is problematic on the dendritic side of the neuron, because a single active impulse on any dendritic branch would then fire the entire structure, precluding the possibility of integrating the incoming information.

Such concerns evaporate, however, when one realizes that an active fiber requires a threshold charge to be supplied before an impulse emerges, and this condition may fail to be satisfied where a dendritic fiber branches. Although the concept of threshold charge was understood in the early 1970s and independently suggested by several researchers as a basis for dendritic logic, the inertia of collective belief is strong. Only recently have the possibilities of information processing in dendritic structures become fashionable among neuroscientists.

What are the lessons to be learned from such examples of scientific oversight? Can we assume that the list is complete? Do no more surprises lie in wait? Is "science finished" as was believed a hundred years ago and some are saying again? How are young scientists to set their sails for the challenges of the present century?

Perhaps the most important message to carry away from the experiences of nonlinear science over the past three decades is this. Do not be overly impressed by theorists. Always think things through for yourself. Be wary of procrustean tendencies to stretch or truncate the facts of nature until they fit within the confines of some narrow doctrine. As the theorists lovingly unfold their formulations, maintain a jaundiced eye.

Although the list of errors in widely held scientific opinions may be complete and science ended, it seems imprudent to assume so. Furthermore, I don't believe it. My intuition suggests that there is yet much to learn in the realm of nonlinear science, where the whole is more than the sum of its parts. Many aspects of the natural world remain mysterious.

Among the foremost of these mysteries is an understanding of the phenomenon of life. Just as the twentieth century can properly be described as the "century of physics," during which—for better and worse—numerous secrets of the inanimate world have been prized from nature's store, many predict that the twenty-first will become the "century of biology." How will the life sciences of this new century differ from those of the past?

To comprehend the nature of life, one must begin by considering what is known about the dynamic structure of a living organism, and we have noted in Chapter 4 that the arrangement is hierarchical, with relevant knowledge organized more or less as in the following diagram:

Environment
Species
Organism
Physiology
Cytology
Replication of biomolecules
Genetic transcription
Biochemical cycles
Biomolecular dynamics
Chemistry
Atomic physics

This diagram is not intended to be definitive or complete; it is merely a sketch to present the general idea of the hierarchical structure of biological organization. One may well suggest the inclusion of additional levels, branchings, between (say) the plant and animal kingdoms, or whatever corresponds to the facts of the matter. Furthermore, there is nothing original about the suggestion of a hierarchical structure of biology; it has been observed by many and is obvious to anyone who considers how the books are arranged in a scientific library [1].

Nonetheless, there are some comments to make about the diagram that may not be obvious. Note first that each of these levels is thought to be characterized by some order parameters obeying nonlinear rules or regularities of interaction that the scientists devoted to that level are working to discover. These rules may be nonlinear differential equations of the sort considered in this book, or they may be of a more subtle nature, yet unknown. At some levels of description (psychology and cultural anthropology come to mind), there may be few quantifiable rules, or none at all [9].

From the nonlinear dynamics at each level—as we have seen throughout this book—there emerge stable entities (new things), providing a basis for the dynamics at higher levels. Guided by the Born–Oppenheimer potential energy surface, atomic elements organize themselves into molecules. Out of the nonlinear interactions of chemical molecules emerge the proteins, acting as catalysts (enzymes) in biochemical cycles. Biochemical cycles support the replication of biomolecules, underlying cellular reproduction. In their turn, cells comprise organs, which join together to form organisms. These go on to interact with each other (nonlinearly, of course) to become components of species and the biosphere.

Although the dynamics at each level is clearly influenced by those at adjacent higher and lower levels, effects of longer range are also expected, examples of which abound. An environmental change that alters the chemical composition of the atmosphere can have profound influence at levels related to the biochemistry of a particular species. Subconscious fear can alter the chemistry of one's blood. A cultural obligation to ingest a recreational drug can change the dynamics of certain membrane proteins. And so on.

Fledgling scientists and learned philosophers alike should be aware that nonlinear dynamic hierarchies are more intricate than anything hitherto conceived by science. Upon consideration of the several surprises that have emerged from seemingly benign differential equations, it is evident that the behavioral bounds of living systems are presently unknown. Those who make pronouncements about the ultimate nature of biological organisms are merely expressing their opinions—often based upon experience with inanimate matter—rather than presenting scientific knowledge.

With this hierarchical structure in mind, we can begin to ask: What is life? Where does it reside? What is its physiological substrate? How does one deal with such phenomena?

To illustrate the difficulties of comprehending the dynamics of a living organism, think of each chain of causal implications as a very long piece of spaghetti, threading its way through the hierarchical structure, branching and converging (as causal implications do), folding back on itself to form closed causal loops, engaged in endless involutions, implying immense numbers of possible future configurations, boggling the mind. Where—in this slithering nest of ourobori—is the physiological substrate of life?

If my heart begins to fibrillate, I die. Does this mean that life is located in the heart? If the oxygen is removed from the atmosphere, we all die. Does this imply that life resides in the biosphere? If the iron atoms in the hemoglobin of my blood are changed into atoms of nickel, I die yet again. Is life then situated at the level of biochemistry? Or is life a phenomenon that emerges from all levels of an organism's dynamic hierarchy?

A central feature of life is the ability to pursue attentive and goal oriented behavior. Some call this property "autopoietic," while others use the term "complex adaptive system." For human life this ability has evolved to abstract thinking and self-awareness in the context of our cultural configurations. Collectively called "mind," such phenomena provide the bases for our lives, the feelings that color our hopes and dreams, our joy and pain. Although poorly understood, complex adaptive systems point to some of the most challenging problems of modern mathematics [9].

As was noted at the close of Chapter 5, there are reasons to suppose that mind emerges from the hierarchical organization of the brain [15, 16], which can be sketched as follows:

>Human culture
>Phase sequence
>Complex assembly
>$\vdots$
>Assembly of assemblies of assemblies
>Assembly of assemblies
>Assembly of neurons
>Neuron
>Nerve fiber
>Membrane protein

Comparison of the hierarchy of life with that of a brain leads one to ask: What is the relationship between life and mind? The importance of this question suggests that the study of interacting dynamic hierarchies will become a fruitful area of research in applied mathematics. Others have expressed related ideas.

1. Erwin Schrödinger, the Austrian physicist, philosopher, and sometime poet (whose formulation of quantum mechanics we studied in the previous chapter) was deeply interested in understanding the natures of life and mind [13, 14]. In 1944, he published an influential book entitled *What is Life?*, proposing (from measurements of mutation rates of fruit flies induced by gamma rays) that the genetic code is embodied in a single molecule [11]. Among the several conclusions drawn from this idea was the following:

   From all we have learned about the structure of living matter, we must be prepared to find it working in a manner that cannot be reduced to the ordinary laws of physics. And that is not on the ground that there is any new force directing the behaviour of the single atoms within a living organism, or because the laws of chemistry do not apply, but because life at the cellular level is more ornate, more elaborate than anything we have yet attempted in physics.

2. In his development of the meta-science of "synergetics," the German condensed matter physicist Hermann Haken has emphasized the importance of using descriptions adapted to the levels being investigated, and defining appropriate order parameters to describe the emergent entities at each level [6, 7]. With respect to dynamic hierarchies in the brain, he notes [3]:

   While in physics or chemistry it is not too difficult to define the microscopic and macroscopic levels, with respect to the brain we must ask what adequate intermediate levels we have to choose between "microscopic" and "macroscopic." ... Probably there is not only one intermediate level at the neuronal level, and we may ask what are the appropriate intermediate levels to be chosen, which are already macroscopic with respect to the neuronal level but still microscopic to more complex functions of the brain.

3. The German biochemists Manfred Eigen and Peter Schuster have sketched a scheme of life based upon a hierarchy of cycles of dynamic activity, working together to form "hypercycles" [5]. In their words:

> The hypercycle is the tool for integrating length-restricted self-replicative entities into a new stable order, which is able to evolve coherently. No other kind of organization, such as mere compartmentation, or non-cyclic networks could fulfill simultaneously all three of the following conditions
>
> - to maintain competition among the wild-type distribution of every self-replicative entity in order to preserve their information,
> - to allow for a coexistence of several (otherwise competitive) entities and the mutual distributions, and
> - to unify these entities into a coherently evolving unit, where advantages of one individual can be utilized by all members and where this unit as a whole remains in sharp competition with any unit of alternative composition.

4. Introducing her recent book entitled *What is Life?*—written in tribute to Schrödinger—the American evolutionary biologist Lynn Margulis emphasizes the immense intricacy of the interrelations among living organisms that have developed throughout the course of evolution [10]. Pressing is the need for science to find a middle path between the Scylla of determinism and the Charybdis of animism in order to deal with the realities of reproducing and attentive organisms. Following a suggestion of Schrödinger [13], she comments thus on the relation between life and mind:

> If we grant our ancestors even a tiny fraction of the free will, consciousness, and culture we humans experience, the increase in complexity on Earth over the last several thousand million years becomes easier to explain: life is the product not only of blind physical forces but also of selection in the sense that organisms choose. All autopoietic beings have two lives, as the country song goes, the life we are given and the life we make.

Although such ideas may seem far-fetched to those trained in the analysis of inanimate matter, they are beginning to be taken up by members of the nonlinear science community [2, 4, 12, 17]. But the task of understanding the nature of life will not be easy for several reasons. First of all, there is a very large number of parameters to be determined in any realistic model of even the simplest living organism. Second, the phenomenon of emergence leads to the appearance of new dynamic entities in phase spaces of uncharted dimension, and these entities may be not only unknown but unknowable [9]. Third, it is difficult to distinguish between cause and effect in a system with myriad closed causal loops interwoven through many dynamic levels, often making the determination of even a single parameter problematic. Fourth, time scales tend to change by an order of magnitude or more between adjacent levels of life's hierarchy, rendering numerical studies challenging if not impractical. Finally, the involution of life and mind introduces profound philosophical problems that are yet to be sorted out [8, 9, 14–16].

The task of science, however, is to deal with nature as it is, not as we would have it be. From a perspective that appreciates the hierarchical nature of emergent phenomena and the immense number of ways that life and mind may interweave, science is far from finished, and researchers of the present century face challenges worthy of their very best efforts.

REFERENCES

1. P W Anderson. More is different: Broken symmetry and the nature of the hierarchical structure of science. *Science* 177 (1972) 393–396.
2. N A Baas. Emergence, hierarchies, and hyperstructures. In *Artificial Life III*, C G Langton, ed., Addison-Wesley, Reading, MA, 1994.
3. E Basar, H Flohr, H Haken, and A J Mandell (eds.). *Synergetics of the Brain*. Springer-Verlag, Berlin, 1983.
4. M C Boerlijst and P Hogeweg. Spiral wave structure in pre-biotic evolution: Hypercycles stable against parasites. *Physica D* 48 (1991) 17–28.
5. M Eigen and P Schuster. *The Hypercycle: A Principle of Natural Self-organization*. Springer-Verlag, Berlin, 1979.
6. H Haken. *Advanced Synergetics*. Springer-Verlag, Berlin, 1983.
7. H Haken. *Synergetics*. 3rd edition. Springer-Verlag, Berlin, 1983.
8. H Haken. *Principles of Brain Functioning: A Synergetic Approach to Brain Activity, Behavior and Cognition*. Springer-Verlag, Berlin, 1996.
9. S A Levin. Complex adaptive systems: Exploring the known, the unknown and the unknowable. *Bull. Am. Math. Soc.* 40 (2002) 3–19.
10. L Margulis and D Sagan. *What is Life?* Simon & Schuster, New York, 1995.
11. W Moore. *Schrödinger: Life and Thought*. Cambridge University Press, Cambridge, 1989.
12. J S Nicolis. *Dynamics of Hierarchical Systems: An Evolutionary Approach*. Springer-Verlag, Berlin, 1986.
13. E Schrödinger. *What is Life?* Cambridge University Press, Cambridge, 1944 (republished 1967).
14. E Schrödinger. *Mind and Matter*. Cambridge University Press, Cambridge, 1958 (republished 1967).
15. A C Scott. *Stairway to the Mind*. Springer-Verlag (Copernicus), New York, 1995.
16. A C Scott. *Neuroscience: A Mathematical Primer*. Springer-Verlag, New York, 2002.
17. B H Voorhees. Axiomatic theory of hierarchical systems. *Behavioral Sci.* 28 (1983) 24–34.

APPENDIX A

CONSERVATION LAWS AND CONSERVATIVE SYSTEMS

Assume a system that is uniform in the x-direction ($-\infty < x < +\infty$) and conserves some quantity, Q. In other words, Q is neither created nor destroyed in the course of the dynamics under investigation. Then it is convenient to introduce the following definitions.

- $F(x,t)$ is the flow of the conserved quantity, or the rate that it passes the point x at time t.
- $D(x,t)$ is the density of the conserved quantity, or the amount of Q per unit of x.

In the context of these definitions,

$$\Delta x \frac{d}{dt} D(x + \Delta x/2, t) = F(x,t) - F(x + \Delta x, t) + O(\Delta x^2).$$

Taking the limit of this expression as $\Delta x \to 0$ yields the conservation law:

$$\frac{\partial D}{\partial t} + \frac{\partial F}{\partial x} = 0. \tag{A.1}$$

The total quantity

$$Q = \int_{-\infty}^{\infty} D \, dx$$

is evidently conserved because

$$\frac{dQ}{dt} = \int_{-\infty}^{\infty} \frac{\partial D}{\partial t} dx = -\int_{-\infty}^{\infty} \frac{\partial F}{\partial x} dx$$
$$= F(-\infty, t) - F(+\infty, t).$$

Thus
$$Q = \text{constant}$$
if the flow into the system at $x = -\infty$ equals the flow out at $x = +\infty$. This condition is satisfied if the dynamic variables approach zero as $x \to \pm\infty$.

There are many physical examples of conserved quantities, including the number of automobiles in a study of highway traffic flow, water in a river, minority carriers in a semiconductor, electromagnetic energy in a pulse of radio wave transmission, and mechanical energy in an elastic wave.

In nonlinear science, it is useful to have a means of distinguishing between systems or subsystems that include the effects of energy conservation and those that do not. The former are often referred to in the engineering literature as "conservative" or "lossless" and are considered to be constructed from elements such as inductors and capacitors or their mechanical analogs, masses and springs, excluding resistors or "dashpots" (mechanical resistors). Systems that do not conserve energy are called "dissipative" or "open" and require the presence of batteries or amplifying devices (e.g., transistors) to maintain dynamic activity.

For some arbitrary collection of terms in a PDE, however, it is not always clear whether energy is conserved. In such a situation, one can proceed by checking whether the system can be derived from a Lagrangian density $\mathcal{L}$, which is a function of the dependent variable $u(x,t)$ and certain of its derivatives [1–3].

Basic to this perspective is the assumption that the action integral

$$I = \int_{x_1}^{x_2} \int_{t_1}^{t_2} \mathcal{L}(u, u_x, u_t, \cdots) \, dx \, dt$$

takes a maximum or minimum value along the true solution $u(x,t)$. In other words, the variation of I (written δI) is equal to zero when $\mathcal{L}$ is evaluated on $u(x,t)$.

Now let

$$\mathcal{L} = \mathcal{L}(u, u_x, u_t)$$

and choose

$$\delta u(x,t)$$

to be a small change of $u(x,t)$ that is zero at x_1, x_2, t_1, and t_2. Under these assumptions

$$\delta I = \int_{x_1}^{x_2} \int_{t_1}^{t_2} \left[\frac{\partial \mathcal{L}}{\partial u} \delta u + \frac{\partial \mathcal{L}}{\partial u_x} \delta u_x + \frac{\partial \mathcal{L}}{\partial u_t} \delta u_t \right] dx \, dt$$

$$= \int_{x_1}^{x_2} \int_{t_1}^{t_2} \left[\frac{\partial \mathcal{L}}{\partial u} - \frac{\partial}{\partial x} \frac{\partial \mathcal{L}}{\partial u_x} - \frac{\partial}{\partial t} \frac{\partial \mathcal{L}}{\partial u_t} \right] \delta u(x,t) \, dx \, dt,$$

after integrating by parts using the boundary conditions assumed for $\delta u(x,t)$.

Evidently the condition

$$\delta I = 0$$

requires that $u(x,t)$ must satisfy the Lagrange–Euler equation[1]

$$\frac{\partial \mathcal{L}}{\partial u} - \frac{\partial}{\partial x}\frac{\partial \mathcal{L}}{\partial u_x} - \frac{\partial}{\partial t}\frac{\partial \mathcal{L}}{\partial u_t} = 0. \tag{A.2}$$

As a simple example, note that a Lagrangian density for the linear wave equation

$$\frac{\partial^2 u}{\partial x^2} - \frac{\partial^2 u}{\partial t^2} = 0$$

is

$$\mathcal{L} = \tfrac{1}{2}(u_x^2 - u_t^2).$$

Systems of partial differential equations associated with a Lagrangian density in the manner just described are said to be conservative for the following reasons.

1. One can define a momentum density as

$$\pi \equiv \frac{\partial \mathcal{L}}{\partial u_t}.$$

2. Then, an energy density (often called a Hamiltonian density) can be defined through the transformation

$$\mathcal{H}(u, u_x, \pi) \equiv \mathcal{L}(u, u_x, u_t) - \pi u_t.$$

3. Direct calculation shows that

$$E = \int_{-\infty}^{\infty} \mathcal{H}\, dx$$

is a conserved quantity obeying the conservation law

$$\frac{\partial \mathcal{H}}{\partial t} + \frac{\partial \mathcal{P}}{\partial x} = 0,$$

where

$$\mathcal{P} \equiv -\frac{\partial \mathcal{L}}{\partial u_x} u_t$$

is a flow of energy, or a power.

[1] If $\mathcal{L}$ depends upon higher derivatives of u, the Lagrange–Euler equation includes additional terms that are obtained through integration by parts in a similar manner. Thus for

$$\mathcal{L} = \mathcal{L}(u, u_x, u_t, u_{xx}),$$

the corresponding Lagrange–Euler equation is

$$\frac{\partial \mathcal{L}}{\partial u} - \frac{\partial}{\partial x}\frac{\partial \mathcal{L}}{\partial u_x} - \frac{\partial}{\partial t}\frac{\partial \mathcal{L}}{\partial u_t} + \frac{\partial^2}{\partial x^2}\frac{\partial \mathcal{L}}{\partial u_{xx}} = 0,$$

and so on.

For the wave equation,

$$\pi = -u_t,$$

$$\mathcal{H} = \tfrac{1}{2}(u_x^2 + \pi^2),$$

and

$$\mathcal{P} = -u_x u_t.$$

Not all PDE systems can be formulated in this manner. For example, the linear diffusion equation

$$\frac{\partial^2 u}{\partial x^2} - \frac{\partial u}{\partial t} = 0,$$

which conserves the integral of u, does not have a Lagrangian density. Thus the linear diffusion equation is not conservative of energy in the present sense.

Interestingly, the nonlinear diffusion equation

$$\frac{\partial^2 u}{\partial x^2} - \frac{\partial u}{\partial t} = f(u),$$

which plays a central role in flame propagation and nerve impulse dynamics, has no conserved quantities at all. The same is true for the FitzHugh–Nagumo, Morris–Lecar, and Hodgkin–Huxley equations of nerve impulse conduction.

As a final example, consider the damped wave equation

$$\frac{\partial^2 u}{\partial x^2} - \frac{\partial^2 u}{\partial t^2} = \alpha \frac{\partial u}{\partial t}.$$

As $\alpha \to 0$, this reduces to the linear wave equation. For $\alpha \gg 1$, on the other hand, it approaches the linear diffusion equation.

Note that the left-hand side of this equation can be derived from a Lagrangian density corresponding to the energy density $\mathcal{H} = (u_x^2 + u_t^2)/2$ and total energy

$$E = \int_{-\infty}^{\infty} \frac{1}{2}(u_x^2 + u_t^2)\,dx.$$

Differentiating this energy integral with respect to time, integrating the first term by parts (assuming wave amplitudes go to zero as $x \to \pm\infty$), and substituting the damped wave equation yields

$$\frac{dE}{dt} = \int_{-\infty}^{\infty} (u_x u_{xt} + u_t u_{tt})\,dx$$

$$= \int_{-\infty}^{\infty} u_t(-u_{xx} + u_{tt})\,dx$$

$$= -\alpha \int_{-\infty}^{\infty} u_t^2\,dx.$$

For $\alpha > 0$ it is seen that E decreases with time, justifying the designation of α as a damping parameter.

By putting terms that can be generated from a Lagrangian density (conservative terms) on one side of the equation and all other terms (dissipative terms) on the other side, any PDE—linear or nonlinear—can be analyzed in this manner.

REFERENCES

1. G D Birkhoff. *Dynamical Systems* (revised edition). American Mathematical Society, Providence, Rhode Island, 1966. (First published in 1927.)
2. H Goldstein. *Classical Mechanics*. Addison–Wesley, Reading, Massachusetts, 1951.
3. H Jeffreys and B S Jeffreys. *Methods of Mathematical Physics*. Cambridge University Press, Cambridge, 1962.

APPENDIX B

MULTISOLITON FORMULAS

In this appendix are recorded N-soliton formulas for a few of the most important nonlinear systems.

B.1 The KdV equation

Gardner et al. (and also Hirota) have shown that the equation

$$\frac{\partial u}{\partial t} - 6u\frac{\partial u}{\partial x} + \frac{\partial^3 u}{\partial x^3} = 0$$

has the N-soliton [4, 5]

$$u_N(x,t) = -2\frac{\partial^2}{\partial x^2}\log f(x,t), \tag{B.1}$$

where

$$f(x,t) = \det(M), \tag{B.2}$$

and $M = [m_{ij}]$ is an $N \times N$ matrix with elements

$$m_{ij} = \frac{2\sqrt{\kappa_i \kappa_j}}{\kappa_i + \kappa_j} \exp\left[\frac{1}{2}(\kappa_i + \kappa_j)x - \frac{1}{2}(\kappa_i^2 + \kappa_j^3)t + \alpha_{ij}\right] + \delta_{ij}. \tag{B.3}$$

In these expressions, κ_i, κ_j, α_{ij}, and δ_{ij} are real constants, and

$$\kappa_i \neq \kappa_j \quad \text{for } i \neq j.$$

With $t = 0$, these equations were first obtained by Kay and Moses in 1956 as a class of reflectionless potentials for Schrödinger's equation [8].

B.2 The SG equation

Several authors have obtained N-soliton solutions for the equation

$$\frac{\partial^2 u}{\partial x^2} - \frac{\partial^2 u}{\partial t^2} = \sin u,$$

with a variety of analytic forms. Using methods related to those of Hirota, it has been shown by Caudry et al. that [3]

$$u_N(x,t) = \arccos\left[1 + 2\left(\frac{\partial^2}{\partial t^2} - \frac{\partial^2}{\partial x^2}\right)\log f(x,t)\right], \tag{B.4}$$

is an exact solution, where
$$f(x,t) = \det(M).$$

Elements of the $N \times N$ matrix $M = [m_{ij}]$ are given by
$$m_{ij} = \frac{2}{a_i + a_j} \cosh\left(\frac{\theta_i + \theta_j}{2}\right),$$

with
$$\theta_i = \frac{1}{2}\left[\left(a_i - \frac{1}{a_i}\right)t + \left(a_i + \frac{1}{a_i}\right)x\right] + \delta_i,$$

where the a_i and δ_i are constants. Because $u_N(x,t)$ is assumed to be real, these constants are either real or they must occur in complex conjugate pairs. These same authors show that this result is equivalent to a more complex expression obtained previously by Hirota, and they present closely related formulas for the reduced Maxwell–Bloch equations of nonlinear optics [2, 6].

At about the same time, Ablowitz et al. used an inverse scattering transform to show that an N-soliton solution can also be written as follows [1]. First introduce the transformation
$$\xi = (x+t)/2$$
$$\tau = (x-t)/2$$
$$\tilde{u}(\xi, \tau) = u(x,t)$$

under which the sine–Gordon (SG) equation takes the form
$$\frac{\partial^2 \tilde{u}}{\partial \xi \partial \tau} = \sin \tilde{u}.$$

Then
$$\left(\frac{\partial \tilde{u}_N(x,t)}{\partial \xi}\right)^2 = 4\frac{\partial^2}{\partial \xi^2}[\log \det(BB^* + I)], \tag{B.5}$$

where $B = [b_{ij}]$ is an $N \times N$ matrix with elements
$$b_{ij} = \frac{\sqrt{c_i c_j^*}}{\lambda_i - \lambda_j^*} e^{i(\lambda_i - \lambda_j^*)\xi}$$
$$c_i(t) = e^{-(i/2\lambda_i)\tau + \gamma_i},$$

with λ_i and γ_i constants. For $u_N(x,t)$ to be real, these constants must be purely imaginary or occur in pairs for which
$$\lambda_1 = -\lambda_2^* \quad \text{and} \quad \gamma_1 = -\gamma_2^*.$$

In these equations, the λ_i locate upper half plane poles of the reflection and transmission coefficients. When purely imaginary, they correspond to the kink speeds in the laboratory (x,t) system given by Equation (6.76) as

$$v_i = \frac{1 + 4\lambda_i^2}{1 - 4\lambda_i^2}.$$

B.3 The NLS equation

Using an inverse scattering transform method, Zakharov and Shabat showed that the equation

$$\mathrm{i}\frac{\partial u}{\partial t} + \frac{\partial^2 u}{\partial x^2} + 2|u|^2 u = 0$$

has an N-soliton solution for which [9]

$$|u_N(x,t)|^2 = 2\frac{\partial^2}{\partial x^2}[\log \det(BB^* + I)], \qquad (B.6)$$

where $B = [b_{ij}]$ is an $N \times N$ matrix with elements

$$b_{ij} = \frac{\sqrt{c_i c_j^*}}{\lambda_i - \lambda_j^*} e^{i(\lambda_i - \lambda_j^*)x}$$

$$c_i(t) = e^{4i\lambda_i^2 t + \gamma_i},$$

with λ_i and γ_i being complex constants. As for the SG equation, the λ_i locate upper half plane poles of the transmission and reflection coefficients.

B.4 The Toda lattice

Toda's infinite chain of unit masses connected by the potential

$$\phi(r) = \frac{a}{b}e^{-br} + ar + \text{constant}$$

is governed by the difference–differential equations (Newton's second law)

$$\frac{d^2 r_n}{dt^2} = \left(\frac{d\phi(r_{n+1})}{dr_{n+1}} - \frac{d\phi(r_n)}{dr_n}\right) - \left(\frac{d\phi(r_n)}{dr_n} - \frac{d\phi(r_{n-1})}{dr_{n-1}}\right).$$

Under the transformation

$$r_n = -\frac{1}{b}\log(1 + V_n), \qquad (B.7)$$

V_n obeys the equation

$$\frac{d^2}{dt^2}\log(1 + V_n) = ab(V_{n+1} - 2V_n + V_{n-1}),$$

which describes the propagation of voltage through a nonlinear electric filter. Redefining the timescale as

$$\tau = \sqrt{ab}\, t,$$

Hirota has shown that [7]

$$V_n(t) = \frac{d^2}{d\tau^2} f_n(\tau), \tag{B.8}$$

where

$$f_n(\tau) = \sum_{\mu=0,1} \exp\left(\sum_{i<j}^{(N)} B_{ij}\mu_i\mu_j + \sum_{i=1}^{N} \mu_i X_i\right),$$

$$X_i = \omega_i \tau - \kappa_i n + \gamma_i$$

$$\omega_i = \pm 2\sinh\kappa_i/2$$

and

$$e^{B_{ij}} = -\frac{(\omega_i - \omega_j)^2 - 4\sinh^2(\kappa_i - \kappa_j)/2}{(\omega_i + \omega_j)^2 - 4\sinh^2(\kappa_i + \kappa_j)/2}.$$

In these formulas, the κ_i and γ_i are real constants, and the symbol

$$\sum_{\mu=0,1}$$

implies summation over all possible combinations of $\mu_1 = 0, 1$; $\mu_2 = 0, 1; \ldots$; $\mu_N = 0, 1$. Finally the symbol

$$\sum_{i<j}^{(N)}$$

indicates summation over all possible pairs chosen from N elements.

REFERENCES

1. M J Ablowitz, D J Kaup, A C Newell, and H Segur. Method for solving the sine–Gordon equation. *Phys. Rev. Lett.* 30 (1973) 1262–1264.
2. P J Caudry, J C Eilbeck, and J D Gibbon. Exact multisoliton solution of the reduced Maxwell–Bloch equations of non-linear optics. *J. Inst. Math. Applic.* 14 (1974) 375–386.
3. P J Caudry, J D Gibbon, J C Eilbeck, and R K Bullough. Exact multisoliton solutions of the self-induced transparency and sine–Gordon equations. *Phys. Rev. Lett.* 30 (1973) 237–238.
4. C S Gardner, J M Greene, M D Kruskal, and R M Miura. Method for solving the Korteweg–de Vries equation. *Phys. Rev. Lett.* 19 (1967) 1095–1097.
5. R Hirota. Exact solutions of the Korteweg–de Vries equation for multiple collisions of solitons. *Phys. Rev. Lett.* 27 (1971) 1192–1194.

6. R Hirota. Exact solution of the sine–Gordon equation for multiple collisions of solitons. *J. Phys. Soc. Jpn.* 33 (1972) 1459–1463.
7. R Hirota. Exact N-soliton solutions of a nonlinear lumped network equation. *J. Phys. Soc. Jpn.* 35 (1973) 286–288.
8. I Kay and H E Moses. Reflectionless transmission through dielectrics and scattering potentials. *J. Appl. Phys.* 27 (1956) 1503–1508.
9. V E Zakharov and A B Shabat. Exact theory of two-dimensional self-modulation of waves in nonlinear media. *Sov. Phys. JETP* 34 (1972) 62–69.

APPENDIX C

ELLIPTIC FUNCTIONS

Early on in undergraduate studies, every student of physical science learns that the linear ODE $\dot{x} + x = 0$ is solved by the exponential function $x = e^{-t}$, which is defined by a converging infinite series. Similarly, the linear ODE $\ddot{x} + x = 0$ is solved by a combination of trigonometric functions (sines and cosines), which are also defined by converging infinite series. During our high school years, we were forced to memorize identities involving these functions, which we later found useful in working out the details of particular calculations.

Less well known is the fact that many nonlinear ODEs can be similarly solved in terms of "elliptic functions" which are easily evaluated as quotients of converging series. To see how this goes, consider Equation (3.10), where it is desired to find solutions of the nonlinear ODE

$$\tilde{u}' = \pm \sqrt{P(\tilde{u})}. \tag{C.1}$$

In this equation,
$$P(\tilde{u}) = 2(\tilde{u} - a_1)(\tilde{u} - a_2)(\tilde{u} - a_3)$$

is the cubic polynomial shown in Figure 3.1. As the three roots are real with $a_3 > a_2 > a_1$, $P(\tilde{u})$ will be positive in the interval $a_1 < \tilde{u} < a_2$, allowing Equation (C.1) to have a real and bounded solution. Integration of Equation (C.1) over this range yields

$$\xi = \xi_1 \pm \int_{a_1}^{\tilde{u}} \frac{dy}{\sqrt{2(y - a_1)(y - a_2)(y - a_3)}},$$

where $\xi = \xi_1$ for $\tilde{u} = a_1$. At this point, it is convenient to introduce the substitution

$$y = a_1 + (a_2 - a_1) \sin^2 \theta$$

to obtain

$$\xi = \xi_1 \pm \sqrt{\frac{2}{a_3 - a_1}} \int_0^\phi \frac{d\theta}{\sqrt{1 - k^2 \sin^2 \theta}} \tag{C.2}$$

with

$$k^2 = (a_2 - a_1)/(a_3 - a_1) \tag{C.3}$$

and ϕ related to $\tilde{u}$ by

$$\tilde{u} = a_1 + (a_2 - a_1) \sin^2 \phi = a_2 - (a_2 - a_1) \cos^2 \phi. \tag{C.4}$$

For $k^2 < 1$, the integrand of the integral

$$\zeta = \int_0^\phi \frac{d\theta}{\sqrt{1 - k^2 \sin^2 \theta}} \tag{C.5}$$

is always positive. This implies that ζ is a monotone increasing function of ϕ; thus inverse functions can be defined as

$$\begin{aligned} \mathrm{sn}(\zeta, k) &= \sin \phi, \\ \mathrm{cn}(\zeta, k) &= \cos \phi. \end{aligned} \tag{C.6}$$

These are the Jacobi elliptic (sn and cn) functions with modulus k. (Lawden suggests calling them "ess en" and "cee en" or "san" and "can" [3].) In the context of these functions, Equations (C.2), (C.4), and (C.6) yield

$$\tilde{u}(\xi) = a_2 - (a_2 - a_1)\mathrm{cn}^2[(\xi - \xi_1)\sqrt{(a_3 - a_1)/2}, k], \tag{C.7}$$

as a solution of Equation (C.1). The qualitative character of this solution is discussed in Section 3.1.3.

From Equation (C.3), note that the square of the modulus is restricted to the range $0 \leq k^2 < 1$. Letting $k = 0$ in Equation (C.5) implies that $\phi = \zeta$, so

$$\begin{aligned} \mathrm{sn}(\zeta, 0) &= \sin \zeta, \\ \mathrm{cn}(\zeta, 0) &= \cos \zeta. \end{aligned}$$

For $0 \leq k < 1$, $\mathrm{sn}(\zeta, k)$ and $\mathrm{cn}(\zeta, k)$ are periodic functions with period $4K(k)$, where

$$K(k) = \int_0^{\pi/2} \frac{d\theta}{\sqrt{1 - k^2 \sin^2 \theta}} \tag{C.8}$$

is called the "complete elliptic integral of the first kind," which is plotted vs. k^2 in Figure C.1. Evidently the period of sn and cn reduces to 2π at $k = 0$. From Figure C.2, it is seen that sn (cn) looks like a distorted sine (cosine) function with a quarter period, $K(k)$, that increases without bound as $k \to 1$.

From Equations (C.6), the identity

$$\mathrm{sn}^2(\zeta, k) + \mathrm{cn}^2(\zeta, k) = 1$$

follows from a corresponding identity for trigonometric functions. In a similar way, a third basic Jacobi elliptic function can be defined as

$$\mathrm{dn}^2(\zeta, k) + k^2 \mathrm{sn}^2(\zeta, k) = 1, \tag{C.9}$$

which is plotted in Figure C.3 for selected values of k. (Note that dn has a period of $2K$ rather than $4K$ as for sn and cn.) In terms of dn, the complete elliptic integral of the second kind can be defined as

$$E(k) = \int_0^K \mathrm{dn}^2(\zeta, k)\, d\zeta,$$

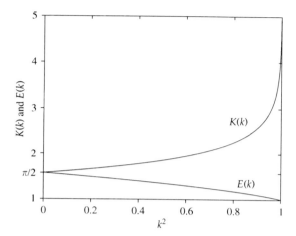

FIG. C.1. *Plots of the complete elliptic integral of the first kind $K(k)$ and the complete elliptic integral of the second kind $E(k)$ vs. k^2.*

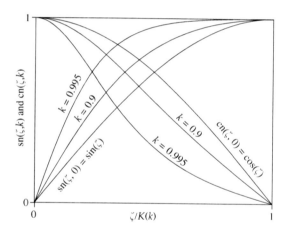

FIG. C.2. *Plots of $\mathrm{sn}(\zeta, k)$ and $\mathrm{cn}(\zeta, k)$ as functions of $\zeta/K(k)$.*

which is also plotted vs. k^2 in Figure C.1.

Three additional elliptic functions are defined as ratios of sn, cn, and dn. Thus

$$\mathrm{sc}(\zeta, k) = \frac{\mathrm{sn}(\zeta, k)}{\mathrm{cn}(\zeta, k)}, \quad \mathrm{cd}(\zeta, k) = \frac{\mathrm{cn}(\zeta, k)}{\mathrm{dn}(\zeta, k)}, \quad \text{and} \quad \mathrm{ds}(\zeta, k) = \frac{\mathrm{dn}(\zeta, k)}{\mathrm{sn}(\zeta, k)}.$$

Limiting forms and values for these elliptic functions in the trigonometric limit ($k = 0$) and the hyperbolic limit ($k = 1$) are given in Table C.1. Finally, six

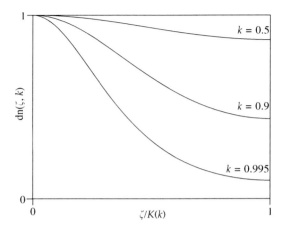

FIG. C.3. *Plots of* $\mathrm{dn}(\zeta, k)$ *as functions of* $\zeta/K(k)$. *(Note that* $\mathrm{dn}(\zeta, 0) = 1$.*)*

Table C.1. *Some limiting forms and values for elliptic functions.*

Function	$k = 0$	$k = 1$
$K(k)$	$\pi/2$	∞
$E(k)$	$\pi/2$	1
$\mathrm{sn}(\zeta, k)$	$\sin \zeta$	$\tanh \zeta$
$\mathrm{cn}(\zeta, k)$	$\cos \zeta$	$\mathrm{sech}\, \zeta$
$\mathrm{dn}(\zeta, k)$	1	$\mathrm{sech}\, \zeta$
$\mathrm{sc}(\zeta, k)$	$\tan \zeta$	$\sinh \zeta$
$\mathrm{cd}(\zeta, k)$	$\cos \zeta$	1
$\mathrm{ds}(\zeta, k)$	$\mathrm{cosec}\, \zeta$	$\mathrm{cosech}\, \zeta$

additional functions are defined as: $\mathrm{ns} = 1/\mathrm{sn}, \mathrm{nc} = 1/\mathrm{cn}, \mathrm{nd} = 1/\mathrm{dn}, \mathrm{cs} = 1/\mathrm{sc}, \mathrm{dc} = 1/\mathrm{cd}$, and $\mathrm{sd} = 1/\mathrm{ds}$.

As this appendix provides only the briefest introduction to elliptic functions, students of nonlinear science are encouraged to spend time with handbooks and texts devoted to the subject [1–4]. In these sources will be found many identities, power series, addition theorems, derivative expressions, properties of functions with imaginary arguments, integral formulas, techniques for reducing nonlinear ordinary differential equations to standard forms, and numerical methods for evaluating elliptic functions, amid a variety of applications.

REFERENCES

1. P F Byrd and M D Friedman. *Handbook of Elliptic Integrals for Engineers and Scientists.* Springer-Verlag, New York, 1971.
2. A Cayley. *Elementary Treatise on Elliptic Functions.* Dover, New York, 1961.
3. D F Lawden. *Elliptic Functions and Applications.* Springer-Verlag, New York, 1989.
4. E H Neville. *Elliptic Functions: A Primer.* Pergamon Press, Oxford, 1971.

APPENDIX D

STABILITY OF NERVE IMPULSES

Under traveling-wave analysis, one assumes that solutions are functions only of the variable $\xi = x - vt$, where v is a wave speed that enters into the resulting ODE system as an adjustable parameter. To study the stability of a traveling wave, however, it is necessary to allow an explicit dependence upon time. To include this dependence, note that traveling-wave analysis is a special case of the independent variable transformation

$$u(x, t) \longrightarrow \tilde{u}(\xi, \tau),$$

where ξ and τ are related to the original independent variables (x and t) by

$$\xi = x - vt \quad \text{and} \quad \tau = t.$$

In the new (ξ, τ) system, ξ is measured on a distance scale (or meter stick) moving with velocity v in the x-direction, and time τ is measured on the same time scale (the same clock) as in the laboratory frame of reference. Thus the transformation is from the stationary (or laboratory) system (x, t) to a moving system (ξ, τ).

It is convenient to assign a different symbol for time in the moving system because partial derivatives transform as

$$\frac{\partial u(x,t)}{\partial x} \longrightarrow \frac{\partial \tilde{u}(\xi, \tau)}{\partial \xi} \frac{\partial \xi}{\partial x} + \frac{\partial \tilde{u}(\xi, \tau)}{\partial \tau} \frac{\partial \tau}{\partial x}$$

and

$$\frac{\partial u(x,t)}{\partial t} \longrightarrow \frac{\partial \tilde{u}(\xi, \tau)}{\partial \xi} \frac{\partial \xi}{\partial t} + \frac{\partial \tilde{u}(\xi, \tau)}{\partial \tau} \frac{\partial \tau}{\partial t}.$$

Because $\partial \xi / \partial x = 1$, $\partial \tau / \partial x = 0$, $\partial \xi / \partial t = -v$, and $\partial \tau / \partial t = 1$, the partial derivatives transform as

$$\partial/\partial x \longrightarrow \partial/\partial \xi \quad \text{and} \quad \partial/\partial t \longrightarrow \partial/\partial \tau - v \partial/\partial \xi.$$

As a specific example, consider the Z–F equation

$$\frac{\partial^2 u}{\partial x^2} - \frac{\partial u}{\partial t} = f(u),$$

which transforms to
$$\frac{\partial^2 u}{\partial \xi^2} - \frac{\partial u}{\partial \tau} + v\frac{\partial u}{\partial \xi} = f(u) \tag{D.1}$$
where the tildes have been dropped for notational convenience. Assuming that u is independent of τ in the moving system, this PDE reduces to the ODE
$$\frac{d^2 u_0}{d\xi^2} + v\frac{du_0}{d\xi} = f(u_0) \tag{D.2}$$
of traveling-wave analysis. In general, however, u depends upon both ξ and τ; thus we must study Equation (D.1) to learn about the stability of a traveling wave.

To investigate the stability of $u_0(\xi)$, write
$$u(\xi, \tau) = u_0(\xi) + \phi(\xi, \tau),$$
where $\phi(\xi, \tau)$ is an alteration of the traveling wave. Then from Equations (D.1) and (D.2), $\phi(\xi, \tau)$ satisfies the nonlinear PDE
$$\frac{\partial^2 \phi}{\partial \xi^2} + v\frac{\partial \phi}{\partial \xi} - \frac{\partial \phi}{\partial \tau} = f(u_0 + \phi) - f(u_0).$$
To this point, no approximations have been made—we have merely transformed the independent and dependent variables which is a matter of bookkeeping.

A linear-stability analysis for this system was first carried out by Zeldovich and Barenblatt in 1959 [15].[1] Following these authors, we assume
$$|\phi(\xi, 0)| \ll |u_0(\xi)|$$
and approximate
$$f(u_0 + \phi) - f(u_0) \doteq G[u_0(\xi)]\,\phi(\xi, \tau),$$
where
$$G[u_0(\xi)] \equiv G(\xi) \equiv \left.\frac{df(u)}{du}\right|_{u=u_0(\xi)},$$
and ϕ satisfies the linear equation
$$\frac{\partial^2 \phi}{\partial \xi^2} + v\frac{\partial \phi}{\partial \xi} - \frac{\partial \phi}{\partial \tau} \doteq G(\xi)\,\phi. \tag{D.3}$$

This linear PDE has been obtained by linearizing the transformed nonlinear PDE—Equation (D.1)—about the traveling-wave solution $u_0(\xi)$.

At the price of assuming that $\phi(\xi, 0)$ is a sufficiently small initial alteration of the traveling wave, in other words, we have obtained a linear PDE for the

---
[1] Nonlinear stability analyses of this equation have been published by Lindgren and Buratti [8] and by Maginu [9, 12].

evolution of $\phi(\xi,\tau)$ in time. From the perspective of linear stability analysis, we can say that our system is:

(1) asymptotically stable if all solutions of Equation (D.3) approach zero as $\tau \to \infty$,
(2) unstable if any solution of Equation (D.3) grows as $\tau \to \infty$, and
(3) stable otherwise.

Equation (D.3) is conveniently analyzed by separating variables. Thus $\phi(\xi,\tau)$ is expressed as a generalized sum of elementary products of the form $\Phi(\xi)T(\tau)$. Here
$$T(\tau) = e^{\lambda\tau},$$
λ is complex, and $\Phi(\xi)$ is a solution of the stability equation
$$\frac{d^2\Phi}{d\xi^2} + v\frac{d\Phi}{d\xi} - [\lambda + G(\xi)]\Phi = 0. \qquad (D.4)$$

Each bounded solution of this equation is an eigenfunction and the corresponding value of λ is an eigenvalue. All of the values of λ for which Equation (D.4) has bounded solutions are referred to as its spectrum. Depending upon the boundary conditions imposed as $\xi \to \pm\infty$, there are two types of eigenvalues.

Continuous eigenvalues. If it is required that solutions of Equation (D.4) be bounded by some finite value as $\xi \to \pm\infty$, then $\Phi(\xi)$ has the asymptotic form
$$\Phi(\xi) \sim e^{\pm ik\xi}.$$

Such eigenfunctions represent radiation from the underlying traveling wave $u_0(\xi)$, and if the radiation grows with time, the underlying impulse is unstable.

As $\xi \to +\infty$, $G(\xi)$ approaches a positive constant, $G_1 > 0$. Similarly, as $\xi \to -\infty$, $G(\xi) \to G_2 > 0$. Substitution of $\Phi(\xi) = \exp(\pm ik\xi)$ into Equation (D.4), therefore, leads to the eigenvalues
$$\lambda = -(G_1 + k^2) \pm ikv$$
as $\xi \to +\infty$. This set of eigenvalues is said to be continuous because it contains elements for all real values of k.

To see that these continuous eigenvalues correspond to eigenfunctions representing radiation from the traveling wave, note that in the $+\xi$-direction, $\phi(\xi,\tau)$ has the form
$$\phi_{\rm rad}(\xi,\tau) \sim e^{-(G_1+k^2)\tau}\cos(k\xi + kv\tau),$$
with a corresponding expression for radiation in the $-\xi$-direction.

For all real values of k, this radiative eigenfunction is damped with time, falling exponentially to zero as $\tau \to +\infty$. Thus the continuous spectrum does not contribute to instability of the underlying traveling wave.

Discrete (or "point") eigenvalues. With the boundary conditions

$$\Phi(\xi) \to 0 \quad \text{as } \xi \to \pm\infty,$$

the corresponding eigenvalues occur at discrete (isolated) locations on the real axis of the complex λ-plane. These "point" eigenvalues are the particular values of λ for which the asymptotic behavior of Equation (D.4) as $\xi \to -\infty$ goes smoothly over into a solution of this same equation as $\xi \to +\infty$.

If Equation (D.4) has any eigenfunction with a positive eigenvalue ($\lambda > 0$), then the corresponding traveling-wave solution is unstable. If all eigenvalues are negative, then the traveling wave is asymptotically stable.

It is immediately evident that Equation (D.4) always has an eigenfunction for the eigenvalue $\lambda = 0$. To see this, differentiate Equation (D.2) with respect to ξ; thus

$$\frac{d^2}{d\xi^2}\left(\frac{du_0}{d\xi}\right) + v\frac{d}{d\xi}\left(\frac{du_0}{d\xi}\right) = G(\xi)\left(\frac{du_0}{d\xi}\right),$$

which is identical to Equation (D.4) with

$$\lambda = 0 \quad \text{and} \quad \Phi(\xi) = du_0/d\xi.$$

Such a striking property has physical significance, which can be understood as follows. Suppose we start with a traveling-wave solution $u_0(\xi)$ and add a small amount of its derivative $du_0/d\xi$. The result is

$$u_0(\xi) + \epsilon du_0/d\xi = u_0(\xi + \epsilon) + \mathrm{O}(\epsilon^2).$$

Thus adding the derivative of a traveling wave with respect to its traveling-wave variable merely translates the original solution in the ξ-direction. Because a translated traveling wave is still an exact solution of the system, such a disturbance has no dependence on time.

Although we have obtained this result for a special case, it is generally true for all linear-stability analyses of traveling waves that the derivative of a traveling-wave with respect to its traveling-wave variable is an eigenfunction with zero eigenvalue of the corresponding stability equation.

Armed with this knowledge, the task is to determine whether Equation (D.4) has a negative eigenvalue. To this end, it is convenient to introduce the dependent-variable transformation

$$\Phi(\xi) = \psi(\xi)\,e^{-v\xi/2},$$

changing Equation (D.4) to

$$\frac{d^2\psi}{d\xi^2} - \left[\lambda + \frac{v^2}{4} + G(\xi)\right]\psi = 0. \tag{D.5}$$

Equation (D.5) is a second-order self-adjoint operator equation $d\psi/d\xi - [\lambda + U(\xi)]\psi = 0$ with[2]

$$U(\xi) = \frac{v^2}{4} + G(\xi).$$

Such systems share the following properties [13].

- Because $U(\xi)$ has the shape of a potential well, the eigenfunction (ψ_0) with the most positive eigenvalue (λ_0) has no finite zero crossing.
- The eigenfunction (ψ_1) with the next largest eigenvalue (λ_1) has one zero crossing.
- The eigenfunction (ψ_2) with the next largest eigenvalue (λ_2) has two zero crossings.
- And so on.

Thus the discrete eigenvalues of Equation (D.5) can be ordered as

$$\lambda_0 > \lambda_1 > \lambda_2 > \cdots > \lambda_n > \text{etc.},$$

where eigenfunction ψ_n has n finite zero crossings.[3]

From these results, the following conclusions can be drawn. First consider the stability of a monotone decreasing leading-edge solution $u_0(\xi)$ as shown in Figure 4.2(c). The function $\psi(\xi) = e^{v\xi/2} du_0/d\xi$ is the eigenfunction of Equation (D.5) corresponding to the eigenvalue $\lambda = 0$. Because this eigenfunction has no zero crossings, $\lambda_0 = 0$ is the most positive eigenvalue. In other words, all eigenvalues are less than or equal to zero, and the traveling-wave solution $u_0(\xi)$ is stable. It is not asymptotically stable because the perturbation can include components with the most positive eigenvalue, which do not decay with time. Because this argument uses only the qualitative shape of $u_0(\xi)$, the same conclusion holds for any monotone increasing or decreasing (level change) traveling-wave solution of the nonlinear diffusion equation.

As we have seen in Section 4.1.1, the Z–F equation also has a solution that is impulse-shaped, with a maximum amplitude (u_m) at some finite value of ξ, corresponding to homoclinic trajectories in the (u, w) phase plane of Figure 4.3(b). Because the ξ-derivative of such a function has a zero crossing at a finite value

---

[2]To check that the transformation from $\Phi(\xi)$ to $\psi(\xi)$ does not violate the null boundary condition on $\Phi(\xi)$ as $\xi \to -\infty$, assume that $G(\xi) \to G_2 > 0$ as $\xi \to -\infty$. This implies

$$\psi(\xi) \sim \exp\left(\xi\sqrt{G_2 + v^2/4 + \lambda}\right)$$

and therefore

$$\Phi(\xi) \sim \exp\left[\xi\left(-v/2 + \sqrt{G_2 + v^2/4 + \lambda}\right)\right] \quad \text{as } \xi \to -\infty.$$

Thus for $\lambda \geq 0, \psi(\xi) \to 0$ as $\xi \to -\infty$, implying $\Phi(\xi) \to 0$ as $\xi \to -\infty$.

[3]An intuitive way to establish these results is to study Equation (D.5) in the $(\psi, d\psi/d\xi)$ phase plane.

of ξ, $\lambda = 0$ is not the most positive eigenvalue of Equation (D.5). In other words, $\lambda_0 > 0$, so the solution of Figure 4.3(c) is unstable.

Augmentations of the above method to general systems that include both the Hodgkin–Huxley (H–H) and FitzHugh–Nagumo (F–N) formulations of nerve impulse dynamics have been described by Evans [1]–[4] and by Sattinger [11], and there have been several more detailed analyses of the F–N system [5]–[7, 10, 14]. Among the conclusions of these studies are the following.

- A sufficient condition for instability is that one mode of the linearized PDE has an eigenvalue with positive real part.
- Because there is always an eigenfunction of the linearized PDE for $\lambda = 0$, impulses are at most stable.
- Necessary conditions for stability are that no eigenvalues have positive real parts and that the zero eigenvalue ($\lambda = 0$) is nondegenerate.

To appreciate the requirement that the $\lambda = 0$ eigenvalue must be nondegenerate for stability, consider a Green function $\mathcal{G}(\xi, \tau)$ for the linearized PDE from which the total disturbance of the traveling-wave solution can be computed as [13]

$$\phi(\xi, \tau) = \int_0^\tau d\tau' \int_{-\infty}^{+\infty} \mathcal{G}(\xi - \xi', \tau - \tau') F(\xi', \tau') \, d\xi'.$$

In this formulation, $F(\xi', \tau')$ represents an arbitrary disturbance of the traveling wave, and $\mathcal{G}(\xi, \tau)$ is the response of the linearized PDE to a disturbance that is delta-function localized in both space and time. Because F is arbitrary, the properties of $\mathcal{G}(\xi, \tau)$ indicate whether any disturbance will grow with time.

To see how this comes about, consider the Laplace transform of $\mathcal{G}(\xi, \tau)$, which is defined as

$$\tilde{\mathcal{G}}(\xi, \lambda) \equiv \int_0^\infty \mathcal{G}(\xi, \tau) e^{-\lambda \tau} d\tau.$$

From this function of ξ and λ, $\mathcal{G}(\xi, \tau)$ can be recovered through the inverse transform

$$\mathcal{G}(\xi, \tau) = \int_C \tilde{\mathcal{G}}(\xi, \lambda) e^{\lambda \tau} d\lambda,$$

where for $\tau > 0$ the integration is over a closed curve (C) in the complex λ-plane that encloses all of the singularities of $\tilde{\mathcal{G}}(\xi, \lambda)$.

Notice that the inverse transform is a generalized sum of terms of the form $e^{\lambda \tau}$, which appears in the separation of variables for the linearized PDE. Thus, the singularities of $\tilde{\mathcal{G}}(\xi, \lambda)$ comprise the spectrum of the linearized PDE. The implications of a degenerate eigenvalue at $\lambda = 0$ can now be appreciated by recalling the Laplace transform pair

$$\frac{\tau^{n-1}}{(n-1)!} \longleftrightarrow \frac{1}{\lambda^n}.$$

Thus if the $\lambda = 0$ eigenvalue is doubly degenerate, $\tilde{\mathcal{G}}(\xi,\lambda)$ contains the factor $1/\lambda^2$, implying a corresponding temporal response that is proportional to τ. Linear growth with τ implies instability.

With the exception of the translation-mode eigenvalue at $\lambda = 0$, all eigenvalues of Equation (D.4) have negative real parts of finite magnitude for the leading edge solution of Figure 4.2(c). Thus if the eigenvalues of a F–N system are assumed to depend continuously on ε, there is a parameter range of some $\varepsilon > 0$ over which the upper curve in Figure 4.12 is stable [6, 7, 14]. Similarly for an H–H system with some

$$\tau_\mathrm{m} > 0,$$
$$1/\tau_\mathrm{h} > 0,$$
$$1/\tau_\mathrm{n} > 0,$$

the faster impulse should be stable.

To extend an analytic proof of stability out to the critical values of these time constants (beyond which traveling waves are not found), it is necessary to show that no pair of eigenvalues at $\lambda = \sigma \pm i\omega$ has a real part (σ) that becomes positive. If so, the impulse would have an unstable internal mode of oscillation with the temporal behavior $e^{\sigma\tau}\cos\omega\tau$. It is not known presently whether such unstable internal modes can occur for some values of F–N or H–H parameters.

REFERENCES

1. J W Evans. Nerve axon equations: I. Linear approximations. *Indiana Univ. Math. J.* 21 (1972) 877–885.
2. J W Evans. Nerve axon equations: II. Stability at rest. *Indiana Univ. Math. J.* 22 (1972) 75–90.
3. J W Evans. Nerve axon equations: III. Stability of the nerve impulse. *Indiana Univ. Math. J.* 22 (1972) 577–593.
4. J W Evans. Nerve axon equations: IV. The stable and unstable impulse. *Indiana Univ. Math. J.* 24 (1975) 1169–1190.
5. J A Feroe. Temporal stability of solitary impulse solutions of a nerve equation. *Biophys. J.* 21 (1978) 103–110.
6. C K R T Jones. Some ideas in the proof that the FitzHugh–Nagumo pulse is stable. In *Nonlinear Partial Differential Equations*, J Smoller, ed., Contemporary Mathematics 17, American Mathematical Society, Providence, 1984, pp. 287–292.
7. C K R T Jones. Stability of the travelling wave solution of the FitzHugh–Nagumo system. *Trans. Am. Math. Soc.* 286 (1984) 431–469.
8. A G Lingren and R J Buratti. Stability of waveforms on active nonlinear transmission lines. *Trans. IEEE Circuit Theor.* CT-16 (1969) 274–279.
9. K Maginu. On asymptotic stability of waveforms on a bistable transmission line. *IECE Professional Group on Nonlinear Problems*. NLP, Institute of Electrical and Computer Engineering, Tokyo, Japan, pp. 70–24 (in Japanese), 1971.

10. K Maginu. Stability of periodic travelling wave solutions of a nerve conduction equation. *J. Math. Biol.* 6 (1978) 49–57.
11. D H Sattinger. On the stability of waves of nonlinear parabolic systems. *Adv. Math.* 22 (1976) 312–355.
12. A C Scott. The electrophysics of a nerve fiber. *Rev. Mod. Phys.* 11 (1975) 487–533.
13. I Stackgold. *Green's Functions and Boundary Value Problem.* John Wiley & Sons, New York, 1979.
14. E Yanagida. Stability of fast travelling pulse solutions of the FitzHugh–Nagumo equations. *J. Math. Biol.* 22 (1985) 81–104.
15. Y B Zeldovich and G I Barenblatt. Theory of flame propagation. *Combust. Flame* 3 (1959) 61–74.

APPENDIX E

PERIODIC TODA-LATTICE SOLITONS

Finding a solitary-wave solution for the Toda lattice is not as straightforward as were the examples considered in Chapter 3, so it is instructive to go through the details. Following Toda [1] and Whitham [2], we begin by writing Equation (5.1) as the first-order system

$$\frac{ds_n}{dt} = -\frac{d\phi(r_n)}{dr_n} = ae^{-br_n} - a$$
$$\frac{dr_n}{dt} = 2s_n - s_{n+1} - s_{n-1},$$
(E.1)

and noting that

$$\frac{d^2 s_n}{dt^2} = -abe^{-br_n}\frac{dr_n}{dt} = -b\left(a + \frac{ds_n}{dt}\right)\frac{dr_n}{dt},$$

whereupon

$$\frac{\ddot{s}_n}{b(a + \dot{s}_n)} = s_{n+1} - 2s_n + s_{n-1},$$

with the dots indicating time derivatives.

For a traveling wave, assume that $s_n(t) = \tilde{s}(\theta)$, where

$$\theta \equiv \omega t - \beta n$$

so $\tilde{s}(\theta)$ is governed by

$$\frac{\omega^2 \tilde{s}''}{b(a + \omega \tilde{s}')} = \tilde{s}(\theta + \beta) - 2\tilde{s}(\theta) + \tilde{s}(\theta - \beta).$$
(E.2)

At this point, the challenge is to find a functional form for $\tilde{s}(\theta)$ that can satisfy Equation (E.2). To this end, Toda suggested the identity

$$\mathrm{dn}^2(\theta + \beta) - \mathrm{dn}^2(\theta - \beta) = -2k^2 \frac{d}{d\beta}\left(\frac{\mathrm{sn}\,\theta\,\mathrm{cn}\,\theta\,\mathrm{dn}\,\theta\,\mathrm{sn}^2\beta}{1 - k^2 \mathrm{sn}^2\theta\,\mathrm{sn}^2\beta}\right),$$
(E.3)

where k is the modulus of the Jacobi elliptic functions (see Appendix C). As Jacobi's epsilon function is defined by

$$E(u) = \int_0^u \mathrm{dn}^2 y\, dy,$$

Equation (E.3) can be integrated to

$$E(\theta + \beta) - 2E(\theta) + E(\theta - \beta) = -2k^2 \frac{\mathrm{sn}\theta\, \mathrm{cn}\theta\, \mathrm{dn}\theta\, \mathrm{sn}^2\beta}{1 - k^2\mathrm{sn}^2\theta\, \mathrm{sn}^2\beta}, \quad (\text{E.4})$$

which is close to Equation (E.2). Noting further that

$$E'(\theta) = \mathrm{dn}^2\theta = 1 - k^2\mathrm{sn}^2\theta$$
$$E''(\theta) = -2k^2\mathrm{sn}\theta\, \mathrm{cn}\theta\, \mathrm{dn}\theta,$$

Equation (E.4) can be written as

$$\frac{E''}{\alpha + E'} = E(\theta + \beta) - 2E(\theta) + E(\theta - \beta)$$

with

$$\alpha \equiv \frac{1}{\mathrm{sn}^2\beta} - 1.$$

$E(\theta)$ is not a periodic function, but it is related to the periodic Jacobi zeta function

$$Z(\theta) \equiv E(\theta) - \frac{E(k)}{K(k)}\theta,$$

where $K(k) = K$ and $E(k) = E$ are complete elliptic integrals of the first and second kind, respectively, and the period of $Z(\theta)$ is $2K$.

Normalizing the phase so a period corresponds to a unit increase in θ, $\dot{s}_n(t)$ can be written as

$$\dot{s}_n(t) = \frac{(2K\omega)^2}{b}\left[\mathrm{dn}^2[2K(\omega t - \beta n)] - E/K\right],$$

Finally, from the first of Equations (E.1),

$$r_n = -\frac{1}{b}\log\left(1 + \frac{\dot{s}_n}{a}\right). \quad (\text{E.5})$$

In the hyperbolic limit (see Table C.1), $k \to 1$, $E(k) \to 1$, $K(k) \to \infty$, and $\mathrm{dn}(\zeta, k) \to \mathrm{sech}(\zeta)$. Thus it is convenient to define κ such that

$$2K\beta \to \kappa \quad 2K\omega \to \sqrt{ab}\sinh\kappa,$$

leading to the solitary wave of Equation (5.5).

REFERENCES

1. M Toda. *Theory of Nonlinear Lattices*. Springer-Verlag, Berlin, 1981.
2. G B Whitham. *Linear and Nonlinear Waves*. John Wiley, New York, 1974.

APPENDIX F

ANALYTIC APPROXIMATIONS FOR LONG LATTICE SOLITARY WAVES

In Equation (5.8), a general equation for long lattice solitary waves is given as

$$v^2 R''(z) = U'[R(z+1)] - 2U'[R(z)] + U'[R(z-1)], \qquad (F.1)$$

where $r_n(t) = R(n - vt) \equiv R(z)$ is the separation of adjacent lattice sites, and $U[R(z)]$ is the corresponding spring potential.

Analytic approximations to numerical solutions of such equations can be obtained in the following manner [1]. Writing

$$F(z) \equiv U'[R(z)]$$

and indicating Fourier transforms as

$$\mathcal{F}(k) = \int_{-\infty}^{\infty} F(z) e^{-ikz} dz$$

and

$$\mathcal{R}(k) = \int_{-\infty}^{\infty} R(z) e^{-ikz} dz,$$

the transform of Equation (F.1) becomes

$$v^2 \mathcal{R}(k) = \left(\frac{4 \sin^2(k/2)}{k^2} \right) \mathcal{F}(k) \equiv \Lambda(k) \mathcal{F}(k).$$

Approximating $\Lambda(k)$ as

$$\Lambda(k) \approx \frac{1}{1 + k^2/12} \qquad (F.2)$$

implies the differential equation

$$v^2 \left(1 - \frac{1}{12} \frac{d^2}{dz^2} \right) R(z) \doteq U'[R(z)],$$

which can be written as

$$R'' \doteq \frac{12}{v^2} \left(v^2 R(z) - U'[R(z)] \right). \qquad (F.3)$$

For many forms of the function $U'[R(z)]$, this equation can be integrated in terms of elliptic or hyperbolic functions.

REFERENCE

1. D B Duncan, J C Eilbeck, H Feddersen, and J A D Wattis. Solitons on lattices. *Physica D* 68 (1993) 1–11.

APPENDIX G

MULTIPLE-SCALE ANALYSIS OF A DAMPED-HARMONIC OSCILLATOR

Dynamics of a damped-harmonic oscillator are governed by the equation

$$\frac{d^2x}{dt^2} + \varepsilon \frac{dx}{dt} + x = 0, \tag{G.1}$$

and we seek a solution that evolves from the initial conditions $x(0) = 1$ and $\dot{x}(0) = 0$. In a multiple-scale analysis of such a second-order ODE system, one assumes scaled time variables of the form [1–4]

$$T_0 = t, \quad T_1 = \varepsilon t, \quad T_2 = \varepsilon^2 t, \text{ etc.}, \tag{G.2}$$

where ε is a small parameter. A solution of the ODE can then be written as

$$x(t) = x(T_0, T_1, T_2).$$

Under this formulation, the first and second derivatives with respect to time t become

$$\frac{d}{dt} = \frac{\partial}{\partial T_0} + \varepsilon \frac{\partial}{\partial T_1} + \varepsilon^2 \frac{\partial}{\partial T_2},$$

$$\frac{d^2}{dt^2} = \frac{\partial^2}{\partial T_0^2} + 2\varepsilon \frac{\partial^2}{\partial T_0 \partial T_1} + \varepsilon^2 \left(\frac{\partial^2}{\partial T_1^2} + 2 \frac{\partial^2}{\partial T_0 \partial T_2} \right).$$

Inserting $x(T_0, T_1, T_2)$ into Equation (G.1), for example, and assuming that the scaled variables are independent gives:

$$\frac{\partial^2 x}{\partial T_0^2} + x + \varepsilon \left(\frac{\partial x}{\partial T_0} + 2 \frac{\partial^2 x}{\partial T_0 \partial T_1} \right)$$
$$+ \varepsilon^2 \left(\frac{\partial x}{\partial T_1} + \frac{\partial^2 x}{\partial T_1^2} + 2 \frac{\partial^2 x}{\partial T_0 \partial T_2} \right) + O(\varepsilon^3) = 0. \tag{G.3}$$

The next step is to introduce the truncated series expansion

$$x = x_0(T_0, T_1, T_2) + \varepsilon x_1(T_0, T_1, T_2) + \varepsilon^2 x_2(T_0, T_1, T_2)$$

The author thanks M.P. Sørensen for writing the first draft of this appendix.

into (G.3). Ordering according to powers of ε, yields the following system of equations:

$$\frac{\partial^2 x_0}{\partial T_0^2} + x_0 = 0,$$

$$\frac{\partial^2 x_1}{\partial T_0^2} + x_1 = -\frac{\partial x_0}{\partial T_0} - 2\frac{\partial^2 x_0}{\partial T_0 \partial T_1},$$

$$\frac{\partial^2 x_2}{\partial T_0^2} + x_2 = -\frac{\partial x_1}{\partial T_0} - 2\frac{\partial^2 x_1}{\partial T_0 \partial T_1} - \frac{\partial x_0}{\partial T_1}$$

$$- \frac{\partial^2 x_0}{\partial T_1^2} - 2\frac{\partial^2 x_0}{\partial T_0 \partial T_2}.$$

(G.4)

The solution of the first equation is

$$x_0 = A(T_1, T_2)e^{iT_0} + B(T_1, T_2)e^{-iT_0}.$$

The fact that A and B depend on T_1 and T_2 expresses the slow variation of these parameters on time, while the fast variation enters only through the exponentials.

The function x_0 can now be inserted into the right hand side of the second of Equations (G.4), to obtain

$$\frac{\partial^2 x_1}{\partial T_0^2} + x_1 = -i\left(A + 2\frac{\partial A}{\partial T_1}\right)e^{iT_0} + i\left(B + 2\frac{\partial B}{\partial T_1}\right)e^{-iT_0}.$$

Terms on the right hand side that are proportional to $e^{\pm iT_0}$ will produce secular terms proportional to $te^{\pm iT_0}$ in the solution for x_1. In order to avoid secular terms in the solution, it is sufficient that the coefficients of $e^{\pm iT_0}$ vanish. This requirement can be satisfied by the T_1 dependence of A and B. In other words

$$A + 2\frac{\partial A}{\partial T_1} = 0 \Rightarrow A(T_1, T_2) = A_1(T_2)e^{-T_1/2},$$

and similarly

$$B(T_1, T_2) = B_1(T_2)e^{-T_1/2}.$$

The solution of the second of Equations (G.4) now reads

$$x_1 = C(T_1, T_2)e^{iT_0} + D(T_1, T_2)e^{-iT_0}.$$

Insertion of x_0 and x_1 into the last of Equations (G.4) leads to

$$\frac{\partial^2 x_2}{\partial T_0^2} + x_2 = \left(\frac{A_1}{4} - 2i\frac{dA_1}{dT_2}\right)e^{iT_0 - T_1/2} + \left(\frac{B_1}{4} + 2i\frac{dB_1}{dT_2}\right)e^{-iT_0 - T_1/2}$$

$$- i\left(C + 2\frac{\partial C}{\partial T_1}\right)e^{iT_0} + i\left(D + 2\frac{\partial D}{\partial T_1}\right)e^{-iT_0}.$$

Again secular terms are avoided by demanding that coefficients to $e^{\pm iT_0}$ vanish. Accordingly

$$2i\frac{dA_1}{dT_2} - \frac{A_1}{4} = 0 \Rightarrow A_1 = A_{10}e^{-iT_2/8}$$

and

$$2\frac{\partial C}{\partial T_1} + C = 0 \Rightarrow C = C_1(T_2)e^{-T_1/2}.$$

Similarly $B_1 = B_{10}e^{iT_2/8}$ and $D = D_1(T_2)e^{-T_1/2}$.

To order ε^2 the solution then becomes

$$x = \left[A_{10}e^{i(T_0-T_2/8)} + B_{10}e^{-i(T_0-T_2/8)}\right]e^{-T_1/2}$$
$$+ \varepsilon\left[C_1(T_2)e^{iT_0} + D_1(T_2)e^{-iT_0}\right]e^{-T_1/2}.$$

With the definitions of T_0, T_1, and T_2 from Equations (G.2), this is

$$x(t) = \left[A_{10}e^{it(1-\varepsilon^2/8)} + B_{10}e^{-it(1-\varepsilon^2/8)}\right]e^{-\varepsilon t/2}$$
$$+ \varepsilon\left[C_1(\varepsilon^2 t)e^{it} + D_1(\varepsilon^2 t)e^{-it}\right]e^{-\varepsilon t/2} + O(\varepsilon^3). \qquad (G.5)$$

Although the dependencies of C_1 and D_1 upon $\varepsilon^2 t$ (or T_2) have not been determined, there is no loss of generality in writing

$$C_1(\varepsilon^2 t) = \tilde{C}_1(\varepsilon^2 t)e^{-i\varepsilon^2 t/8}$$
$$D_1(\varepsilon^2 t) = \tilde{D}_1(\varepsilon^2 t)e^{+i\varepsilon^2 t/8},$$

which makes the sinusoidal frequency of the second square bracket equal to that of the first. Then setting $A_{10} = B_{10} = 1/2$ and $\tilde{C}_1 = -\tilde{D}_1 = 1/4i$ satisfies the initial conditions $x(0) = 1$ and $\dot{x}(0) = 0$, reducing Equation (G.5) to

$$x(t) = e^{-\varepsilon t/2}\cos[(1 - \tfrac{1}{8}\varepsilon^2)t] + \tfrac{1}{2}\varepsilon e^{-\varepsilon t/2}\sin[(1 - \tfrac{1}{8}\varepsilon^2)t] + O(\varepsilon^3),$$

in agreement with Equation (7.21).

REFERENCES

1. J Kevorkian and J D Cole. *Multiple Scale and Singular Perturbation Methods*. Springer-Verlag, New York, 1996.
2. A H Nayfeh. *Perturbation Methods*. John Wiley & Sons, New York, 1973.
3. A H Nayfeh and D T Mook. *Nonlinear Oscillations*. John Wiley & Sons, New York, 1979.
4. A H Nayfeh. *Introduction to Perturbation Techniques*. John Wiley & Sons, New York, 1981.

APPENDIX H

GREEN FUNCTIONS FOR SOLITON RADIATION

Consider the task of calculating the radiation that is generated by a multifluxon waveform as it responds to a perturbation of the pure soliton equation [3]. The primary tool in such a calculation is a Green function $\mathcal{G}$ for calculating $\mathbf{W}_1$ in Equation (7.84). In terms of this Green function,

$$\mathbf{W}_1(x,t) = \int_0^t \int_{-\infty}^{\infty} \mathcal{G}(x,t|x',t')\mathcal{F}(x',t')\,dx'\,dt', \tag{H.1}$$

where the matrix kernel $\mathcal{G}(x,t|x',t')$ is determined in the following manner.

First, the expression for $\mathbf{W}_1(x,t)$ given in Equation (H.1) is substituted into the left-hand side of Equation (7.84), giving the condition

$$\mathcal{F}(x,t) = \int_0^t \int_{-\infty}^{\infty} L\mathcal{G}(x,t|x',t')\mathcal{F}(x',t')\,dx'\,dt'.$$

Thus as a function of x and t, $\mathcal{G}(x,t|x',t')$ is defined by the homogeneous initial value problem

$$\begin{bmatrix} \partial_t & -1 \\ (-\partial_{xx} + \cos u_0) & \partial_t \end{bmatrix} \mathcal{G}(x,t|x',t') = 0, \tag{H.2}$$

where $t > t' \geq 0$, and

$$\mathcal{G}(x,t|x',t') \to \begin{bmatrix} 1 & 0 \\ 0 & 1 \end{bmatrix} \delta(x-x')$$

as $t \to t'$.

Second, the expression for $\mathcal{F}(x',t')$ from Equation (7.84) is substituted into Equation (H.1), whereupon

$$\mathbf{W}_1(x,t) = \int_0^t \int_{-\infty}^{\infty} \mathcal{G}(x,t|x',t')L_{x',t'}\mathbf{W}_1(x',t')\,dx'\,dt',$$

where $L_{x',t'}$ indicates the operator L written for the primed variables. After integrating by parts, one finds that $\mathcal{G}^T$ satisfies an adjoint problem in the (x',t')-coordinates:

$$-\begin{bmatrix} \partial_{t'} & (\partial_{x'x'} - \cos u_0) \\ 1 & \partial_{t'} \end{bmatrix} \mathcal{G}^T(x,t|x',t') = 0, \tag{H.3}$$

where $t > t' \geq 0$, and

$$\mathcal{G}^T(x,t|x',t') \to \begin{bmatrix} 1 & 0 \\ 0 & 1 \end{bmatrix} \delta(x-x')$$

as $t' \to t$.

Thus the columns of the matrix $\mathcal{G}$ as functions of x and t belong to the null space of L, $\mathcal{N}(L)$, while the rows of $\mathcal{G}$ as functions of x' and t' belong to the null space of $L^\dagger$, $\mathcal{N}(L^\dagger)$. With these facts in hand, it is possible to construct $\mathcal{G}(x,t|x',t')$.

In Section 7.5.1, it was assumed that $\mathbf{W}_0$ is an exact multisoliton wave with $2(N+M+2L)$ speed and location parameters—the p_j—that modulate in response to the perturbation εR. The first-order correction—$\mathbf{W}_1$—describes the radiation field that is generated by these modulating fluxons. Because the operator L arises from linearization of the nonlinear equation about the multisoliton waveform $\mathbf{W}_0$, some members of $\mathcal{N}(L)$ and $\mathcal{N}(L^\dagger)$ can be generated by differentiating $\mathbf{W}_0$ with respect to each of its velocity and location parameters. But this finite collection of parameters cannot generate all members of the infinite dimensional null spaces; thus it is necessary to enlarge this collection of parameters to an infinite set that generates a basis for each null space. With this basis, an explicit representation of the Green function can be constructed [1, 2].

The Bäcklund transform provides a means for generating multisoliton formulas that include additional parameters related to the radiation field. To see how this goes, consider the most simple case: no solitons at all. Then the unperturbed problem involves only radiation, and at low amplitude each radiation component is a solution of the linear (Klein–Gordon) equation $(\partial_{tt} - \partial_{xx} + 1)u_r = 0$. Thus

$$u_r(x,t;k,A) = A e^{\pm i(kx+\omega t)}, \tag{H.4}$$

where $\omega^2 = k^2 + 1$, $-\infty < k < \infty$, and $|A| \ll 1$.

For each value of k, the vector

$$\mathbf{W}_0 = \begin{bmatrix} u_r \\ u_{r,t} \end{bmatrix}$$

resides in the null space of L, so a Fourier representation of the zero-soliton Green function becomes

$$\mathcal{G}_0(x,t|x',t')$$
$$= \frac{1}{2\pi} \int_{-\infty}^{\infty} dk \begin{bmatrix} \cos\omega(t-t') & (1/\omega)\sin\omega(t-t') \\ -\omega\sin\omega(t-t') & \cos\omega(t-t') \end{bmatrix} e^{-ik(x-x')}. \tag{H.5}$$

Next, consider a single kink, so $\mathbf{W}_0$ is as in Equation (7.83). Now the null space of L is composed of two distinct subspaces: (i) a discrete subspace $\mathcal{N}_d(L)$ that is spanned by the two derivatives of $\mathbf{W}_0$ with respect to v and x_0, and

(ii) a continuous subspace $\mathcal{N}_c(L)$, spanned by the radiative null vectors of L. Consequently the Green function is composed of two parts

$$\mathcal{G}(x,t|x',t') = \mathcal{G}_d(x,t|x',t') + \mathcal{G}_c(x,t|x',t'),$$

where the columns of $\mathcal{G}_d(x,t|x',t')$, as functions of x and t, lie in $\mathcal{N}_d(L)$, and the columns of $\mathcal{G}_c(x,t|x',t')$, as functions of x and t, lie in $\mathcal{N}_c(L)$. Similarly the rows of $\mathcal{G}_d(x,t|x',t')$, as functions of x' and t', lie in $\mathcal{N}_d(L^\dagger)$, and the rows of $\mathcal{G}_c(x,t|x',t')$, as functions of x' and t', lie in $\mathcal{N}_c(L^\dagger)$.

From Equation (H.1), the discrete Green function makes no contribution to $\mathbf{W}_1$. Why? Because the time variation of the parameters—governed by Equations (7.90)—was chosen to ensure that $\mathcal{F}(x',t')$ is orthogonal to the rows of $\mathcal{G}_d(x,t|x',t')$, as functions of x' and t'. This condition prevents linear growth of $\mathbf{W}_1$ with time. Thus Equation (H.1) may as well be written as

$$\mathbf{W}_1(x,t) = \int_0^t \int_{-\infty}^\infty \mathcal{G}_c(x,t|x',t')\mathcal{F}(x',t')\,dx'\,dt'. \tag{H.6}$$

In order to construct $\mathcal{G}_c$, we require sets of functions that span both $\mathcal{N}_c(L)$ and $\mathcal{N}_c(L^\dagger)$. We cannot find them from $\mathbf{W}_0$ because it contains no parameters related to the radiation field. Thus it is necessary to augment $\mathbf{W}_0$ to a solution of the pure soliton equation with parameters that are related to the radiation field, and such an augmented solution can be found using the Bäcklund transform.

Starting with the zero-soliton radiative solutions $u_r(x,t;k)$, defined in Equation (H.4), new solutions are generated as

$$u_r(x,t;k,A) \xrightarrow{\text{BT}} u(x,t;k,A,\zeta).$$

Here ζ is a constant introduced by the Bäcklund transform, and $u(x,t;k,A,\zeta)$ represents a kink of velocity

$$v = \frac{16\zeta^2 + 1}{16\zeta^2 - 1} \tag{H.7}$$

that is riding on top of radiation of wave number k and amplitude A.

Because u is an exact solution of the SG equation, the derivative of u with respect to A is a null function of the operator $\partial_{tt} - \partial_{xx} + \cos u$. Thus

$$\boldsymbol{\psi}(x,t;k,\zeta) \equiv \lim_{A\to 0} \frac{\partial}{\partial A}\begin{bmatrix} u(x,t;k,A,\zeta) \\ u_t(x,t;k,A,\zeta) \end{bmatrix} \in \mathcal{N}_c(L),$$

and $\mathcal{N}_c(L)$ is spanned by all such vector functions. Likewise

$$\tilde{\boldsymbol{\psi}}(x',t';k,\zeta) \equiv \lim_{A\to 0} \frac{\partial}{\partial A}\begin{bmatrix} u_{t'}(x',t';k,A,\zeta) \\ -u(x',t';k,A,\zeta) \end{bmatrix} \in \mathcal{N}_c(L^\dagger),$$

and $\mathcal{N}_c(L^\dagger)$ is similarly spanned.

From ψ and $\tilde{\psi}$, the single-kink radiative Green function is constructed as

$$\mathcal{G}_c(x,t|x',t') = \frac{1}{4\pi i}\int_{-\infty}^{\infty} d\lambda G \frac{\exp\{-i[k(\lambda)(x-x')+\omega(\lambda)(t-t')]\}}{\lambda(\zeta^2-\lambda^2)^2}, \quad (H.8)$$

where

$$G = \begin{bmatrix} g_{11} & g_{12} \\ g_{21} & g_{22} \end{bmatrix}$$

and

$$g_{11} \equiv (\zeta^2+\lambda^2+2\zeta\lambda\tanh\Theta)[i\omega(\lambda)(\zeta^2+\lambda^2-2\zeta\lambda\tanh\Theta') - 2i\zeta\omega(\zeta)\operatorname{sech}\Theta']$$
$$g_{12} \equiv -(\zeta^2+\lambda^2+2\zeta\lambda\tanh\Theta)(\zeta^2+\lambda^2-2\zeta\lambda\tanh\Theta')$$
$$g_{21} \equiv [-i\omega(\lambda)(\zeta^2+\lambda^2+2\zeta\lambda\tanh\Theta) + 2i\zeta\lambda\omega(\zeta)\operatorname{sech}^2\Theta]$$
$$\quad \times [i\omega(\lambda)(\zeta^2+\lambda^2-2\zeta\lambda\tanh\Theta') - 2i\zeta\omega(\zeta)\operatorname{sech}\Theta']$$
$$g_{22} \equiv [i\omega(\lambda)(\zeta^2+\lambda^2+2\zeta\lambda\tanh\Theta) - 2i\zeta\lambda\omega(\zeta)\operatorname{sech}^2\Theta](\zeta^2+\lambda^2-2\zeta\lambda\tanh\Theta').$$

Also

$$\omega(\lambda) \equiv 2\lambda + \frac{1}{8\lambda}$$

$$\omega(\zeta) \equiv 2\zeta + \frac{1}{8\zeta}$$

$$k(\lambda) \equiv 2\lambda - \frac{1}{8\lambda}$$

$$\Theta \equiv \frac{x - \int_0^t v(\alpha)\,d\alpha - x_0(t)}{\sqrt{1-v^2(t)}}$$

$$\Theta' \equiv \frac{x' - \int_0^{t'} v(\alpha)\,d\alpha - x_0(t')}{\sqrt{1-v^2(t')}},$$

where ζ is related to the kink velocity through Equation (H.7) and to the BT parameter a, introduced in Section 3.2.5, by $\zeta = i/4a$.

This procedure can be repeated any number of times in the BT sequence

$$u(x,t;k,A) \xrightarrow{\zeta_1} u_1 \xrightarrow{\zeta_2} u_2 \ldots \xrightarrow{\zeta_{N+M+2L}}$$
$$u_{N+M+2L}(x,t;k,A,\zeta_1,\ldots,\zeta_{N+M+2L}),$$

after which elements of $\mathcal{N}_c(L)$ and $\mathcal{N}_c(L^\dagger)$ can be obtained by differentiating u_{N+M+2L} with respect to A and then setting $A = 0$. Evidently, these expressions for

$$\psi(x,t;k,\zeta_1,\ldots,\zeta_{N+M+2L}) \in \mathcal{N}_c(L)$$

and
$$\tilde{\psi}(x', t'; k, \zeta_1, \ldots, \zeta_{N+M+2L}) \in \mathcal{N}_c(L^\dagger)$$
will be quite intricate, but it may be possible to construct them, and the corresponding Green function, with the aid of an algebraic computer code. This is a challenge for the computationally gifted.

REFERENCES

1. J P Keener and D W McLaughlin. Solitons under perturbations. *Phys. Rev. A* 16 (1977) 777–790.
2. J P Keener and D W McLaughlin. A Green function for a linear equation associated with solitons. *J. Math. Phys.* 18 (1977) 2008–2013.
3. D W McLaughlin and A C Scott. Perturbation analysis of fluxon dynamics. *Phys. Rev. A* 18 (1978) 1652–1680.

INDEX

Ab initio calculations 347–350
Ablowitz, Kaup, Newell, and Segur, *see* AKNS
Ablowitz–Ladik,
  see AL
Ablowitz, M.J. 193, 258, 265, 377, 439
Absolute instability 40
Absorption, infrared 10, 351–354, 391, 400
Acetanilide, *see* ACN
Acids, amino 20, 203, 206, 386, 387
ACN (acetanilide) 20, 198, 390–400, 425–426
  Birge–Sponer relation for 396–397
  coherent state for 397
  Davydov's analysis for 393–394
  infrared absorption spectrum 391, 400
  local modes in 399
  overtone spectra of 396–397
  Peierls barrier in 394
  self-trapping in 391, 399
  soliton 394–396
    binding energy of 393, 396
    effective mass of 396
  structure of 390–391
  temperature dependence 397–398
Acoustic vs. optical phonons 399
Action integral 434
Action potential 126
Active node 215, 216
Active sites in protein 207
Adelman, W.J., Jr. 123
Adenosine triphosphate 386
Adiabatic approximation 205, 390, 393
Adjoint
  matrix 289
  operator 36, 303, 312, 463
AKNS (Ablowitz–Kaup–Newell–Segur)
  formulation of ISM 250–252
AL (Ablowitz–Ladik)
  equation 193–195, 377–380, 408
    quantum theory 339–347

Alexander, Denise 400
All-or-nothing propagation 143, 150, 426
  response of a cell assembly 222
Alpha-helix 203
  in protein 204–205, 386–390
  parameters for 390
  soliton, 204–205, 390
    additional interactions 205
    thermal effects 205
Altenberger, R. 152, 153
Altruism 161, 430
Ambiguous perceptions 214–215, 222
Amide-I (CO) band 203, 204, 387, 390–391, 398–399, 425–426
Analog to digital converters 187
AND branch 153
Anderson, Carl 8
Anderson localization 207, 399
Anharmonic
  extrinsic vs. intrinsic 197
  limit of DST chaos 201–202
  potential 179, 182, 185
Annihilation, kink–antikink 81–82, 314–317
Anticommutators 381–382
Assemblies of neurons 221–223, 429
Aubry, Serge 186, 193, 207, 402
Autocatalytic reactions, *see* reactions
Autopoietic organisms 428, 430
Axon, giant, of squid 9, 12–13, 118–124

Bäcklund transform, *see* BT
Bäcklund, Albert 6
Bambusi, D. 193
Barenblatt, G.I. 138, 449
Bazin, M.H. 55
Belousov, Boris 13, 157
Belousov–Zhabotinsky reaction 13, 157
Benjamin, T.B. 77
Benjamin–Feir instability 77, 90
Benzene 10, 11, 364–365, 425
  experimental parameters 365

Benzene (cont.)
  overtone bands 365
  quantum theory 364–365
Bernstein, Lisa 19
Bessel, Friedrich Wilhelm 398
Bethe equations 405, 406, 408
Bethe, Hans 405
Beurle, R.L. 223
Bifurcation diagrams 198–200
Bifurcations 133, 356
Bilinear differential operators 67, 102
Binczak, Stephane 219
Biological
  form 159–160, 425
  molecules 202–203, 205–206, 207–209, 425–426
  solitons 205, 207–210
  species 211, 213
Birge–Sponer relation 347, 396–397
Birkhoff normal form 184
Black body 339–340
Blockage of nerve impulse 148, 150–153
Block-diagonalization of matrices 363–364, 367, 404, 408
Blow-up 97, 191, 193
Boerlijst, M.C. 160–161
Boesch, R. 189
Bohm, David 21
Bohr, Niels 340, 402
Bohr–Sommerfeld quantum condition 402, 410
Boltzmann, Ludwig 339
Boltzmann's constant 339, 340
Born, Max 8–9, 21, 341, 347
Born–Oppenheimer potential 348–349
Bose–Einstein condensate 193
Bose, Satyendra Nath 343
Bosons 343–345, 403
Bound states 16, 42–43, 240–242, 256, 259
  normalization 243, 245, 250
Boussinesq, Joseph 2, 55
Brain, organization of 176, 221–223, 429
Bray, W.C. 157
Breathers 80, 84–85, 186, 189, 271
  DSG 187–189, 402
  molecular crystal 426
  stationary 189, 198–199
BT (Bäcklund transformation) 6–8, 466–467
  history 6–8
  KdV 61–67

NLS 94–95
permutative property of 65, 83
SG 83–85
Toda lattice 178, 225
Buratti, R.J. 147, 449
Burgers, equation of 116–117, 164
Burgers, J.M. 116
Burgess, A. 170

Canal de Bourgogne 2
Canal, Union 1
Candle 4, 110–111, 162
Careri, Giorgio 20, 390–391, 398
Carlile, C.J. 402
Carrier wave 85, 90, 91, 92, 93, 299
Carr, J. 199
Catalytic hypercycle 160–161
Cauchy, Augustin Louis 47
Cauchy's integral theorem 47, 247, 262–263
Caudry, P.J. 438
Causality 34
  in pseudotime 44, 46, 47
Causal loops 3, 4, 20, 204, 428
Cause and effect 42, 242
Cell assemblies 221–223, 429
Cellular automata 176
Chaos
  and coherence 190, 201–202
  low dimensional 424
Characteristic admittance 152
Chemical diffusion 159
  oscillators 154
Chen, H.H. 93
Chizmadzhev, Y.A. 127
Christiansen, P.L. 55
Citric acid cycle 160
Coherent states 376–377, 397
Coherent structure(s)
  local modes 10–11, 19–20
  nerve impulses 4–6, 9, 11–12, 124–127
  ring waves 12–14, 154–156
  scroll waves 156–159
  solitons 14–19, 55–56
  spiral waves 155–156
Cole–Hopf transformation 117
Cole, Kenneth 9, 121
Collapse, dynamic 97
  of a wave packet 342, 360
Collective coordinates 306
Collins, M.A. 179
Collision of nerve impulses 12

of solitons 2, 14–15, 64, 82, 253–254, 270, 314–317
Complete elliptic integrals 444, 445
Computation, dendritic 151–153, 426
Condon, Edward Uhler 399
Conductance
 membrane 9, 119
 nonlinear 211
Conservation law(s) 273–277, 433
 approximate, for nerve 149
Conservative systems 434–437
Conserved densities
 for matrix scattering 276–277
 for Schrödinger operator 274–275
Continuous eigenvalues 450
Convective intability 39–40
Coupled nerves 322–326
Cowan, J. 213
Cross-derivative condition 264, 266
Cross, Michael 98
Cruzeiro-Hansson, Leonor 399
Crystal dislocations 9–10, 56, 187
Crystals, molecular 19–20, 22, 187, 390–391
Current, transmembrane 9, 111
Curvature of reaction-diffusion front 155, 157
Cusps in wave intersections 13, 154, 156

Damped harmonic oscillator 290–293, 460–462
Dauxois, T. 209
Davydov, A.S. 20, 203, 386, 387, 391, 399
De Broglie, Louis 21
Decremental conduction 143–147
Deift, P. 181
Delta function scattering 42–43, 255–256
Dendritic
 field 150–153
 information processing 151–153, 426
Density, Lagrangian, see Lagrangian
Deoxyribonucleic acid, see DNA
Derrick, G.H. 87
de Vries, Hendrik 2
Dichloromethane 353, 360
Differential operators, bilinear 67, 102
Diffusion
 constants 159–160
 genetic 9
 linear 5, 31–32, 436
 nonlinear 5, 12, 111–117, 154–159, 424–425

Dihalomethane
 antisymmetric mode 355
 classical analysis 354–356
 experimental results 360–361
 local mode 355–356
 quantum analysis 356–360
 self-trapping in 359–360
 symmetric mode 355
Dirac, P.A.M. 350
Dirac's notation 350–351
Direct problem 16, 40–43, 240–249, 258–263
Discrete
 breathers 186–187
 eigenvalues 451
 kinks 188
 nonlinear Schrödinger, see DNLS
 sine–Gordon, see DSG
Dislocations, crystal 9–10, 22, 56, 187
Dispersion relation 30–31
 nonlinear 18, 77, 79, 80, 81, 89
Displaced harmonic oscillator 394
Dissipative lattices 210–215
DNA (deoxyribonucleic acid)
 linear models 208
 nonlinear models 208
 solitons 207–210
 structure 208
DNLS (discrete nonlinear Schrödinger)
 equation 190–193, 365–370
 for alpha-helix soliton 205
Doglov, A.S. 182, 185
Domain walls 15, 22, 56
Doppler, Christian Johann 319
Drazin, P.G. 61, 243, 254
Dressing term 302, 318
DSG (discrete sine–Gordon)
 breathers 189
 equation 187–189
 kink 188
DST (discrete self-trapping)
 chaos 201–202
 dimer, QISM analysis of 406–407
 equation 197–202, 361–365
 for globular protein 205–207
 Hamiltonian 199
 Hartree approximation for 375–376
 quantum 361–365
 stability of stationary solutions 199–201
 stationary solutions 198–199

Edler, J. 399
Effective mass 366–367, 369, 370, 379
Eigen, Manfred 160, 430
Eigenvalues, bound state 16, 41, 241–242, 256
Eilbeck, J.C. 179, 188, 199, 200
Einstein, Albert 8, 12, 20
Einstein oscillator 392, 393
Electron and positron 8, 73
Electronic
  charge 8, 354
  cloud 197, 347–350
Elementary particles 7–8, 20–21, 73, 87–88
  of thought 12
Elliptic functions 443–447
Emergence 3, 4, 12, 161, 223, 429
  of form 219–220
  in biology 159–161, 429–430
Emergent structures
  cell assemblies 221–223, 429
  history 1–23
  local modes 10–11, 19–20, 186–187, 199, 200, 355–356, 359–360, 391, 400
  nerve impulses 124–127
  scroll waves 156–159
  social assemblies 221
  spiral waves 155–156
  solitons 55–98
  Turing patterns 159–160
Energy analysis
  damped oscillator 290–293, 460–462
  KdV 294–296
  NLS 299–301
  SG 296–299
Energy operator
  acetanilide 392
  Davydov model 387
  displaced harmonic oscillator 394
  DST 362
  fermionic 382
  Hubbard model 384, 385
  Salerno's equation 380
Enol'skii, V.Z. 407
Envelope wave(s) 85, 90, 91, 92, 93, 299
Ephaptic coupling of nerves 322–326
Equation(s)
  Ablowitz–Ladik (AL) 193–195, 377–380
  Burgers 116–117
  diffusion 31–33, 436

  discrete NLS 190–193
  discrete self-trapping (DST) 197–202, 361–365
  FitzHugh–Nagumo (F–N) 129–134, 320–322
  Gel'fand–Levitan 43–48, 245–249
  Hodgkin–Huxley (H–H) 117–127
  Klein–Gordon 73
  Korteweg–de Vries (KdV) 2–3, 55, 57–70, 250–258, 294–296
  Markin–Chizmadzhev 127–129
  modified KdV 61
  Morris–Lecar 134–138
  nonlinear Schrödinger (NLS) 88–97, 271–273
  Schrödinger 341, 342, 345, 348
  sine–Gordon (SG) 71–88, 266–271, 296–299, 301–306, 310–319
  Swift–Hohenberg 98
  wave 29, 435
  Zeldovich–Frank-Kamenetsky 111–116
Erneux, T. 218
Estabrook, F.B. 61
Evanescence 39
Evans function 142, 143, 453
Evans, J.W. 142, 453
Exponential taper 147–148

Faddeev, L.D. 258
Failure of nerve propagation 218–219
Fairy rings of mushrooms 154, 155
Faraday, Michael 110, 161
Feddersen, H. 189, 206
Feir, J.E. 77
Fermi, Enrico 14, 177
Fermionic polaron model 381–384
Fermions 381–382, 403–404
Fermi–Pasta–Ulam recurrence 14, 15
Fermi's Golden Rule 352
Fibers, optical 22, 96, 98, 308–309
Fibrillation 158–159, 428
Filamentation 96–97
Filip, A.M. 181
Fillaux, F. 401, 402
FitzHugh, Richard 12, 129, 132–133
FitzHugh–Nagumo, see F–N
Flach, S. 186
Flame of candle 4, 110–111, 162
Flesch, R. 179
Flip-flop 210, 214
Flux, magnetic
  loops 87

quantum 74
Fluxon(s)
  and antifluxons 72–74
  dynamics 297–299
  interacting 314–317
  oscillator 297–298
  radiation 317–319
F–N (FitzHugh–Nagumo)
  coupled 322–326
  decremental conduction 145–147
  equation 129–134, 320–322
  impulses 130–133
  periodic solutions 133–134
  three dimensional 154, 159
  traveling waves 130–133
Form, differential 6
Fourier analysis 30–31, 32, 43, 257–258
Fourier, Jean Baptiste Joseph 30
4-methyl-pyridine 187, 401–403
Franck–Condon factor 399
Franck, James 399
Frank-Kamenetsky, David 9, 111
Frankenstein, Victor 4
Franklin, Ben 4
Fredholm
  condition 303, 304, 313, 316
  theorem 35–37
    for matrices 289, 290
Fredholm, Erik 36
Frenkel, Yakov 9–10, 15, 176
Freud, Sigmund 221
Friesecke, G. 181
Friesecke–Wattis theorem 181–182
Fundamental equation of
  chemistry 342

Galvani, Luigi 4
Gardner, C.S. 239, 438
Gel'fand–Levitan equation 43–48, 245–249
Genetic code 207–208
Geometric ratio 152
Globular protein(s) 205–207
Golkany, I.M. 20
Green function 33–35
  for multisoliton radiation 310–314, 317–319, 453, 463, 463–467, 464–467
Green, George 34
Greene, P.H. 213, 214
Group velocity 77, 90, 92, 367, 379–380

Gudermann, Christoph 7
Gudermannian 7, 10, 72–73

Haken, Hermann 14, 215, 219, 223, 429–430
Hamiltonian 191, 195, 196, 294, 297, 338
Hamm, Peter 391, 399, 400
Harmonic balance 184
Harmonic oscillator 36–37, 290, 337–338, 342–345
Hartree approximation 186, 372–376
  accuracy of 376
  normalization 374
  wave function 374, 389
Hartree, Douglas R. 372
Heart failure 158–159
Heat conduction 33, 36
Hebb, Donald 221, 222
'Hebbian' synapses 221
Heisenberg picture 366
Heisenberg uncertainty 341, 371–372
Heisenberg, Werner 21, 340
Helmholtz, Hermann 4–5
Hermite, Charles 344
Hermite polynomials 344
H–H (Hodgkin–Huxley)
  equations 117–127
  threshold effects 149–150
Hierarchy(ies)
  biological 160–161, 427–428
  brain 222–223, 429
  BT 6–7, 62–67, 83–84, 93–95
Hirota, R. 67, 438, 439, 441
Hirota's method 67–70
Hochstrasser, D. 179
Hodgkin, Alan 12, 118
Hodgkin–Huxley, see H–H
Homoclinic trajectories 116, 125, 132
Hubbard model 384–386
Human brain 176, 214–215, 221–223, 429
Huxley, Andrew 12, 118
Hydrodynamic solitons 1–3, 57–58, 256–257
Hydrogen bond 203–204, 349–350, 386, 387
  in DNA 207
Hypercycles 160–161, 430

Impulse, nerve 4–6, 9
  propagation 126
  velocity 320–322

Inductance, phenomenological 146
Infeld, Leopold 8
Instability
  absolute 40
  Benjamin–Feir 77, 90
  convective 39–40
  modulational 77, 90
  traveling wave 138–143, 448–455
Intensities of overtones 360–361, 396–397
Intrinsic gap modes 183, 184
  localized modes 182–183
Invariance, Lorentz 8, 73, 85, 86
Inverse problem, see Scattering
Inverse scattering method, see ISM
Ion current(s) 119–123
Iooss, G. 181
ISM (inverse scattering method)
  general 15–17, 238–239
  KdV 250–257
  NLS 271–273
  SG 266–271
Izergin, A.G. 193

James, G. 185
James's theorem 185
Johansson, M. 193
Johnson, R.S. 61, 243, 254
Jones, C.K.R.T. 142–143
Josephson junction(s) 71–72
  discrete arrays 187, 189
  long 71–74, 297–299
  oscillators 298, 317–319
Jost functions 243, 260

Kaup, D.J. 258
KdV (Korteweg–de Vries)
  delta function initial condition 255–256
  energy density for 294
  equation 3, 16, 55, 57–70
  ISM for 239, 250–252
  Lagrangian density for 294
  N-soliton formulas 76–70, 254–256, 438
  perturbation theory 294–296
  soliton(s) 58–59, 252–253
  square-well initial conditions 256–257
Keener, J.P. 309
Keller, J. 133
Kenkre, Nitant 381
Khodorov, Boris 152, 153
Kink(s) and antikink(s) 73–85
  collision 15, 81–82, 270–271, 314–317
  oscillation 80–81, 270–271
Kirchhoff, Gustav Robert 71
Kirchhoff's laws 71, 117
Kivshar, Yuri 315, 327
Klein–Gordon equation 73
Knickerbocker, C.J. 295
Kolmogoroff, Andrej 9
Konno, K. 93
Kontorova, T. 9–10, 15, 176
Kopidakis, G. 207
Korepin, V.E. 193, 404
Korteweg–de Vries, see KdV
Korteweg, Diederik 2
Krebs cycle 160–161
Krebs, Hans Alfred 160
Krumhansl, J.A. 209, 400
Kruskal, M.D. 14, 15, 55, 188, 193

Ladik, J.F. 193, 377
Lagrange–Euler equation 435
Lagrangian
  averaged 17–18, 61, 74–77, 306
  density 75, 146, 306, 434
Laguerre, Edmond Nicolas 395
Lamb, George 6
Lamb, Horace 2
Landau, Lev 8
Langmuir, Irving 97
Laplace, Pierre Simon de 31
Laplace transform 31–32, 44, 453
Laser, multimode 210–211, 214
Lattice NLS equation, see DNLS
Lattice solitary waves 179–180, 458–459
  existence of 180–182
  with NL on-site potentials 185–202
Lattice soliton, see Soliton
Lattice, Toda 17, 19, 178–179, 456–457
Lawden, D.F. 444
Lax, Peter 16
Leading edge charge 149–150
Lecar, Harold 134
Legendre, Adrien-Marie 254
Lichens 170
"Light cone" coordinates 266
Lindgren, A.G. 147, 449
Linearity vs. nonlinearity 28
Line widths 361
Liouville–Arnold theorem 191
Localization, nonlinear vs. Anderson 207
Local mode(s) 199, 353

history of 10–11, 19–20, 186
  in biomolecules 205–207, 386–390
  in molecules 197–201, 361–365, 425
  in molecular crystals 390–398
Logic, dendritic 151–153, 426
Loop, closed causal 3, 4, 20, 204
Lorentz, Hendrik Antoon 73
Lorentz transform 73, 85, 86–87
Los Alamos Laboratories 14, 176, 347
Lotka, Alfred 12
Lowering operators 343–344, 351, 378, 381–382
Luther, Robert 5–6, 12, 157

MacKay–Aubry theorem 186–193
MacKay, Robert 186, 193
Maginu, K. 449
Makhankov, V.G. 377
Malomed, Boris 315, 327
Manakov, S.V. 95
MANIAC computer 14, 176
Margulis, Lynn 430
Markin, V.S. 127
Matrices, large 326, 327
  perturbation theory for 288–290
Mauro, Alex 146
Matter, theory of 7–8, 20–21, 56, 87–88
McLaughlin, David W. 309
Mecke, Reinhard 10, 19, 186
Membrane resonance 146
Messenger RNA 209
Methane 199–200, 201
Method of
  averaging 17–18, 61, 74–77, 306–309
  harmonic balance 211–212
  number states (NSM) 357–358, 403–404
Methylated pyridines 187, 401–403
Metropolous, Nick 14
Mie, Gustav 7
'Missing mass' for KdV soliton 296
Miura, Robert 18, 61, 275
Model proteins 390–400
Modes, local (*see* Local modes)
Modified KdV equation 61
Modulated traveling waves 17–18, 61, 74–77
Molecular crystal(s) 19–20, 22, 176, 187, 190–193, 390–392

Molecular dynamics codes 205–206, 349–350
Mollenauer, L.F. 93
Momentum density 435
Morphogenesis, biological 159–160
Morris, Catherine 134
Mortensen, Ole Sonnich 360
Multimode lasers 210–211, 214
Multiple scale analysis 460–462
  damped oscillator 291–293
  SG 301–306
Multisoliton
  perturbation theory 309–314, 463–467
  wave(s) 67–70, 254–255, 438–442
Mushrooms, fairy rings of 154
Myelinated nerve axon 215–219

Nagumo, Jin-ichi 12, 129
Narcotization of nerve 144–145
Naugolnykh, K. 55
Necker cube 214, 222
Neocortex 214–215, 221–223
Nernst, Walther 6
Nerve(s) 117–127
  impulse
    decremental 143–147
    propagation 4–6
    stability 138–143
  model
    F–N 129–134, 320–322
    H–H 117–127
    M–C 127–129
    M–L 134–138
    Z–F 111–116
  myelinated 215–219
  parallel 322–326
Neuristor 130, 147
Newell, Alan 18, 258, 295
Newton, Isaac 1
Newton's equations 177, 338, 350, 440
Nicolis, G. 218
NLS (nonlinear Schrödinger)
  equation(s) 18, 56, 88–95
    coupled 95–96
    ISM for 271–273
    Lagrangian density for 299, 306
    lattices 190–197
    N-soliton solution 93–95, 440
    transverse phenomena 95–97
    vector 95–96

NLS (nonlinear Schrödinger) (contd.)
    soliton(s) 91, 272–273
        averaged Lagrangian for 306–309
        damping rates 301, 309
        energy density for 300
        envelope velocity of 91, 273
        power balance for 299–301
Node(s) of Ranvier 215, 216
Nonergodicity 14, 176
Nonlinear(ity) 3
    definition of 28
    diffusion
        in more space dimensions 154–159
        in nerve 117–127
    dynamic hierarchies 160–161, 222–223, 427, 429
    electric filter 180, 440–441
    optics 95–98
    soft 338
    standing waves 77–81
Nonlinear Schrödinger, see NLS
Normalization of bound states 41, 243, 250
NSM and QISM, comparison of 407–409
N-soliton formula(s)
    KdV 67–70, 254–255, 438
    NLS 440
    nonlinear filter 440–441
    SG 438–440
    Toda lattice 440–441
Nucleic acids 202, 207–208
Null space 35, 289, 304, 312–313, 316
Number operator
    bosonic 343–344, 356, 362, 366
    fermionic 382
    Hubbard model 385
    q-deformed 378, 380
Number
    propagation 31, 354
    wave 353–354
Number state method (NSM) 357–358, 403–404

Operator(s)
    bosonic 344
    fermionic 381–382
    ISM
        Lax 16, 238–239, 312
        matrix 258, 264
        Schrödinger 340–341
    linear 35–36

    number (see Number operator)
    q-deformed 378, 380
    quantum 340–341
    trace 405, 406
    transfer 404–405
    translation 364
Oppenheimer, J. Robert 342, 347
Optical vs. acoustic modes 182–184, 399
Optics, nonlinear 22, 93–94, 95, 308–309
OR branch 153
Order parameters 209, 214, 429–430
Organizing center 13–14, 155
Oscillator, classical nonlinear 337–338
Ostrovsky, L.A. 55, 77
Ostwald, William 5
Ott, E. 295
Ourobori 3, 428
Overtone(s) 186, 347, 360–361, 396–397

Page, J.B. 182
Parker-Rhodes, A.F. 154, 155
Parmentier, R.D. 189
Particle(s)
    and antiparticle 73
    elementary 7–8, 20–21, 85–88
    of thought 12
Pasta, John 14
Pedersen, N.F. 315
Pego, R.L. 181
Peierls barrier 188, 205, 380, 394
Peierls, Rudolph E. 188
Perelomov, A.M. 377
Periodic trains of
    nerve impulses 133–134
    solitons 59–61, 74–77, 92
Permeability, ionic 119–121
Permutation property of BT 65–66, 83, 94
Perring, J.K. 15, 83
Perturbation(s)
    general 287–288, 326–327
    theory 309–314, 463–467
        based on ISM 309
        Born–Oppenheimer 347–350
Perturbed
    KdV equation 294–296
    kink-antikink collision 314–317
    NLS equation 299–301, 306–309
    SG equation 296–299, 301–306
    matrices 288–290
Peyrard, M. 188, 189, 209

Phase space for traveling waves 113, 116, 124–125, 130–132
Pinning to lattice 188, 205, 380, 394
Planck, Max 339, 340, 341
Planck's constant 19, 72, 339, 340, 341
Plane waves of F–N 155
Plasma
  collisionless 55, 97
  frequency 319
  oscillations 79
Plastic deformation 9–10, 176, 187
Pnevmatikos, Stephanos 198
Poisson bracket 191, 195, 196
  nonstandard 195, 196, 377
Poisson, Simeon 191
Polaron(s) 8, 203, 386–400, 425
Poles of scattering functions
  simple 42–43, 242, 249–250
  multiple 273
Polysaccharides 202
Potential,
  Born–Oppenheimer 348–349
  Toda 17, 178, 440, 456
Power balance
  candle 4, 110–111
  nerve 146–147
  NLS soliton 309
  SG fluxon 297–298
Prairie fires 154, 425
Product wave function
  assumption in ACN 393
  Born–Oppenheimer 348
  Davydov 387–388
  Hartree 374
Propagation number 31, 354
Propagator variable 156, 159–160
Protein(s) 20, 202–203, 205–207
  alpha-helix 202–203, 386–387
  structure 203, 205–206, 386–387
Pseudotime 43–44
Pulse
  locking on coupled nerves 322–326
  narrowing, optical 93–94
Pulson 87
Pump–probe measurements 351–353, 399
Purkinje cell 150–151

Quack, M. 361
Quality factor 213
Quantum
  discrete NLS (QDNLS) 365–370
  field theory 369, 418–420
  inverse scattering method (QISM) 404–407
  lattice sine–Gordon equation 401–403
  linear oscillator 342–345
  soliton, behavior of 370–372, 409
  theory, origins of 339–342
  trace operator 405, 406
  uncertainty principle 341, 371–372, 380, 409
Quasidegeneracy of eigenvalues 19, 186–187, 358, 359, 365, 425
Quasiharmonic lattices 210–214

Radiation field 256, 309, 319
  intensity 339, 340
  steady state 319
Raising operator(s) 343–344, 351, 378, 381–381
Raman, Chandrasekhra Venkata 353
Raman data for ACN 391, 400
  scattering 353, 372
Rasmussen, Jens Juul 97
Reaction(s)
  autocatalytic 5–6, 159–160, 424–425
  Belousov–Zhabotinsky (BZ) 13–14
Reciprocity theorem 197
Recovery variable 130, 156–157, 159–160
Reflection coefficient 42, 241, 252, 258
  time dependence of 251–252, 267, 272
Reflectionless potential(s) 48, 249–250
Remoissenet, Michel 56, 179
Replication 219–220
  of DNA 207, 209
Residue theorem 46, 47, 247, 262–263
Resonator, nonlinear 77–81, 297–299
Rinzel, John 133
RNA polymerase 209
Rosenau, P. 179
Rotating wave
  amplitudes 190, 197, 342–343
  approximation 184, 186, 190, 345–347
Russell, John Scott 1–2, 10, 22–23, 110, 182, 256–257
Rydpal, K. 97

Safety factor, for nerve 150
Salerno, Mario 196, 380, 381, 407
Salerno's equation 195–197
  application to molecular crystals 380–381

Salerno's equation (*cont.*)
  AL and DNLS limits 196, 380–381
  quantum analysis of 380–381
Saltatory conduction 218–219
Satsuma, J. 94
Sattinger, D.H. 453
Saturation of optical fiber 308–309
Scaled variables 291–293
  for SG 301–302
Scanning laser microscopy 189
Scattering
  data 16, 43, 252
  operator(s) 16, 239, 240, 251, 258
  potential 241, 249–250, 259
  problem 40–43, 240–242, 258–263
    multiple poles 273
    simple poles 242–245
  states 16, 40–41, 240–242, 258–260
Schrödinger('s)
  equation 21, 40, 340, 342, 345
  operator 340–341
  picture 366, 372
Schrödinger, Erwin 9, 21, 56, 340, 429, 430
Schuster, Peter 160, 430
Scroll ring 157–159
Seeger, A. 83
Segur, Harvey 258, 265
Self-dual lattices 179
Semiconductor laser 98, 214
Separation of variables, nonlinear 77–79
SG (sine–Gordon)
  equation 10, 37, 56, 71–85
    adjoint null space 312–314, 463–467
    boundary conditions for 298–299
    breathers 84–85, 271
    energy density for 297
    ISM for 239, 266–271
    kinks and antikinks 73, 267–268
    laboratory coordinates 266
    Lagrangian density for 296
    mechanical model of 74
    N-soliton formulas 438–440
    periodic waves 74–77
    perturbation theory 296–299, 301–306
    square well initial conditions 268–270
    three-dimensional 87–88
    two-dimensional 85–87

kink
  dynamics of 297–298
  energy equation 297, 304, 316
  Green function for 463–465
  power balance for 298
  radiation from 317–319
  perturbed fluxon 296–299
  pulson 87
Shabat, A.B. 18, 96, 239, 258, 264, 273, 440
Shallow water waves 1–3, 57–58, 256–257, 295
Shelf, emergence of 296
Shelley, Mary 4
"Shooting" method 113, 124–125, 188, 192
Sievers, A.J. 182, 184, 185
Signaling problem 39
Sine–Gordon, *see* SG
Six-twelve potential 349
Skyrme, T.H.R. 15, 83
Smets, D. 181
Social amobae (slime molds) 154, 160, 425
Social assemblies 221
Soliton(s)
  AL 193–195
  biological 202–210
  classical 55–98
  history 1–3, 6–7, 14–19
  KdV 57–70, 252–257
  lattice 178–179, 190–193
  NLS 88–98, 271–273
  perturbation theory for 293–310
  quantum 370–372, 409
  SG 71–88, 266–271
  Toda 178–179, 456–457
Sommerfeld, Arnold 402
Sørensen, M.P. 238, 287, 460
Spontaneous analysis 256–257
Spring-mass lattice 177–185
Spring-mass system 337–338
Squid axon 118
  impulse speed 4–5, 126
  threshold 149–150
Stability 37–40
  asymptotic 37
  linear 38–40
  nerve 138–143
  neutral 37
  traveling wave 138–143, 448–455
Steuerwald, Rudolph 10

INDEX

Stretching oscillation
  CH 10–11, 185, 353, 360–361, 364–365
  CO 20, 203–206, 386–387, 390–392, 398–400
  NH 198, 199, 400
Struik, Dirk 34
Sudan, R.N. 295
Sudden cardiac death 158–159
Swift–Hohenberg equation 98
Swihart velocity 72, 86, 319
Symmetric ordering of boson operators 346
Symmetry breaking 360
Synergetics 14, 215, 429–430

Tachyon(s) 73, 75
Takeno, S. 182, 185, 399
Takhtajan, L.A. 239, 258
Tapered fibers 147–148
Temperature
  effects on protein soliton 205, 391, 397–398, 399, 400
  factor for nerve 120, 132–133, 145
Terahertz range 20, 22, 299, 319, 403
Theory of the double solution 21
Thermalization 14, 339–340
Thermodynamic limit 408
Thompson, D'Arcy W. 159
Threshold
  charge 150, 426
  condition 116, 127, 149–150
  ignition of an assembly 222
Toda lattice 17, 178–179, 181, 456–457
  BT for 224–225
  N-soliton formula for 440–441
Toda, Morikazu 17, 178, 180, 185, 456
Transform(ation)
  Bäcklund 6–7, 19, 61–67, 83–84, 93–95, 465–467
  Cole–Hopf 117
  Fourier 30–31, 43, 257–258
  Galilean 57–58, 448
  inverse scattering 15–17, 238–239, 250–252, 264–265
  Laplace 32, 44
  Lorentz 8, 73, 85, 86
Transmission line, superconducting, see Josephson junctions
Tunnel diode arrays 210–215
Tunneling time 359, 360
Turing, Alan 159
Turing patterns 159–160

Turn-off (on) variable 119
Two-component scattering 258–263
  integral equation 263
  ISM for 264–265
Two-soliton collision
  KdV 63–67, 253–254
  SG 81–83, 270–271, 314–317

Ulam, Stan 14

van der Pol, B. 211, 291
van der Waals, Johannes Diderik 350
Variational analysis of NLS 306–309
Varicosities of nerve 148
Vector NLS solitons 95–96, 283–284
Velarde, M.G. 220
Vella, D. 193
Venakides, S. 181
Volta Alessandro 4
Volterra, Vito 213
Vortex
  hydrodynamic 5
  line 155–156
  ring 158–159
Vvedensky, D.D. 243

Wadati, M. 93, 224–225
Wahlquist, H.D. 61
Wattis, J.A.D. 179, 180, 181, 189, 402
Wave(s)
  equation 29, 435
  evanescent 39
  number 353–354
  of neural information 223
  of translation 1–3, 59, 256–257
  packet 19, 186, 341, 342, 370–372, 409
  packet, nonlinear 88–90
  ring 13, 154–155
  solitary 1–3
  spherical 157
  spiral 155–156, 169, 170
  tank 2, 256–257
Whitham G.B. 17, 61, 75, 456
Whitham's method 17–18, 74–77, 100–101, 181, 306
Winfree, A.T. 14, 155, 156, 158, 159, 170
Willem, M. 181
Willis, C.R. 186
Wronskian 244–245
Wronski, Josef Maria 244

Yajima, N. 94
Yakushevich, L. 208
Yanagida, E. 143

Zabusky, N.J. 14, 15, 64, 66, 238, 254
Zaikin, A.N. 13, 154

Zakharov, V.E. 18, 96, 239, 258, 264, 273, 440
Zeldovich–Frank–Kamenetsky equation 111–116
  stability of 143, 448, 452
Zeldovich, Yakov 9, 111, 138, 449
Zero-soliton Green function 464
Zhabotinsky, A.M. 13, 154, 157

Lightning Source UK Ltd.
Milton Keynes UK
18 May 2010

154328UK00009B/86/A